Research on Soil Management and Conservation

Research on Soil Management and Conservation

Editor

Luis Eduardo Akiyoshi Sanches Suzuki

Basel • Beijing • Wuhan • Barcelona • Belgrade • Novi Sad • Cluj • Manchester

Editor
Luis Eduardo Akiyoshi
Sanches Suzuki
Federal University of Pelotas
Pelotas
Brazil

Editorial Office
MDPI AG
Grosspeteranlage 5
4052 Basel, Switzerland

This is a reprint of articles from the Special Issue published online in the open access journal *Soil Systems* (ISSN 2571-8789) (available at: https://www.mdpi.com/journal/soilsystems/special_issues/S01JI34XAA).

For citation purposes, cite each article independently as indicated on the article page online and as indicated below:

Lastname, A.A.; Lastname, B.B. Article Title. *Journal Name* **Year**, *Volume Number*, Page Range.

ISBN 978-3-7258-1965-2 (Hbk)
ISBN 978-3-7258-1966-9 (PDF)
doi.org/10.3390/books978-3-7258-1966-9

Contents

About the Editor

Luis Eduardo Akiyoshi Sanches Suzuki

Luis Eduardo Akiyoshi Sanches Suzuki has been a professor at the Center of Technology Development, Federal University of Pelotas, Brazil, since 2009. His research and extensionist topics include soil physics, compaction, physical–hydric and mechanical properties, management, and conservation, with over 24 years of experience publishing national and international papers. His teaching area of expertise also includes soil science, a field which he has publicized and promoted for over twelve years. He graduated in Agronomy from the São Paulo State University, Brazil, and Geography from the Federal University of Pelotas, Brazil, with a Master's Degree in Soil Science and a Ph.D. in Forestry Engineering, both from the Federal University of Santa Maria, Brazil.

soil systems

MDPI

Editorial

Research on Soil Management and Conservation

Luis Eduardo Akiyoshi Sanches Suzuki

Center of Technology Development, Federal University of Pelotas, Pelotas 96010-610, Rio Grande do Sul State, Brazil; dusuzuki@gmail.com

The soil is the base of a sustainable agricultural system; it is the key for food and energy production, a reservoir of water and nutrients [1–3], and a filter for water and contaminants [4]. However, inadequate soil management may significantly negatively impact the environment [5–7], crop development and yield [8–12], natural resources such as air and water [13–15], and human and animal health [16,17]. Soil management practices that favor soil and water conservation and the improvement of soil functions and structure are preferable. The diversity of soil uses and types [18–20], climate [21], relief, and origin materials make the study of better soil management practices a worldwide challenge. Thus, the Special Issue "Soil Management and Conservation" addresses topics such as soil tillage, the influence of machinery on soil structure, erosion processes, control practices, the influence of plants on soil structure, and practices used for soil conservation and the improvement of soil structure. Strategies to avoid soil structure degradation, such as studies on precompression stress and the compression index, were called upon to present the new findings on that subject.

In this sense, we had the response of the scientific community, presenting their findings in this Special Issue. The transition from irrigated cropland to irrigated management-intensive grazing (MiG) as a proposal to sustainably intensify agroecosystems was evaluated through soil health indicators using the Soil Management Assessment Framework (SMAF) and verified to have significant improvements in biological soil health indicators such as β-glucosidase, microbial biomass carbon, and potentially mineralizable nitrogen, while nutrient status was relatively stable (Trimarco et al., Contribution 1). On the other hand, the authors verified that soil compaction, based on bulk density, reduced soil physical health. In this context, the soil structure can be very susceptible to compaction, requiring attention to machinery traffic and animal trampling, especially for loads larger than the precompression stress that can overcompact the soil, despite the fact that loose soil is more compressive, presenting greater deformation than a preserved soil structure (Suzuki et al., Contribution 2).

Our Special Issue also brought an alert to the effects of polyethylene polymer on soil properties and plant growth, showing that the presence of polyethylene polymer on microplastics in the soil may increase the amount of available Zn and Cd in highly contaminated soils and lettuce and edible leaves, posing a risk to the environment and human health (Bethanis and Golia, Contribution 3). We highlighted the challenges in the management of environmentally fragile sandy soils, showing these soils play an important role in aquifer recharge, but they are soils of low suitability for agricultural use and with a high risk of leaching and aquifer contamination (Suzuki et al., Contribution 4).

Studies on soil management and conservation were presented in the Special Issue, showing the no-tillage system maintains soil carbon in the deepest soil layers, but the loss of carbon rate in the topsoil is greater than the input due to soil erosion and organic matter mineralization (Thomaz and Kurasz, Contribution 5), while green manure together with grass is an economical and environmentally sound strategy to restore the macrofauna of an anthropogenically degraded soil (Bonini et al., Contribution 6). The adoption of conservation agriculture principles such as conservation tillage, permanent plant cover, and crop

Citation: Suzuki, L.E.A.S. Research on Soil Management and Conservation. *Soil Syst.* **2024**, *8*, 42. https://doi.org/10.3390/soilsystems8020042

Received: 12 March 2024
Accepted: 3 April 2024
Published: 5 April 2024

diversification can contribute to the mitigation of climate change without compromising food security on local and global scales, but knowledge about biophysical, technical, socio-economic, cultural, and political barriers is still necessary since the success of the adoption of conservation agriculture practices will depend on environmental–socio-economic factors (Francaviglia et al., Contribution 7). In this line, conventional and organic management did not significantly affect the decomposition rate of soil organic matter in cranberry soils but varied with soil layer, incubation time, and temperature, with the rate of CO_2 emissions decreasing with elapsed time and the soil subsurface contributing increasingly to CO_2 emissions (Dossou-Yovo et al., Contribution 8). Combinations of multiple conservation management practices, such as crop residues, cover crop planting and termination timing, seeding rate or species selection, tillage practices, and organic amendments such as manure and compost, which affect soil water and nutrient dynamics, may improve soil water functions (Ghimire et al., Contribution 9).

Models based on machine learning algorithms were used to understand the geographical distribution of gully erosion and identify regions more prone to gully formation to ensure appropriate management (Eloudi et al., Contribution 10). We also presented that the use of oat straw mulching decreased soil runoff, and the cost to replace the available nutrients via mineral fertilizer varies from USD 75.4 (no mulching) to USD 2.70 per hectare (8 Mg ha^{-1} oat straw mulching) (Suzuki et al., Contribution 11), showing the efficiency of conservation practices in decreasing soil runoff. In this same way, it was verified that the integration of GIS-based analyses with hydrological modeling at watershed scales provides additional capabilities to quantify the effect of conservation practices on sediment loads by spatially characterizing different types of conservation practices and scenarios and their relative impact on sediment reduction (ElKadiri et al., Contribution 12). The water erosion processes caused by intensive soil trampling on the tourist trails were also contemplated in this Special Issue, indicating strategies for appropriate management and recovery of the degraded trails and avoiding accidents involving visitors (Lima et al., Contribution 13).

In terms of soil functioning and behavior, we should consider that continuous wetting and drying soil cycles may influence water retention, water holding capacity, and water movement (Pires, Contribution 14). Moreover, the soil amendments with biochar and silica have the potential to reduce the adverse effect of salt stress on cucumber (Al-Toobi et al., Contribution 15), which is a strategy to improve the soil environment for crop growth and yield.

Future research was pointed out in this Special Issue, such as quantifying net carbon accumulation in cranberry soils through litter burial by sanding and the development of methodologies for site sampling and monitoring of the cranberry production system to meet sustainable development goals for emissions (Dossou-Yovo et al., Contribution 8). The need to develop region-specific, stakeholder-driven approaches for a more reliable estimate of soil health in water-limited environments like arid and semi-arid regions of the USA, since the indicators or weighting of them may differ from more humid regions or according to climate context (Ghimire et al., Contribution 9). The soil organic carbon content can be a good indicator of the effectiveness of the adoption of a certain conservation agriculture practice, considering its agro-environmental benefits and its potential for climate change mitigation, despite the fact that studies are necessary to accurately assess soil organic carbon (SOC) gains and address the limitations of soil organic carbon sequestration, especially the uncertainty associated with SOC estimations at the farm level (Francaviglia et al., Contribution 7).

Soil and site-specific studies are necessary to define the soil moisture needed to remove animals from wet soils to reduce or avoid soil compaction caused by hoof pressure (Trimarco et al., Contribution 1), as well as the knowledge of the precompression stress values and their relationship with soil structure to avoid additional compaction caused by machinery traffic and animal trampling (Suzuki et al., Contribution 2).

Further studies are still necessary to understand the mechanisms that catalyze the synergistic toxicity of the coexistence of metals and microplastics in soils, which may pose risks to soil, plants, and human health, and to find appropriate appointments to reduce such risks (Bethanis and Golia, Contribution 3).

Our Special Issue also reinforces the following: challenging application of study results to farmers, extension agents, and stakeholders, such as conservationist agriculture; mulching on the ground is kept; a crop rotation system is planned and not only a crop succession (e.g., wheat–soybean); replacement of fallow by cover crops; and usage of contour tillage and terraces wherever necessary (Thomaz and Kurasz, Contribution 5; Suzuki et al., Contribution 11); and challenges in the management of sandy soils (Suzuki et al., Contribution 4).

Models based on machine learning algorithms for determining the gully formation need further studies to test their performance under many subdivisions of the input data in order to improve the prediction (Eloudi et al., Contribution 10), as well as their applicability for identification of different types of practices, controlling parameters, and their location in the watershed to reduce sediment load (ElKadiri et al., Contribution 12).

Our Special Issue achieved, directly or indirectly, at least some of the 17 sustainable development goals (SDGs) [22], especially the goals (2) zero hunger, (6) clean water and sanitation, (13) climate action, and (15) life on land. Goal (2) zero hunger: The papers present strategies to improve the soil environment for crop growth and yield, although more food yield does not mean equal distribution and access. Goal (6) clean water and sanitation: Includes improving soil quality, reducing the input of sediments into the water, and maintaining clean water. Goal (13) climate action: Some papers approached carbon sequestration, which meets the climate action to combat climate change and its impacts. Goal (15) life on land: Some papers met the actions to protect, restore, and promote sustainable use of terrestrial ecosystems.

The Special Issue "Soil Management and Conservation" presented some advances and contributions to the knowledge of some topics addressed and indicated the possibility for future research.

Conflicts of Interest: The author declares no conflicts of interest.

List of Contributions

1. Trimarco, T.; Brummer, J.E.; Buchanan, C.; Ippolito, J.A. Tracking soil health changes in a management-intensive grazing agroecosystem. *Soil Syst.* **2023**, *7*, 94. https://doi.org/10.3390/soilsystems7040094.
2. Suzuki, L.E.A.S.; Reinert, D.J.; Secco, D.; Fenner, P.T.; Reichert, J.M. Soil structure under forest and pasture land-uses affecting compressive behavior and air permeability in a subtropical soil. *Soil Syst.* **2022**, *6*, 98. https://doi.org/10.3390/soilsystems6040098.
3. Bethanis, J.; Golia, E.E. Revealing the combined effects of microplastics, Zn, and Cd on soil properties and metal accumulation by leafy vegetables: a preliminary investigation by a laboratory experiment. *Soil Syst.* **2023**, *7*, 65. https://doi.org/10.3390/soilsystems7030065.
4. Suzuki, L.E.A.S.; Pedron, F.d.A.; Oliveira, R.B.d.; Rovedder, A.P.M. Challenges in the management of environmentally fragile sandy soils in Southern Brazil. *Soil Syst.* **2023**, *7*, 9. https://doi.org/10.3390/soilsystems7010009.
5. Thomaz, E.L.; Kurasz, J.P. Long term of soil carbon stock in no-till system affected by a rolling landscape in Southern Brazil. *Soil Syst.* **2023**, *7*, 60. https://doi.org/10.3390/soilsystems7020060.
6. Bonini, C.d.S.B.; Maciel, T.M.d.S.; Moreira, B.R.d.A.; Chitero, J.G.M.; Henrique, R.L.P.; Alves, M.C. Long-term integrated systems of green manure and pasture significantly recover the macrofauna of degraded soil in the brazilian savannah. *Soil Syst.* **2023**, *7*, 56. https://doi.org/10.3390/soilsystems7020056.
7. Francaviglia, R.; Almagro, M.; Vicente-Vicente, J.L. Conservation agriculture and soil organic carbon: principles, processes, practices and policy options. *Soil Syst.* **2023**, *7*, 17. https://doi.org/10.3390/soilsystems7010017.
8. Dossou-Yovo, W.; Parent, S.-É.; Ziadi, N.; Parent, L.E. CO_2 emissions in layered cranberry soils under simulated warming. *Soil Syst.* **2023**, *7*, 3. https://doi.org/10.3390/soilsystems7010003.
9. Ghimire, R.; Thapa, V.R.; Acosta-Martinez, V.; Schipanski, M.; Slaughter, L.C.; Fonte, S.J.; Shukla, M.K.; Bista, P.; Angadi, S.V.; Mikha, M.M.; et al. Soil health assessment and management framework for water-limited environments: examples from the great plains of the USA. *Soil Syst.* **2023**, *7*, 22. https://doi.org/10.3390/soilsystems7010022.

10. Eloudi, H.; Hssaisoune, M.; Reddad, H.; Namous, M.; Ismaili, M.; Krimissa, S.; Ouayah, M.; Bouchaou, L. Robustness of optimized decision tree-based machine learning models to map gully erosion vulnerability. *Soil Syst.* **2023**, *7*, 50. https://doi.org/10.3390/soilsystems7020050.
11. Suzuki, L.E.A.S.; Amaral, R.d.L.d.; Almeida, W.R.d.S.; Ramos, M.F.; Nunes, M.R. Oat straw mulching reduces interril erosion and nutrient losses caused by runoff in a newly planted peach orchard. *Soil Syst.* **2023**, *7*, 8. https://doi.org/10.3390/soilsystems7010008.
12. ElKadiri, R.; Momm, H.G.; Bingner, R.L.; Moore, K. Spatial optimization of conservation practices for sediment load reduction in ungauged agricultural watersheds. *Soil Syst.* **2023**, *7*, 4. https://doi.org/10.3390/soilsystems7010004.
13. Lima, G.M.d.; Guerra, A.J.T.; Rangel, L.d.A.; Booth, C.A.; Fullen, M.A. Water Erosion Processes on the Geotouristic Trails of Serra da Bocaina National Park Coast, Rio de Janeiro State, Brazil. *Soil Syst.* **2024**, *8*, 24. https://doi.org/10.3390/soilsystems8010024.
14. Pires, L.F. Changes in soil water retention and micromorphological properties induced by wetting and drying cycles. *Soil Syst.* **2023**, *7*, 51. https://doi.org/10.3390/soilsystems7020051.
15. Al-Toobi, M.; Janke, R.R.; Khan, M.M.; Ahmed, M.; Al-Busaidi, W.M.; Rehman, A. Silica and Biochar amendments improve cucumber growth under saline conditions. *Soil Syst.* **2023**, *7*, 26. https://doi.org/10.3390/soilsystems7010026.

References

1. Mengel, K.; Kirkby, E.A.; Kosegarten, H.; Appel, T. The soil as a plant nutrient medium. In *Principles of Plant Nutrition*; Mengel, K., Kirkby, E.A., Kosegarten, H., Appel, T., Eds.; Springer: Dordrecht, The Netherlands, 2001; pp. 15–110. Available online: https://link.springer.com/chapter/10.1007/978-94-010-1009-2_2 (accessed on 10 March 2024).
2. Suzuki, L.E.A.S.; Lima, C.L.R.; Reinert, D.J.; Reichert, J.M.; Pilon, C.N. Estrutura e armazenamento de água em um Argissolo sob pastagem cultivada, floresta nativa e povoamento de eucalipto no Rio Grande do Sul. *Rev. Bras. Ciênc. Solo* **2014**, *38*, 94–106. [CrossRef]
3. FAO—Food and Agriculture Organization of the United Nations. *Soils for Nutrition: State of the Art*; FAO: Rome, Italy, 2022; 78p, Available online: https://www.fao.org/documents/card/en/c/cc0900en (accessed on 10 March 2024).
4. Cheng, K.; Xu, X.; Cui, L.; Li, Y.; Zheng, J.; Wu, W.; Sun, J.; Pan, G. The role of soils in regulation of freshwater and coastal water quality. *Philos. Trans. R. Soc. B* **2021**, *B376*, 20200176. Available online: https://royalsocietypublishing.org/doi/10.1098/rstb.2020.0176 (accessed on 10 March 2024). [CrossRef] [PubMed]
5. Lal, R. Soil erosion and sediment transport research in tropical Africa. *Hydrol. Sci. J.* **1985**, *30*, 239–256. [CrossRef]
6. Didoné, E.J.; Minella, J.P.G.; Merten, H. Quantifying soil erosion and sediment yield in a catchment in southern Brazil and implications for land conservation. *J. Soils Sediments* **2015**, *15*, 11. [CrossRef]
7. Global Soil Partnership—GSP. Global Soil Partnership Endorses Guidelines on Sustainable Soil Management. 2016. Available online: https://www.fao.org/global-soil-partnership/resources/highlights/detail/en/c/416516/ (accessed on 10 March 2024).
8. Kaiser, D.R.; Reinert, D.J.; Reichert, J.M.; Collares, G.L.; Kunz, M. Intervalo hídrico ótimo no perfil explorado pelas raízes de feijoeiro em um latossolo sob diferentes níveis de compactação. *Rev. Bras. Ciênc. Solo* **2009**, *33*, 845–855. [CrossRef]
9. Moraes, M.T.; Debiasi, H.; Franchini, J.C.; Mastroberti, A.A.; Levien, R.; Leitner, D.; Schnepf, A. Soil compaction impacts soybean root growth in an Oxisol from subtropical Brazil. *Soil Tillage Res.* **2020**, *200*, 104611. [CrossRef]
10. Nunes, M.R.; Pauletto, E.A.; Denardin, J.E.; Suzuki, L.E.A.S.; van Es, H.M. Dynamic changes in compressive properties and crop response after chisel tillage in a highly weathered soil. *Soil Tillage Res.* **2019**, *186*, 183–190. [CrossRef]
11. Suzuki, L.E.A.S.; Reinert, D.J.; Alves, M.C.; Reichert, J.M. Critical limits for soybean and black bean root growth, based on macroporosity and penetrability, for soils with distinct texture and management systems. *Sustainability* **2022**, *14*, 2958. [CrossRef]
12. Botta, G.F.; Antille, D.L.; Nardon, G.F.; Rivero, D.; Bienvenido, F.; Contessotto, E.E.; Ezquerra-Canalejo, A.; Ressia, J.M. Zero and controlled traffic improved soil physical conditions and soybean yield under no-tillage. *Soil Tillage Res.* **2022**, *215*, 105235. [CrossRef]
13. Minella, J.P.G.; Merten, G.H.; Reichert, J.M.; Santos, D.R. Identificação e implicações para a conservação do solo das fontes de sedimentos em bacias hidrográficas. *R. Rev. Bras. Ciênc. Solo* **2007**, *31*, 1637–1646. Available online: https://www.scielo.br/j/rbcs/a/RGKRtjfsRZjByjJGzjxWh7D/?format=pdf&lang=pt (accessed on 10 March 2024). [CrossRef]
14. Tiecher, T.; Minella, J.P.G.; Caner, L.; Evrard, O.; Mohsin Zafar, M.; Capoane, V.; Gall, M.L.; Santos, D.R. Quantifying land use contributions to suspended sediment in a large cultivated catchment of Southern Brazil (Guaporé River, Rio Grande do Sul). *Agric. Ecosyst. Environ.* **2017**, *237*, 95–108. Available online: https://hal.science/hal-01686523/document (accessed on 10 March 2024). [CrossRef]
15. Tiecher, T.; Minella, J.P.G.; Evrard, O.; Caner, L.; Merten, G.H.; Capoane, V.; Didoné, E.J.; Santos, D.R. Fingerprinting sediment sources in a large agricultural catchment under no-tillage in Southern Brazil (Conceição River). *Land Degrad. Dev.* **2018**, *29*, 939–951. [CrossRef]
16. Dorici, M.; Costa, C.W.; Moraes, M.C.P.; Piga, F.G.; Lorandi, R.; Lollo, J.A.; Moschini, L.E. Accelerated erosion in a watershed in the southeastern region of Brazil. *Environ. Earth Sci.* **2016**, *75*, 1301. [CrossRef]

17. Lense, G.H.E.; Servidoni, L.E.; Parreiras, T.C.; Santana, D.B.; Bolleli, T.M.; Ayer, J.E.B.; Spalevic, V.; Mincato, R.L. Modeling of soil loss by water erosion in the Tietê River Hydrographic Basin, São Paulo, Brazil. *Semina Ciênc. Agrár.* **2022**, *43*, 1403–1422. [CrossRef]
18. Soil Survey Staff. *Keys to Soil Taxonomy*, 12th ed.; USDA—Natural Resources Conservation Service: Washington, DC, USA, 2014; 142p. Available online: https://www.nrcs.usda.gov/wps/portal/nrcs/detail/soils/survey/class/taxonomy/?cid=nrcs142p2 _053580 (accessed on 8 December 2021).
19. IUSS Working Group WRB. *World Reference Base for Soil Resources 2014, Update 2015 International Soil Classification System for Naming Soils and Creating Legends for Soil Maps*; World Soil Resources Reports No. 106; FAO: Rome, Italy, 2015.
20. Santos, H.G.; Jacomine, P.K.T.; Anjos, L.H.; Oliveira, V.A.; Lumbreras, J.F.; Coelho, M.R.; Almeida, J.A.; Araujo Filho, J.C.; Oliveira, J.B.; Cunha, T.J.F. *Sistema Brasileiro de Classificação de Solos*; Embrapa: Brasília, Brasil, 2018; Available online: https://ainfo.cnptia.embrapa.br/ digital/bitstream/item/199517/1/SiBCS-2018-ISBN-9788570358004.pdf (accessed on 29 November 2021).
21. IBGE—Instituto Brasileiro de Geografia e Estatística. Mapa Brasil Climas—Escala 1:5.000.000. 1978. Available online: https://geoftp. ibge.gov.br/informacoes_ambientais/climatologia/mapas/brasil/Map_BR_clima_2002.pdf (accessed on 26 February 2023).
22. Department of Economic and Social Affairs, United Nations. *Sustainable Development. The 17 Goals*. 2024. Available online: https://sdgs.un.org/goals (accessed on 12 March 2024).

 *soil systems*

Article

Soil Structure under Forest and Pasture Land-Uses Affecting Compressive Behavior and Air Permeability in a Subtropical Soil

Luis Eduardo Akiyoshi Sanches Suzuki [1,*], Dalvan José Reinert [2], Deonir Secco [3], Paulo Torres Fenner [4] and José Miguel Reichert [2]

[1] Center of Technological Development, Federal University of Pelotas, Pelotas 96010-610, RS, Brazil
[2] Soils Department, Federal University of Santa Maria, Santa Maria 97105-9000, RS, Brazil
[3] Centro de Ciências Exatas e Tecnológicas, Universidade Estadual do Oeste do Paraná-UNIOESTE, Cascavel 85819-110, PR, Brazil
[4] Department of Forest Sciences, UNESP Universidade Estadual Paulista, Botucatu 18610-307, SP, Brazil
* Correspondence: luis.suzuki@ufpel.edu.br

Abstract: Machinery traffic and animal trampling can deform the soil and, consequently, impair soil pore functioning. This study aimed to evaluate how soil structure affects the compressibility, physical properties and air permeability of a Typic Paleudalf under forest, pasture and eucalyptus. Soil samples with preserved structure were used to determine soil physical (bulk density, porosity, degree of water saturation at 33 kPa-tension, air permeability) and mechanical properties (soil deformation, precompression stress, compressibility index). After these evaluations, each soil sample was fragmented, sieved, and the metal rings filled with structureless soil, and underwent the same determinations as the samples with preserved structure. For loads greater than the precompression stress (load greater than 200 kPa), soil with non-preserved structure had the largest deformation. An increase in bulk density decreased macropores linearly ($R^2 = 0.77$ and 0.87, respectively, to preserved and non-preserved soil structure) and air flow exponentially. The soil with preserved structure was less susceptible to further compaction. Air flow was greatest in soils with lower bulk density, microporosity and water saturation degree, and a high volume of macropores. Soil structure (preserved and non-preserved) had more significant differences in microporosity, compressibility index, soil deformation, and bulk density at the end of the compression test.

Keywords: compressibility; precompression stress; soil compaction; soil permeability

Citation: Suzuki, L.E.A.S.; Reinert, D.J.; Secco, D.; Fenner, P.T.; Reichert, J.M. Soil Structure under Forest and Pasture Land-Uses Affecting Compressive Behavior and Air Permeability in a Subtropical Soil. *Soil Syst.* **2022**, *11*, 98. https://doi.org/ 10.3390/soilsystems6040098

Academic Editor: Heike Knicker

Received: 1 November 2022
Accepted: 10 December 2022
Published: 15 December 2022

Publisher's Note: MDPI stays neutral with regard to jurisdictional claims in published maps and institutional affiliations.

1. Introduction

Soils are responsible for many processes essential to life [1], serving as a substrate to support plant growth, a reservoir of nutrients [2,3], and the site for many biological processes involving the decomposition and cycling of animal and plant compounds [4–6]. Soil influences air and water quality through interactions with the atmosphere, and as a system for storing [7] and purifying water flowing through the soil profile [8].

Soil structural quality is essential for proper pore functioning for water and air flow and biological activity, all important for the maintenance of life. In cultivated soil, its structure is affected by machinery traffic, animal trampling and soil tillage, for example. Management practices that alter the classes of pores with larger diameters will directly affect flows of air and water in the soil [9]. Soil tillage alters the mechanical strength of soil aggregates, pore continuity, and hydraulic, gas and heat fluxes [10]. Soil as a three-phase system (solid–air–water) has limited resilience and sustainability and, when stress-supporting limits are exceeded, soil properties and functions are affected, particularly pore size and distribution, affecting the flux of water, gas and heat [11].

Soil precompression stress and the compression index are useful indicators to apprise, respectively, soil load bearing capacity and susceptibility to compaction [12,13]. When pressure applied to the soil is lower than the precompression stress, elastic (recoverable) deformation occurs in the soil, and physical properties undergo minor changes. However, with pressure greater than the load-bearing capacity of the soil, plastic deformation (non-recoverable) occurs, and soil physical properties change considerably [12,13].

A soil with good physical quality allows infiltration, retention and availability of water to plants, streams and subsurface; responds to management and resists degradation; allows exchange of heat and gases with the atmosphere and plant roots; and enables root growth [14]. Water, oxygen, temperature and root penetration resistance directly affect plant growth. These properties are affected by bulk density, aggregation, aggregate stability and pore size distribution, all indirectly related to crop growth and yield [15]. Precompression stress and compressibility index, soil mechanical parameters related to machinery traffic and to animal trampling, are associated with soil structure [16–18] and plants [18], and their knowledge may help maintain soil structure adequate for its functioning [19]. The knowledge on the transition from elastic to plastic properties and changes of function of the soil is essential to increase or at least maintain soil functions such as fluxes, root penetrability, filtering and buffering [4].

Some studies were conducted to understand the influence of soil tillage (no-tillage, plowing, chiseling) and land use (annual crops, forest, pasture) on the physical properties of soil [19–34]. However, the knowledge on soil structure, especially porosity (arrangement and continuity, for example), related to soil resistance to support loads and air flow is still incipient.

Thus, for soil physical, chemical and biological processes to contribute to improved environmental quality, soil structure must allow adequate aeration, infiltration and retention of water and exchanges of gases and heat with the atmosphere. Furthermore, field operations that involve soil rupture (e.g., by tillage) and/or machinery traffic and animal trampling can substantially change soil structure (soil loosening or compaction), modifying the conditions that determine root and plant growth and yield, and water and air flows [18,34–36]. Remolded soil or non-preserved structure has been used in compressibility tests, especially in engineering tests to demonstrate the soil supportability of buildings [37–42], but few studies [43] have focused on soil function.

Considering the importance of soil structure, this study aimed to evaluate the effect of soil structure on its compressibility, physical properties, and air permeability. Our hypothesis was that a loose soil has lower load bearing capacity and is more susceptible to soil compaction than a structured soil; and even with lower bulk density and greater macroporosity, the loose soil has lower air permeability due to decreased pore continuity. Therefore, this study contributes towards a better understanding of the relationship between soil structure, compressibility, and permeability.

2. Materials and Methods

2.1. Study Site and Treatments

The study area is located in the municipality of Butiá, in the physiographic region of the Southeast Mountain Range (Rio-Grandense Shield) of the Rio Grande do Sul State, southern Brazil, with geographic coordinates of $30°06'06''$ south latitude and $51°52'18''$ west longitude (Figure 1). According to the Köppen system of climatic classification, the climate in the region is "Cfa" type—subtropical, humid, without drought. Based on 30 years of data (1981 to 2020), the minimum and maximum temperature of the hottest (January) and least hot (July) month is, respectively, 19.4 and 30.8 °C and 9.4 and 19.2 °C, and the rainfall varies from 99.6 mm (March) to 149.7 mm (June), with an annual average of 124.37 mm [44].

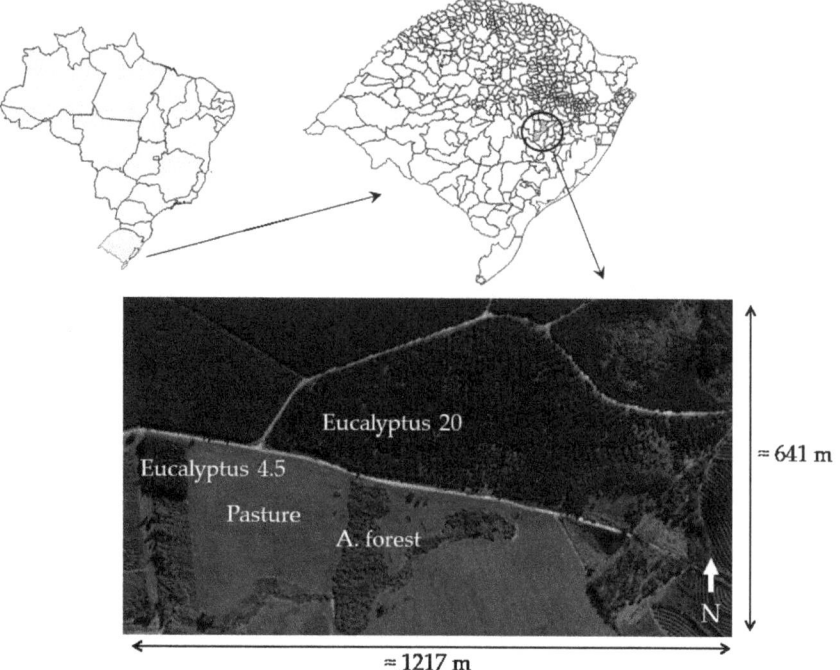

Figure 1. Map of Brazil, with Rio Grande do Sul State shown as hatched; map of Rio Grande do Sul State with Butiá shown as hatched; and image from Google Earth with the land uses studied. Image of Google Earth dated 5 September 2005.

The soil in the area is classified as Typic Paleudalf [45], Umbric Rhodic Acrisol [46] or "Argissolo Vermelho Distrófico" by the Brazilian Soil Classification System [47], with low-activity clays, moderate A horizon (i.e., not included in other categories of A horizon), medium texture in the horizon A/clay in the horizon B with gravel, smooth undulated and undulated relief, and the soil parent material is granite. The uses or treatments in this study were in contiguous areas and were as follows:

(1) Anthropized forest: forest composed by tree and shrub species with a height of approximately 4 m, used as shelter for cattle. Due to the possibility of cattle gaining access to this sampling point in the driest periods, this area was called anthropized forest;

(2) Pasture: 5-y-old pasture, consisting of brachiaria brizanta (*Brachiaria brizantha*) intercropped with Pensacola (*Paspalum lourai*) and clover (*Trifolium* sp.). The pasture was installed in an area of 1200 ha under conventional tillage (plowing and harrowing) in 2001. Before the pasture, there was natural forest and pasture, and soybean intermittently;

(3) Eucalyptus 20: a 20-y-old *Eucalyptus saligna* stand, with conventional tillage used to plant the stand in 1986. Before the eucalyptus, the area consisted of pasture;

(4) Eucalyptus 4.5: clonal *Eucalyptus saligna* in a second rotation, with 4.5 years of age. The original planting occurred in 1993, with soil tillage in strip and a three-stem chisel. The harvesting of eucalyptus in the first cycle, at 8.5 years of age, was performed manually with a chainsaw, and the wood extraction was carried out with a Forwarder Valmet 890 with a load capacity of 18 Mg, without burning the crop residue. The traffic for the harvesting of eucalyptus in the first cycle was at random, with number of passes reaching up to 16. The second planting of eucalyptus was carried out between the rows in 2002. Before the first planting in 1993, the area was used for soybean and pasture.

Soil particle size distribution and total organic carbon content in soils are presented in Table 1.

Table 1. Mean values of gravel, particle size distribution and total organic carbon for the studied land uses and six soil layers.

Layer (m)	Gravel (20–2 mm)	Sand			Silt (0.05–0.002 mm)	Clay (<0.002 mm)	* Total Organic Carbon
		Total (2–0.05 mm)	Coarse (2–0.2 mm)	Fine (0.2–0.05 mm)			
				g kg^{-1}			g dm^{-3}
			Anthropized Forest				
0.00–0.05	8	407	245	162	191	402	34
0.05–0.10	12	385	210	175	193	422	21
0.10–0.20	12	379	213	166	187	434	17
0.20–0.40	23	345	198	147	179	476	14
0.40–0.60	48	293	171	122	165	542	14
0.60–1.00	47	277	167	110	144	579	11
			Pasture				
0.00–0.05	38	362	206	156	193	445	27
0.05–0.10	21	355	200	155	199	446	24
0.10–0.20	36	334	193	141	185	481	19
0.20–0.40	41	301	175	126	165	534	16
0.40–0.60	75	300	186	114	137	563	14
0.60–1.00	68	282	167	115	130	588	12
			Eucalyptus 20				
0.00–0.05	30	374	212	162	161	465	32
0.05–0.10	40	371	213	158	161	468	18
0.10–0.20	75	385	220	165	157	458	17
0.20–0.40	274	353	206	147	156	491	17
0.40–0.60	110	302	185	117	134	564	13
0.60–1.00	97	285	176	109	120	595	11
			Eucalyptus 4.5				
0.00–0.05	14	475	272	203	200	325	34
0.05–0.10	14	460	265	195	194	346	16
0.10–0.20	19	426	240	186	192	382	16
0.20–0.40	55	376	226	150	162	462	15
0.40–0.60	47	314	188	126	151	535	14
0.60–1.00	37	288	171	117	141	571	9

* Source: [48].

2.2. Soil Sampling and Analyses

Samples of soil with preserved or undisturbed structure, as it was in the field, were collected in September 2006. For this purpose, three trenches in each use were opened and, in each trench, two samples per soil layer were collected, totaling six replicates per layer. The samples were collected in metal cylinders of 2.5 cm height and 6.1 cm diameter, in the 0.025 to 0.05 m, 0.10 to 0.125 m, and 0.20 to 0.225 m soil layers. These were saturated by capillarity and, later, positioned on a tension table at 0.60 m of water column to determine the microporosity [49]. Soil macroporosity was calculated by the difference between the total porosity and the microporosity. The total porosity was calculated by the equation:

$$Tp = 1 - (Bd/Pd) \tag{1}$$

where Tp is the total porosity (m^3 m^{-3}), Bd is the bulk density (Mg m^{-3}), and Pd is the particle density (Mg m^{-3}). Soil particle density was determined by the method proposed by Gubiani et al. [50], in soil samples with non-preserved structure collected in September 2006 in three trenches within each use, in the 0.00–0.05; 0.10–0.20 and 0.20–0.40 m soil layers.

The soil samples with preserved structure were re-saturated, equilibrated at a tension of 33 kPa using Richards pressure chambers [51] and, then, submitted to the air permeability test using an air permeameter. Permeability was calculated as:

$$K = \rho * g * [(\Delta v * L) / (\Delta t * \Delta p * A)] \qquad (2)$$

where K is the air permeability (m s^{-1}), ρ is the air density in the moment of measurement (kg m^{-3}), g is the acceleration of gravity (m s^{-2}), Δv is the reading on the flowmeter (m^3), L is the height of the cylinder (m), Δt is the time (minutes), Δp is the air pressure applied (hPa) and A is the area of the cylinder (m^2). We used $\rho = 1.169$ kg m^{-3}, g = 9.81 m s^{-2}, $\Delta t = 1$ min, and $\Delta p = 1$ hPa. Air density was calculated as:

$$\rho = \rho_n * [(T_n * p) / (p_n * T)] \qquad (3)$$

where ρ is the air density in the moment of measuring (kg m^{-3}), ρ_n is the standard air density (kg m^{-3}), T_n is the standard temperature ($^\circ$K), p is the atmospheric pressure in the measurement (mbar), p_n is standard atmospheric pressure (mbar), and T is the temperature in the measurement ($^\circ$K). We used: ρ_n (atmospheric pressure of 1013 mbar and temperature of 273.15 $^\circ$K = 0 $^\circ$C) = 1293 kg m^{-3}, T_n (0 $^\circ$C) = 273.15 $^\circ$K, p = 1000 mbar, p_n = 1013 mbar, and T (25 $^\circ$C) = 298.15 $^\circ$K.

After the air permeability test, the soil samples were submitted to a uniaxial compression test in the laboratory, with a five minutes application of successive static loads of 12.5, 25, 50, 100, 200, 400, 800 and 1600 kPa in the Terraload model S−450 (Durham Geo-Enterprises) consolidator, with pressure applied by compressed air. Maximum soil deformation was determined by following the methodology of Silva et al. [52], without considering pore water pressure changes during the test, since our apparatus had no such capability. Although this loading time might be considered a short interval in the multistep loading because of water pressure, as discussed in Rosa et al. [53], with the possibility of saturation and prevention, this loading time allows more than 99% of soil deformation. After the compression test, the soil samples were oven dried at 105 $^\circ$C.

Before the compression test, soil bulk density (Bd) and degree of water saturation at 33 kPa matric tension (Sd) were calculated. Based on the vertical displacement measured in the laboratory by the consolidometer after the application of each load, the deformation (Def) of the soil at the end of the test was calculated. The compressibility index (Ci) and the precompression stress (Pcs) were calculated using Casagrande's method [54]. Soil compression curves were plotted relating the observed bulk density to the applied pressure in the uniaxial compression test.

After performing the determinations of macroporosity, microporosity, total porosity, air permeability, bulk density and compressibility with the soil samples with preserved structure, the samples from each ring were unstructured so that the particles passed through a 2 mm mesh sieve. Then the rings were filled with their respective soil (particles > and <2 mm), suffering a slight compaction so that all the soil filled the ring, maintaining the original bulk density of the soil sample. That soil was named non-preserved structure. The samples went through the same processes (saturated by capillarity, submitted at 0.60 m tension on a tension table and at 33 kPa tension using Richards pressure chambers, and oven dried at 105 $^\circ$C) and determinations (bulk density, macroporosity, microporosity, total porosity, air permeability, degree of water saturation at 33 kPa matric tension and compressibility) of the samples with preserved structure.

A completely randomized design was used, comparing samples with preserved and non-preserved structure for each soil layer and land use. The analysis of variance and the Tukey test of means were performed considering 5% significance, as well as regression analysis considering the properties evaluated.

3. Results

Soil macroporosity, total porosity and initial bulk density were not significantly ($p > 0.05$) influenced by soil structure (Tables 2 and 3), whereas microporosity was significantly ($p < 0.05$) influenced by soil structure (preserved and non-preserved), with an increase in the soil with non-preserved structure in the eucalyptus areas. However, this increase in microporosity was not reflected in significant differences in total porosity ($p > 0.05$). The unstructured, sieving and reorganization of soil particles during sample accommodation in the cylinder with non-preserved structure may have contributed to the increase in the microporosity and decrease (not statistically significative) in the macroporosity.

Table 2. Coefficient of variation (cv) and mean values of macroporosity, microporosity and total porosity, for soil with preserved (Pres) and non-preserved (NPres) structure under different land uses and layers.

Layer, m	Macroporosity, m^3 m^{-3}			Microporosity, m^3 m^{-3}			Total Porosity, m^3 m^{-3}		
	Pres	NPres	cv, %	Pres	NPres	cv, %	Pres	NPres	cv, %
				Forest					
0.025–0.05	0.109 a	0.149 a	41.73	0.367 a	0.337 b	5.93	0.475 a	0.486 a	7.54
0.10–0.125	0.159 a	0.183 a	17.10	0.347 a	0.335 a	6.42	0.506 a	0.518 a	2.64
0.20–0.225	0.150 a	0.146 a	18.66	0.336 a	0.348 a	5.06	0.486 a	0.495 a	2.54
				Pasture					
0.025–0.05	0.093 a	0.094 a	35.96	0.356 a	0.370 a	4.03	0.449 a	0.463 a	4.90
0.10–0.125	0.105 a	0.107 a	26.10	0.358 a	0.366 a	3.84	0.463 a	0.473 a	4.25
0.20–0.225	0.140 a	0.126 a	46.87	0.342 a	0.363 a	9.21	0.482 a	0.489 a	6.72
				Eucalyptus 20					
0.025–0.05	0.354 a	0.333 a	10.25	0.237 b	0.258 a	4.78	0.591 a	0.591 a	5.00
0.10–0.125	0.226 a	0.204 a	52.43	0.287 a	0.315 a	17.59	0.513 a	0.519 a	11.95
0.20–0.225	0.205 a	0.196 a	26.27	0.303 a	0.319 a	11.66	0.508 a	0.515 a	4.13
				Eucalyptus 4.5					
0.025–0.05	0.082 a	0.068 a	54.58	0.299 b	0.339 a	3.08	0.381 a	0.407 a	9.97
0.10–0.125	0.127 a	0.099 a	51.04	0.286 b	0.330 a	6.24	0.413 a	0.429 a	11.67
0.20–0.225	0.120 a	0.085 b	17.09	0.311 b	0.355 a	4.20	0.432 a	0.440 a	4.08

Means followed by same letters in a given line, for each physical property, do not differ statistically from each other by Tukey's test at 5% significance.

Although the initial bulk density was equal for soil with preserved and non-preserved structure, the latter soil reached the highest values at the end of the compression test and, consequently, the largest soil deformation (Table 3, Figures 2–5). As the initial bulk density increased, there was a decrease in soil deformation, and this decrease was more pronounced in soil with preserved structure (Figure 6a). Macropores decreased as bulk density increased ($R^2 = 0.77$ and 0.87, respectively, for preserved and non-preserved soil structure) (Figure 6b); therefore, the soil became less compressive, i.e., lower deformation occurred ($R^2 = 0.88$ and 0.32, respectively, for preserved and non-preserved soil structure) (Figure 6c). With an increase in the initial bulk density, there was an increase in the range of the final bulk density between soil with preserved and non-preserved structure ($R^2 = 0.46$ and 0.74, respectively, for preserved and non-preserved soil structure) (Figure 6d).

Table 3. Coefficient of variation (cv) and mean values of bulk density in the beginning and in the end of the uniaxial compression test, and deformation, for soil with preserved (Pres) and non-preserved (NPres) structure under different land uses and layers.

Layer, m	Bulk Density Initial, Mg m^{-3}			Bulk Density Final, Mg m^{-3}			Deformation, mm		
	Pres	NPres	cv, %	Pres	NPres	cv, %	Pres	NPres	cv, %
				Forest					
0.025–0.05	1.28 a	1.25 a	6.93	1.64 b	1.80 a	3.31	0.551 b	0.759 a	12.41
0.10–0.125	1.25 a	1.23 a	2.69	1.72 a	1.81 b	1.61	0.673 b	0.792 a	8.76
0.20–0.225	1.30 a	1.27 a	2.41	1.73 b	1.85 a	1.72	0.624 b	0.783 a	6.11
				Pasture					
0.025–0.05	1.38 a	1.32 a	3.39	1.73 b	1.89 a	3.03	0.513 b	0.744 a	7.91
0.10–0.125	1.36 a	1.33 a	3.77	1.72 b	1.88 a	3.61	0.538 b	0.728 a	6.24
0.20–0.225	1.29 a	1.27 a	6.67	1.69 a	1.78 a	6.14	0.593 a	0.711 a	17.38
				Eucalyptus 20					
0.025–0.05	1.03 a	0.99 a	6.33	Not determined			Not determined		
0.10–0.125	1.19 a	1.21 a	13.24	1.79 a	1.89 a	4.87	0.639 a	0.756 a	15.03
0.20–0.225	1.23 a	1.21 a	4.35	1.78 a	1.78 a	3.65	0.769 a	0.702 a	11.46
				Eucalyptus 4.5					
0.025–0.05	1.50 a	1.47 a	6.82	1.87 b	1.99 a	2.88	0.483 b	0.650 a	16.10
0.10–0.125	1.47 a	1.43 a	8.48	1.86 a	1.96 a	5.77	0.539 a	0.656 a	16.70
0.20–0.225	1.44 a	1.42 a	3.19	1.82 b	1.96 a	2.38	0.530 b	0.695 a	8.25

Means followed by same letters in a given line, for each physical property, do not differ statistically from each other by Tukey's test at 5% significance.

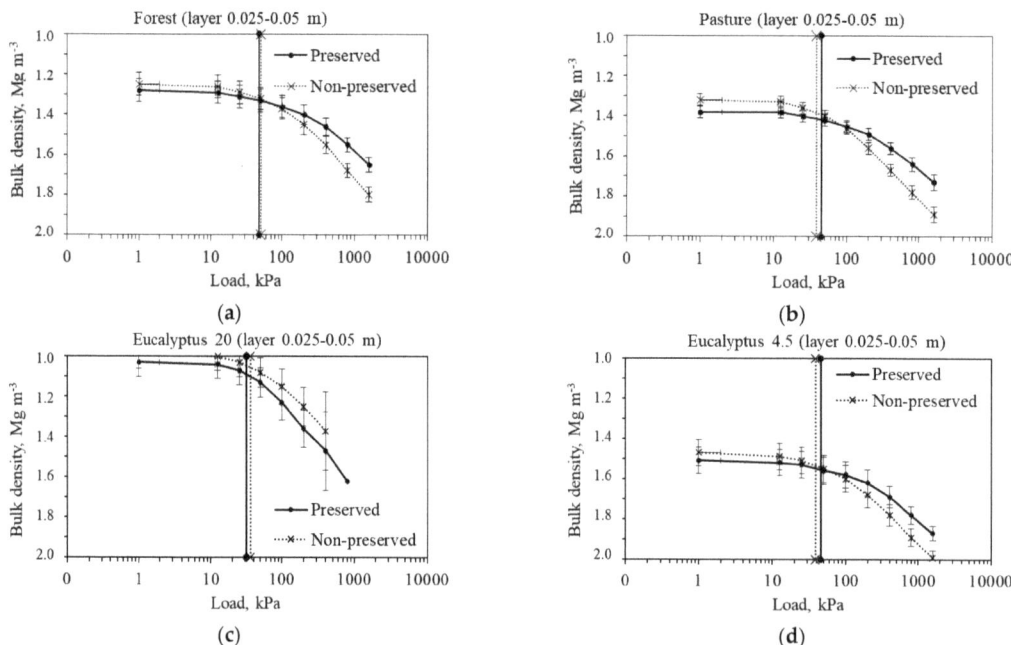

Figure 2. Soil compression curve for soil with preserved and non-preserved structure in the 0.025–0.05 m soil layer for four land uses. Vertical error bars for each pressure indicate the least significance difference, while vertical bars that accompany the superior and inferior axes in the figure indicate the precompression stress value for the preserved and non-preserved soil structures.

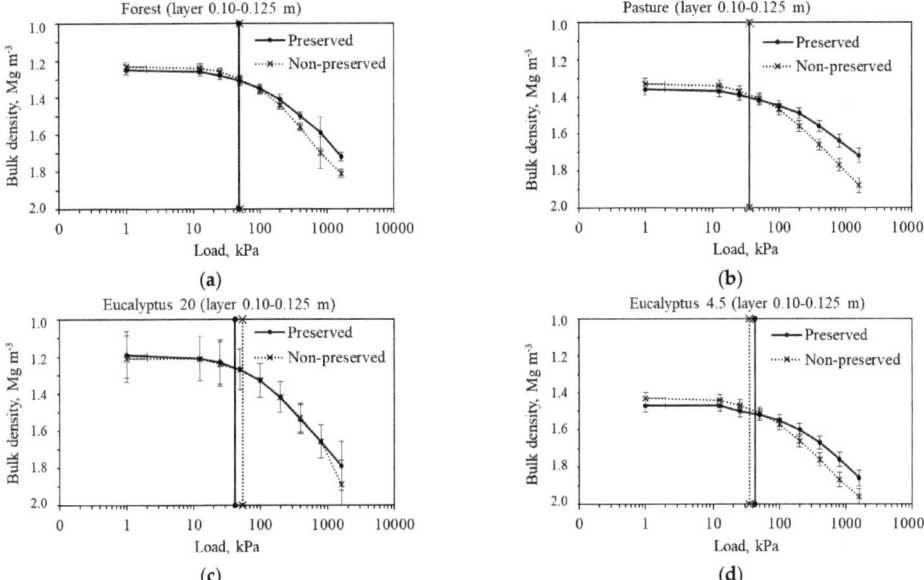

Figure 3. Soil compression curve for soil with preserved and non-preserved structure in the 0.10–0.125 m soil layer for four land uses. Vertical error bars for each pressure indicate the least significance difference, while vertical bars that accompany the superior and inferior axes in the figure indicate the precompression stress value for the preserved and non-preserved soil structures.

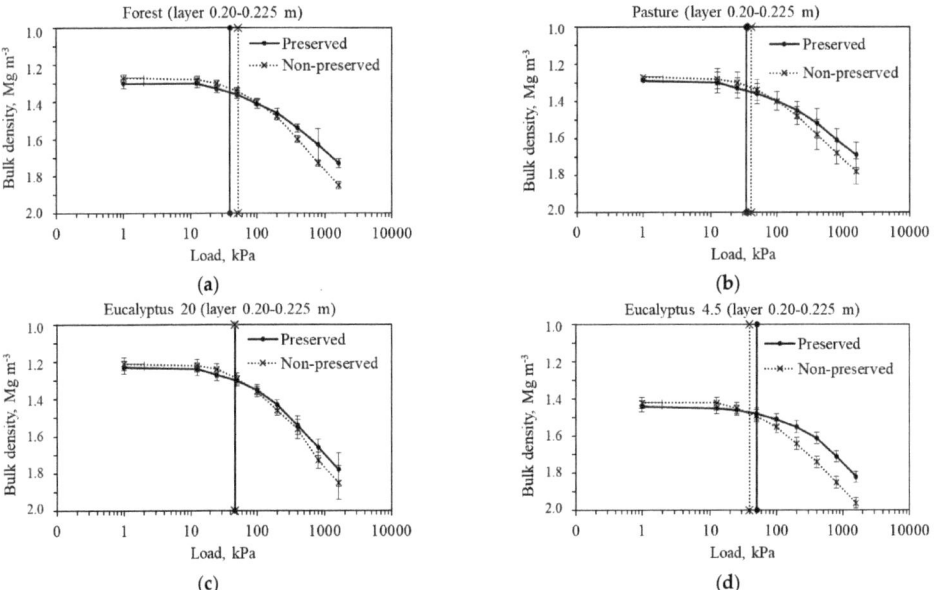

Figure 4. Soil compression curve for soil with preserved and non-preserved structure in the 0.20–0.225 m soil layer for four land uses. Vertical error bars for each pressure indicate the least significance difference, while vertical bars that accompany the superior and inferior axes in the figure indicate the precompression stress value for the preserved and non-preserved soil structures.

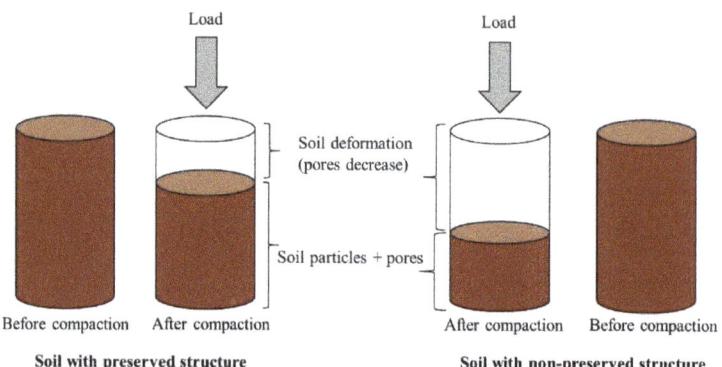

Figure 5. Soil with preserved and non-preserved structure with the same bulk density before the uniaxial compression test and the differences in soil deformation and pores decrease at the end of the test.

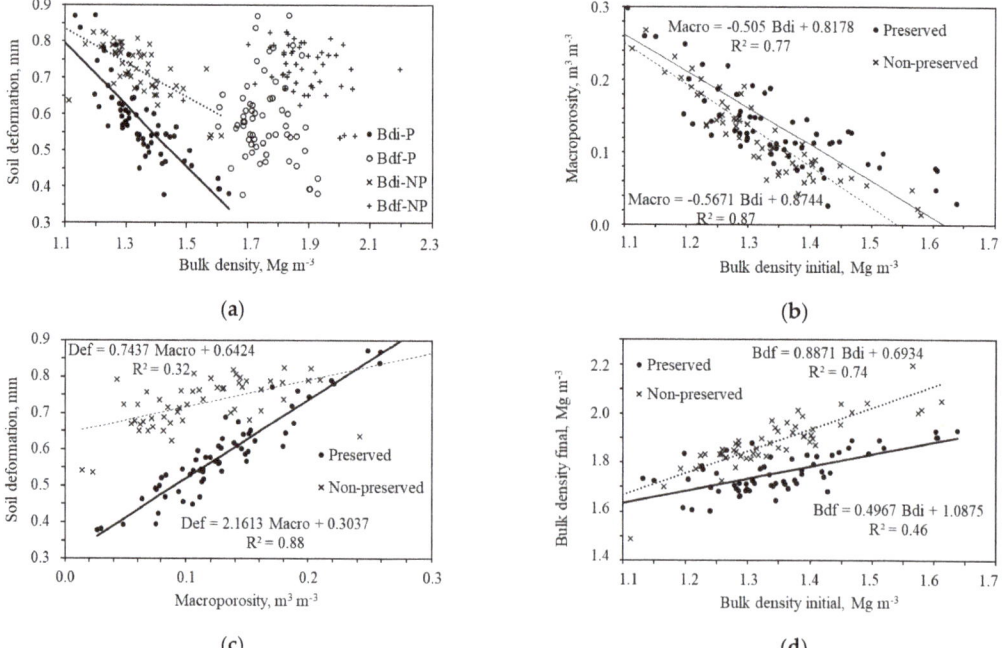

Figure 6. Regression between physical properties for soil with preserved (P) and non-preserved (NP) structure. Macro = macroporosity; Bdi and Bdf = bulk density in the beginning and in the final of the uniaxial compression test, respectively; Def = soil deformation in the final of the uniaxial compression test.

Soil precompression stress was similar for both types of soil structure ($p > 0.05$) (Table 4), while differences between soil structure occurred at loads greater than the precompression stress, i.e., greater than 200 kPa (Figures 2–4). Soil compressibility index was affected by soil structure ($p < 0.05$) for forest and pasture uses, where the non-preserved structure presented highest values (Table 4), i.e., the soil was more susceptible to compaction. The increase in bulk density ($R^2 = 0.79$ and 0.76, respectively, for preserved and non-preserved soil structure) (Figure 7a) and the degree of water saturation ($R^2 = 0.78$

and 0.65, respectively, for preserved and non-preserved soil structure) (Figure 6b) was associated with a decrease in the compressibility index.

Table 4. Coefficient of variation (cv) and average values of precompression stress, compressibility index, degree of water saturation and air permeability for soil with preserved (Pres) and non-preserved (NPres) structure under different land uses and layers.

Layer, m	Precompression Stress, kPa			Compressibility Index			Degree of Water Saturation, %			Air Permeability, mm h⁻¹		
	Pres	NPres	cv, %	Pres	NPres	cv, %	Pres	NPres	cv, %	Pres	NPres	cv, %
						Forest						
0.025–0.05	47.53 a	49.85 a	18.46	0.25 b	0.39 a	18.69	66.52 a	52.71 b	16.67	17.29 a	27.71 a	87.15
0.10–0.125	48.10 a	49.85 a	20.18	0.33 a	0.38 a	16.67	57.07 a	50.37 b	7.84	34.55 a	30.69 a	80.60
0.20–0.225	39.35 b	51.92 a	17.42	0.28 b	0.40 a	7.89	61.73 a	54.04 b	9.56	19.03 a	19.78 a	95.27
						Pasture						
0.025–0.05	44.56 a	38.47 a	20.31	0.21 b	0.34 a	10.51	69.62 a	64.21 a	6.74	26.09 a	10.07 a	69.75
0.10–0.125	35.53 a	35.50 a	22.58	0.22 b	0.32 a	9.83	67.39 a	62.34 a	11.93	15.09 a	10.21 a	53.41
0.20–0.225	34.42 a	40.76 a	39.80	0.25 a	0.33 a	21.35	61.94 a	60.04 a	17.74	16.10 a	14.75 a	65.28
						Eucalyptus 20						
0.025–0.05	31.24 a	35.85 a	28.57	0.60 a	0.58 a	16.43	36.65 a	33.92 a	8.58	Not determined		
0.10–0.125	42.20 a	54.40 a	32.70	0.43 a	0.45 a	26.76	45.87 a	47.98 a	29.49	192.70 a	110.99 a	78.48
0.20–0.225	46.47 a	46.62 a	27.51	0.38 a	0.45 a	15.23	50.33 a	47.42 a	13.60	66.33 a	119.39 a	67.66
						Eucalyptus 4.5						
0.025–0.05	46.00 a	38.65 a	18.27	0.18 a	0.25 a	24.57	68.45 a	67.83 a	15.84	19.16 a	7.37 a	117.84
0.10–0.125	42.27 a	34.85 a	28.26	0.21 a	0.28 a	24.41	63.23 a	59.95 a	18.08	27.53 a	17.18 a	107.33
0.20–0.225	50.92 a	39.33 a	25.42	0.21 b	0.29 a	12.01	62.02 a	61.27 a	5.11	26.35 a	9.67 a	78.51

Means followed by same letters in a given line, for each physical property, do not differ statistically from each other by Tukey's test at 5% significance.

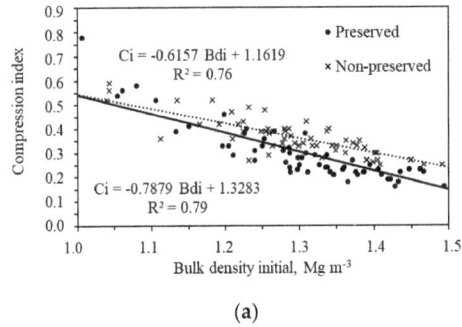

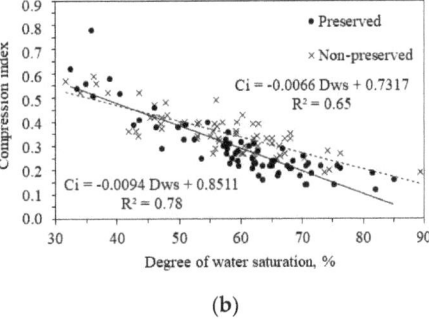

(a)　　　　　　　　　　　　　　　　　　　　(b)

Figure 7. Regression between physical properties for soil with preserved and non-preserved structure. Bdi = bulk density in the beginning of the uniaxial compression test; Dws = degree of water saturation; Ci = compressibility index.

The degree of water saturation was affected by soil structure type (preserved and non-preserved) ($p < 0.05$) only in forest (Table 4). By decreasing macroporosity ($R^2 = 0.87$ and 0.91, respectively, for preserved and non-preserved soil structure) (Figure 8a) and increasing microporosity ($R^2 = 0.47$ and 0.71, respectively, for preserved and non-preserved soil structure) (Figure 8b), there was an increase in the degree of water saturation. Air permeability did not differ statistically ($p > 0.05$) between soil structure types (Table 4).

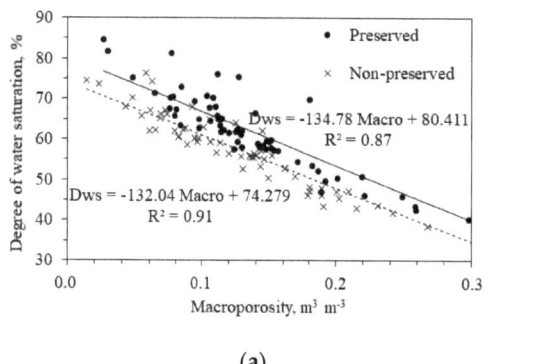 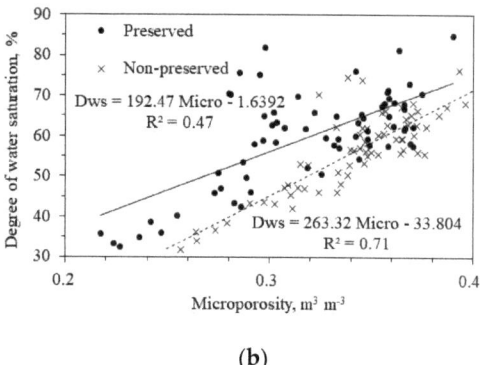

(a) (b)

Figure 8. Regression between physical properties for soil with preserved and non-preserved structure. Macro = macroporosity; Micro = microporosity; Dws = degree of water saturation.

Increase in air permeability was associated with an exponential decrease in bulk density (Figure 9a), degree of water saturation (Figure 9b), microporosity (Figure 9d), and an increase in macropores (Figure 9c). This behavior shows that the air flow occurred mainly in the macropores. By increasing macroporosity (Figure 9c) and reducing the degree of water saturation (Figure 9b), more pores were available for air flow.

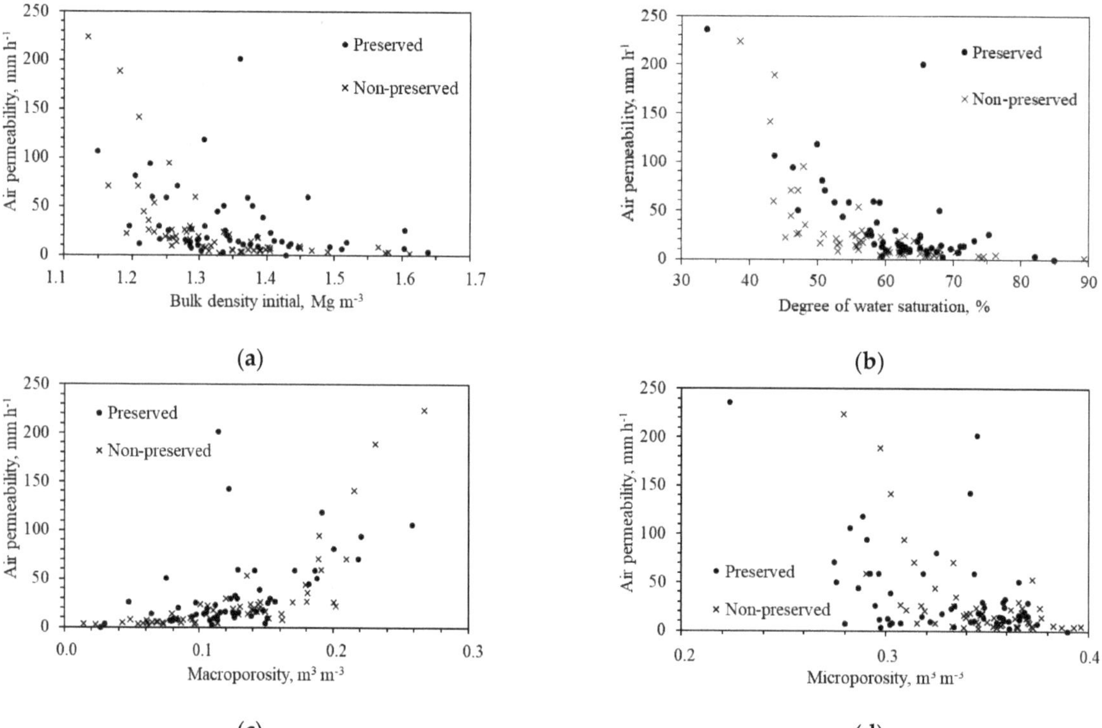

(a) (b)

(c) (d)

Figure 9. Regression between physical properties for soil with preserved and non-preserved structure.

Soil structure condition (preserved or not) had few influences on bulk density, macroporosity, total porosity, air permeability, and precompression stress ($p > 0.05$). However, for loads greater than the precompression stress (load greater than 200 kPa), the soil with non-preserved structure had greater deformation. The results show that compaction reduced macropores and air flow; as a consequence the soil experienced less deformation with further loading and was less susceptible to additional compaction.

4. Discussion

We observed that soil structure (preserved and non-preserved) had significant influence, especially on microporosity, compressibility index, soil deformation, and bulk density at the end of the compression test. Increasing bulk density and degree of water saturation decreased air permeability, soil deformation, macropores and compression index (Figure 10).

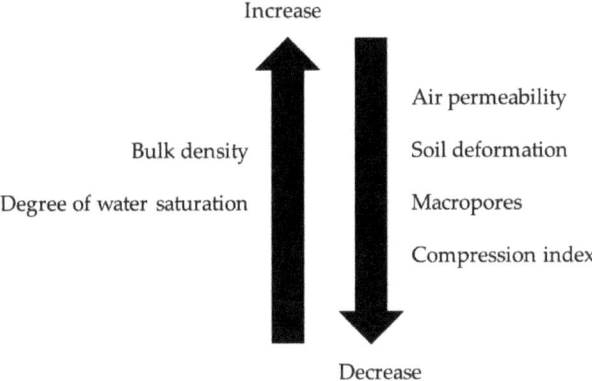

Figure 10. Scheme showing alterations on bulk density and degree of water saturation, with changes on physical and compressive properties and air permeability.

Chiseling and/or harrowing disaggregate the soil and modify the relation of mass/volume in the field [19,34,55]. In our case, in the laboratory the soil mass was the same for both preserved and non-preserved structure, justifying the similar values of bulk density and total porosity in both structural conditions. However, when the soil sample with non-preserved structure was disrupted, sieved and rearranged in the metal ring, the relation between microporosity and macroporosity was modified, with an increase in the microporosity and a decrease in the macroporosity. Pores in the soil with preserved structure were more continuous, formed by the decomposition of roots and biological activity [56,57], while pores in the soil with non-preserved structure were randomly distributed in the soil mass.

We expected greater values of precompression stress for undisturbed soil samples because of the history of loads applied by machinery traffic and animal trampling, and differences in the precompression stress values when comparing the soil structure (preserved and non-preserved), but our expectations were not confirmed. For instance, the pressure applied on the soil by forest machines and by horse-hoof can exceed 300 kPa [58]. When comparing soil tillage treatments (no-tillage, chisel plow, conventional tillage), Veiga et al. [43] obtained differences using undisturbed soil samples, but less difference in the precompression stress between treatments when using remolded soil samples, and suggested that remolding soil samples eliminates the effect of age hardening and soil aggregation.

Furthermore, in our study, we did not observe significative differences in the porosity and bulk density that could differently influence the precompression stress values when comparing preserved and non-preserved soil structure. For instance, Suzuki et al. [59] demonstrated the negative and positive correlation between precompression stress with, respectively, total porosity and bulk density, and Nunes et al. [18] showed correlation of precompression stress with macroporosity and bulk density.

Our precompression stress values (31.24 to 54.40 kPa) were low, suggesting a possible effect of mineralogy and contents of gravel and sand. The studied soil was derived from granite substrate and may have contained micas in its mineralogical composition, including in the clay fraction 1:1 clay minerals, such as kaolinite dominant in the clay fraction, and feldspars in the sand and silt fractions [60]. Horn and Lebert [61] stated that soil compressibility depends on soil strength, particle size distribution, type of clay mineral, content and type of organic substances, root distribution, soil bulk density, soil distribution, pore size and pore continuity in soil and simple aggregates, and water content and/or water potential. The resistance of the soil to decrease its volume when subjected to pressure is less pronounced in sandy soils and less aggregated. The increment in clay content increased the precompression stress, while the compression index decreased in denser soils and increased in clay soils and with higher void index, except in higher soil moisture [62]. Sandy soils retain less water on their surfaces [63–67], and present greater friction resistance between soil particles, which makes it difficult for particles to move to close-together positions [68].

Other studies [69,70] have shown higher precompression stress values (77 to 183 kPa for natural forest, annual crop and pasture areas in Oxisols, and values larger than 230 kPa for non-irrigated and irrigated grazing systems in Hapludalf) than those obtained in our study (31.24 to 54.40 kPa). However, Capurro et al. [71] showed similar values (35 to 47 kPa) in Vertissol under grazing cattle to those of our study, while Horn et al. [58] found values lower than 60 kPa in Inceptisol under forest. Suzuki et al. [59] verified precompression stress values ranging from 57.09 to 232.42 kPa depending on the sampling position (wheel line, interline planting, line planting and near the peach plant) and soil depth in a peach orchard, and that values were correlated positively with bulk density and negatively with total porosity.

As shown, there is a wide range of precompression stress values in the literature, in different soil types, use and management, and in our study the lower values may have been associated to mineralogy, gravel and sand influence. Corroborating with this suggestion, in the same site of the present study, Suzuki et al. [7] verified that soil texture (sand, silt and clay) and organic matter presented greater correlation with mean weight-diameter of aggregates than with properties related to soil structure, such as porosity and bulk density. The authors also found that, even in a small amount, gravel decreased the mean weight–diameter of aggregates because its low reactivity and greater diameter hindered the formation of stable aggregates.

A classification for precompression stress was proposed by Horn and Fleige [72], considering pF values of 1.8 and 2.5, (respectively, for soil when macropores are drained, and soil at field capacity), density and shear strength parameters. The authors classified precompression stress as very low (<30 kPa), low (30–60 kPa), medium (60–90 kPa), high (90–120 kPa), very high (120–150 kPa), and extremely high (>150 kPa). In our study, the range of precompression stress (31.24 to 54.40 kPa) was considered low according to the proposed classification.

Although the initial bulk density was equal for soil with preserved and non-preserved structure, the latter soil reached the highest values of bulk density at the end of the compression test and, consequently, the largest soil deformation. When the internal soil strength is high, the rigidity of the pore system will be more pronounced, and the more elastic the soil will be within the recompression load range [73]. As the initial bulk density increased, there was a decrease in soil deformation and macropores, making the soil less compressive. Suzuki et al. [17,59] also observed that soil with larger bulk density had smaller deformation and was less susceptible to soil compaction (larger load bearing capacity) when submitted to external loads. Powers et al. [74] observed that soil bulk density augmented with increased compaction, particularly in soils with low or moderate initial bulk density, while for soils with higher bulk density this increase was small. This behavior was attributed to the difficulty in compressing smaller pores, caused by high bulk density and pores filled with water. Soil deformation occurs when particles are able to separate and move towards each other, having their movements limited by friction and bonds between particles. Therefore, the more compact the soil and the closer the particles, the greater the friction forces, which are responsible for resistance [75].

With increasing bulk density, the soil becomes less compressive and less susceptible to compaction. Additionally, increasing the degree of water saturation increases moisture, firstly in micropores, resulting in the pore–pressure effect. Water in the micropores receives the applied load and, as the drainage of these pores is very slow, decreases soil susceptibility to compaction. Pore-pressure is the pressure exerted by water that occupies the pore space of the soil and corresponds to a force that can delay the consolidation of a cohesive soil [13].

When soil aggregates (from homogeneous via prismatic to subangular blocky and finally crumbly structure) are formed, the accessibility of particle and pore surfaces is better and maintains site productivity and biodiversity. However, soil compaction and deformation result in a platy rigid structure that is difficult for the roots to access water, ions and gas and change flux directions, and this occurs within the virgin compression stress range [4]. Mentges et al. [76] mention that the type of soil structure (prismatic, massive, for example) should be considered in studies that relate elastic parameters of soil.

Soil with non-preserved structure presented the highest values of compressibility index, while the increase in bulk density and degree of water saturation decreased the compressibility index. Other authors [17,59,77] also observed that the increase in bulk density decreased the compressibility index, while Reichert et al. showed that by increasing moisture, the compressibility index increased as well [77].

Although the types of soil structure did not show statistical differences for macroporosity and air permeability, a greater permeability was expected in samples with preserved structure due to greater pore continuity associated with the activity and root decomposition, while in samples with non-preserved structure there were possible less-continuous pores due to soil disruption and rearrangement. Mechanical deep-ploughing or soil loosening result in less dense soil layers, but they deprive soils of their internal strength and destroy pore continuity and the increased sensitivity to further soil settlement [78]. Even with lower total porosity than in conventional tillage, soil under no-tillage in agricultural areas generally conducts water more efficiently [79], due to bioporosity [57,80]. Mando et al. [56] found the efficiency of biological pores in increasing water infiltration.

We observed that increase in air permeability was associated with a decrease in bulk density, microporosity and degree of water saturation, and an increase in macropores, demonstrating that the air flow occurred mainly in the macropores; while increasing macroporosity and reducing degree of water saturation caused more pores to be available for air flow. During compaction, the larger pores responsible for soil aeration decreased and were replaced by smaller pores, mainly pores that retain water.

This decrease in aeration porosity can be 1.5–2 times greater than the decrease in total pore space. The decrease in the oxygen diffusion coefficient, however, will depend on the geometry and stability of aeration pore channels and deformation degree during compaction [81]. Horn et al. [58] found that soils with low bulk density generally have high air permeability. Soil compaction caused by a tractor changed the pore orientation that persisted two years after the traffic event in a Typic Argiudoll [82]. A long-term no-tillage (around 25 years old), increased soil bulk density and reduced air-filled porosity and macroporosity, but created a continuous and stable pore organization system, which is one of the most important properties for gas transport through soils [83].

With reduction in soil moisture, there was an increase in air permeability because of a greater amount of water-free and continuity of pores available for air flow [84,85]. Mentges et al. [84] also found that, in areas under no-tillage for annual crops, the increase in permeability is greater in sandy soils than in clayey ones. In an area with eucalyptus, the variation in soil saturated hydraulic conductivity and in air permeability was related to pore size distribution, especially for the >300 μm diameter pores [8]. When a load is applied to the soil surface, the stress is transmitted three-dimensionally through the solid, liquid and gas phases. If air permeability in the soil is high enough to allow the immediate deformation of the pores filled with air, the air flow can be interrupted by changes in water content or pore-pressure [13].

5. Conclusions

Our results contribute towards a better understanding of the relationship between soil structure, compressibility and air permeability in a Typic Paleudalf under forest, eucalyptus and pasture, with gravel and clay content ranging from 325 to 595 g kg^{-1}. Total porosity and initial soil bulk density were not influenced by soil structure (preserved and non-preserved), but the relation between macroporosity and microporosity was influenced; moreover, by increasing bulk density, there was a decrease in macropores and in deformation of soil under loading.

Precompression stress was low (<54.40 kPa) and similar between soil structure (condition preserved and non-preserved), refuting one of our hypotheses that preserved structure would have a larger precompression stress due to the history of loads applied by machinery traffic and animal trampling. Structure effect occurred for loads above the precompression stress (load larger than 200 kPa), where non-preserved structure presented a larger deformation. Compressibility index was highest for non-preserved soil under forest and pasture uses. With an increase in bulk density and degree of water saturation, the compressibility index decreased.

Air permeability was not affected by soil structure (preserved and non-preserved) in this soil with presence of gravel, and increase in air permeability was associated with a decrease in bulk density, microporosity and degree of water saturation, and increase of macropores, refuting our hypothesis since we expected lower air permeability in the loose soil due to the absence of pore continuity.

Soil structure (preserved and non-preserved) significantly influenced microporosity, compressibility index, soil deformation and bulk density at the end of the compression test.

In terms of farm management, both soil structures (preserved and non-preserved) require greater care in machinery traffic and animal trampling because of compaction susceptibility, especially for loads larger than the precompression stress, that can over-compact the soil, increasing bulk density and decreasing macroporosity and air flow. However, loose soil (non-preserved soil structure) requires more care, especially for loads greater than 200 kPa, when the soil becomes more compressive (greater deformation) than preserved structure.

Author Contributions: Conceptualization, L.E.A.S.S. and D.J.R.; formal analysis, L.E.A.S.S., D.J.R., D.S., P.T.F. and J.M.R.; investigation, L.E.A.S.S. and D.J.R.; methodology, L.E.A.S.S., D.J.R., D.S., P.T.F. and J.M.R.; resources, D.J.R.; writing—original draft, L.E.A.S.S. and D.J.R.; writing—review and editing, D.S., P.T.F. and J.M.R. All authors have read and agreed to the published version of the manuscript.

Funding: This research received no external funding.

Institutional Review Board Statement: Not applicable.

Informed Consent Statement: Informed consent was obtained from all subjects involved in the study.

Data Availability Statement: Data available on request from the authors.

Acknowledgments: We thank technicians and student helpers from the "Embrapa Clima Temperado" of Pelotas-RS, for their assistance in field and laboratory tasks. We also thank Capes for the scholarship during the first author's doctorate degree (finance code 001), CNPq for research productivity grants, and landowners for allowing access to the field sites.

Conflicts of Interest: The authors declare no conflict of interest.

References

1. Wienhold, B.J.; Andrews, S.S.; Karlen, D.L. Soil quality: A review of the science and experiences in the USA. *Environ. Geochem. Health* **2004**, *26*, 89–95. [CrossRef] [PubMed]
2. Mengel, K.; Kirkby, E.A.; Kosegarten, H.; Appel, T. The soil as a plant nutrient medium. In *Principles of Plant Nutrition*; Mengel, K., Kirkby, E.A., Kosegarten, H., Appel, T., Eds.; Springer: Dordrecht, The Netherlands, 2001; pp. 15–110. [CrossRef]
3. FAO—Food and Agriculture Organization of the United Nations. *Soils for Nutrition: State of the Art*; FAO: Rome, Italy, 2022; 78p. [CrossRef]
4. Mary, B.; Recous, S.; Darwis, D.; Robin, D. Interactions between decomposition of plant residues and nitrogen cycling in soil. *Plant Soil* **1996**, *181*, 71–82. [CrossRef]
5. Piano, J.T.; Egewarth, J.F.; Frandoloso, J.F.; Mattei, E.; de Oliveira, P.S.R.; Rego, C.A.R.M.; de Herrera, J.L. Decomposition and nutrients cycling of residual biomass from integrated crop-livestock system. *Aust. J. Crop Sci.* **2019**, *13*, 739–745. [CrossRef]
6. Henneron, L.; Kardol, P.; Wardle, D.A.; Cros, C.; Fontaine, S. Rhizosphere control of soil nitrogen cycling: A key component of plant economic strategies. *New Phytol.* **2020**, *228*, 1269–1282. [CrossRef] [PubMed]
7. Suzuki, L.E.A.S.; Lima, C.L.R.; Reinert, D.J.; Reichert, J.M.; Pilon, C.N. Estrutura e armazenamento de água em um Argissolo sob pastagem cultivada, floresta nativa e povoamento de eucalipto no Rio Grande do Sul. *Rev. Bras. Ciênc. Solo* **2014**, *38*, 94–106. [CrossRef]
8. Cheng, K.; Xu, X.; Cui, L.; Li, Y.; Zheng, J.; Wu, W.; Sun, J.; Pan, G. The role of soils in regulation of freshwater and coastal water quality. *Philos. Trans. R. Soc. B* **2021**, *B376*, 20200176. [CrossRef]
9. Prevedello, J.; Vogelmann, E.S.; Kaiser, D.R.; Reinert, D.J. A funcionalidade do sistema poroso do solo em floresta de eucalipto sob Argissolo. *Sci. For.* **2013**, *41*, 557–566. Available online: https://www.ipef.br/PUBLICACOES/SCIENTIA/nr100/cap13.pdf (accessed on 31 October 2022).
10. Horn, R. Soil structure formation and management effects on gas emission. *J. Agric. Mach. Sci.* **2008**, *4*, 13–18. [CrossRef]
11. Horn, R.; Blum, W.E.H. Effect of land-use management systems on coupled physical and mechanical, chemical and biological soil processes: How can we maintain and predict soil properties and functions? *Front. Agric. Sci. Eng.* **2020**, *7*, 243–245. [CrossRef]
12. Holtz, R.D.; Kovacs, W.D. *An Introduction to Geotechnical Engineering*; Prentice-Hall: Hoboken, NJ, USA, 1981.
13. Reichert, J.M.; Reinert, D.J.; Suzuki, L.E.A.S.; Horn, R. Mecânica do solo. In *Física do Solo*; van Lier, Q.J., Ed.; Sociedade Brasileira de Ciência do Solo: Viçosa, Brazil, 2010; pp. 29–102.
14. Reichert, J.M.; Reinert, J.M.; Braida, J.A. Qualidade dos solos e sustentabilidade de sistemas agrícolas. *Ciência Ambiente* **2003**, *27*, 29–48.
15. Letey, J. Relationship between soil physical conditions and crop production. *Adv. Soil Sci.* **1985**, *1*, 277–293. [CrossRef]
16. Lima, C.L.R.; Reinert, D.J.; Reichert, J.M.; Suzuki, L.E.A.S. Compressibilidade de um Argissolo sob plantio direto escarificado e compactado. *Ciênc. Rural* **2006**, *36*, 1765–1772. [CrossRef]
17. Suzuki, L.E.A.S.; Reinert, D.J.; Reichert, J.M.; Lima, C.L.R. Estimativa da suscetibilidade à compactação e do suporte de carga do solo com base em propriedades físicas de solos do Rio Grande do Sul. *Rev. Bras. Ciênc. Solo* **2008**, *32*, 963–973. [CrossRef]
18. Nunes, M.R.; Pauletto, E.A.; Denardin, J.E.; Suzuki, L.E.A.S.; van Es, H.M. Dynamic changes in compressive properties and crop response after chisel tillage in a highly weathered soil. *Soil Tillage Res.* **2019**, *186*, 183–190. [CrossRef]
19. Suzuki, L.E.A.S.; Reinert, D.J.; Fenner, P.T.; Secco, D.; Reichert, J.M. Prevention of additional compaction in eucalyptus and pasture land uses, considering soil moisture and bulk density. *J. S. Am. Earth Sci.* **2022**, *120*, 104113. [CrossRef]
20. Collares, G.L.; Reinert, D.J.; Reichert, J.M.; Kaiser, D.R. Compactação de um latossolo induzida pelo tráfego de máquinas e sua relação com o crescimento e produtividade de feijão e trigo. *Rev. Bras. Ciênc. Solo* **2008**, *32*, 933–942. [CrossRef]
21. Flores, C.A.; Reinert, D.J.; Reichert, J.M.; Albuquerque, J.A.; Pauletto, E.A. Recuperação da qualidade estrutural, pelo sistema plantio direto, de um Argissolo Vermelho. *Ciênc. Rural* **2008**, *38*, 2164–2172. [CrossRef]
22. Suzuki, L.E.A.S.; Lima, C.L.R.; Reinert, D.J.; Reichert, J.M.; Pilon, C.N. Condição estrutural de um Argissolo no Rio Grande do Sul, em floresta nativa, em pastagem cultivada e em povoamento com eucalipto. *Ciênc. Florest.* **2012**, *22*, 833–843. [CrossRef]
23. Kunz, M.; Gonçalves, A.D.M.A.; Reichert, J.M.; Guimarães, R.M.L.; Reinert, D.J.; Rodrigues, M.F. Compactação do solo na integração soja-pecuária de leite em Latossolo argiloso com semeadura direta e escarificação. *Rev. Bras. Ciênc. Solo* **2013**, *37*, 6. [CrossRef]
24. Prevedello, J.; Kaiser, D.R.; Reinert, D.J.; Vogelmann, E.S.; Fontanela, E.; Reichert, J.M. Manejo do solo e crescimento inicial de Eucalyptus grandis Hill ex Maiden em Argissolo. *Ciênc. Florest.* **2013**, *23*, 129–138. [CrossRef]
25. Prevedello, J.; Vogelmann, E.S.; Kaiser, D.R.; Fontanela, E.; Reinert, D.J.; Reichert, J.M. Agregação e matéria orgânica de um Argissolo sob diferentes preparos do solo para plantio de Eucalipto. *Pesq. Florest. Bras.* **2014**, *34*, 149–158. [CrossRef]
26. Reichert, J.M.; Bervald, C.M.P.; Rodrigues, M.F.; Kato, O.R.; Reinert, D.J. Mechanized land preparation in eastern Amazon in fire-free forest-based fallow systems as alternatives to slash-and-burn practices: Hydraulic and mechanical soil properties. *Agric. Ecosyst. Environ.* **2014**, *192*, 47–60. [CrossRef]
27. Suzuki, L.E.A.S.; Reichert, J.M.; Reinert, D.J.; de LIMA, C.L.R. Degree of compactness and mechanical properties of a subtropical Alfisol with eucalyptus, native forest, and grazed pasture. *Forest. Sci.* **2015**, *61*, 716–722. [CrossRef]
28. Reichert, J.M.; Brandt, A.A.; Rodrigues, M.F.; da Veiga, M.; Reinert, D.J. Is chiseling or inverting tillage required to improve mechanical and hydraulic properties of sandy clay loam soil under long-term no-tillage? *Geoderma* **2017**, *301*, 72–79. [CrossRef]

29. Holthusen, D.; Brandt, A.A.; Reichert, J.M.; Horn, R. Soil porosity, permeability and static and dynamic strength parameters under native forest/grassland compared to no-tillage cropping. *Soil Tillage Res.* **2018**, *177*, 113–124. [CrossRef]
30. Holthusen, D.; Brandt, A.A.; Reichert, J.M.; Horn, R.; Fleige, H.; Zink, A. Soil functions and in situ stress distribution in subtropical soils as affected by land use, vehicle type, tire inflation pressure and plant residue removal. *Soil Tillage Res.* **2018**, *184*, 78–92. [CrossRef]
31. Reichert, J.M.; Fontanela, E.; Awe, G.O.; Fasinmirin, J.T. Is cassava yield affected by inverting tillage, chiseling or additional compaction of no-till sandy-loam soil? *Rev. Bras. Ciênc. Solo* **2021**, *45*, e0200134. [CrossRef]
32. Reichert, J.M.; Morales, C.A.S.; Lima, E.M.; Bastos, F.; Sampietro, J.A.; Araújo, E.F.; Srinivasan, R. Best tillage practices for early-growth of clonal eucalyptus in soils with distinct granulometry, drainage and profile depth. *Soil Tillage Res.* **2021**, *212*, 105038. [CrossRef]
33. França, J.S.; Reichert, J.M.; Holthusen, D.; Rodrigues, M.F.; Araújo, E.F. Subsoiling and mechanical hole-drilling tillage effects on soil physical properties and initial growth of eucalyptus after eucalyptus on steeplands. *Soil Tillage Res.* **2021**, *207*, 104860. [CrossRef]
34. Suzuki, L.E.A.S.; Reinert, D.J.; Alves, M.C.; Reichert, J.M. Medium-term no-tillage, additional compaction, and chiseling as affecting clayey subtropical soil physical properties and yield of corn, soybean and wheat crops. *Sustainability* **2022**, *14*, 9717. [CrossRef]
35. Nunes, M.R.; Denardin, J.E.; Pauletto, E.A.; Faganello, A.; Pinto, L.F.S. Effect of soil chiseling on soil structure and root growth for a clayey soil under no-tillage. *Geoderma* **2015**, *259–260*, 149–155. [CrossRef]
36. Suzuki, L.E.A.S.; Reinert, D.J.; Alves, M.C.; Reichert, J.M. Critical limits for soybean and black bean root growth, based on macroporosity and penetrability, for soils with distinct texture and management systems. *Sustainability* **2022**, *14*, 2958. [CrossRef]
37. Habibbeygi, F.; Nikraz, H.; Verheyde, F. Determination of the compression index of reconstituted clays using intrinsic concept and normalized void ratio. *Int. J. Geotec. Const. Mat. Env.* **2017**, *13*, 54–60. [CrossRef]
38. Ma, J.; Qian, M.; Yu, C.; Yu, X. Compressibility evaluation of reconstituted clays with various initial water contents. *J. Perform. Constr. Facil.* **2018**, *32*, 04018077. [CrossRef]
39. Xu, B.; Zhang, N. Determination of compression curve of in-situ soil considering soil disturbance. *Adv. Eng. Res.* **2018**, *162*, 213–216. [CrossRef]
40. Oluwaseun, A.A.; Yinusa, A.A.; Siyan, M. Compressibility characteristics of remoulded residual soils under loading. *J. Appl. Geol. Geophys.* **2018**, *6*, 52–57. [CrossRef]
41. Zhang, C.; Liu, E.; Tang, Y. Investigation on mechanical properties of artificially structured soils under different stress paths. *IOP Conf. Ser. Earth Environ. Sci.* **2019**, *267*, 042089. [CrossRef]
42. Zheng, J.; Yang, Z.; Gao, H.; Lai, X.; Wu, X.; Huang, Y. Experimental study on microstructure characteristics of saturated remolded cohesive soil during consolidation. *Sci. Rep.* **2022**, *12*, 18378. [CrossRef]
43. Veiga, M.; Horn, R.; Reinert, D.J.; Reichert, J.M. Soil compressibility and penetrability of an Oxisol from southern Brazil, as affected by long-term tillage systems. *Soil Tillage Res.* **2007**, *92*, 104–113. [CrossRef]
44. Somar Meteorologia. Médias Climatológicas de Butia. Available online: https://irga.rs.gov.br/medias-climatologicas (accessed on 5 May 2022).
45. Soil Survey Staff. *Keys to Soil Taxonomy*, 12th ed.; USDA-Natural Resources Conservation Service: Washington, DC, USA, 2014. Available online: https://www.nrcs.usda.gov/wps/portal/nrcs/detail/soils/survey/class/taxonomy/?cid=nrcs142p2_053580 (accessed on 8 December 2021).
46. IUSS Working Group WRB. *World Reference Base for Soil Resources 2014, Update 2015 International Soil Classification System for Naming Soils and Creating Legends for Soil Maps*; World Soil Resources Reports No. 106; FAO: Rome, Italy, 2015.
47. Santos, H.G.; Jacomine, P.K.T.; Anjos, L.H.; Oliveira, V.A.; Lumbreras, J.F.; Coelho, M.R.; Almeida, J.A.; Araujo Filho, J.C.; Oliveira, J.B.; Cunha, T.J.F. *Sistema Brasileiro de Classificação de Solos*; E-book: Il. Color; Embrapa: Brasília, Brazil, 2018; Available online: https://www.embrapa.br/solos/sibcs (accessed on 5 October 2022).
48. Pilon, C.N.; Santos, D.C.; Lima, C.L.R.; Antunes, L.O. Carbono e nitrogênio de um Argissolo Vermelho sob floresta, pastagem e mata nativa. *Ciênc. Rural* **2011**, *41*, 447–453. [CrossRef]
49. Teixeira, P.C.; Donagemma, G.K.; Fontana, A.; Teixeira, W.G. (Eds.) *Manual de Métodos de Análise de Solo*, 3rd ed.; e-Book: Il. Color.; Embrapa: Brasília, Brazil, 2017; 212p. Available online: http://www.infoteca.cnptia.embrapa.br/infoteca/handle/doc/1085209 (accessed on 10 November 2021).
50. Gubiani, P.I.; Reinert, J.; Reichert, J.M. Método alternativo para a determinação da densidade de partículas do solo—Exatidão, precisão e tempo de processamento. *Ciênc. Rural* **2006**, *36*, 664–668. [CrossRef]
51. Klute, A. Water retention: Laboratory methods. In *Methods of Soil Analysis: Physical and Mineralogical Methods*, 2nd ed.; Klute, A., Ed.; American Society of Agronomy and Soil Science Society of America: Madison, WI, USA, 1986; pp. 635–660.
52. Silva, V.R.; Reinert, D.J.; Reichert, J.M. Suscetibilidade à compactacão de um latossolo vermelho-escuro e de um podzólico vermelho-amarelo. *Rev. Bras. Ciênc. Solo* **2000**, *24*, 239–249. [CrossRef]
53. Rosa, D.P.; Reichert, J.M.; Lima, E.M.; Rosa, V.T. Chiselling and wheeling on sandy loam long-term no-tillage soil: Compressibility and load bearing capacity. *Soil Res.* **2021**, *59*, 488–500. [CrossRef]
54. Casagrande, A. The determination of the pre-consolidation load and its practical significance. In Proceedings of the International Conference on Soil Mechanics and Foundation Engineering; Harvard University: Cambridge, MA, USA, 1936; pp. 60–64.

55. Alves, M.C.; Suzuki, L.E.A.S.; Hipólito, J.L.; Castilho, S.R. Propriedades físicas e infiltração de água de um Latossolo Vermelho Amarelo (Oxisol) do noroeste do estado de São Paulo, Brasil, sob três condições de uso e manejo. *Cad. Lab. Xeolóxico Laxe* **2005**, *30*, 167–180. Available online: https://ruc.udc.es/dspace/bitstream/handle/2183/6311/CA-30-10.pdf?sequence=1&isAllowed=y (accessed on 31 October 2022).
56. Mando, A.; Stroosnijder, L.; Brussaard, L. Effects of termites on infiltration into crusted soil. *Geoderma* **1996**, *74*, 107–113. [CrossRef]
57. Abreu, S.L.; Reichert, J.M.; Reinert, D.J. Escarificação mecânica e biológica para a redução da compactação em argissolo franco-arenoso sob plantio direto. *Rev. Bras. Ciênc. Solo* **2004**, *28*, 519–531. [CrossRef]
58. Horn, R.; Vossbrink, J.; Becker, S. Modern forestry vehicles and their impacts on soil physical properties. *Soil Tillage Res.* **2004**, *79*, 207–219. [CrossRef]
59. Suzuki, L.E.A.S.; Reisser Júnior, C.; Miola, E.C.C.; Rostirolla, P.; Strieder, G.; Scherer, V.S.; Pauletto, E.A. Variabilidade da compressibilidade e do grau de compactação de um Argissolo cultivado com pessegueiro. *Sci. Rural* **2021**, *23*, 60–75. Available online: http://www.cescage.com.br/revistas/index.php/ScientiaRural/article/view/1642 (accessed on 31 October 2022).
60. Castro, P.P.; Curi, N.; Furtini Neto, A.E.; Resende, Á.V.; Guilherme, L.R.G.; Menezes, M.D.; Araújo, E.F.; Freitas, D.A.F.; Mello, C.R.; Silva, S.H.G. Química e mineralogia de solos cultivados com Eucalipto (*Eucalyptus* sp.). *Sci. For.* **2010**, *38*, 645–657. Available online: http://www.bibliotecaflorestal.ufv.br:80/handle/123456789/16762 (accessed on 31 October 2022).
61. Horn, R.; Lebert, M. Soil compactability and compressibility. In *Soil Compaction in Crop Production*; Soane, B.D., van Ouwerkerk, C., Eds.; Elsevier: Amsterdam, The Netherlands, 1994; pp. 45–69.
62. Braga, F.V.A.; Reichert, J.M.; Mentges, M.I.; Vogelmann, E.S.; Padrón, R.A.R. Propriedades mecânicas e permeabilidade ao ar em topossequência Argissolo-Gleissolo: Variação no perfil e efeito de compressão. *Rev. Bras. Ciênc. Solo* **2015**, *39*, 1025–1035. [CrossRef]
63. Vaz, C.M.P.; Freitas Iossi, M.; Mendonça Naime, J.; Macedo, Á.; Reichert, J.M.; Reinert, D.J.; Cooper, M. Validation of the Arya and Paris water retention model for Brazilian soils. *Soil Sci. Soc. Am. J.* **2005**, *69*, 577–583. [CrossRef]
64. Klein, V.A.; Reichert, J.M.; Reinert, D.J. Água disponível em um Latossolo Vermelho argiloso e murcha fisiológica de culturas. *Rev. Bras. Eng. Agrícola Ambient.* **2006**, *10*, 646–650. [CrossRef]
65. Awe, G.O.; Reichert, J.M.; Timm, L.C.; Wendroth, O.O. Temporal processes of soil water status in a sugarcane field under residue management. *Plant Soil* **2015**, *387*, 395–411. [CrossRef]
66. Reichert, J.M.; Albuquerque, J.A.; Kaiser, D.R.; Reinert, D.J.; Urach, F.L.; Carlesso, R. Estimation of water retention and availability in soils of Rio Grande do Sul. *Rev. Bras. Ciênc. Solo* **2009**, *33*, 1547–1560. [CrossRef]
67. Reichert, J.M.; Albuquerque, J.A.; Solano Peraza, J.E.; Costa, A. Estimating water retention and availability in cultivated soils of southern Brazil. *Geoderma Reg.* **2020**, *21*, e00277. [CrossRef]
68. Lima, C.L.R.; Silva, A.P.; Imhoff, S.; Lima, H.V.; Leão, T.P. Heterogeneidade da compactação de um Latossolo Vermelho-Amarelo sob pomar de laranja. *Rev. Bras. Ciênc. Solo* **2004**, *28*, 409–414. [CrossRef]
69. Kondo, M.K.; Dias Junior, M.S. Compressibilidade de três Latossolos em função da umidade e uso. *Rev. Bras. Ciênc. Solo* **1999**, *23*, 211–218. [CrossRef]
70. Lima, C.L.R.; Silva, A.P.; Imhoff, S.; Leão, T.P. Compressibilidade de um solo sob sistemas de pastejo rotacionado intensivo irrigado e não irrigado. *Rev. Bras. Ciênc. Solo* **2004**, *28*, 945–951. [CrossRef]
71. Capurro, E.P.G.; Secco, D.; Reichert, J.M.; Reinert, D.J. Compressibilidade e elasticidade de um Vertissolo afetado pela intensidade de pastejo bovino. *Ciênc. Rural* **2014**, *44*, 283–288. [CrossRef]
72. Horn, R.; Fleige, H. A method for assessing the impact offload on mechanical stability and on physical properties of soils. *Soil Tillage Res.* **2003**, *73*, 89–99. [CrossRef]
73. Horn, R. Effect of land-use management systems on coupled hydraulic mechanical soil processes defining the climate-food-energy-water nexus. *Bulg. J. Soil Sci.* **2019**, *4*, 3–15. Available online: https://www.bsss.bg/issues/Issue1_2019/BJSS_2019_1_1.pdf (accessed on 31 October 2022).
74. Powers, R.F.; Scott, D.A.; Sanchez, F.G.; Voldseth, R.A.; Page-Dumroese, D.; Elioff, J.D.; Stone, D.M. The North American long-term soil productivity experiment: Findings from the first decade of research. *For. Ecol. Manag.* **2005**, *220*, 31–50. [CrossRef]
75. Guérif, J. Effects of compaction on soil strength parameters. In *Soil Compaction in Crop Production*; Soane, B.D., van Ouwerkerk, C., Eds.; Elsevier: Amsterdam, The Netherlands, 1994; pp. 191–214.
76. Mentges, M.I.; Reichert, J.M.; Gubiani, P.I.; Reinert, D.J.; Xavier, A. Alterações estruturais e mecânicas de solo de várzea cultivado com arroz irrigado por inundação. *Rev. Bras. Ciênc. Solo* **2013**, *37*, 221–231. [CrossRef]
77. Reichert, J.M.; Mentges, M.I.; Rodrigues, M.F.; Cavalli, J.P.; Awe, G.O.; Mentges, L.R. Compressibility and elasticity of subtropical no-till soils varying in granulometry organic matter, bulk density and moisture. *Catena* **2018**, *165*, 345–357. [CrossRef]
78. Horn, R.; Mordhorst, A.; Fleige, H.; Zimmermann, I.; Burbaum, B.; Filipinski, M.; Cordsen, E. Soil type and land use effects on tensorial properties of saturated hydraulic conductivity in northern Germany. *Eur. J. Soil Sci.* **2020**, *71*, 179–189. [CrossRef]
79. Wu, L.; Swan, J.B.; Paulson, W.H.; Randalla, G.W. Tillage effects on measured soil hydraulic- properties. *Soil Tillage Res.* **1992**, *25*, 17–33. [CrossRef]
80. Costa, F.S.; Albuquerque, J.A.; Bayer, C.; Fontoura, S.M.V.; Wobeto, C. Propriedades físicas de um Latossolo Bruno afetadas pelos sistemas plantio direto e preparo convencional. *Rev. Bras. Ciênc. Solo* **2003**, *27*, 527–535. [CrossRef]
81. Boone, F.R.; Veen, B.W. Mechanisms of crop responses to soil compaction. In *Soil Compaction in Crop Production*; Soane, B.D., van Ouwerkerk, C., Eds.; Elsevier: Amsterdam, The Netherlands, 1994; pp. 237–264.

82. Soracco, C.G.; Lozano, L.A.; Villarreal, R.; Palancar, T.C.; Collazo, D.J.; Sarli, G.O.; Filgueira, R.R. Effects of compaction due to machinery traffic on soil pore configuration. *Rev. Bras. Ciênc. Solo* **2015**, *39*, 408–415. [CrossRef]
83. Talukder, R.; Plaza-Bonilla, D.; Cantero-Martínez, C.; Wendroth, O.; Castel, J.L. Soil gas diffusivity and pore continuity dynamics under different tillage and crop sequences in an irrigated Mediterranean area. *Soil Tillage Res.* **2022**, *221*, 105409. [CrossRef]
84. Mentges, M.I.; Reichert, J.M.; Rodrigues, M.F.; Awe, G.O.; Mentges, L.R. Capacity and intensity soil aeration properties affected by granulometry, moisture, and structure in no-tillage soils. *Geoderma* **2016**, *263*, 47–59. [CrossRef]
85. Ambus, J.V.; Reichert, J.M.; Gubiani, P.I.; Carvalho, P.C.F. Changes in composition and functional soil properties in long-term no-till integrated crop-livestock system. *Geoderma* **2018**, *330*, 232–243. [CrossRef]

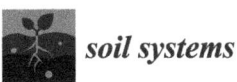 *soil systems*

Communication

CO$_2$ Emissions in Layered Cranberry Soils under Simulated Warming

Wilfried Dossou-Yovo [1,2,*], Serge-Étienne Parent [1], Noura Ziadi [2] and Léon E. Parent [1,*]

1   Department of Soils and Agri-Food Engineering, Université Laval, Quebec, QC G1V 0A6, Canada
2   Agriculture and Agri-Food Canada, Quebec Research and Development Center, 2560 Hochelaga Boulevard, Quebec, QC G1V 2J3, Canada
*   Correspondence: wilfrieddossouyovo16@gmail.com (W.D.-Y.); leon-etienne.parent@fsaa.ulaval.ca (L.E.P.)

Abstract: Sanding to bury the overgrowth of uprights and promote new growth results in alternate sand and organic sublayers in the 0–30 cm layer of cranberry soils contributing to global carbon storage. The aim of this study was to measure CO$_2$ emission rates in cranberry soil sublayers under simulated warming. Soil samples (0–10, 10–20 and 20–30 cm) were incubated in jars for up to 105 days at 10, 20 and 30 °C. The CO$_2$ emission rate was measured biweekly by gas chromatography. The CO$_2$ emission rate increased with temperature and decreased in deeper soil sublayers. Linear regression relating CO$_2$ efflux to soil sublayer and temperature returned $R^2 = 0.87$. Sensitivity of organic matter decomposition to temperature was estimated as activation energy and as Q_{10} coefficient, the increase in reaction rate per 10 °C. Activation energy was 50 kJ mol^{-1}, 59 kJ mol^{-1} and 71 kJ mol^{-1} in the in the 0–10, 10–20 and 20–30 cm sublayers, respectively, indicating higher molecular-weight compounds resisting to decomposition in deeper sublayers. The Q_{10} values were significantly higher ($p < 0.01$) in the 10–30 cm (2.79 ± 0.10) than the 0–10 cm (2.18 ± 0.07) sublayers. The 20–30 cm sublayer where less total carbon was stored was the most sensitive to higher temperature. Cranberry soils could be used as sensitive markers of global warming.

Keywords: carbon accumulation; cranberry soils; activation energy; temperature-dependent CO$_2$ emissions rate

Citation: Dossou-Yovo, W.; Parent, S.-É.; Ziadi, N.; Parent, L.E. CO$_2$ Emissions in Layered Cranberry Soils under Simulated Warming. *Soil Syst.* 2023, 7, 3. https://doi.org/10.3390/soilsystems7010003

Academic Editor: Luis Eduardo Akiyoshi Sanches Suzuki

Received: 9 November 2022
Revised: 28 December 2022
Accepted: 30 December 2022
Published: 9 January 2023

1. Introduction

Terrestrial carbon (C) is three times greater than atmospheric C [1]. Soil C sequestration is the conversion of atmospheric CO$_2$ into long-lived C pools [2]. While soil C sink capacity of managed ecosystems is the estimated historic cumulative C loss of 55–78 Gt, the attainable capacity is only 50–66% of that potential [3]. Cranberry agroecosystems are exceptions to this general perspective [4]. The fate of organic matter in soils depends primarily on its intrinsic decomposability and on protection mechanisms such as soil aggregation [5] in silty or clayey agricultural soils [6].

In North America, conventionally and organically managed cranberry agroecosystems are mostly established on acid sandy soils arranged as flat beds in low-lying positions to facilitate water transfer [7]. Beds are diked, then capped with 0.3–1.0 m of sand. Native soil C is accumulated in dikes, beds, and the subsoil. The seasonal C flux of leaf and stem litterfall was estimated at 2.15–2.57 Mg C ha^{-1} (153-d)$^{-1}$ in Wisconsin [8]. The belowground vegetative biomass may contribute up to 2/3 of total vegetative C stocks. The overgrowth of uprights is buried every 2–5 years by spreading two to five cm of sand onto frozen soil to promote new growth [9]. This results in alternate layers of sand and organic matter in the root zone [10] and high potential for C storage due to physical protection through anthropic surface sanding [4].

The composition and biomass of the microbial community generally differ between upper and lower soil layers [11–13]. The decreasing rates of CO$_2$ emissions in deeper soil layers [14,15] have been attributed to the vertical distribution of soil organic carbon in terms of amount and quality [13–16]. Indeed, the C:N ratio is narrower, and soil organic matter (SOM) is more decomposed, in deeper layers of cranberry soils [10]. Temperature at depth should also be considered. The threshold temperature for mineralization activity of cranberry soil was set at 13 °C [8].

The effect of temperature on organic matter decomposition is crucial to understand the global C cycle and potential feedbacks to the climate system [17]. The largest C stocks have been found at high latitude [18,19]. While the seasonal CO_2 emission of Quebec cranberry soils has been estimated at 2.7–3.4 t CO_2 eq ha^{-1} [20], the effect of global warming on CO_2 emission in the soil profile as a function of temperature has not been established. The activation energy of decomposition and the Q_{10} coefficient as increase in reaction rate per 10 °C [17,21,22] can reflect the differential contribution of soil layers to CO_2 emissions in cranberry agroecosystems in areas of rapid climate change such as Eastern Canada.

We hypothesized that activation energy of soil organic matter (SOM) decomposition and Q_{10} differ in alternate sand and organic matter layers due to the differential C/N ratio and decomposition degree of organic matter in two differently managed cranberry soils. The aim of this study was to measure the decomposition rate of SOM in layered cranberry soils as a function of management (conventional vs. organic), soil layer, incubation time and temperature under controlled environments to assess the differential effects of global warming on soil C storage in cranberry soils.

2. Materials and Methods

2.1. Soil Sampling and Analysis

Sites were selected to cover the two main management practices in south-central Quebec. Site #45 (46°16′34.7″ N, 71°51′30.0″ W, elevation 112 m) was conventionally managed, and site #A9 (46°14′16.5″ N, 72°02′13.4″ W, elevation 92 m) was organically farmed. Sites #45 and #A9 have been planted to cultivar "Stevens" in 1999 and 2004, respectively. The climate of the region is sub-humid temperate and continental with cold winters and hot summers. Soil series were the Saint-Jude series at site #45 and Sainte-Sophie series at site #9, both classified as Humo-Ferric Podzols in the Canadian System Haplorthods in the U.S. Soil Taxonomy, and Orthic Podzols in the World Reference Base for Soil Resources. The soil contained 937 g sand kg^{-1}, 37 g silt kg^{-1}, and 26 g clay kg^{-1} at site #45, and 915 g sand kg^{-1}, 49 g silt kg^{-1}, and 36 g clay kg^{-1} at site #9 [4].

Fields received 40 kg N ha^{-1} yr^{-1} as ammonium sulfate (site #45) or granules of poultry manure (site #9). The source of phosphorus was mono-ammonium phosphate at site #45 as recommended locally from soil and tissue tests, and granules of poultry manure at site #9 from a N-based recommendation. Potassium was applied at a rate of 100 kg K ha^{-1} as KCl, sul-po-mag and/or granules of poultry manure. Micro-nutrients were applied at need depending on the results of tissue testing. Fields were sprinkler-irrigated at need.

Soil samples were collected for physical analyses in spring 2018. Three soil layers were sampled (0–10; 10–20; 20–30 cm) at four places per site using cylinders (diameter = 5.5 cm, height = 7.6 cm). Samples were sealed in plastic bags and stored at 4 °C until use within a week. Soil samples were air dried, and 2 mm sieved before analysis. Soil pH was measured in 0.01 M $CaCl_2$ (soil to solution ratio of 1:2 *v:v*). Soil carbon and nitrogen were quantified by combustion [23] using the Leco CNS model 630-300-200 (Leco Corporation, Saint-Joseph, MI, USA). Soil bulk density was determined as the mass of air-dry soil divided by the volume of the cylinder. Soil carbon content and porosity were computed in each 10 cm thick layers as follows [24]:

$$C_s = P_b \times C_c \tag{1}$$

$$C_{layer} = C_s \times [layer\ thickness] \times area \tag{2}$$

$$TP = \left(1 - P_b/P_p\right) \times 100 \tag{3}$$

$$P_d = \frac{100}{\frac{\%Organic\ matter}{1.55} + \frac{100 - \%Organic\ matter}{2.65}} \tag{4}$$

where Cs = carbon stock (kg m^{-3}), Cc = carbon content (%), TP is total soil porosity (volumetric fraction); P_b = soil bulk density (g cm^{-3}); P_d = soil particle density assuming 1.55 g cm^{-3} as the particle density of organic matter and 2.65 g cm^{-3} as the particle density of mineral matter [25].

2.2. CO$_2$ Emission

Soils were sampled at three depths (0–10 cm, 10–20 cm, 20–30 cm) and four locations in each field (785 mL per sample using cylinders 10 cm in height and 10 cm Ø) and introduced in plastic bags. The 72 fresh soil samples were 6 mm sieved, filled to 1/3 of the volume of 250-mL Mason jars with 100 g dry-based material, and placed in temperature-controlled chambers following a completely randomized design with four replications per site. There were three (3) soil layers (0–10; 10–20; 20–30 cm), three (3) temperatures (10, 20 and 30 °C), four (4) replicates and two (2) sites. Soil water content was adjusted twice a week with distilled water to water-filled pore space (WFPS) close to 0.50–0.70 as volumetric fraction [24,26]. Water content was assessed by weighing the jars, assuming a density of one g cm^{-3}.

Soil CO$_2$ flux was measured using a close chamber protocol [26]. At sampling time taken biweekly during 105 d, jars were capped with a lid containing two male slips. One slip was fitted to a septum for headspace sampling using a 20 mL polypropylene syringe. The other slip was used to equilibrate the jar internal pressure during sampling. Air samples were taken at 0 and 24 h, then transferred into pre-evacuated 12 mL glass vials (Exetainer, Labco, High Wycombe, UK). Gas samples were analyzed for CO$_2$ using a gas chromatograph fitted to a Ni-NO$_3$ (10%) catalyst column and a flame ionization detector (Model 3800, Varian Inc., Walnut Creek, CA, USA), equipped with a headspace auto-injector (Combi Pal, CTC Analytics, Zurich, Switzerland). The CO$_2$ flux (Fc, μg g^{-1} h^{-1}) was measured as [26].

$$F_c = \frac{dc}{dt} \times \frac{v}{M_v} \times \frac{M_m}{W}, \tag{5}$$

where dc/dt (μL L^{-1} h^{-1}) is change rate of headspace CO$_2$ concentration in dry air samples estimated at time = 0 and time = 24 h, assuming that CO$_2$ emissions vary linearly through time; v (L) is pot headspace volume; Mv (L mol^{-1}) is molecular volume at the pre-deployment air temperature (22–24 °C); Mm (μg mol^{-1}) is molecular mass of CO$_2$ (44,000,000); and W (g) is dry soil mass.

2.3. Statistical Analysis

2.3.1. First Order Kinetics

The decomposition rate constant (k) was computed as follows [27]:

$$k = \frac{ln([C_{initial} - CO_2\_C_t]/C_{initial})}{t} \tag{6}$$

where $C_{initial}$ (mg kg^{-1}) is initial soil carbon content, and $CO_2\_C_t$ (mg kg^{-1}) is cumulative CO$_2$ released during incubation period t.

2.3.2. Q$_{10}$ and Activation Energy

The increase in reaction rate per 10 °C was reported as follows [21,22]:

$$Q_{10} = \frac{k \times (t + 10)}{k(t)} \tag{7}$$

where $k_{(t)}$ is k at temperature t (°K) and $k_{(t+10)}$ is k at temperature $t + 10$ (°K).

Activation energy was derived from the Arrhenius equation [19,22] as follows [21]:

$$k = A \times exp(-E_a/RT) \tag{8}$$

where A is pre-exponential factor and Ea is activation energy assumed to be independent of temperature, R is the universal gas constant, and T is absolute temperature ($^{\circ}$K).

2.3.3. Statistical Analysis

Statistical analyses were performed in the R environment version 4.1.0 [28]. The difference between conventional and organic farming systems were tested using a mixed model. The CO_2 emission rates were fitted to elapsed time, temperature and soil layers across farming systems using the *lm* linear regression model as follows [29]:

$$y = \beta_0 + \beta_1 x_1 + \ldots + \beta_k x_k + \varepsilon \tag{9}$$

where y (CO_2 emission rate) is the predicted value, x_1 through x_k are k independent variables or predictors (time, temperature in $^{\circ}$K, soil layers), β_0 is the value of y where independent variables take zero values, and β_1 through β_k are estimated regression coefficients. The R codes and dataset are available online at https://bit.ly/3gbi6Ov (accessed on 8 January 2023).

3. Results

3.1. Soil Properties

Soil properties are presented in Figure 1. Soil carbon content varied from 1.67 to 30.9 Mg C ha^{-1}, being larger in the 0–10 cm (16.55 ± 1.15), than in the 10–20 cm (13.63 ± 2.95) and the 20–30 cm (6.09 ± 1.44) layers (Figure 1A). The C:N ratios were 20.08 ± 1.05, 16.01 ± 1.91 and 9.02 ± 1.96 in 0–10, 10–20 and 20–30 cm layers, respectively (Figure 1B). Soil bulk density increased in lower layers as a result of sand accumulation and organic matter decomposition while biomass production reduced bulk density in the upmost layer. Lower pH values in upper layers under conventional farming are attributable to soil acidification by elemental sulfur amendment and ammonium sulfate fertilization. In organic farming, high-ammonium poultry manure granules likely acidified the upper soil layer in the first place. Soil porosity, water content and bulk density followed the same trends as inter-related properties.

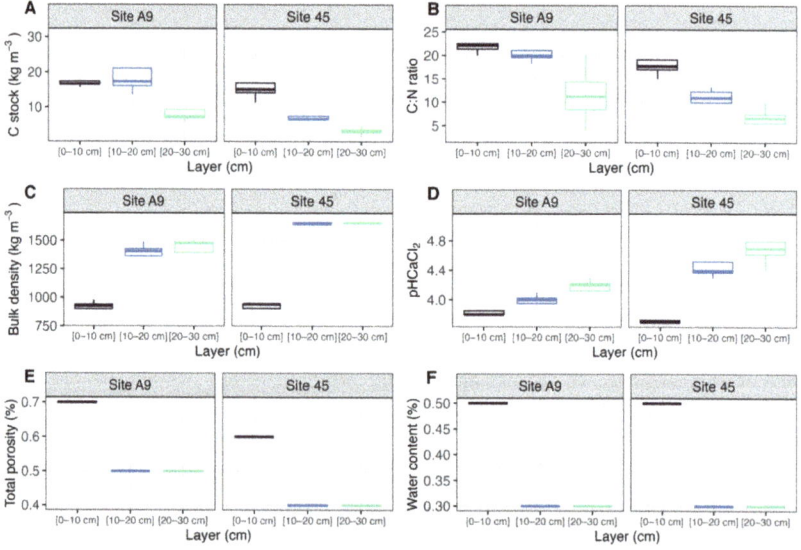

Figure 1. Soil properties in layers of cranberry soils: (**A**) C stock, (**B**) C:N ratio, (**C**) bulk density, (**D**) pH$_{CaCl_2}$, (**E**) total porosity (volumetric fraction), (**F**) water content (volumetric fraction).

3.2. CO_2 Emission Rate

The CO_2 emission rates did not differ significantly between sites (p-value > 0.05), decreased (p-value ≤ 0.05) through time and soil depth, and increased (p-value ≤ 0.05) with temperature (Figure 2). The soil layer showed the largest effect followed by temperature and incubation time. The CO_2 emissions are presented in Figure 3 for significant treatments.

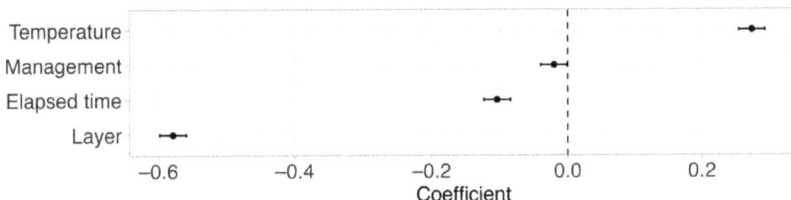

Figure 2. Effect of temperature, management, elapsed time and soil layer on CO_2 emissions. Probability for significance is 0.05.

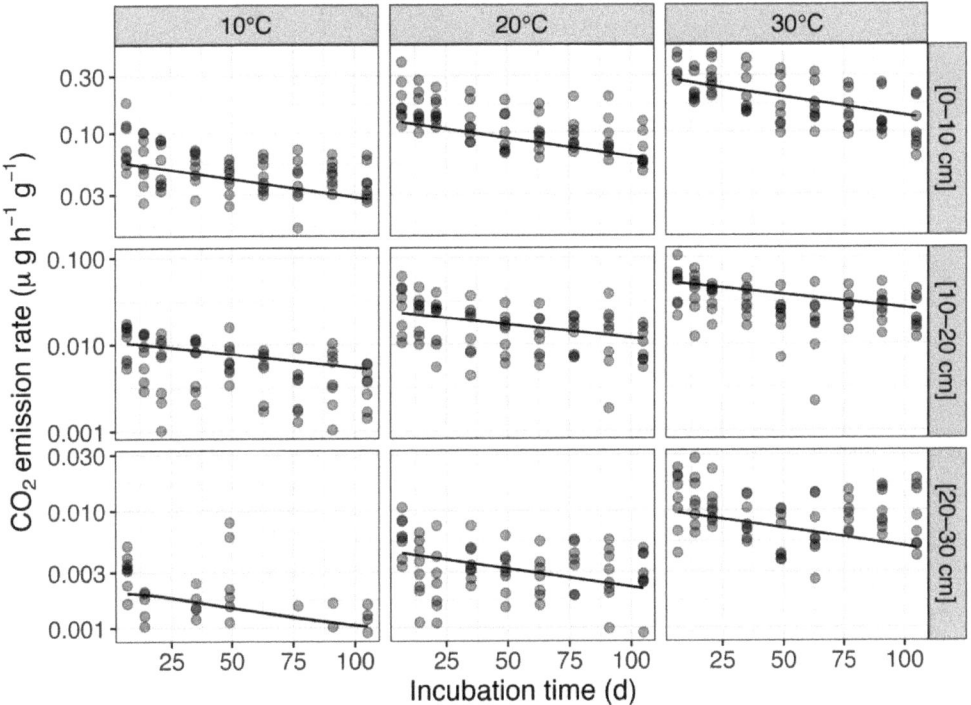

Figure 3. Influence of soil layer, temperature and time on CO_2 emission rate.

The CO_2 emission rate was highest in the 0–10 cm layer at 30 °C and lowest in the 20–30 cm layer at 10 °C. At 10 °C, 87.2% of the total across-layer CO_2 emissions during the incubation period occurred in the 0–10 cm layer compared to 12.1% in the 10–20 cm layer and 0.7% in the 20–30 cm layer. At 20 °C, 83.1% of total across-layer CO_2 emission occurred in the 0–10 cm layer compared to 14.3% in the 10–20 cm layer and 2.5% in the 20–30 cm layer. At 30 °C, 82.8% of total across-layer CO_2 emission occurred in the 0–10 cm layer compared to 13.5% in the 10–20 cm layer and 3.8% in the 20–30 cm layer. The R^2 value of the equation relating CO_2 emission to temperature, soil depth and elapsed time was 0.87 and root-mean-square-error was 0.24 (Figure 3).

The effect of temperature on SOM decomposition rates indicated large differences between layers in biochemical composition of the organic materials and SOM resistance to decomposition. The slope of the Arrhenius equation ($-Ea/R$) was highest in the (0–10 cm) layer, intermediate in the 10–20 cm layer and lowest in the 20–30 cm layer (p-value < 0.001) (Figure 4). The activation energy required to decompose SOM was 50 kJ mol^{-1} in the 0–10 cm layer, 59 kJ mol^{-1} in 10–20 cm layer and 71 kJ mol^{-1} in 20–30 cm layer (Figure 5). The Q_{10} (mean $\pm$ SE) was 2.79 $\pm$ 0.10 in the 20–30 cm layer and 2.18 $\pm$ 0.07 in the 0–10 and 10–20 cm layers (p-value < 0.001) (Figure 6).

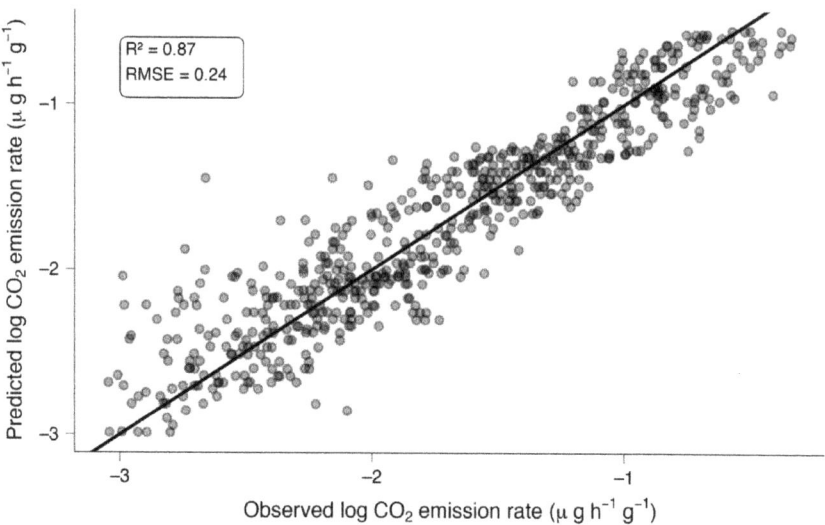

Figure 4. Relationship between observed and predicted CO_2 emission rates.

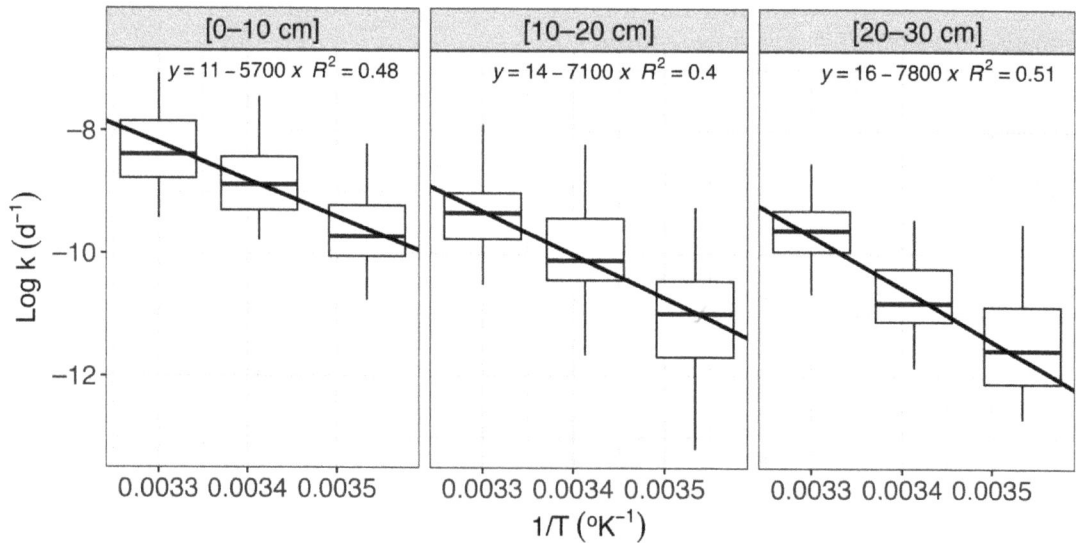

Figure 5. Experimental data fitted to the Arrhenius equation (p-value < 0.001).

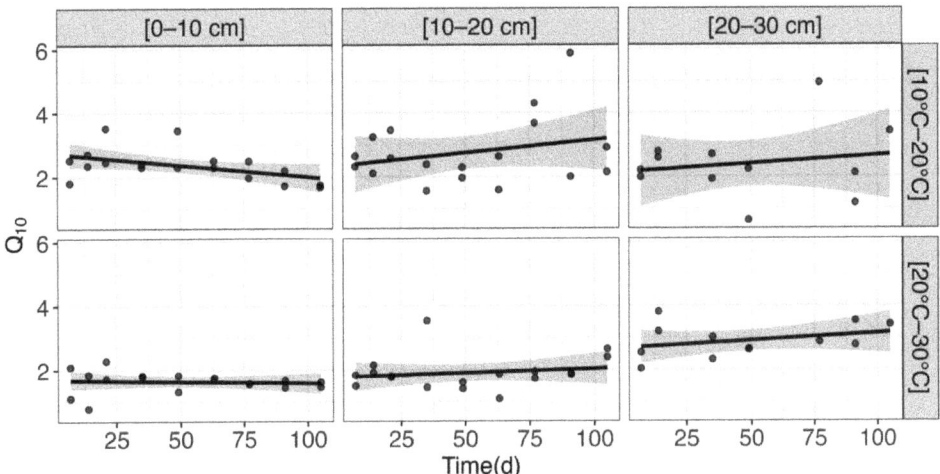

Figure 6. Time variation of Q_{10} across soil layers (*p*-value < 0.001).

4. Discussion

Several factors such as climate, the amount and quality of plant residues, soil management, mineralogy, texture [30–33], bulk density [34], as well as layer location in the soil profile [10,15,16,35,36] control organic matter turnover in soils. Organic matter decomposition rate also depends on the spatial distribution of organic matter, and the site-specific microbial community impacted by land use, temperature, rainfall, soil type and bulk density [12,37]. In cranberry soils documented in Figure 1, differences in soil parameters should be further addressed in relation with carbon accumulation and microbe abundance and diversity.

Cranberry agroecosystems were shown to contribute to CO_2 emissions much less (2.7–3.4 t CO_2 eq ha^{-1}) compared to other horticultural cropping systems [20]. Indeed, slowly decomposing carbon can accumulate in large amounts in layered cranberry soils after burial of organic matter through regular sanding. This paper quantified layer × temperature interactions regulating CO_2 emissions in cranberry soils.

4.1. Dependency of CO_2 Emission on Soil Depth

The decreasing CO_2 emission rate in deeper soil layers results in part from the vertical distribution of soil organic carbon (SOC) in terms of amount and quality [16]. The biochemical composition varies considerably among cranberry soil layers [4]. The biochemical quality of plant species and that of the soil are the main factors driving litter decomposition under otherwise similar conditions of temperature and rainfall [38–40]. For example, litter quality differs considerably among tundra, grassland, and boreal, conifer, deciduous, and tropical forest biomes [38]. Litter decomposability is associated with species' ecological strategy within different ecosystems globally [39]. The effect of climate on litter mass loss can be offset by differences in soil parameters as mediated by soil microbial populations [40].

Fresh sources of SOC such as shoot litter, senescent roots, and root exudates [41,42] are directly available to soil microbes in upper layers [43]. Fresh organic matter decreases deeper in the soil profile [36,44,45]. There is abundant young fast-cycling C in upper layers compared to ancient slow-cycling C in the subsoil [36]. As a result, soil respiration is greater near soil surface (0–5 and 5–10 cm layers) compared to lower layers [46]. The cranberry litter deposited on the floor of cranberry beds contains approximately 80% of lignocellulose while 89% of the particles are larger than 2 mm in size and the C:N ratio is 55 in average [4]. Lignocellulose is a compact material made of strongly bound cellulose, lignin, and hemicellulose in structural networks in stems and roots [47]. Lignocellulose is broken down by a suite of extracellular enzymes [48]. Due to structural and biochemical constraints, litter is slowly decomposed in cranberry soils [4].

4.2. Temperature Sensitivity on CO_2 Emission Rates

The activation energy required to decompose SOM in cranberry agroecosystems was lower in upper layers (50–59 kJ mol^{-1}) than in the 20–30 cm layer (71 kJ mol^{-1}) where high-molecular-weight phenolic compounds abound [4]. In comparison, the *Ea* values for enzyme activities were 75 kJ mol^{-1} for phenol oxidase and, 40–45 kJ mol^{-1} for β-glucosidase, cellobiohydrolase and peroxidase in the A horizons of three temperate biomes [48], compared to that of pyrophosphatase that averaged 22–33 kJ mol^{-1} in Histosols and 33–43 kJ mol^{-1} in mineral soils [49].

Low-quality organic C limits the energy available to the microbial community [43,50]. Humic substances, complex organic molecules and recalcitrant SOM as shown by higher activation energy requirements in lower soil layers resist microbial attack and may combine with minerals to reduce microbial degradability even more [51]. As a result, higher temperatures showed disproportionate impacts on the depolymerization of high-molecular weight constituents of SOM [40]. More decomposed soil organic matter in the deepest layer is shown by the lower C:N ratios compared upper layers (Figure 1). In upper layers, cranberry plant residues show high C:N ratio of 66.7 ± 5.7 [4], indicating a decreasing gradient of C:N ratios from litter in upper soil layers and to more decomposed materials in the lower layer of cranberry soils [7].

The Q_{10} values were higher (2.79 ± 0.10) in the 20–30 cm than upper (2.18 ± 0.07) layers in the range of 283–303 K in the present study. In comparison, the Q_{10} values varied between 1.2 and 2.8 within temperature range of 278–308 K in the 0–29 cm layer of cropland, grassland, deciduous forest, and coniferous forest ecosystems [52]. The Q_{10} values of Massachusetts forest soils in the 278–303 K range were 2.43–5.00 for hemlock, 2.62–3.77 for young birch, and 2.59–5.23 for mature birch [53]. Indeed, compared to commonly used values of 1.5–2.0, the Q_{10} values may vary widely from 1 to 12 depending on land use, C:N ratio and degradability of SOM, soil class, moisture content, texture, and acidity [54]. Mycorrhizae may impact Q_{10} values of carbon sources. Ericoid mycorrhizal (ErM) foliar litters, fine roots, fungal biomass and the necromass generally decomposes slower than those of arbuscular and ecto-mycorrhizal fungi, which could contribute to organic matter accumulation in sites where ErM plants occur [55]. This aspect could be further examined in cranberry agroecosystems.

5. Conclusions

The the decomposition rate of SOM in cranberry soils did not vary significantly with management (conventional vs. organic), but varied with soil layer, incubation time and temperature. The rate of CO_2 emissions decreased with elapsed time. Activation energy was 50–59 kJ mol^{-1} in upper layers (0–20 cm) compared to 71 kJ mol^{-1} in the 20–30 cm layer required to decompose high-molecular-weight materials. The Q_{10} values were 2.9–3.1 in the deepest layer compared to 1.9–2.3 for the Q_{10} values in upper layers. Temperature sensitivity of C decomposition rate in layered cranberry soils thus impacted differentially the C storage capacity of cranberry agroecosystems. Activation energy and Q_{10} increased deeper in the soil, indicating higher temperature sensitivity of the most recalcitrant sources of SOM. Despite their smaller contribution to total C storage compared to upper layers of cranberry soils, the 20–30 cm soil layer would contribute increasingly to CO_2 emissions in the context of global warming.

Future research could quantify net C accumulation in cranberry soils through litter burial by sanding since the establishment of cranberry beds and the management practices required to promote C storage as ecosystem service. This will require developing a methodology for site sampling and monitoring covering several aspects of the cranberry production system to meet sustainable development goals, addressing the destruction of native ecosystems as well as the carbon already accumulated in dikes and the subsoil. Organic layers alternating with sand layers in cranberry beds and showing disproportionate contributions to CO_2 emissions could be used as sensitive markers of the impact of global warming on soil C storage capacity over decades.

Author Contributions: Conceived and designed the experiment: W.D.-Y. and N.Z.; performed the experiment: W.D.-Y.; analyzed the data: W.D.-Y. and S.-É.P.; wrote the first draft of the manuscript: W.D.-Y., S.-É.P. and L.E.P.; all authors provided critical feedback on the manuscript. All authors have read and agreed to the published version of the manuscript.

Funding: This collaborative research project was funded by Les Atocas de l'Érable Inc., Les Atocas Blandford Inc., La Cannebergière Inc., the Natural Sciences and Engineering Research Council of Canada (RDCPJ-469358-14) and Agriculture Agri-food Canada (AAFC-1555).

Institutional Review Board Statement: Not applicable.

Informed Consent Statement: Not applicable.

Data Availability Statement: Not applicable.

Conflicts of Interest: The authors declare that they have no known competing financial interests or personal relationships that could have appeared to influence the work reported in this paper.

References

1. Post, W.M.; Peng, T.-H.; Emanuel, W.R.; King, A.W.; Dale, V.H.; De Angelis, D.L. The global carbon cycle. *Am. Scient.* **1990**, *78*, 310–326.
2. Lal, R. Soil carbon stocks under present and future climate with specific reference to European ecoregions. *Nutr. Cycl. Agroecosyst.* **2008**, *81*, 113–127. [CrossRef]
3. Lal, R. Soil carbon sequestration impacts on global climate change and food security. *Science* **2004**, *304*, 1623–1628. [CrossRef] [PubMed]
4. Dossou-Yovo, W.; Parent, S.-É.; Ziadi, N.; Parent, É.; Parent, L.E. Tea Bag Index to Assess Carbon Decomposition Rate in Cranberry Agroecosystems. *Soil Syst.* **2021**, *5*, 44. [CrossRef]
5. Angers, D.A.; Chenu, C. Dynamics of soil aggregation and C sequestration. In *Soil Processes and the Carbon Cycle*; Lal, R., Kimble, J.M., Follett, R.F., Stewart, B.A., Eds.; CRC Press: Boca Raton, FL, USA, 1997; pp. 199–206.
6. Hassink, J.; Whitmore, A.P.; Kuba, J. Size and density fractionation of soil organic matter and the physical Size and density fractionation of soil organic matter and the physical capacity of soils to protect organic matter. *Eur. J. Agron.* **1997**, *7*, 189–199. [CrossRef]
7. Kennedy, C.D.; Wilderotter, S.; Payne, M.; Buda, A.R.; Kleinman, P.J.A.; Bryant, R.B. A geospatial model to quantify mean thickness of peat in cranberry bogs. *Geoderma* **2018**, *319*, 122–131. [CrossRef]
8. Stackpoole, S.M.; Kosola, K.R.; Workmaster, B.A.A.; Guldan, N.M.; Browne, B.A.; Jackson, R.D. Looking beyond fertilizer: Assessing the contribution of nitrogen from hydrologic inputs and organic matter to plant growth in the cranberry agroecosystem. *Nutr. Cycl. Agroecosyst.* **2011**, *91*, 41–54. [CrossRef]
9. Sandler, H.; DeMoranville, C. *Cranberry Production: A Guide for Massachusetts. Summary Edition*; University of Massachusetts Cranberry Station: Amherst, MA, USA, 2008; pp. 1–198.
10. Kosola, K.R.; Workmaster, B.A.A. Mycorrhizal colonization of cranberry: Effects of cultivar, soil type, and leaf litter composition. *J. Am. Soc. Hortic. Sci.* **2007**, *132*, 134–141. [CrossRef]
11. Eilers, K.G.; Debenport, S.; Anderson, S.; Fierer, N. Digging deeper to find unique microbial communities: The strong effect of depth on the structure of bacterial and archaeal communities in soil. *Soil Biol. Biochem.* **2012**, *50*, 58–65. [CrossRef]
12. Fierer, N.; Schimel, J.P.; Holden, P.A. Variations in microbial community composition through two soil depth profiles. *Soil Biol. Biochem.* **2003**, *35*, 167–176. [CrossRef]
13. Kramer, C.; Gleixner, G. Soil organic matter in soil depth profiles: Distinct carbon preferences of microbial groups during carbon transformation. *Soil Biol. Biochem.* **2008**, *40*, 425–433. [CrossRef]
14. Balesdent, J.; Basile-Doelsch, I.; Chadoeuf, J.; Cornu, S.; Derrien, D.; Fekiacova, Z.; Hatté, C. Atmosphere–soil carbon transfer as a function of soil depth. *Nature* **2018**, *559*, 599–602. [CrossRef] [PubMed]
15. Beier, C.; Rasmussen, L. Organic matter decomposition in an acidic forest soil in Denmark as measured by the cotton strip assay. *Scand. J. For. Res.* **1994**, *9*, 106–114. [CrossRef]
16. Li, J.; Yan, D.; Pendall, E.; Pei, J.; Noh, N.J.; He, J.S.; Li, B.; Nie, M.; Fang, C. Depth dependence of soil carbon temperature sensitivity across Tibetan permafrost regions. *Soil Biol. Biochem.* **2018**, *126*, 82–90. [CrossRef]
17. Davidson, E.A.; Janssens, I.A. Temperature sensitivity of soil carbon decomposition and feedbacks to climate change. *Nature* **2006**, *440*, 165–173. [CrossRef] [PubMed]
18. Bird, M.I.; Chivas, A.R.; Head, J. A latitudinal gradient in carbon turnover times in forest soils. *Nature* **1996**, *381*, 143–146. [CrossRef]

19. Von Lützow, M.; Kögel-Knabner, I. Temperature sensitivity of soil organic matter decomposition-what do we know? *Biol. Fertil. Soil* **2009**, *46*, 1–15. [CrossRef]
20. Lloyd, K.; Madramootoo, C.A.; Edwards, K.P.; Grant, A. Greenhouse gas emissions from selected horticultural production systems in a cold temperate climate. *Geoderma* **2019**, *349*, 45–55. [CrossRef]
21. Lee, C.G.; Suzuki, S.; Inubushi, K. Temperature sensitivity of anaerobic labile soil organic carbon decomposition in brackish marsh. *Soil Sci. Plant Nutr.* **2018**, *64*, 443–448. [CrossRef]
22. Sierra, C.A. Temperature sensitivity of organic matter decomposition in the Arrhenius equation: Some theoretical considerations. *Biogeochemistry* **2012**, *108*, 1–15. [CrossRef]
23. Kowalenko, C.G. Assessment of Leco CNS-2000 analyzer for simultaneously measuring total carbon, nitrogen, and sulphur in soil. *Commun. Soil Sci. Plant Anal.* **2001**, *32*, 2065–2078. [CrossRef]
24. Linn, D.M.; Doran, J.W. Effect of Water-Filled Pore Space on Carbon Dioxide and Nitrous Oxide Production in Tilled and Nontilled Soils. *Soil Sci. Soc. Am. J.* **1984**, *48*, 1267–1272. [CrossRef]
25. Verdonck, O.F.; Cappaert, I.M.; De Boodt, M.F. Physical characterization of horticutural substrates. *Acta Hortic.* **1978**, *178*, 191–200. [CrossRef]
26. Gagnon, B.; Ziadi, N.; Rochette, P.; Chantigny, M.H.; Angers, D.A.; Bertrand, N.; Smith, W.N. Soil-surface carbon dioxide emission following nitrogen fertilization in corn. *Can. J. Soil Sci.* **2016**, *96*, 219–232. [CrossRef]
27. Newton, L.S.J.; Lopes, A.; Spokas, K.; Archer, D.W.; Reicosky, D. First-order decay models to describe soil C-CO$_2$ loss after rotary tillage. *Sci. Agric.* **2009**, *66*, 650–657.
28. Verzani, J. *Using R for Introductory Statistics*; CRC Press: Boca Raton, FL, USA, 2018.
29. Dalgaard, P. Multiple Regression. In *Introductory Statistics with R*; Springer Science + Business Media, LLC.: Berlin/Heidelberg, Germany, 2008; pp. 185–194.
30. Cai, A.; Feng, W.; Zhang, W.; Xu, M. Climate, soil texture, and soil types affect the contributions of fine-fraction-stabilized carbon to total soil organic carbon in different land uses across China. *J. Environ. Manag.* **2016**, *172*, 2–9. [CrossRef]
31. Fissore, C.; Jurgensen, M.F.; Pickens, J.; Miller, C.; Page-Dumroese, D.; Giardina, C.P. Role of soil texture, clay mineralogy, location, and temperature in coarse wood decomposition-A mesocosm experiment. *Ecosphere* **2016**, *7*, e01605. [CrossRef]
32. McInerney, M.; Bolger, T. Temperature, wetting cycles and soil texture effects on carbon and nitrogen dynamics in stabilized earthworm casts. *Soil Biol. Biochem.* **2000**, *32*, 335–349. [CrossRef]
33. Sugihara, S.; Funakawa, S.; Kilasara, M.; Kosaki, T. Effects of land management on CO2 flux and soil C stock in two Tanzanian croplands with contrasting soil texture. *Soil Biol. Biochem.* **2012**, *46*, 1–9. [CrossRef]
34. Ngao, J.; Epron, D.; Delpierre, N.; Bréda, N.; Granier, A.; Longdoz, B. Spatial variability of soil CO2 efflux linked to soil parameters and ecosystem characteristics in a temperate beech forest. *Agric. For. Meteorol.* **2012**, *154–155*, 136. [CrossRef]
35. De Graaff, M.A.; Jastrow, J.D.; Gillette, S.; Johns, A.; Wullschleger, S.D. Differential priming of soil carbon driven by soil depth and root impacts on carbon availability. *Soil Biol. Biochem.* **2014**, *69*, 147–156. [CrossRef]
36. Fontaine, S.; Barot, S.; Barré, P.; Bdioui, N.; Mary, B.; Rumpel, C. Stability of organic carbon in deep soil layers controlled by fresh carbon supply. *Nature* **2007**, *450*, 277–280. [CrossRef] [PubMed]
37. Hobley, E.U.; Wilson, B. The depth distribution of organic carbon in the soils of eastern Australia. *Ecosphere* **2016**, *7*, e01214. [CrossRef]
38. Bonan, G.B.; Hartman, M.D.; Parton, W.J.; Wieder, W.R. Evaluating litter decomposition in earth system models with long-term litterbag experiments: An example using the Community Land Model version 4 (CLM4). *Glob. Change Biol.* **2013**, *19*, 957–974. [CrossRef]
39. Cornwell, W.K.; Cornelissen, H.C.; Dorrepaal, E.; Eviner, V.T.; Godoy, O.; Hobbie, S.E.; Hoorens, B.; Van Bodegom, P. Plant species traits are the predominant control on litter decomposition rates within biomes worldwide. *Ecol. Lett.* **2008**, *11*, 1065–1071. [CrossRef]
40. Duboc, O.; Zehetner, F.; Djukic, I.; Tatzber, M.; Berger, T.W.; Gerzabek, M.H. Decomposition of European beech and black pine foliar litter along an Alpine elevation gradient: Mass loss and molecular characteristics. *Geoderma* **2012**, *189*, 522–531. [CrossRef]
41. Jobbágy, E.G.; Jackson, R.B. The vertical distribution of soil organic carbon and its relation to climate and vegetation. *Ecol. Appl.* **2000**, *10*, 423–436. [CrossRef]
42. Schrumpf, M.; Kaiser, K.; Guggenberger, G.; Persson, T.; Kögel-Knabner, I.; Schulze, E.-D. Storage and stability of organic carbon in soils as related to depth, occlusion within aggregates, and attachment to minerals. *Biogeosciences* **2013**, *10*, 1675–1691. [CrossRef]
43. Bosatta, E.; Ågren, G.I. Soil organic matter quality interpreted thermodynamically. *Soil Biol. Biochem.* **1999**, *31*, 1889–1891. [CrossRef]
44. Hicks Pries, C.E.; Sulman, B.N.; West, C.; O'Neill, C.; Poppleton, E.; Porras, R.C.; Castanha, C.; Zhu, B.; Wiedemeier, D.B.; Torn, M.S. Root litter decomposition slows with soil depth. *Soil Biol. Biochem.* **2018**, *125*, 103–114. [CrossRef]
45. Kuzyakov, Y. Priming effects: Interactions between living and dead organic matter. *Soil Biol. Biochem.* **2010**, *42*, 1363–1371. [CrossRef]
46. St-Luce, M.; Ziadi, N.; Chantigny, M.H.; Braun, J. Long-term effects of tillage and nitrogen fertilization on soil C and N fractions in a corn–soybean rotation. *Can. J. Soil Sci.* **2022**, *102*, 277–292. [CrossRef]
47. Andlar, M.; Rezić, T.; Marđetko, N.; Kracher, D.; Ludwig, R.; Šantek, B. Lignocellulose degradation: An overview of fungi and fungal enzymes involved in lignocellulose degradation. *Eng. Life Sci.* **2018**, *18*, 768–778. [CrossRef] [PubMed]

48. Steinweg, J.M.; Jagadamma, S.; Frerichs, J.; Mayes, M.A. Activation energy of extracellular enzymes in soils from different biomes. *PLoS ONE* **2013**, *8*, e59943. [CrossRef] [PubMed]
49. Parent, L.E.; Mackenzie, A.F. Rate of Pyrophosphate Hydrolysis in Organic Soils. *Can. J. Soil Sci.* **1985**, *65*, 497–506. [CrossRef]
50. Fierer, N.; Craine, J.M.; McLauchlan, K.K.; Schimel, J.P. Litter quality and the temperature sensitivity of decomposition. *Ecology* **2005**, *86*, 320–326. [CrossRef]
51. Gerke, J. Concepts and misconceptions of humic substances as the stable part of soil organic matter: A review. *Agronomy* **2018**, *8*, 76. [CrossRef]
52. Meyer, N.; Welp, G.; Amelung, W. The temperature sensitivity (Q_{10}) of soil respiration: Controlling factors and spatial prediction at regional scale based on environmental soil classes. *Glob. Biogeochem. Cycl.* **2018**, *32*, 306–323. [CrossRef]
53. Ignace, D.D. Determinants of temperature sensitivity of soil respiration with the decline of a foundation species. *PLoS ONE* **2019**, *14*, e0223566. [CrossRef]
54. Meyer, N.; Meyer, H.; Welp, G.; Amelung, W. Soil respiration and its temperature sensitivity (Q10): Rapid acquisition using mid-infrared spectroscopy. *Geoderma* **2018**, *323*, 31–40. [CrossRef]
55. Ward, E.B.; Duguid, M.C.; Kuebbing, S.E.; Lendemer, J.C.; Warren II, R.J.; Bradford, M.A. Ericoid mycorrhizal shrubs alter the relationship between tree mycorrhizal dominance and soil carbon and nitrogen. *J. Ecol.* **2021**, *109*, 3524–3540. [CrossRef]

soil systems

Article

Spatial Optimization of Conservation Practices for Sediment Load Reduction in Ungauged Agricultural Watersheds

Racha ElKadiri [1],*, Henrique G. Momm [1], Ronald L. Bingner [2] and Katy Moore [1]

[1] Department of Geosciences, Middle Tennessee State University, Murfreesboro, TN 37132, USA
[2] National Sedimentation Laboratory, United States Department of Agriculture, Oxford, MS 38655, USA
* Correspondence: racha.elkadiri@mtsu.edu

Abstract: Conservation practices (CPs) are used in agricultural watersheds to reduce soil erosion and improve water quality, leading to a sustainable management of natural resources. This is especially important as more pressure is applied on agricultural systems by a growing population and a changing climate. A challenge persists, however, in optimizing the implementation of these practices given their complex, non-linear, and location-dependent response. This study integrates watershed modeling using the Annualized Agricultural Non-Point-Source model and a GIS-based field scale localization and characterization of CPs. The investigated practices are associated with the implementation of riparian buffers, sediment basins, crop rotations, and the conservation reserve program. A total of 33 conservation scenarios were developed to quantify their impact on sediment erosion reduction. This approach was applied in an ungauged watershed as part of the Mississippi River Basin initiative aiming at reducing one of the largest aquatic dead zones in the globe. Simulation results indicate that the targeted approach has a significant impact on the overall watershed-scale sediment load reduction. Among the different evaluated practices, riparian buffers were the most efficient in sediment reduction. Moreover, the study provides a blueprint for similar investigations aiming at building decision-support systems and optimizing the placement of CPs in agricultural watersheds.

Keywords: AnnAGNPS; ungauged watershed; conservation practices; agricultural watershed; soil erosion; sediment load; watershed modeling; GIS; riparian buffer; sediment basin

Citation: ElKadiri, R.; Momm, H.G.; Bingner, R.L.; Moore, K. Spatial Optimization of Conservation Practices for Sediment Load Reduction in Ungauged Agricultural Watersheds. *Soil Syst.* **2023**, *7*, 4. https://doi.org/10.3390/soilsystems7010004

Academic Editor: Luis Eduardo Akiyoshi Sanches Suzuki

Received: 19 November 2022
Revised: 29 December 2022
Accepted: 30 December 2022
Published: 13 January 2023

1. Introduction

Soil erosion is a global challenge causing the siltation of waterways and dams, the degradation of water quality and the reduction of soil fertility and crop yield among other environmental and socio-economic issues [1–4]. These problems are projected to increase with a changing climate characterized by a higher frequency and intensity of extreme precipitation events [5,6], hence threatening further, the future of food security. Presently, the U.S. agricultural sector loses about $44 billion per year from soil erosion impacts on fertile soil quantity and water quality [7]. In fact, 80% of freshwater bodies in the U.S. are impacted by nonpoint source pollution [8,9], which is primarily caused by soil movement from agricultural fields [10].

Over the last few decades, conservation practices have proven to be an effective measure for preventing or minimizing soil erosion and its negative impacts on crops, water, and soils [11–13]. These practices include, but are not limited to, minimum tillage, permanent soil cover, riparian vegetation buffer, crop rotation, terraces, sedimentation basins, strip cropping, and intercropping [14,15]. The implementation of these practices has shown results in terms of water quality improvement, soil erosion reduction, soil fertility increase, soil moisture retention, long-term yield increase, and food security [16].

However, conservation practices range broadly in effectiveness, since water, nutrients and soil particles are transported along various pathways and controlled by multiple natural and anthropogenic processes [17–23]. One of the main factors impacting conservation practices effectiveness is their spatial placement within the watershed. The non-point source pollution reduction results vary as a function of the upstream area drainage patterns, soil characteristics, topography, vegetation density and land management among other physical factors [23–27]. Hence, the need for investigations to determine the best implementation conditions of these conservation practices while accounting for their complex, nonlinear, and location dependent response. Often, the evaluation of conservation practices at the watershed-scale is conducted by two types of studies.

The first type involves the identification of candidate locations for conservation practices implementation in agricultural watersheds using geographic information system (GIS) analyses [28,29]. These efforts support the development of decision support systems to aid watershed scale conservation plans based on ranking location suitability for specific conservation practices. However, despite the significant contribution of these studies in identifying potential sites for practice implementation, they are not designed to generate temporal sediment yield and load estimates like in the case of comparable watershed models.

The second type involves investigations conducted to evaluate the impact of conservation practices using watershed-scale hydrological models. In this type of study, characterization and simulation of conservation practices is often performed indirectly. Input parameters controlling flow and/or infiltration are adjusted to estimate the impact of the conservation practice within the basic modeling unit (e.g., Hydrological Response Unit (HRU)). For example, in studies based on the Soil and Water Assessment Tool (SWAT) [30], contour farming has been described by adjustments to curve numbers (CNs) [11], conservation tillage has been described by adjustments to CNs and Manning's roughness coefficient (n) [31], and vegetative practices has been simulated by adjusting CN, n, and universal soil loss equation cover management factor (USLE–C) [32]. In addition to an indirect representation of these conservation practices, their location is also under-characterized since the location of HRUs within the sub-catchment is not considered in the calculations [31]. These factors increase the uncertainties in the evaluation of most conservation practices in which efficiency is dependent on location and physical characteristics (e.g., width of the riparian buffer, size of the sediment basin).

In this study, a comprehensive evaluation of conservation practices efficiency was conducted through the integration of both hydrological modeling and GIS-based field scale localization and characterization of these practices in a wide range of scenarios. Four sets of conservation practices are targeted in this study: riparian buffers, sediment basins, crop rotations, and the conservation reserve program. These practices were selected due to their high efficiency in reducing sediment detachment and/or transport at field and/or watershed scales [33–39], in addition to the suitability of the developed integrated methodology in quantifying their location-dependent impact [40–48]. This overarching goal can be further subdivided into the following specific objectives: (1) evaluation of conservation scenarios based on location of conservation practices at field-scale informed by hydrological model results and characterized at raster-grid scale by GIS tools, (2) quantification of type and placement of conservation practices efficiency locally, where they were implemented, and their contribution to the overall watershed's sediment load, and (3) assessment of the tradeoff between gains in sediment load reduction and potential loss in productive area based on a multi-objective function.

2. Materials and Methods

2.1. Study Area

The North Fork Forked Deer River watershed used in this study is in the northern portion of the Lower Mississippi hydrological region in West Tennessee (Figure 1). The total drainage area is 631.31 km^2 flowing into the Forked Deer River. The average annual precipitation in the watershed is 1437 mm, with about 70% occurring during the growing season (March–October), March being the wettest month of the year (i.e., 167 mm on average), and August the driest month (i.e., 84 mm on average). Soils in the watershed include 55 classes as defined by the USDA Web Soil Survey [49], with silt loam being the most dominant soil texture in the watershed.

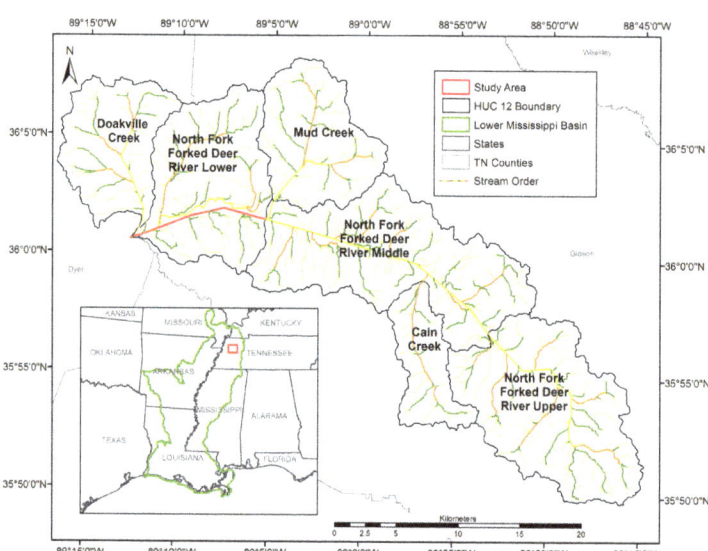

Figure 1. Location of the study area in the lower Mississippi water resource region, including six USGS HUC-12 sub-watersheds.

The watershed consists of six USGS 12-digits hydrologic-unit code (HUC-12), referred to as sub-watersheds by the U.S. Geological Survey (Figure 1). The North Fork Forked Deer River Upper covers a total area of 147.80 km^2 with an average slope of 7%. Predominant land uses include agricultural cropland (38%), forest (36%), pastures (19%), and developed land (7%). Cain Creek is in the southernmost region with a total area of 43.56 km^2 with average slopes of 6.0%. Predominant land uses include agricultural cropland (59%), forest (18%), pastures (10%), and developed land (13%). Mud Creek is in Gibson County. It covers a drainage area of 84.64 km^2 and has an average slope of 4.89%. Predominant land uses within this sub-watershed include agricultural cropland (78%), forest (11%), pastures (4%), and developed land (7%). The North Fork Forked Deer River middle is the largest of the HUC-12 divisions with a total area of 161.13 km^2 and an average slope of 5.6%. Predominant land uses include agricultural cropland (51%), forest (33%), pastures (8%), and developed land (7%). The Doakville Creek is the western-most HUC 12 division and is in Dyer County. It is the second smallest sub-watershed, with a total area of 77.49 km^2. Predominant land uses within this sub-catchment include agricultural cropland (80%), forest (7%), pastures (7%), and developed land (6%). The North Fork Forked Deer River lower covers a drainage area of 115.38 km^2 and has an average slope of 3.84%. Land uses within this sub-watershed include agricultural cropland (73%), forest (19%), pastures (4%), and developed land (4%).

The lack of streamflow observations in most catchments around the world and the nation, especially in small catchments in rural areas has created a setback for hydrological and conservation investigations in these regions [50]. This ungauged watershed represents a study case of this challenge, where conservation practices are needed but cannot be supported by field measured streamflow and sediment data. Methods to minimize the uncertainties introduced by the lack of field gauges are discussed in Section 3.1.

2.2. The AnnAGNPS Model

The USDA Annualized Agricultural Non-Point Source (AnnAGNPS) [40–44] watershed pollution model was designed to evaluate the impact of integrated long-term effects of farming and conservation practices on water quality within ungauged agricultural watersheds. The model contains components to describe processes controlling pollutants sources and sinks, their corresponding movement throughout the watershed, and their interrelated contribution to the watershed total pollutant load. A description of the AnnAGNPS model's components and mathematical formulation has been provided in previous studies [40–44], and only a summary of the AnnAGNPS model's key characteristics relevant to this study is provided.

The watershed is internally represented within AnnAGNPS as cells connected to reaches. Reaches are designated to simulate physical processes resulting from concentrated flow (channels) while cells (often referred to as sub-catchments, fields, or AnnAGNPS cells to distinguish from raster grid cells) are designed to simulate physical processes occurring at upland areas that drain into reaches. Sub-catchments are hierarchically connected to reaches to describe surface and shallow subsurface flow, sediment detachment, transport and deposition processes, and pollutants transport throughout the overland flow areas of the watershed. Upland erosion processes include sheet and rill, tillage-induced ephemeral gullies, and classical and edge-of-field gullies. Reaches and sub-catchments are individually described in terms of topography, weather, soil, and management. Sizes of sub-catchments are often selected based on field sizes to enhance the spatial characterization of practices [45]. Specifically, management input databases are designed to describe farming practices on high temporal resolution (up to daily) to capture unique farming management schedules and their associated operations at the field scale.

2.3. Characterization and Modeling of Riparian Buffers

Riparian buffers are defined as either natural or planted vegetation located at the edge of fields or along reaches. They are designed to reduce the delivery of eroded sediment from fields or sub-catchments (cell-located buffers) into reaches or the delivery of sediments from one reach to another (reach-located buffers). The potential maximum sediment trapping efficiency input for each AnnAGNPS cell containing a riparian buffer was determined using a GIS approach available from the AGNPS Buffer Utility Feature (AGBUF) software package [46]. Using a user-provided GIS layer describing spatially the location of the buffer and the vegetation type, the software analyzes raster grid cells within the sub-catchment to calculate the potential maximum sediment trapping efficiency based on slope, drainage area, and vegetation type [42]. The actual sediment trapped by riparian buffers for each of the five AnnAGNPS sediment particle size classes is determined using relationships involving the potential maximum sediment trapping efficiency from AGBUF and daily surface flow [47]. The advantage of the approach is the scale at which calculations are conducted (3 m raster grid cell) to describe and place the riparian buffer within each sub-catchment (i.e., AnnAGNPS cell).

2.4. Characterization and Modeling of Constructed Sediment Basins

Similar to the riparian buffer component, sediment basins are optimally located within the watershed and physically characterized using GIS-based analyses and the AGNPS Wetland Feature (AGWET) software package [47]. The latter is designed to record information of each sediment basin's surface area, barrier height, presence/absence of vegetation, location in the watershed, and upstream drainage area are calculated using the user-provided GIS layers [48]. The development of the sediment basin input databases for the AnnAGNPS model is generated using AGWET. The AnnAGNPS watershed pollution model is then used to determine the change in energy between inflows and outflows and the respective impact on water quality processes for each sub-catchment. The conservation of mass is applied to both hydrology and pollutant balances. The integration of GIS analyses including the spatial distribution of constructed sediment basins with the sub-catchment, and the description of soil, land use, and topography at raster grid scales enhance the characterization of each sediment basin and its location driven impacts on water and sediment.

2.5. Baseline Conditions

2.5.1. Topography

Topographic information was obtained from LiDAR datasets available in the public data repository of the Tennessee Department of Finance and Administration [51]. Datasets were provided as raster grids with one-meter spatial resolution. Raster grids were mosaiced, reprojected, scaled, resampled to three-meter spatial resolution, and hydrologically enforced. Topographic analyses were performed with the GIS TopAGNPS software package. The TopAGNPS technology is built based on the Topographic Parameterization (TOPAZ) software [52,53]. In addition to standard GIS operations for processing DEM, TopAGNPS has the tools to sub-divide the watershed into reaches and sub-catchments. An iterative approach was applied where datasets generated by the TopAGNPS computer program were compared to high-resolution imagery and auxiliary GIS layers to determine whether manmade obstructions would hinder the flow routing algorithm. The latter could cause structures to work as pseudo-dams resulting in an incorrect surface flow network and/or increase ponding beyond normal levels. A custom computer program has been developed to modify user-selected regions in the DEM to enforce surface flow. Two user-provided parameters control how the watershed is subdivided into sub-catchments and reaches. A critical source area (CSA) value of five hectares and a maximum source channel length (MSCL) value of 250 m were selected, yielding a total of 12,573 sub-catchments and 5047 reaches (Table 1 and Figure 2).

Table 1. Discretization of the watershed into AnnAGNPS cells.

HUC-12 Name	AnnAGNPS Cells (Sub-Catchments)			
	Number	Average Area (ha)	Average Slope (%)	Average Flow Length (m)
North Fork Forked Deer River Upper	2946	5.00	8%	246.45
Cain Creek	904	4.81	8%	245.89
Mud Creek	1682	5.04	6%	255.84
North Fork Forked Deer River Middle	3165	5.10	5%	259.92
Doakville Creek	1565	4.95	6%	243.84
North Fork Forked Deer River Lower	2311	4.99	4%	259.56
Total	12,573			

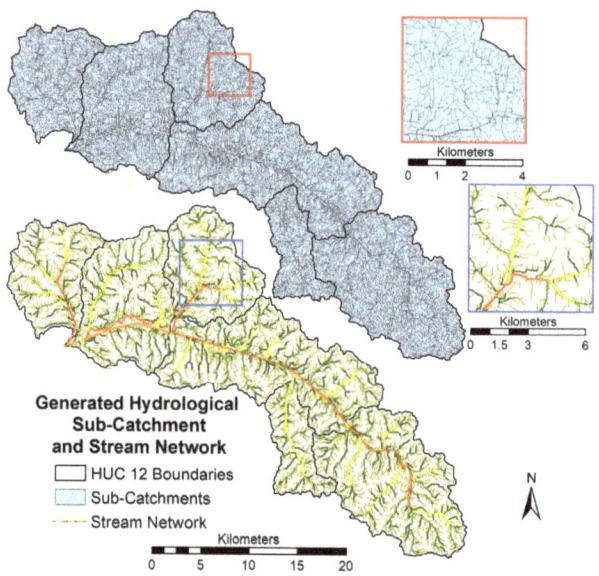

Figure 2. Discretization of the watershed into sub-catchments representing fields (upper map) and reaches representing concentrated flow (lower map).

2.5.2. Climate

Climate datasets at daily temporal scale from 2008 to 2018 were obtained from the U.S. National Oceanic and Atmospheric Administration (NOAA) [54] for Gibson, Dyer, Weakley, Carroll, Crockett, and Madison Counties. These datasets, including daily precipitation and minimum and maximum air temperature information, were pre-processed, quality controlled, and analyzed to identify the weather stations with the greatest temporal data coverage and their location in relation to the watershed. However, most of these stations were located outside of the study area boundaries (Figure 3). Hence, neighboring stations from outside the watershed were used to fill data gaps in the stations located within the study area (referred to as secondary stations). Decision on which station to draw information from was based on spatial zones of influence using Thiessen polygons. The primary station was defined to be in the centroid of the study area (Figure 3) and to be the average of all stations. In the AnnAGNPS watershed model, the primary climate station is only used as a backup to fill a missing data point in the secondary stations. The secondary station records were also filled with data from the nearest neighboring stations (Figure 3) and were evaluated to remove data anomalies. In the AnnAGNPS model, each sub-catchment (i.e., AnnAGNPS cell) is assigned to a secondary climate station.

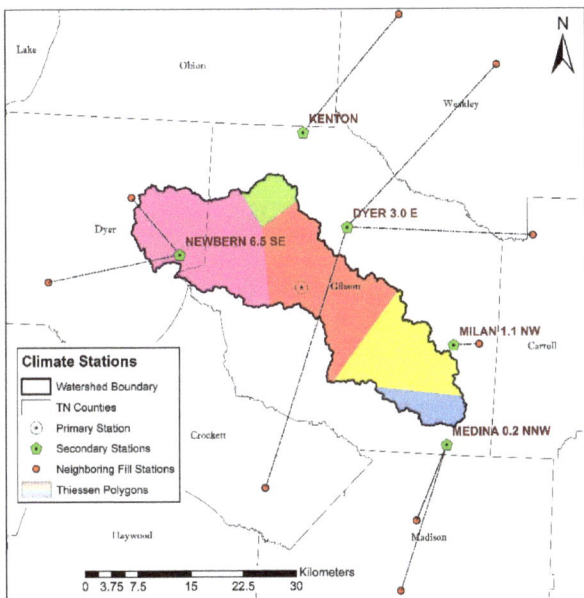

Figure 3. Climate station locations used in the study.

Weather characteristics not available from the historic observations were also generated using AGNPS Generation of weather Elements for Multiple applications (AgGEM) software package [55], including dew point, sky cover, wind speed, and solar radiation. The AgGEM software package generates synthetic data for these four parameters based on long-term statistics derived from records of different regions in the US.

2.5.3. Soil

Soil spatial distribution data was retrieved from the Web Soil Survey (WSS) using the Natural Resources Conservation Service's (NRCS) website [56] for Gibson and Dyer counties. Complementary soil description of physical and chemical properties in tabular format were retrieved from the USDA Soil Data Access website [57]. A custom SQL script was used to query and retrieve soil information needed for the input soil database (e.g., hydrologic soil group, erodibility factor, impervious depth, specific gravity, clay, silt, sand

and rock ratios, and number of soil layers). These datasets were post processed using the National Soil Information System (NASIS) Import to AnnAGNPS (NITA) software package (i.e., a component of the AGNPS modeling system) to ensure the accuracy of the soil characteristics table. Once the data were quality controlled, the soil characteristics data table was joined with the attribute table of the original soil shapefile. Using GIS analysis, each sub-catchment was assigned to a soil type based on spatial majority analysis. A total of 80 unique soil types were used.

2.5.4. Management

Field management within the AnnAGNPS model is represented by a multi-year temporal sequence of operations describing key farming activities and their potential impact on soil cover, surface runoff/infiltration, fertilizer application, irrigation strategy, and soil disturbance by equipment. Each sub-catchment in the watershed is assigned to a management ID. This procedure ensures that farming practices, land cover, and their impact on soil detachment and transportation are characterized in time and space. Land use/land cover information were obtained from the U.S. Department of Agriculture—National Agricultural Statistics Service—CropScape website [58]. Agricultural land use information, referred to as crop data layer (CDL), were downloaded as annual raster grids at 30 m spatial resolution describing crop types and main non-agricultural land covers [59]. In this study, raster grids from 2008 to 2018 were used. Statistical analyses were performed to determine dominant crop types based on datasets for the years 2008, 2013, and 2018. Nine major classes were ascertained from the original land use classes. The nine dominant consistent classes are corn, cotton, winter wheat/soybean, forest, developed, grass/pasture, soybeans, woody wetlands, and water. These classes represent about 99% of the watershed. Crops that were less conventional, such as pumpkins or Christmas trees, were classified under "grass/pasture", given that they represent about 1% of the total watershed area, and their incorporation in the management schedule and operations will significantly increase the processing time and complexity. The original raster grids for all years were re-coded to the nine main land use classes. For each year considered, a majority spatial zonal statistic GIS analysis was used to assign the dominant land use for each sub-catchment (Figure 4).

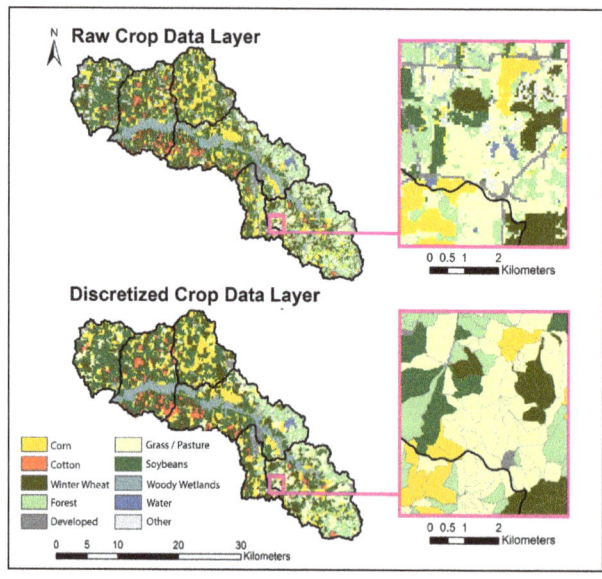

Figure 4. Spatial-temporal characterization of land use. The procedure was applied for all years, but only 2008 is included for illustration purposes.

The input AnnAGNPS farming management database was assembled by integrating three sources of information: (1) spatiotemporal crop type information at sub-watershed scale (from annual discretized CDL datasets), (2) average annual crop type yield information at county scale (from agricultural census downloaded from USDA-NASS [60]), and (3) one-year farming practices (from RUSLE 2 database). The latter represent typical farming operations and schedules for a particular crop type in this region (Table 2). The three datasets were combined using a custom Python script, generating 4135 unique 11-year crop/landuse rotations for the 12,573 sub-catchments in the study area. The output from the custom script are management schedule and operation databases required by the AnnAGNPS model.

Table 2. Example of the annual management schedule for three major crops in the study area.

Crop	Date	Operation
Corn	15 Mar.	Bedder/Lister
	1 Apr.	Disk
	13 Apr.	Fertilizer
	14 Apr.	Disk
	15 Apr.	Sprayer
	16 Apr.	Plant
	9 May	Sprayer
	15 May	Fertilizer
	29 May	Sprayer
	15 Sep.	Harvest
	16 Sep.	Weed Growth
Cotton	17 Apr.	Sprayer
	18 Apr.	Fertilizer
	1 May	Plant
	15 May	Sprayer
	15 June	Fertilizer
	16 June	Sprayer
	15 July	Sprayer
	31 July	Sprayer
	15 Aug.	Sprayer
	29 Aug.	Defoliant
	15 Oct.	Weed Growth
Soybean	20 Mar.	Chisel
	5 May	Fertilizer
	10 May	Disk
	11 May	Plant
	29 May	Sprayer
	20 July	Sprayer
	28 Aug.	Sprayer
	15 Oct.	Harvest
	20 Oct.	Weed Growth

2.6. Conservative Practices Scenarios

2.6.1. Riparian Buffer

A riparian buffer is an area of vegetation designed to slow down and spread surface flow thus promoting infiltration and allowing contaminants and sediment to be deposited. Riparian covers typically include vegetation like grasses, sedges, rushes, and ferns that are tolerant of intermittent flooding or saturated soils. They are established and managed as the dominant vegetation in the transitional zone between upland areas such as agricultural fields and aquatic habitats such as streams [61,62]. The overall sediment trapping efficiency of riparian buffers is impacted by location and buffer physical characteristics (e.g., width, length, vegetation density). In this study we simulate the potential impact of riparian buffers in a wide range of scenarios based on varying their width and location within the watershed.

The use of sediment yield to define potential riparian buffer implementation scenarios, allows to investigate the impact of the conservation practice when is targeted toward hotspot areas only, when is implemented around all streams in the watershed which represents a theoretical maximum reduction, and in-between scenarios. Hence, the first two scenarios involve the implementation of riparian buffers (1) around all streams in the watersheds, and (2) around all streams adjacent to agricultural fields (Figure 5A). In addition, results from the baseline conditions simulation were used to calculate the mean (i.e., 8.94 Mg/ha/year) and the standard deviation (i.e., 5.24 Mg/ha/year) of sediment yield from all agricultural fields in the study area. These values were used in defining a range of location scenarios within the watershed for potential implementation of riparian buffers: (3) around all streams adjacent to agricultural fields that have a sediment yield higher than mean minus standard deviation (referred to as ">medium") (Figure 5B), (4) around all streams adjacent to agricultural fields that have a sediment yield higher than mean plus standard deviation (referred to as ">high") (Figure 5C), and (5) around all streams adjacent to agricultural fields that have a sediment yield higher than mean plus two standard deviations (referred to as ">very high") (Figure 5D).

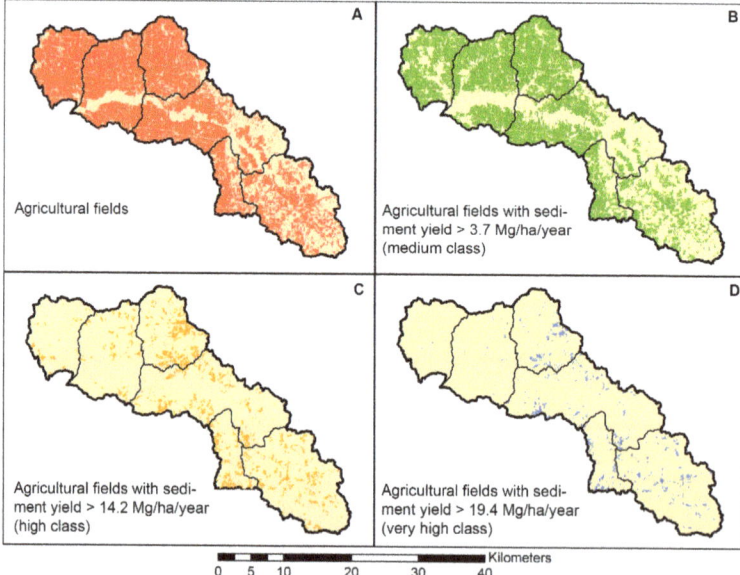

Figure 5. Study area agricultural fields classified by sediment yield ((**A**): All agricultural fields, (**B**): medium class, (**C**): high class, and (**D**): very high class). These four classes are used as scenarios in our final simulations to estimate the contribution of using conservation practices in reducing sediment yield (from least to most restrictive).

The simulated riparian buffer has widths of 10 m, 30 m and 60 m. We selected these buffer widths, given that the EPA defines narrow buffer width as 1–15 m and wide buffer width as broader than 50 m. State and federal guidelines range from seven to 200 m [63,64]. Multiple studies investigated and reported an efficient and cost-effective buffer at widths ranging between 10 and 30 m [65–67]. In addition to these two typical widths (10 and 30 m), we simulated a wide buffer (i.e., 60 m) to quantify the impact of doubling the riparian width from the typical previously investigated scenarios, while still being within the applicable limits [68]. The channel network was buffered by the three buffer widths and the resulting polygon layers were intersected with sub-catchments that meet the sediment yield condition of the respective simulated class.

The combination of these different conditions (i.e., sediment yield class, all or just agricultural areas, buffer width) yielded a total of 15 scenarios (Figure 6a). In addition, a sixteenth scenario was created to represent existing riparian buffer conditions. The existing riparian vegetation (mainly trees) was delineated using a custom script, consisting of a pretrained machine learning canopy detection model based on LiDAR point cloud data [69].

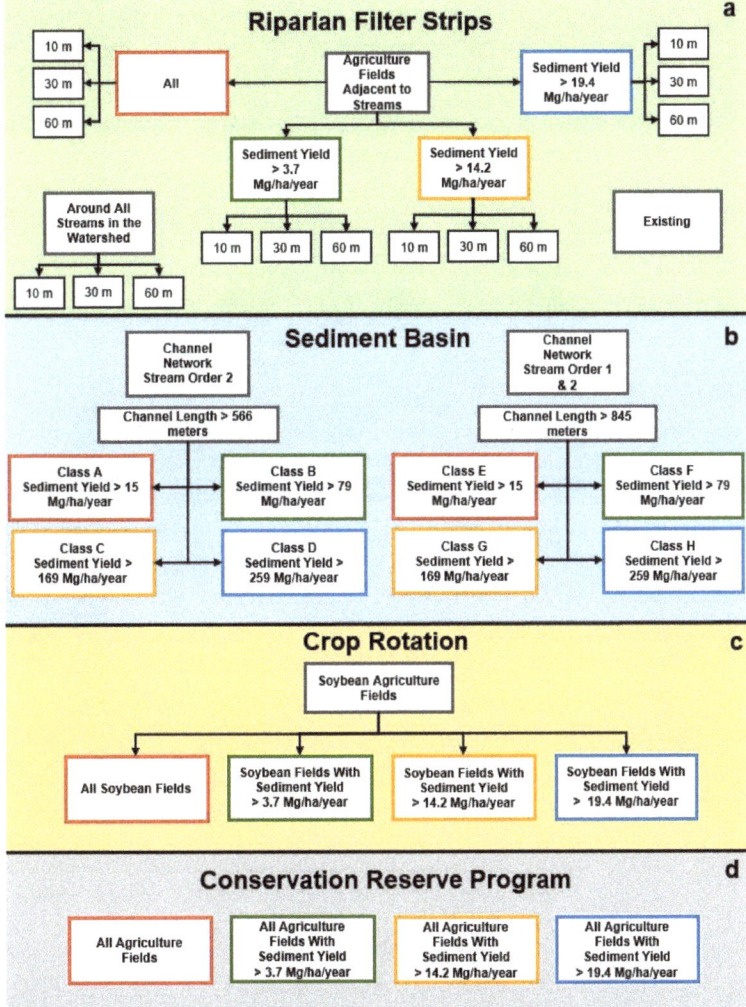

Figure 6. Flowchart summarizing all simulated conservation practice scenarios and their differences (**a–d**).

These 16 riparian buffer GIS layers were further processed at a 3 m spatial resolution using the AGBUF GIS tool (Section 2.3) to evaluate all flow paths through the riparian layer and allowing the determination of a potential sediment trapping efficiency for each sub-catchment. The outcomes of this analysis are used as input databases describing physical properties of riparian zones for each sub-catchment for 16 separate AnnAGNPS simulations.

2.6.2. Sediment Basin

Sediment basins are built to capture runoff, changing the energy of surface flow, and, therefore, allowing the settlement of sediment and other suspended solids [70]. The existing sediment basins were digitally delineated by researching and examining all the water bodies in the watershed using publicly available high resolution remotely sensed imagery, available via Google Earth and ESRI base-maps. As a result, 57 water bodies were identified as sediment basins (Figure 7). These polygons were described using the AGWET tool (Section 2.4) by performing GIS analysis at 3 m spatial resolution to determine area, barrier height, average water depth, and other properties [48,71]. AGWET was used in an iterative mode to delineate the extent of the existing sediment basins and determine the appropriate weir height for each one of them that matches their observed surface area in high resolution satellite imagery. The goal of this process is to find a relationship between sediment basin weir height and upstream area specific to this watershed. This relationship is used to partially determine the morphological characteristics of the proposed new sediment basins in the watershed.

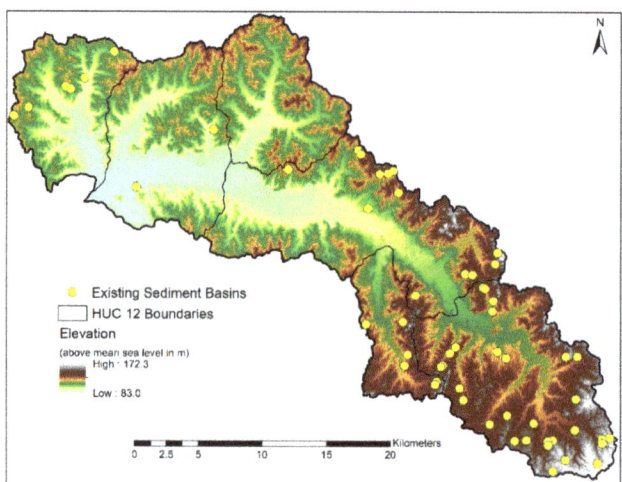

Figure 7. Distribution of existing sediment basins in the study area.

The potential new sediment basin locations were selected based on three criteria: (1) channel stream order to ensure intermittence of flow (i.e., stream order 1 or 2), (2) channel stream length above a statistical threshold (i.e., mean length) to eliminate noisy minuscule channels, and (3) sub-catchment sediment yield categorized into four classes based on its statistical distribution in the targeted sub-catchments as well as on sediment yield in the delineated existing sediment basins, to ensure replication of similar conditions and expansion of impact.

Results from the baseline conditions simulation were used to calculate the mean sediment yield for the AnnAGNPS cells in which the 57 existing sediment basins are located (i.e., 15 Mg/year/ha), and the mean and the standard deviation of all the targeted AnnAGNPS cells by criteria 1 and 2 (i.e., mean: 79 Mg/year/ha and standard deviation: 90 Mg/year/ha). These values were then used in defining a range of location scenarios within the watershed. The considered sediment yield classes are: (1) above mean sediment yield for existing sediment basins (i.e., 15 Mg/year/ha), (2) above mean sediment yield of the considered sub-catchments (i.e., 79 Mg/year/ha), (3) above (mean + standard deviation) sediment yield of the considered sub-catchments (i.e., 169 Mg/year/ha), and (4) above (mean + 2 × standard deviations) sediment yield of the considered sub-catchments (i.e., 259 Mg/year/ha). The use of a statistical distribution to generate simulation classes allow testing the impact of sediment basins based on a spectrum of scenarios ranging from most to least conservative in terms of number of basins, location, and sediment yield, providing multiple options of targeted implementation for stakeholders.

The combination of the aforementioned three criteria (i.e., sediment yield, stream order and channel length) led to eight total scenarios (Figure 8) including 233 sediment basins for class A, 147 sediment basins for class B, 67 sediment basins for class C, 21 sediment basins for class D, 527 sediment basins for class E, 297 sediment basins for class F, 126 sediment basins for class G, and 54 sediment basins for class H. A sediment basin scenario generation flowchart is summarized in Figure 6b.

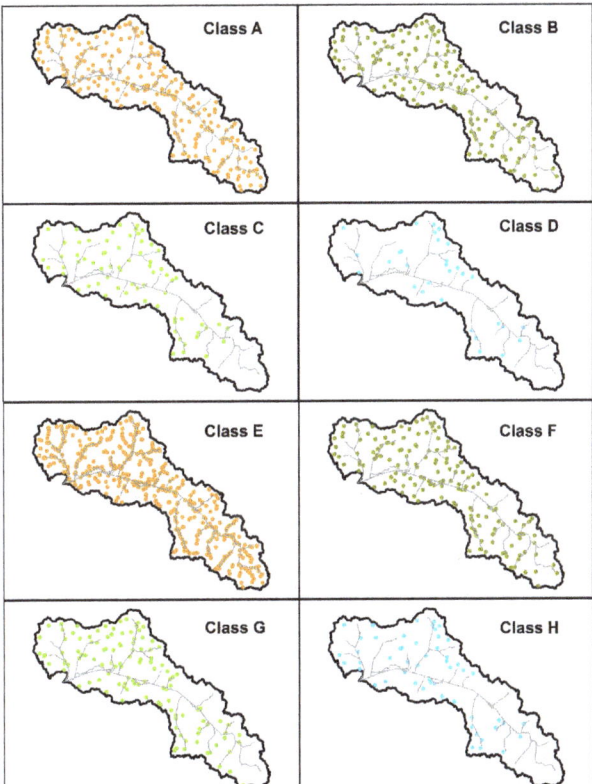

Figure 8. Proposed sediment basin locations corresponding to 8 scenarios (i.e., (**A–H**) described in Section 2.6.2).

Each proposed sediment basin location was analyzed using AgWET. The weir height and surface area of each sediment basin were determined based on the local topography and the average characteristics of the existing delineated sediment basins. These characteristics alongside the location of the basins were inputted into AnnAGNPS to simulate sediment basin impact on sediment reduction in the watershed.

2.6.3. Crop Rotation

Conservation crop rotation was applied as a seasonal sequence of crops grown in the same field yielding a multi-crop rotation cycle. The crops in the rotation should include a high residue producing crop such as wheat or corn along with a low residue producing crop such as soybeans or vegetables. Conservation crop rotation has many benefits which includes reducing sheet, rill, and wind erosion, increasing soil health and organic matter content, and improving soil moisture efficiency [72–75].

The only major crop in our study area that allows rotation is soybeans, so we simulated the soybeans/winter wheat rotations in four location scenarios: (1) all soybean agricultural fields amounting to 7340 AnnAGNPS cells (Figure 5A), (2) soybean agricultural fields that have a sediment yield higher than mean—standard deviation (referred to as ">medium") amounting to 5689 AnnAGNPS cells (Figure 5B), (3) soybean agricultural fields that have a sediment yield higher than mean + standard deviation (referred to as ">high") amounting to a total of 1166 AnnAGNPS cells (Figure 5C), and (4) soybean agricultural fields that have a sediment yield higher than mean + 2 × standard deviation (referred to as ">very high") amounting to a total of 424 AnnAGNPS cells (Figure 5D). The used mean (i.e., 5.25 Mg/ha/year) and standard deviation (i.e., 8.94 Mg/ha/year) values represent the statistical distribution of sediment yield baseline conditions of all agricultural fields in the study area as highlighted in Section 2.6.1.

The simulation of the crop rotation scenarios was conducted by adjusting the management schedule from Soybean to Soybean/Winter Wheat (Table 3) for each AnnAGNPS cell under the respective four crop rotation scenarios. These four scenarios are summarized in Figure 6c.

Table 3. Comparison of annual management schedule for soybean and soybean + winter wheat rotation.

Crop	Date	Operation
Soybean (with no rotation)	20 Mar.	Chisel
	5 May	Fertilizer
	10 May	Disk
	11 May	Plant
	29 May	Sprayer
	20 July	Sprayer
	28 Aug.	Sprayer
	15 Oct.	Harvest
	20 Oct.	Weed Growth
Winter Wheat + Soybean Rotation	30 Oct.	Sprayer
	31 Oct.	Fertilizer
	1 Nov.	Rill or Air Seeder
	15 Nov.	Sprayer
	15 Feb.	Fertilizer
	15 Mar.	Sprayer
	10 May	Disk
	15 May	Sprayer
	10 June	Harvest
	11 June	Sprayer
	13 June	Plant
	27 June	Sprayer
	20 July	Sprayer
	28 Aug.	Sprayer
	15 Oct.	Harvest
	20 Oct.	Weed Growth

2.6.4. Conservation Reserve Program

The Conservation Reserve Program (CRP) is a voluntary program implemented by the USDA Farm Service Agency (FSA) in coordination with agricultural landowners, to remove environmentally fragile land from agricultural production and improve water quality, wildlife habitat, and prevent soil erosion. During the period of the program, the land is not farmed or ranched but planted with native plant species that improve the long-term environmental health and quality of the land. The contract between farmers and FSA lasts between ten to fifteen years with annual rental payments and cost share assistance provided by FSA [76–78].

The simulation period for this study is 11 years (January 2008–December 2018), which fits the typical duration range of CRP. The selected scenarios for this conservation practice are: (1) all agricultural fields (Figure 5A), (2) all agricultural fields that have a sediment yield higher than mean minus standard deviation (referred to as ">medium") (Figure 5B), (3) all agricultural fields that have a sediment yield higher than mean + standard deviation (referred to as ">high") (Figure 5C), and (4) all agricultural fields that have a sediment yield higher than mean + 2 × standard deviation (referred to as ">very High") (Figure 5D). The used mean (i.e., 5.25 Mg/ha/year) and standard deviation (i.e., 8.94 Mg/ha/year) values represent the statistical distribution of sediment yield baseline conditions of all agricultural fields in the study area as highlighted in Section 2.6.1.

We implemented CRP in our simulations by changing the crop type and schedule to grass for the selected cells under each considered scenario. The grass rotation simulates the conditions of land being returned to native plant species with no farming activities. The four CRP scenarios are summarized in Figure 6d.

2.6.5. Optimization and Comparison of Conservation Practice Scenarios

A multi-dimension scoring function was employed to compare all the considered conservation scenarios. This cost function was designed to account for the relative reduction in annual average sediment yield per unit area while, at the same time, consider the potential reduction in productive agricultural land. This analysis was performed for all individual sub-catchments affected by the conservation practice. Both sediment yield per unit area and the area of land converted from agricultural production to conservation were normalized to values between 0 and 1 and the totals from all sub-catchments were calculated. The score for each alternative scenario i was calculated as follows:

$$Score_i = \frac{\left(T_{S_i} \times W_S\right) + \left(T_{L_i} \times W_L\right)}{\left(W_s \times W_L\right)} \tag{1}$$

where T_{S_i} is the total scaled sediment yield per unit area for scenario i, T_{L_i} is the total area converted into a conservation practice, and W_S and W_L are the weighting factors for sediment load and spatial footprint, respectively.

3. Results

3.1. Baseline Conditions

Evaluation of simulation results were performed spatially (Figure 9) and temporally (Figure 10) using two output parameters: sediment yield and sediment load, respectively. Sediment yield is reported as annual average per unit area for each sub-catchment in Mega grams per hectare per year and it represents eroded sediment by inter-rill and rill processes leaving the field into streams. The annual average sediment yield per unit area is intended to describe non-point sources spatially and to serve as a reference to quantify the effect of conservation practices locally (field scale at different parts of the watershed). Sediment load is reported as the annual average at the watershed outlet in Mega grams and is intended to demonstrate how the overall watershed responds to natural drivers (e.g., climate/weather), anthropogenic drivers (e.g., farming management), and to quantify the combined effect of conservation practices on the system. Both parameters are described by key particle sizes: sand, silt, and clay.

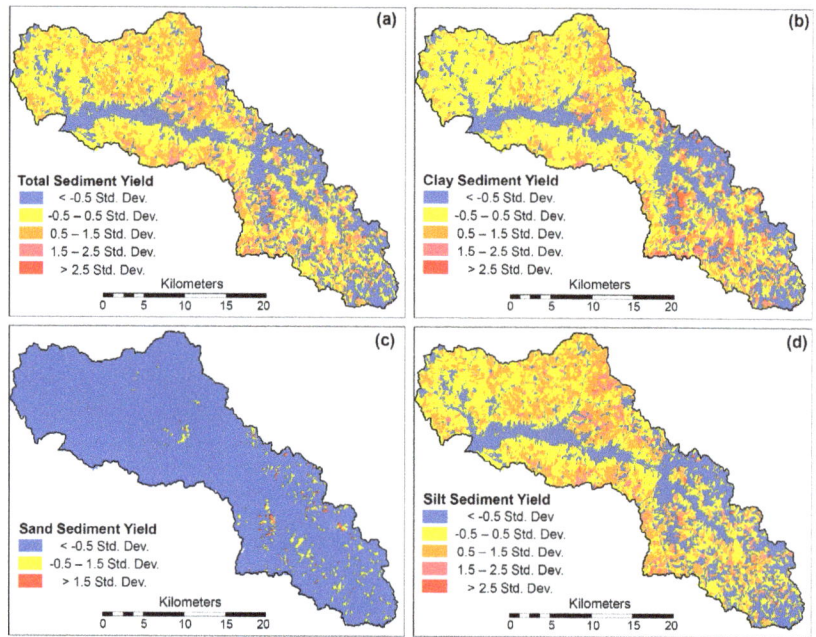

Figure 9. Spatial distribution of sediment yield for different particles sizes: (**a**) clay, (**b**) sand, (**c**) silt, and (**d**) total sediment, estimated using the AnnAGNPS model to describe existing conditions (baseline condition simulation). Graph depicts standardized approach in which values are expressed as deviations from the mean value of all 12,573 sub-catchments.

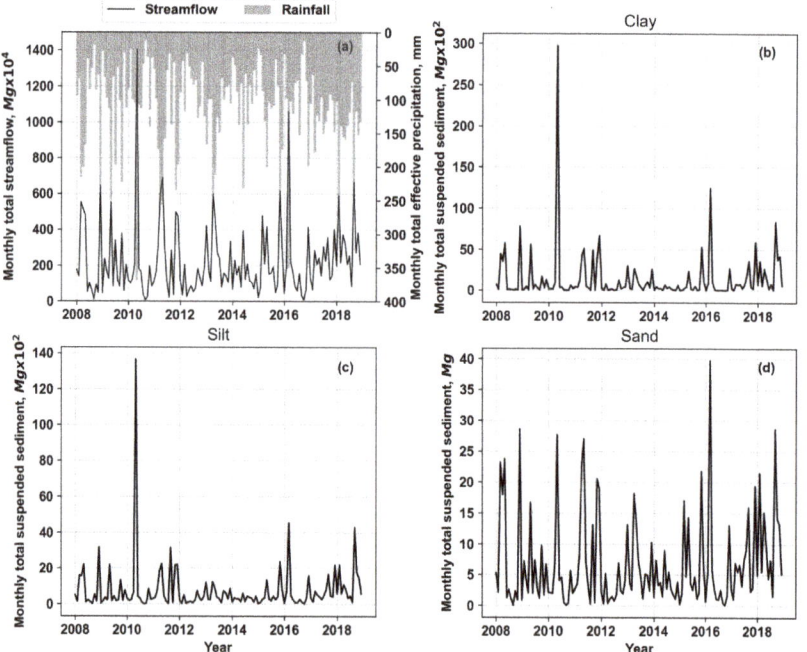

Figure 10. Streamflow (**a**) and suspended sediment load for silt (**b**), clay (**c**), and sand (**d**) sizes at the watershed outlet as simulated by the AnnAGNPS model representing existing conditions.

The lack of streamflow observations in most catchments around the world and the nation, especially in small catchments in rural areas has created a setback for hydrological and conservation investigations in these regions [67]. In this study, the uncertainties caused by the lack of observed data (streamflow and sediment gauges) were minimized by the following. First, the AnnAGNPS watershed pollution model was selected due to its suitable application to agricultural ungauged watersheds based on the detailed spatiotemporal information needed to characterize natural and anthropogenic processes in this type of watersheds [79–85]. Second, relative measurements were used instead of absolute to evaluate conservation scenarios. This was achieved by comparing average sediment load at the outlet of the watershed for each scenario (33 different AnnAGNPS simulations describing conservation practices) to baseline conditions. Absolute sediment load/yield was not used as the basis for any recommendation. Third, comparisons of annual average streamflow with the publicly available web-based Generalized Watershed Loading Function Enhanced (GWLF-E) model [86] indicate that the two estimates are in agreement (i.e., 42.22 cm/year and 42.27 cm/year for AnnAGNPS and GWLF-E, respectively).

Sediment yield was reported as annual average per unit area for each sub-catchment. Results were classified for each particle size per its statistical distribution to facilitate visualization and allow a spatial identification of hot spot sub-catchments in terms of sediment yield (Figure 9). The generated maps were used as the basis for the development of alternative conservation scenarios to target problematic sub-catchments and reduce their sediment yield.

Results from the baseline simulation indicates an overall low sand particle yield relatively to both silt and clay. Sediment yield of both silt and clay are high especially in Mud Creek, North Fork Forked Deer River Lower and Doakville Creek in comparison to Cain Creek and North Fork Forked Deer River Upper (Figure 9). Simulation results demonstrate the combined effect of complex processes driving sediment sources and sinks as well as the spatiotemporal variation of land cover, climate, soil, and farming practices, with agricultural fields driving most of the sediment detachment and transport in the study area.

Temporal evaluation of monthly streamflow at the outlet (Figure 10a) highlights a major rainfall event in 1–2 May 2010, responsible for significant floods in the region. Streamflow peaks correlate with peaks of suspended silt (Figure 10b) and clay (Figure 10c). Sand load peaks does not correlate with streamflow peaks, with the former peaking in 2016 (Figure 10d). It is important to note that estimates of sediment load of sand size particles are two orders of magnitude smaller than silt and clay.

3.2. Conservation Practices Scenarios

In addition to the baseline condition simulation, 33 AnnAGNPS simulations (i.e., 16 for riparian buffer, 9 for sediment basin, 4 for crop rotation and 4 for CRP), with a computer run time of up to 50 h per simulation, were performed to explore the impact of these conservation practices on sediment reduction in a wide range of conditions.

Tables 4–7 summarize the simulation results for the four sets of the investigated conservation practices. Each table includes scenario characteristics, relative reduction of sediment load at the outlet of the watershed from the baseline conditions, as well as the area of land used to implement the practice. The implementation decision is a tradeoff, between maximizing soil erosion reduction and minimizing the economic burden on farmers represented by the area of land that is taken out of production.

Table 4. Riparian buffer scenarios results.

Simulation ID	Location Description	Riparian Buffer Width	Sediment Load (Mg/Year)	Sediment Reduction (%)	Spatial Footprint (Ha)
1	All Streams	10 m	71,312.7	65%	506.6
2	All Streams	30 m	44,970.2	78%	1458.3
3	All Streams	60 m	39,745.7	81%	2728.2
4	All streams adjacent to agricultural fields.	10 m	113,389.3	45%	338.0
5	All streams adjacent to agricultural fields.	30 m	99,602.6	52%	1018.9
6	All streams adjacent to agricultural fields.	60 m	96,321.7	53%	2043.8
7	All streams adjacent to agricultural fields with a sediment yield higher than 3.7 Mg/ha/year	10 m	118,246.5	43%	228.6
8	All streams adjacent to agricultural fields with a sediment yield higher than 3.7 Mg/ha/year	30 m	105,205.8	49%	893.9
9	All streams adjacent to agricultural fields with a sediment yield higher than 3.7 Mg/ha/year	60 m	101,905.8	51%	1823.0
10	All streams adjacent to agricultural fields with a sediment yield higher than 14.2 Mg/ha/year	10 m	178,229.7	13%	60.9
11	All streams adjacent to agricultural fields with a sediment yield higher than 14.2 Mg/ha/year	30 m	175,074.4	15%	175.0
12	All streams adjacent to agricultural fields with a sediment yield higher than 14.2 Mg/ha/year	60 m	174,124.0	15%	357.9
13	All streams adjacent to agricultural fields with a sediment yield higher than 19.4 Mg/ha/year	10 m	195,271.1	5%	19.1
14	All streams adjacent to agricultural fields with a sediment yield higher than 19.4 Mg/ha/year	30 m	194,101.3	6%	52.7
15	All streams adjacent to agricultural fields with a sediment yield higher than 19.4 Mg/ha/year	60 m	193,756.4	6%	106.5
16	Existing riparian buffer	Variable	178,956.2	13%	4683.8
Baseline Conditions	No buffer is integrated into the model	0 m	205,880.2	0%	0

Table 5. Sediment basin scenarios results.

Simulation ID	Scenario Classification (Figure 6b)	Sediment Load (Mg/Year)	Sediment Reduction (%)	Spatial Footprint (Ha)
17	Class A	10,699.1	95%	214.3
18	Class B	70,137.00	66%	200.7
19	Class C	184,607.9	10%	91.2
20	Class D	199,254.0	3%	39.9
21	Class E	164,705.5	20%	252.1
22	Class F	179,666.4	13%	276.4
23	Class G	179,073.7	13%	172.2
24	Class H	198,661.2	4%	79.6
25	Existing Sediment Basins	204,992.0	0%	82.8
	Baseline conditions	205,880.2	0%	0

Table 6. Crop rotation scenarios results.

Simulation ID	Description	Sediment Load (Mg/year)	Sediment Reduction (%)	Spatial Footprint (Ha)
26	Crop rotation is applied to all soybean fields in the watershed	182,278.4	11%	0 *
27	Crop rotation is applied to all soybean fields in the watershed with a sediment yield higher than 3.7 Mg/ha/year	193,476.4	11%	0 *
28	Crop rotation is applied to all soybean fields in the watershed with a sediment yield higher than 14.2 Mg/ha/year	182,969.3	6%	0 *
29	Crop rotation is applied to all soybean fields in the watershed with a sediment yield higher than 19.4 Mg/ha/year	197,905.3	4%	0 *
Baseline Conditions	No additional crop rotation is integrated into the model	205,880.2	0%	0 *

* No land is taken out of production during crop rotation, since winter wheat is planted in the winter when the land is not being used to grow other crops.

Table 7. CRP scenarios results.

Simulation ID	Description	Sediment Load (Mg/Year)	Sediment Reduction (%)	Spatial Footprint (Ha)
30	CRP is applied to all agricultural fields in the watershed	38,598.1	81%	7146.9
31	CRP is applied to all agricultural fields in the watershed with a sediment yield higher than 3.7 Mg/ha/year	44,460.2	78%	6792.5
32	CRP is applied to all agricultural fields in the watershed with a sediment yield higher than 14.2 Mg/ha/year	154,930.8	25%	1519.3
33	CRP is applied to all agricultural fields in the watershed with a sediment yield higher than 19.4 Mg/ha/year	186,290.7	10%	388.5
Baseline Conditions	No CRP is simulated in the model	205,880.2	0%	0

Riparian buffers work by slowing surface flow and promoting infiltration and fine sediment deposition. Our simulations results indicate that sediment reduction of this practice when compared to the baseline condition range between 6% and 81% based on location and buffer width (Table 4). Even though implementation of riparian buffers around every stream in the watershed is not feasible, the value of 81% represents a theoretical maximum potential reduction by this practice. The simulations also indicate that existing riparian buffers provide a reduction of 13% as opposed to a maximum potential of 81%, demonstrating that the watershed is under-served in terms of riparian buffers.

In addition, simulation results indicate that an increase of riparian buffer width from 30 m to 60 m leads to a sediment load reduction at the outlet of the watershed of only 1%, 2%, 0%, and 0% in scenarios 6, 9, 12 and 15, respectively (Table 4). These results highlight a non linear relationship between the buffer width and sediment trapping efficiency, indicating that the expansion of the overall length of the riparian buffer is more effective than increasing the width beyond a certain optimal value [46,87,88].

Sediment basin simulation results indicate a wide range in sediment load reduction from 4% to 95% (Table 5). The overall reduction is a function of not only the number of sediment basins but also their placement throughout the watershed. It can be observed that a similar reduction in sediment load for scenarios 22 and 23 represent a different spatial footprint (i.e., 276.4 and 172.2 hectares, respectively). Additionally, alternative scenarios 17 and 18 that focus on stream order 2 seem significantly more efficient than scenarios 21 and 22. Based on these findings, it is suggested to prioritize the size of the upstream area when deciding where to place sediment basins in the candidate locations in the watershed.

Simulation results indicate that crop rotation between soybean and winter wheat would reduce sediment load at the outlet of the watershed by 4% in case of selecting the top (i.e., mean plus two standard deviations) fields in terms of sediment production and by about 11% in case of implementation in all soybean fields (Table 6), which represent about 27% of the total area of the watershed. This represents a relatively small sediment load reduction, but is a cost-effective alternative since the production areas are not reduced. The only major crop rotation identified in the study area was soybean with winter wheat. A more widespread application of crop rotation could further decrease the sediment yield in the watershed.

Sediment reduction using CRP ranges between 10% and 81% for the study area based on how many fields were removed from production (Table 7). While CRP is one of the most effective tools to address nutrient loss, sediment erosion and wildlife habitat reduction, CRP is one of the costliest practices for the landowners by taking their land completely out of production; hence financial incentives are often provided by the U.S. federal government for this purpose [76,77,89,90].

4. Discussion

An effective implementation of conservation practices designed to promote water quality and minimize soil erosion depends on a comprehensive understanding of watershed processes, farming practices and sediment and agrochemical sources and sinks [48]. This study is an integrated approach using both GIS spatial analysis and watershed modeling to provide stakeholders with a blueprint for targeted conservation plans especially in ungauged watersheds. This investigation allowed us to make four major observations.

4.1. Prioritization of Conservation Practice Location

The definition for the optimal spatial distribution of the proposed conservation practices and their temporal impact on sediment and nutrient detachment, transport, deposition, and trapping is important [28,48]. The description of these spatiotemporal relationships requires the incorporation of specialized technology into watershed modeling, that allows a careful selection of the location and type of practice to optimize implementation [47,48,91,92].

An example corroborating the importance of conservation location is from riparian buffer simulations. Despite an overall positive correlation between the size of the buffer and the amount of sediment reduction, Figure 11 shows that scenarios 4 and 12 use an almost equal buffer surface area (i.e., 338 and 357.9 ha, respectively) but lead to a large difference in sediment load reduction (i.e., 45% and 15%, respectively). Similarly, scenarios 7 and 11 are based on a similar surface area (i.e., 228.6 and 175 ha, respectively) but also lead to a large difference in sediment yield reduction (i.e., 43% and 15%, respectively).

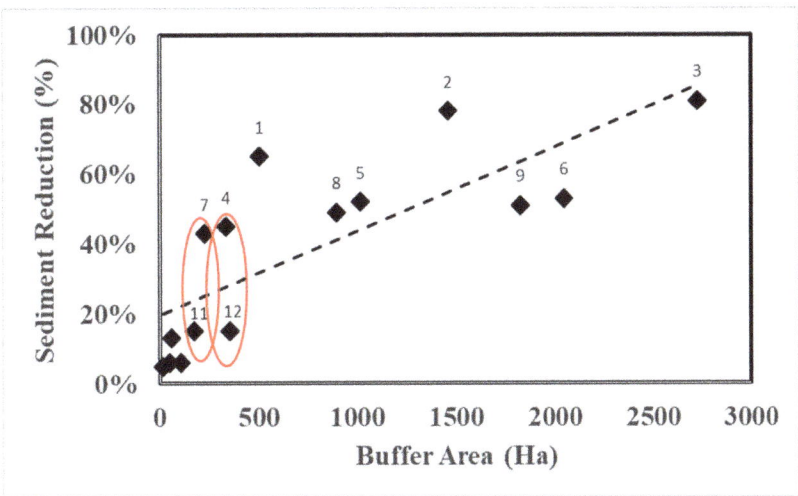

Figure 11. Relative sediment reduction (%) from baseline conditions for riparian buffers in function of their surface area. The numbers refer to scenario ID from Table 4. The two red circles highlight two examples in which a similar riparian buffer area size leads to a large difference in sediment reduction.

A second example is from sediment basin simulations. It was found that proposed scenarios 17 and 18 (i.e., classes A and B of sediment basin) that focus on stream order 2 are significantly more efficient than scenarios 21 and 22 (i.e., classes E and F for sediment basin). In fact, scenario 17 was simulated using optimally located 147 basins, leading to a reduction of 66% in sediment load whereas scenario 22, including more sediment basins (i.e., 297) leads to a reduction of only 13%. Hence, it is suggested to prioritize the size of the upstream area when placing sediment basins in the watershed, while respecting NRCS guidelines.

The selection of where to place the conservation practice within the watershed is important to optimize the available resources and maximize the conservation practices impact on non-point source pollutants.

4.2. Optimization and Comparison of Conservation Practice Scenarios

Three optimization evaluations were performed by varying the weights of both considered parameters: (a) sediment reduction and spatial footprint have equal weight (Figure 12a), (b) a weight of 5 to 1 prioritizing reduction in sediment yield (Figure 12b) and (c) a weight of 5 to 1 prioritizing reduction in spatial footprint (Figure 12c).

When comparing all alternative scenarios using the cost function, riparian buffer seems to be the most efficient. Varying weights can generate different outcomes and this approach can be used by stakeholders to tailor outputs to fit their priorities. Scenarios 1, 2, 3, and 7, representing variations in riparian buffers, were identified as the most efficient conservation practice under all weighting conditions. Conversely, scenarios 21 and 17, representing sediment basins, were scored as least efficient, especially when weights were adjusted to prioritize sediment reduction. Scenarios 30 and 31, which both represent CRP practices, were scored as least efficient when a larger weight was placed on converting land from production to support conservation efforts.

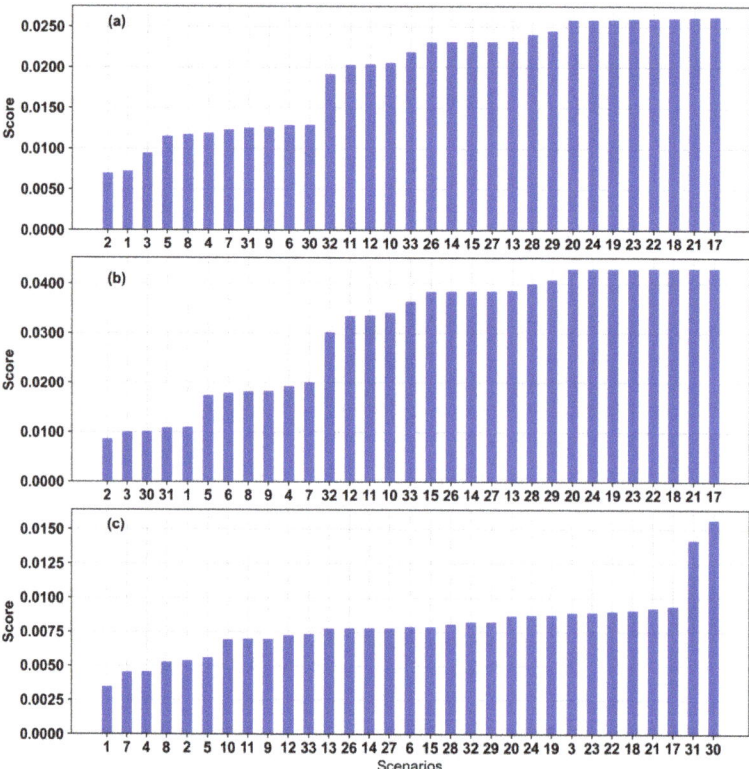

Figure 12. Scoring results comparing alternative conservation scenarios based on sediment yield per unit area and total land converted from agricultural production to a conservation practice. Three evaluations were performed: (**a**) equal weight for sediment yield and spatial footprint, (**b**) prioritize reduction in sediment yield by a factor of 5 to 1, and (**c**) prioritize reduction in spatial footprint by a factor of 5 to 1.

4.3. Building Decision Support Tools Based on Study Results: A Riparian Buffer Example

Simulation results indicate riparian buffers to be the most effective conservation practice for sediment reduction in this watershed, hence they were selected to demonstrate the utility of building decision support tools for stakeholders when implementing practices on the ground designed for optimal reduction of non-point source pollution. Representing simulation results as ranked ratio of accumulated sediment yield per unit of area and accumulated contributing area (Figure 13) illustrates a simple, but effective way, to evaluate individual conservation practices and/or contrast multiple conservation alternatives. For example, based on the AnnAGNPS simulation representing baseline conditions, 40% of the watershed total area has shown to produce 75% of sediment leaving fields into streams. A conservation strategy designed to reduce sediment yield by 25%, could be implemented by targeting specifically those sub-catchments. Alternatively, it is possible to evaluate a wide range of alternative scenarios based on their potential overall reduction. The alternative conservation scenario considering implementing 60 m constructed riparian buffers at sub-catchments classified as "very high" sediment producing locations reduces the overall sediment yield by 7% when compared with baseline conditions. Instead, an alternative scenario considering 10 m constructed riparian buffers implemented in sub-catchments identified as "high" lead to an overall reduction of 27% when compared with baseline conditions. A cluster of alternative scenarios implementing riparian buffers at all agricultural fields and at "medium" classified sub-catchments produce similar reductions

(38–48%) when compared to the baseline scenario. This indicates potential flexibility in selecting where in the watershed and what buffer width to install to obtain similar overall reductions. This decision support tool can assist in the design of conservation practices unique to each watershed.

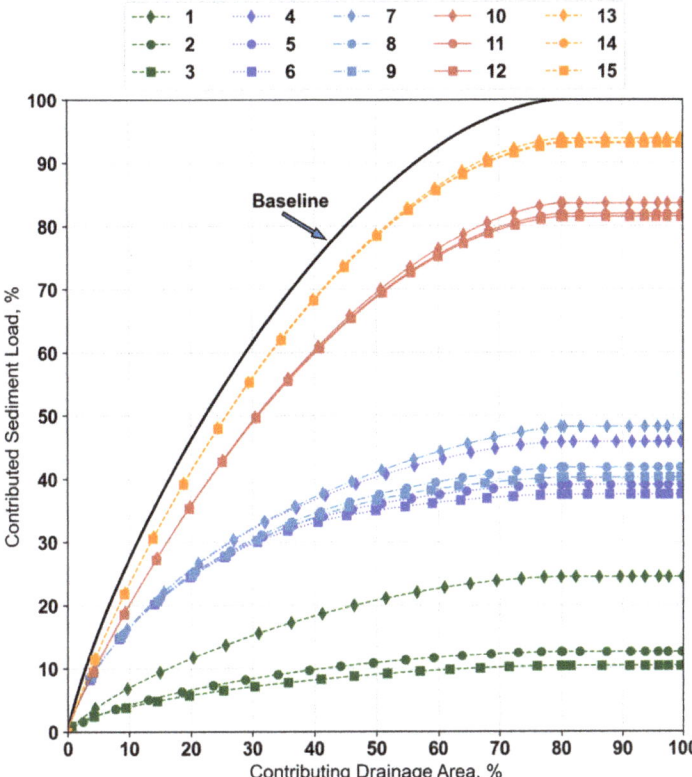

Figure 13. Ranked ratio analysis between accumulated sub-catchment annual average sediment yield per unit area and their corresponding accumulated drainage area.

4.4. Methodology Uncertainties

It is important to recognize that, conservation efforts are often implemented at the watershed scale as a combination of practices varying in type, location, and stakeholder in charge of implementation. In this study, each type of conservation practice was considered separately; this approach was chosen to quantitatively evaluate the effectiveness of individual practices and to provide a platform to compare them. Furthermore, there is an infinite number of possibilities when considering all potential combinations of types of conservation practice, their controlling parameters (width, area, vegetation type, etc.), and, more importantly, their location at the watershed. Additionally, in this investigation, the main quantified sources of sediment are from inter-rill and rill processes, and therefore, channel processes (streambank and streambed erosion) were outside the scope of the study and were kept constant between simulations.

5. Conclusions

Development of conservation plans designed to improve water quality in agricultural watersheds, and in downstream waterways, depends on a detailed spatiotemporal understanding of natural and anthropogenic controlling variables and processes. This constitutes a challenge for conservation stakeholders while managing limited resources, minimizing

Soil Syst. **2023**, *7*, 4

agricultural production loss, and maximizing sediment load reduction. Watershed modeling technology can support the spatial optimization of conservation practices even at ungauged watersheds.

The integration of GIS-based analyses with hydrological modeling at watershed scales provides additional capabilities to quantify the effect of conservation practices to sediment loads by spatially characterizing different types of conservation practices and scenarios and their relative impact on sediment reduction.

The proposed methodology was applied to the North Fork Forked Deer River watershed in west Tennessee as part of the Mississippi River Basin Initiative (MRBI). This watershed was identified as impaired due to high loads of suspended sediment from agricultural sources [93]. Despite the non-availability of continuous runoff and sediment monitoring stations, a detailed characterization of anthropogenic and natural drivers was performed to obtain a relative evaluation of sediment reduction between baseline conditions and potential conservation scenarios. This could serve as a pilot study toward the improvement of non-point source pollution from agricultural activities in ungauged watersheds across the nation and in the Mississippi basin specifically, given that it is responsible for one of the largest aquatic dead zones in the world.

Future directions of this investigation could involve the inclusion of more variables in the scoring function, such as costs of implementation, costs of maintenance, and potential loss of income from a reduced production area. Additionally, the integration of this methodology with machine learning algorithms could aid in the task of selecting and simulating a combination of different types of practices, controlling parameters, and their location in the watershed. This technology could lead to the development of a hybrid customized solutions for impaired watersheds.

Author Contributions: Conceptualization, R.E. and H.G.M.; methodology, R.E. and H.G.M.; formal analysis, R.E., H.G.M. and R.L.B.; data curation, R.E., H.G.M. and K.M.; writing—original draft preparation, R.E. and H.G.M.; writing—review and editing, R.E., H.G.M., R.L.B. and K.M.; visualization, R.E., H.G.M. and K.M.; supervision, R.E. and H.G.M.; project administration, R.E. and H.G.M.; funding acquisition, R.E. and H.G.M. All authors have read and agreed to the published version of the manuscript.

Funding: This research was funded by the National Resources Conservation Service, grant number NR194741XXXXC005 and by the National Institute of Food and Agriculture, award number #: 2020-70001-31278.

Institutional Review Board Statement: Not applicable.

Informed Consent Statement: Not applicable.

Data Availability Statement: Not applicable.

Acknowledgments: Special thanks for John Simpson for assisting with data analysis and processing, and Andrew Osborne for assisting with sediment basin GIS analysis.

Conflicts of Interest: The authors declare no conflict of interest.

References

1. Pimentel, D.; Harvey, C.; Resosudarmo, P.; Sinclair, K.; Kurz, D.; McNair, M.; Crist, S.; Shpritz, L.; Fitton, L.; Saffouri, R.; et al. Environmental and Economic Costs of Soil Erosion and Conservation Benefits. *Science* **1995**, *267*, 1117–1123. [CrossRef] [PubMed]
2. Kagabo, D.M.; Stroosnijder, L.; Visser, S.M.; Moore, D. Soil Erosion, Soil Fertility and Crop Yield on Slow-Forming Terraces in the Highlands of Buberuka, Rwanda. *Soil Tillage Res.* **2013**, *128*, 23–29. [CrossRef]
3. Zhuang, Y.; Du, C.; Zhang, L.; Du, Y.; Li, S. Research Trends and Hotspots in Soil Erosion from 1932 to 2013: A Literature Review. *Scientometrics* **2015**, *105*, 743–758. [CrossRef]
4. Berihun, M.L.; Tsunekawa, A.; Haregeweyn, N.; Dile, Y.T.; Tsubo, M.; Fenta, A.A.; Meshesha, D.T.; Ebabu, K.; Sultan, D.; Srinivasan, R. Evaluating Runoff and Sediment Responses to Soil and Water Conservation Practices by Employing Alternative Modeling Approaches. *Sci. Total Environ.* **2020**, *747*, 141118. [CrossRef]
5. Borrelli, P.; Robinson, D.A.; Panagos, P.; Lugato, E.; Yang, J.E.; Alewell, C.; Wuepper, D.; Montanarella, L.; Ballabio, C. Land Use and Climate Change Impacts on Global Soil Erosion by Water (2015–2070). *Proc. Natl. Acad. Sci. USA* **2020**, *117*, 21994–22001. [CrossRef] [PubMed]

6.   Eekhout, J.P.C.; de Vente, J. Global Impact of Climate Change on Soil Erosion and Potential for Adaptation through Soil Conservation. *Earth Sci. Rev.* **2022**, *226*, 103921. [CrossRef]
7.   Sulaeman, D.; Westhoff, T. The Causes and Effects of Soil Erosion, and How to Prevent It. Available online: https://www.wri. org/insights/causes-and-effects-soil-erosion-and-how-prevent-it (accessed on 15 September 2021).
8.   EPA. *National Nonpoint Source Program: A Catalyst for Water Quality Improvements*; EPA: Cincinnati, OH, USA, 2016.
9.   Jabbar, F.K.; Grote, K.; Tucker, R.E. A novel approach for assessing watershed susceptibility using weighted overlay and analytical hierarchy process (AHP) methodology: A case study in Eagle Creek Watershed, USA. *Environ. Sci. Pollut. Res.* **2019**, *26*, 31981–31997. [CrossRef]
10. EPA. Basic Information about Nonpoint Source (NPS) Pollution. Available online: https://www.epa.gov/nps/basic-information-about-nonpoint-source-nps-pollution#:~{}:TEXT=NONPOINT%20SOURCE%20POLLUTION%20CAN%20 INCLUDE,FOREST%20LANDS%2C%20AND%20ERODING%20STREAMBANKS (accessed on 8 August 2020).
11. Arabi, M.; Frankenberger, J.R.; Engel, B.A.; Arnold, J.G. Representation of Agricultural Conservation Practices with SWAT. *Hydrol. Process.* **2008**, *22*, 3042–3055. [CrossRef]
12. Kassam, A.; Friedrich, T.; Derpsch, R. Global Spread of Conservation Agriculture. *Int. J. Environ. Stud.* **2019**, *76*, 29–51. [CrossRef]
13. Hermans, T.D.G.; Dougill, A.J.; Whitfield, S.; Peacock, C.L.; Eze, S.; Thierfelder, C. Combining Local Knowledge and Soil Science for Integrated Soil Health Assessments in Conservation Agriculture Systems. *J. Environ. Manag.* **2021**, *286*, 112192. [CrossRef]
14. Farooq, M.; Siddique, K.H.M. *Conservation Agriculture: Concepts, Brief History, and Impacts on Agricultural Systems*; Springer International Publishing: Cham, Switzerland, 2015; pp. 3–17.
15. USDA. Conservation Programs. Available online: https://www.ers.usda.gov/topics/natural-resources-environment/ conservation-programs/#:~{}:text=Common%20practices%20include%20nutrient%20management,to%20exclude%20livestock% 20from%20streams (accessed on 12 October 2021).
16. Pittelkow, C.M.; Liang, X.; Linquist, B.A.; van Groenigen, K.J.; Lee, J.; Lundy, M.E.; van Gestel, N.; Six, J.; Venterea, R.T.; van Kessel, C. Productivity Limits and Potentials of The Principles of Conservation Agriculture. *Nature* **2015**, *517*, 365–368. [CrossRef]
17. Brannan, K.M.; Mostaghimi, S.; McClellan, P.W.; Inamdar, S. Animalwaste Bmp Impacts on Sediment and Nutrient Losses in Runoff from the Owl Run Watershed. *Trans. ASABE* **2000**, *43*, 1155–1166. [CrossRef]
18. Inamdar, S.P.; Mostaghimi, S.; McClellan, P.W.; Brannan, K.M. Bmp Impacts on Sediment and Nutrient Yields from an Agricultural Watershed in the Coastal Plain Region. *Trans. ASABE* **2001**, *44*, 1191–1200. [CrossRef]
19. Richards, R.P.; Baker, D.B. Trends in Water Quality in LEASEQ Rivers and Streams (Northwestern Ohio), 1975–1995. *J. Environ. Qual.* **2002**, *31*, 90–96. [CrossRef]
20. Gagnon, S.R.; Makuch, J.; Sherman, T.J. *Implementing Agricultural Conservation Practices: Barriers and Incentives: A Conservation Effects Assessment Bibliography*; Water Quality Information Center, National Agricultural Library: Beltsville, MD, USA, 2004.
21. Gagnon, S.R.; Makuch, J.; Harper, C.Y. *Effects of Agricultural Conservation Practices on Fish and Wildlife [Volume 2]: A Conservation Effects Assessment Bibliography*; Water Quality Information Center, National Agricultural Library: Beltsville, MD, USA, 2008; Volume 7b.
22. Osmond, D.; Meals, D.; Hoag, D.; Arabi, M.; Luloff, A.; Jennings, G.; McFarland, M.; Spooner, J.; Sharpley, A.; Line, D. Improving Conservation Practices Programming to Protect Water Quality in Agricultural Watersheds: Lessons Learned from the National Institute of Food and Agriculture–Conservation Effects Assessment Project. *J. Soil Water Conserv.* **2012**, *67*, 122A–127A. [CrossRef]
23. Her, Y.; Chaubey, I.; Frankenberger, J.; Jeong, J. Implications of Spatial and Temporal Variations in Effects of Conservation Practices on Water Management Strategies. *Agric. Water Manag.* **2017**, *180*, 252–266. [CrossRef]
24. Renschler, C.S.; Harbor, J. Soil Erosion Assessment Tools from Point to Regional Scales—The Role of Geomorphologists in Land Management Research and Implementation. *Geomorphology* **2002**, *47*, 189–209. [CrossRef]
25. Gitau, M.W.; Veith, T.L.; Gburek, W.J.; Jarrett, A.R. Watershed Level Best Management Practice Selection and Placement in the Town Brook Watershed, New York. *J. Am. Water Resour. Assoc.* **2006**, *42*, 1565–1581. [CrossRef]
26. Giri, S.; Nejadhashemi, A.P.; Woznicki, S.A. Evaluation of Targeting Methods for Implementation of Best Management Practices in the Saginaw River Watershed. *J. Environ. Manag.* **2012**, *103*, 24–40. [CrossRef]
27. Xie, H.; Shen, Z.; Chen, L.; Qiu, J.; Dong, J. Time-Varying Sensitivity Analysis of Hydrologic and Sediment Parameters at Multiple Timescales: Implications for Conservation Practices. *Sci. Total Environ.* **2017**, *598*, 353–364. [CrossRef]
28. Tomer, M.D.; Porter, S.A.; James, D.E.; Boomer, K.M.B.; Kostel, J.A.; McLellan, E. Combining Precision Conservation Technologies into a Flexible Framework to Facilitate Agricultural Watershed Planning. *J. Soil Water Conserv.* **2013**, *68*, 113A–120A. [CrossRef]
29. Tomer, M.D.; Porter, S.A.; Boomer, K.M.B.; James, D.E.; Kostel, J.A.; Helmers, M.J.; Isenhart, T.M.; McLellan, E. Agricultural Conservation Planning Framework: 1. Developing Multipractice Watershed Planning Scenarios and Assessing Nutrient Reduction Potential. *J. Environ. Qual.* **2015**, *44*, 754–767. [CrossRef]
30. Arnold, J.G.; Srinivasan, R.; Muttiah, R.S.; Williams, J.R. Large Area Hydrologic Modeling and Assessment Part I: Model Development. *J. Am. Water Resour. Assoc.* **1998**, *34*, 73–89. [CrossRef]
31. Yuan, Y.; Koropeckyj-Cox, L. SWAT Model Application for Evaluating Agricultural Conservation Practice Effectiveness in Reducing Phosphorous Loss from the Western Lake Erie Basin. *J. Environ. Manag.* **2022**, *302*, 114000. [CrossRef] [PubMed]
32. Naseri, F.; Azari, M.; Dastorani, M.T. Spatial Optimization of Soil and Water Conservation Practices Using Coupled SWAT Model and Evolutionary Algorithm. *Int. Soil Water Conserv. Res.* **2021**, *9*, 566–577. [CrossRef]

33. Davie, D.K.; Lant, C.L. The effect of CRP enrollment on sediment loads in two southern Illinois streams. *J. Soil Water Conserv.* **1994**, *49*, 407–412.
34. Carroll, C.; Halpin, M.; Burger, P.; Bell, K.; Sallaway, M.M.; Yule, D.F. The effect of crop type, crop rotation, and tillage practice on runoff and soil loss on a Vertisol in central Queensland. *Aust. J. Soil Res.* **1997**, *35*, 925–940. [CrossRef]
35. Edwards, C.L.; Shannon, R.D.; Jarrett, A.R. Sedimentation Basin Retention Efficiencies for Sediment, Nitrogen, and Phosphorus from Simulated Agricultural Runoff. *Trans. ASAE* **1999**, *42*, 403–409. [CrossRef]
36. Parkyn, S.M.; Davies-Colley, R.J.; Cooper, A.B.; Stroud, M.J. Predictions of stream nutrient and sediment yield changes following restoration of forested riparian buffers. *Ecol. Eng.* **2005**, *24*, 551–558. [CrossRef]
37. Bailey, A.; Deasy, C.; Quinton, J.; Silgram, M.; Jackson, B.; Stevens, C. Determining the cost of in-field mitigation options to reduce sediment and phosphorus loss. *Land Use Policy* **2013**, *30*, 234–242. [CrossRef]
38. Johnson, K.A.; Dalzell, B.J.; Donahue, M.; Gourevitch, J.; Johnson, D.L.; Karlovits, G.S.; Keeler, B.; Smith, J.T. Conservation Reserve Program (CRP) lands provide ecosystem service benefits that exceed land rental payment costs. *Ecosyst. Serv.* **2016**, *18*, 175–185. [CrossRef]
39. Vigiak, O.; Malagó, A.; Bouraoui, F.; Grizzetti, B.; Weissteiner, C.J.; Pastori, M. Impact of current riparian land on sediment retention in the Danube River Basin. *Sustain. Water Qual. Ecol.* **2016**, *8*, 30–49. [CrossRef]
40. Yuan, Y.; Locke, M.A.; Bingner, R.L. Annualized Agricultural Non-Point Source Model Application for Mississippi Delta Beasley Lake Watershed Conservation Practices Assessment. *J. Soil Water Conserv.* **2008**, *63*, 542–551. [CrossRef]
41. Momm, H.G.; Porter, W.S.; Yasarer, L.M.; ElKadiri, R.; Bingner, R.L.; Aber, J.W. Crop Conversion Impacts on Runoff and Sediment Loads in The Upper Sunflower River Watershed. *Agric. Water Manag.* **2019**, *217*, 399–412. [CrossRef]
42. Momm, H.G.; Yasarer, L.M.W.; Bingner, R.L.; Wells, R.R.; Kunhle, R.A. Evaluation of Sediment Load Reduction by Natural Riparian Vegetation in the Goodwin Creek Watershed. *Trans. ASABE* **2019**, *62*, 1325–1342. [CrossRef]
43. Chahor, Y.; Casalí, J.; Giménez, R.; Bingner, R.L.; Campo, M.A.; Goñi, M. Evaluation of The Annagnps Model for Predicting Runoff and Sediment Yield in A Small Mediterranean Agricultural Watershed in Navarre (Spain). *Agric. Water Manag.* **2014**, *134*, 24–37. [CrossRef]
44. Bisantino, T.; Bingner, R.; Chouaib, W.; Gentile, F.; Trisorio Liuzzi, G. Estimation of Runoff, Peak Discharge and Sediment Load at the Event Scale in a Medium-Size Mediterranean Watershed Using the Annagnps Model. *Land Degrad. Dev.* **2015**, *26*, 340–355. [CrossRef]
45. Momm, H.G.; Bingner, R.L.; Emilaire, R.; Garbrecht, J.; Wells, R.R.; Kuhnle, R.A. Automated Watershed Subdivision for Simulations Using Multi-Objective Optimization. *Hydrol. Sci. J.* **2017**, *62*, 1564–1582. [CrossRef]
46. Momm, H.G.; Bingner, R.L.; Yuan, Y.; Locke, M.A.; Wells, R.R. Spatial Characterization of Riparian Buffer Effects on Sediment Loads from Watershed Systems. *J. Environ. Qual.* **2014**, *43*, 1736–1753. [CrossRef]
47. Bingner, R.L.; Theurer, F.D.; Yuan, Y.; Taguas, E.V. *AnnAGNPS Technical Processes*; U.S. Department of Agriculture: Washington, DC, USA, 2018.
48. Momm, H.; Bingner, R.L.; Yuan, Y.; Kostel, J.; Monchak, J.J.; Locke, M.A.; Gilley, A. Characterization and Placement of Wetlands for Integrated Conservation Practice Planning. *Trans. ASABE* **2016**, *59*, 1345–1357. [CrossRef]
49. Web Soil Survey, Natural Resources Conservation Service, United States Department of Agriculture. Available online: http://websoilsurvey.nrcs.usda.gov/ (accessed on 20 May 2020).
50. Wagener, T.; Montanari, A. Convergence of Approaches toward Reducing Uncertainty in Predictions in Ungauged Basins. *Water Resour. Res.* **2011**, *47*. [CrossRef]
51. TN Department of Finance and Administration. State of Tennessee LiDAR Initiative. Available online: https://lidar.tn.gov/ (accessed on 20 May 2020).
52. Garbrecht, J.; Martz, L.W. Digital Landscape Parameterization for Hydrological Applications. *IAHS Publ.-Ser. Proc. Rep.-Intern. Assoc. Hydrol. Sci.* **1996**, *235*, 169–174.
53. Garbrecht, J.; Martz, L.W. The Assignment of Drainage Direction Over Flat Surfaces in Raster Digital Elevation Models. *J. Hydrol.* **1997**, *193*, 204–213. [CrossRef]
54. National Centers for Environmental Information, National Oceanic and Atmospheric Administration. Available online: https://www.ncdc.noaa.gov/cdo-web/search (accessed on 15 June 2020).
55. NRCS. AGNPS Climate Generator GEM. Available online: https://www.nrcs.usda.gov/wps/portal/nrcs/detailfull/national/water/manage/hydrology/?cid=stelprdb1043533 (accessed on 15 June 2020).
56. NRCS. Web Soil Survey. Available online: https://websoilsurvey.sc.egov.usda.gov/App/HomePage.htm (accessed on 19 June 2020).
57. NRCS. Soil Data Access. Available online: https://sdmdataaccess.nrcs.usda.gov (accessed on 10 June 2020).
58. NASS. National Agricultural Statistics Service. Available online: https://nassgeodata.gmu.edu/CropScape/ (accessed on 10 June 2020).
59. Boryan, C.; Yang, Z.; Mueller, R.; Craig, M. Monitoring US agriculture: The US Department of Agriculture, National Agricultural Statistics Service, Cropland Data Layer Program. *Geocarto Int.* **2011**, *26*, 341–358. [CrossRef]
60. U.S. National Agricultural Statistics Service. Available online: https://www.nass.usda.gov (accessed on 10 June 2020).
61. USDA. Riparian Forest Buffers. Available online: https://www.fs.usda.gov/nac/practices/riparian-forest-buffers.php (accessed on 3 February 2020).

62. Lv, J.; Wu, Y. Nitrogen Removal by Different Riparian Vegetation Buffer Strips with different Stand Densities and Widths. *Water Supply* **2021**, *21*, 3541–3556. [CrossRef]
63. Mayer, P.M.; Reynolds, S.K.; McCutchen, M.D.; Canfield, T.J. *Riparian Buffer Width, Vegetative Cover, and Nitrogen Removal Effectiveness: A Review of Current Science and Regulations*; EPA: Cincinnati, OH, USA, 2005.
64. Graziano, M.P.; Deguire, A.K.; Surasinghe, T.D. Riparian Buffers as a Critical Landscape Feature: Insights for Riverscape Conservation and Policy Renovations. *Diversity* **2022**, *14*, 172. [CrossRef]
65. Lee, P.; Smyth, C.; Boutin, S. Quantitative review of riparian buffer width guidelines from Canada and the United States. *J. Environ. Manag.* **2004**, *70*, 165–180. [CrossRef]
66. Clinton, B.D. Stream water responses to timber harvest: Riparian buffer width effectiveness. *For. Ecol. Manag.* **2011**, *261*, 979–988. [CrossRef]
67. King, S.E.; Osmond, D.L.; Smith, J.; Burchell, M.R.; Dukes, M.; Evans, R.O.; Knies, S.; Kunickis, S. Effects of Riparian Buffer Vegetation and Width: A 12-Year Longitudinal Study. *J. Environ. Qual.* **2016**, *45*, 1243–1251. [CrossRef]
68. Wenger, S. *A Review of the Scientific Literature on Riparian Buffer Width, Extent, and Vegetation*; Office of Public Service and Outreach, Institute of Ecology, The University of Georgia: Athens, GA, USA, 1999.
69. Jones, R.S.; ElKadiri, R.; Momm, H. Canopy Classification Using LiDAR: A Generalizable Machine Learning Approach. *Model. Earth Syst. Environ.* **2022**, *58*, 1–14. [CrossRef]
70. Zech, W.C.; Fang, X.; Logan, C. State-Of-The-Practice: Evaluation of Sediment Basin Design, Construction, Maintenance, and Inspection Procedures. *Pract. Period. Struct. Des. Constr.* **2014**, *19*. [CrossRef]
71. Yasarer, L.M.W.; Bingner, R.L.; Momm, H.G. Characterizing Ponds in a Watershed Simulation and Evaluating Their Influence on Streamflow in a Mississippi Watershed. *Hydrol. Sci. J.* **2018**, *63*, 302–311. [CrossRef]
72. Blanco, H.; Lal, R. *Principles of Soil Conservation and Management*; Springer: New York, NY, USA, 2008; Volume 167169.
73. Gonzalez, J.M. Runoff and Losses of Nutrients and Herbicides Under Long-Term Conservation Practices (No-Till and Crop Rotation) in the U.S. Midwest: A Variable Intensity Simulated Rainfall Approach. *Int. Soil Water Conserv. Res.* **2018**, *6*, 265–274. [CrossRef]
74. Shah, K.K.; Modi, B.; Pandey, H.P.; Subedi, A.; Aryal, G.; Pandey, M.; Shrestha, J. Diversified Crop Rotation: An Approach for Sustainable Agriculture Production. *Adv. Agric.* **2021**, *2021*, 8924087. [CrossRef]
75. Shrestha, J.; Subedi, S.; Timsina, K.P.; Subedi, S.; Pandey, M.; Shrestha, A.; Shrestha, S.; Hossain, M.A. Sustainable Intensification in Agriculture: An Approach for Making Agriculture Greener and Productive. *J. Nepal Agric. Res. Counc.* **2021**, *7*, 133–150. [CrossRef]
76. FAPRI-MU. *Estimating Water Quality, Air Quality, and Soil Carbon Benefits of the Conservation Reserve Program*; FAPRI-UMC Report #01-07; Food and Agriculture Policy Research Institute (FAPRI): Reno, NV, USA, 2007.
77. Nagy-Reis, M.B.; Lewis, M.A.; Jensen, W.F.; Boyce, M.S. Conservation Reserve Program is a key element for managing white-tailed deer populations at multiple spatial scales. *J. Environ. Manag.* **2019**, *248*, 109299. [CrossRef]
78. USDA. Conservation Reserve Program. Available online: https://www.fsa.usda.gov/programs-and-services/conservation-programs/conservation-reserve-program/ (accessed on 20 April 2020).
79. Baginska, B.; Milne-Home, W.; Cornish, P.S. Modelling Nutrient Transport in Currency Creek, NSW with AnnAGNPS and PEST. *Environ. Model. Softw.* **2003**, *18*, 801–808. [CrossRef]
80. Polyakov, V.; Fares, A.; Kubo, D.; Jacobi, J.; Smith, C. Evaluation of a Non-Point Source Pollution Model, AnnAGNPS, in a Tropical Watershed. *Environ. Model. Softw.* **2007**, *22*, 1617–1627. [CrossRef]
81. Pease, L.M.; Oduor, P.; Padmanabhan, G. Estimating Sediment, Nitrogen, and Phosphorous Loads from the Pipestem Creek Watershed, North Dakota, Using AnnAGNPS. *Comput. Geosci.* **2010**, *36*, 282–291. [CrossRef]
82. Taguas, E.V.; Yuan, Y.; Bingner, R.L.; Gómez, J.A. Modeling the Contribution of Ephemeral Gully Erosion Under Different Soil Managements: A Case Study in An Olive Orchard Microcatchment using the AnnAGNPS model. *CATENA* **2012**, *98*, 1–16. [CrossRef]
83. Abdelwahab, O.; Bisantino, T.; Milillo, F.; Gentile, F. Runoff and Sediment Yield Modeling in A Medium-Size Mediterranean Watershed. *J. Agric. Eng.* **2013**, *44*. [CrossRef]
84. Li, Z.; Luo, C.; Xi, Q.; Li, H.; Pan, J.; Zhou, Q.; Xiong, Z. Assessment of the AnnAGNPS model in Simulating Runoff and Nutrients in a Typical Small Watershed in The Taihu Lake Basin, China. *CATENA* **2015**, *133*, 349–361. [CrossRef]
85. Nahkala, B.A.; Kaleita, A.L.; Soupir, M.L. Assessment of Input Parameters and Calibration Methods for Simulating Prairie Pothole Hydrology using AnnAGNPS. *Appl. Eng. Agric.* **2021**, *37*, 495–503. [CrossRef]
86. Stroud Water Research Center. Model My Watershed. Available online: https://stroudcenter.org/virtual-learning-resource/model-my-watershed/ (accessed on 13 October 2020).
87. Yuan, Y.; Bingner, R.L.; Locke, M.A. A review of effectiveness of vegetative buffers on sediment trapping in agricultural areas. *Ecohydrology* **2009**, *2*, 321–336. [CrossRef]
88. Zhang, X.; Liu, X.; Zhang, M.; Dahlgren, R.A. A review of vegetated buffers and a meta-analysis of their mitigation efficacy in reducing nonpoint source pollution. *J. Environ. Qual.* **2010**, *39*, 76–84. [CrossRef] [PubMed]
89. Hellerstein, D.M. The US Conservation Reserve Program: The Evolution of an Enrollment Mechanism. *Land Use Policy* **2017**, *63*, 601–610. [CrossRef]

90. Taylor, M.R.; Hendricks, N.P.; Sampson, G.S.; Garr, D. The Opportunity Cost of the Conservation Reserve Program: A Kansas Land Example. *Appl. Econ. Perspect. Policy* **2021**, *43*, 849–865. [CrossRef]
91. Palmeri, L.; Trepel, M. A GIS-Based Score System for Siting and Sizing of Created or Restored Wetlands: Two Case Studies. *Water Resour.* **2002**, *16*, 307–328. [CrossRef]
92. Tanner, C.C.; Kadlec, R.H. Influence of Hydrological Regime on Wetland Attenuation of Diffuse Agricultural Nitrate Losses. *Ecol. Eng.* **2013**, *56*, 79–88. [CrossRef]
93. USDA-NRCS. Mississippi River Basin Healthy Watersheds Initiative. Available online: https://www.nrcs.usda.gov/wps/portal/nrcs/detailfull/national/programs/initiatives/?cid=stelprdb1048200 (accessed on 5 February 2020).

soil systems

Article

Oat Straw Mulching Reduces Interril Erosion and Nutrient Losses Caused by Runoff in a Newly Planted Peach Orchard

Luis Eduardo Akiyoshi Sanches Suzuki [1,*], Rodrigo de Lima do Amaral [1], William Roger da Silva Almeida [1], Mariana Fernandes Ramos [1] and Márcio Renato Nunes [2]

[1] Center of Technological Development, Federal University of Pelotas, Pelotas 96010-610, RS, Brazil
[2] Department of Soil, Water, and Ecosystem Sciences, University of Florida, Gainesville, FL 32611, USA
* Correspondence: luis.suzuki@ufpel.edu.br

Abstract: Soil erosion is one of the major problems in the agricultural areas in the world, and straw mulching is a conservation practice that may reduce soil runoff. How much straw mulching is necessary to reduce soil runoff? The objectives of this study were to quantify and characterize the runoff under different levels of oat straw mulching, as well as to analyze the cost of soil erosion. An experiment was performed in a site with the soil recently tilled for peach orchard implementation. In the ridges in the row of the peach orchard, plots were placed in order to quantify soil and nutrient losses by surface runoff due to interril erosion on the dates 23 August 2015 and 13 March 2016, considering the treatments were composed of different amounts of oat straw mulching (0, 1, 2, 4 and 8 Mg ha^{-1}). The results showed that the use of oat straw mulching decreased soil runoff, especially the doses $\geq$2 Mg ha^{-1}, and the cost to replace the available nutrients P, K, Ca and Mg via mineral fertilizer varies from US$ 75.4 (no mulching) to US$ 2.70 per hectare (8 Mg ha^{-1} oat straw mulching).

Keywords: water erosion; sediment; soil conservation; crop residue; vegetative practices of soil conservation; cost of erosion

Citation: Suzuki, L.E.A.S.; Amaral, R.d.L.d.; Almeida, W.R.d.S.; Ramos, M.F.; Nunes, M.R. Oat Straw Mulching Reduces Interril Erosion and Nutrient Losses Caused by Runoff in a Newly Planted Peach Orchard. *Soil Syst.* **2023**, *7*, 8. https://doi.org/10.3390/soilsystems7010008

Academic Editor: Abdul M. Mouazen

Received: 26 December 2022
Revised: 19 January 2023
Accepted: 27 January 2023
Published: 30 January 2023

1. Introduction

Soil is the foundation for terrestrial life and the sustainability of humankind. However, this natural resource has become increasingly threatened by excessive tillage, limited crop rotations, poor irrigation management, and contaminants [1]. According to FAO [2], one-third of the world's soil resources have been degraded and the remaining topsoil could become unproductive within 60 years if current rates of degradation continue. Agriculture practices may harm the soil due to compaction, acidification, loss of soil organic matter, and soil erosion. Those changes degrade soil physical properties, increase nutrient loss, and reshape fields, ultimately impacting productivity and environmental outcomes [1].

Soil erosion from agricultural fields is estimated to be currently 10 to 20 times (no-tillage) to more than 100 times (conventional tillage) higher than the soil formation rate, and the current levels of global warming are associated with moderate risks from increased soil erosion [3]. Using simulations carried out from 1901 to 1990, soil erosion at the global scale has been increased during the last century, pointing out Brazil as the region with the largest increase, with human activity being the greatest responsible determinant [4]. Yang et al. [4] highlight that in the 2090s, climate change, mainly induced by rainfall increases, is projected to increase soil erosion by around 9% globally, while land use would change about 5%. Higher global temperatures, as impact of climate changes, intensify the hydrological cycle, resulting in more intense rainfall, which is an important driver of soil erosion [5].

Peach orchard production is of great socio-economic importance in southern Brazil, mainly within the Rio Grande do Sul (RS) State, which is the greatest producer in Brazil, accounting for 60.5% of the total Brazilian peach harvest, and it is mainly cultivated by small farmers, with a total area cultivated of 12,468 hectares [6]. However, orchards represent one of the land-uses for which runoff rates and sediment losses may occur if soil and water conservation practices are not adopted, especially in hill slopes.

Runoff and soil erosion have been reported worldwide and are usually associated with (i) the location on hill slopes and disposition of rows along the slope, which makes runoff and erosion stronger [7], (ii) maintenance of bare soil between rows by mechanical or chemical weeding [8–10], and (iii) intense machinery traffic along fixed paths, which promote soil compaction and reduce soil water holding capacity and water infiltration [11].

Runoff and soil erosion in peach orchards tend to be more intense in the first years after plantation, which can be associated with deep and intensive tillage during orchard installation, disaggregating the soil and exposing it to rainfall. A few studies have reported higher runoff and soil erosion rates due to the orchards' installation practices [9,12,13]. Deep tillage is usually applied to incorporate fertilizer and lime and to improve the soil's physical condition prior to plantation; however, it can also decrease soil aggregate stability [14], increase soil organic matter mineralization [15], and promote soil surface crusting [16], as well as decrease water infiltration, which leads to soil erosion [17].

The physical processes of erosion and the control of those events have been studied across a long time and have been established but soil erosion continues to be the greatest threat to soil health and ecosystem services in many regions of the world, having some controversial points that make the establishment of erosion control measures around the world difficult [18].

The use of cover crops and/or mulching in orchards has the potential to reduce runoff and soil erosion [19]. IPCC [3] references growing green manure and cover crops, crop residue retention, reduced/zero tillage, and maintenance of soil covering through improved grazing management as options to reduce vulnerability to soil erosion and nutrient loss. Mulching and cover crops have been proven to be efficient practices to (1) protect the soil from water droplet impact, (2) enhance aggregate stability, (3) improve soil water infiltration, (4) interrupt runoff pathways, (5) improve nutrient cycling and soil water storage, and (6) reduce soil temperature variation and water loss to evaporation [20,21]. Additionally, higher sediment losses and herbicide residues in runoff water were found in bare soil under avocado (*Persea americana* Mill) hillside orchards [9].

Stark and Thorne [12] argued that peach orchards cannot be maintained over a long time without adequate management practices to maintain soil organic matter and to control soil erosion, suggesting the use of cover crops. A literature review by Wolstenholme et al. [8] has also highlighted the benefits of using mulching in avocado orchards. They found that using mulching and/or cover crops decreased tree stress, improved root growth and health, and improved both fruit size and yield compared to avocado under bare soils. These positive effects of mulching within orchards related in past studies were confirmed recently by some other studies around the world [22–24]. However, there is still a gap in knowledge about the effect of cover crop residue as mulch on the triggering of runoff and soil water erosion in peach orchards in southern Brazil, as well as on the exact amount of mulching, considering this involves costs to the farmers. Furthermore, little information exists about the impact of runoff and soil erosion on environmental health (i.e., water contamination), soil losses and their costs for peach farmers in that region.

Therefore, this study aimed to quantify and characterize (i) the runoff and soil erosion under different levels of oat (*Avena sativa*) straw mulching in a commercial peach orchard; (ii) the cost of soil erosion and runoff during the first year after the peach orchard's installation. We hypothesized that straw mulching would decrease the soil runoff and, consequently, reduce nutrient losses and the production costs in the peach production system in southern Brazil.

2. Materials and Methods

2.1. Experimental Area and Treatments

The experimental area is a 0.7 ha commercial peach orchard (variety "sensação") installed in 2015, with 21% slope located in Pelotas City, "Rio Grande do Sul" State, Brazil

(Latitude 31°34′11,76″ S, Longitude 52°30′16,51″ W, 171 m altitude) (Figure 1). The climate is subtropical humid (Cfa) according to the Köppen's Climate Classification System. The mean annual rainfall is 1367 mm at the Pelotas Agroclimatology Station, in the period 1971–2000 [25]. The mean annual temperature is 17.8 °C, January being the hottest month, at 23.2 °C, and July being the coldest, at 12.3 °C (Figure 2).

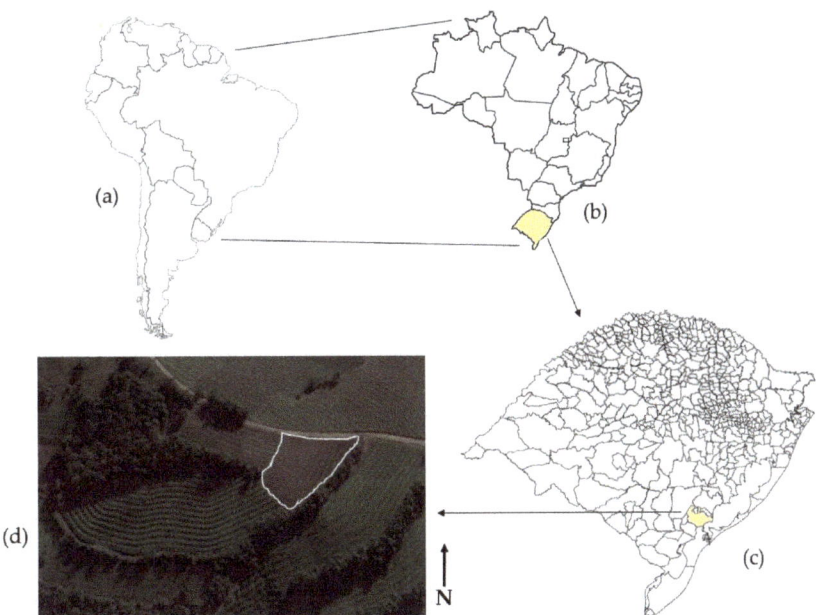

Figure 1. Map from South America (**a**), Brazil and "Rio Grande do Sul" State highlighted (**b**), and "Rio Grande do Sul" State with Pelotas City highlighted (**c**); image from Google Earth with the experimental area surrounded in a white color (**d**). Image of Google Earth dated 7 July 2015.

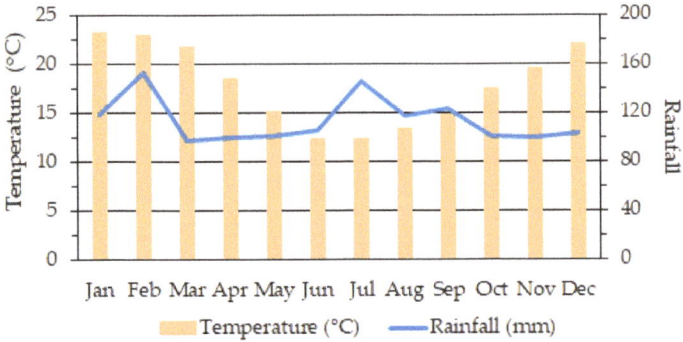

Figure 2. Mean monthly temperature and rainfall of the period 1971–2000. Source: [25].

The soil type at the study region is dominated by Entisols, Mollisols, Ultisols, Inceptisols, Plinthic, Alfisols and Entisols [26] (respectively Neossolos, Chernossolos, Argissolos, Cambissolos, Plintossolos, Planossolos and Gleissolos, accordingly with the Brazilian System of Soil Classification [27]). At the experimental site, the soil was classified as Cambissolo Háplico Tb Distrófico according to the Brazilian System of Soil Classification [27], which corresponds to Inceptisol in Soil Taxonomy [26].

The treatments were composed of different amounts of oat straw (*Avena sativa*) mulching placed on plots with the soil being tilled. The plots were used to measure the soil and nutrient losses by surface runoff.

The site was prepared for planting (peach orchard implementation) in June 2015. For instance, the soil was tilled by plowing (approximately 30 cm deep) followed by harrowing. The ridges in the row were made using the soil from the interrow. The height of the ridges was approximately 0.40 m, having the interrow of the orchard as a reference. The distance between plants in the row was 2.5 m, and in the interrow it was 5.0 m. The soil fertility adjustment was realized in the peach orchard implementation.

On 9 July 2015 (i.e., around one month after orchard implementation), in order to quantify soil and nutrient losses by surface runoff due to interril erosion, plots were placed in the ridges (Figure 3). These plots were constructed using polyvinyl chloride (PVC), with strips of 0.5 m length and 0.15 m height, forming a triangle delimiting an area of 0.11 m^2, and were fitted using PVC pipe with a height of 0.25 m. The strips of the plots were linked through slots, to facilitate assembly, disassembly, and transport of such material. A polyethylene terephthalate (PET) bottle was cut in half and placed in the lower edge of the plot to collect the soil loss by surface runoff. In the field, a hole was opened in the ground for fixing the PET bottle, where its border remained close to the ground surface and the PVC strips connected to the border of the PET bottle. The soil loss by runoff in the delimited area (0.11 m^2) was captured in the PET bottle with a capacity of nearly 1.5 L. Plots larger than 0.11 m^2 were not possible to be used because the area of the ridge was smaller to support larger plots. Because of the plots' size and configuration, only interril erosion was possible to measure, as well as the impact and disaggregation of soil as part of erosion process.

Figure 3. Plots to assess the soil losses by surface runoff, installed in the ridge of the peach's orchard row on 23 August 2015, which was date of application of the treatments with different amounts of oat straw (*Avena sativa*) mulching.

Overall, 15 triangular plots (0.11 m^2) were constructed. The soil within each of those plots was covered by different amounts of oat (*Avena sativa*) straw mulching, which represents the treatments: 0 Mg ha^{-1}, 1 Mg ha^{-1}, 2 Mg ha^{-1}, 4 Mg ha^{-1}, 8 Mg ha^{-1} dry biomass (Figure 4). Each treatment had three replicates. The plots were installed on 9 July

2015 and received the treatments with oat straw mulching on 23 August 2015. When the straw mulching of each treatment was totally or almost totally decomposed in the plots, it was replaced along the experiment, avoiding the zero-straw mulching and being possible to evaluate the period of soil cover and decomposition, considering the different amounts of oat straw mulching. Thus, on 8 November 2015 the oat straw mulching was replaced in the treatments 1 and 2 Mg ha^{-1} and, on 13 January 2016, a new replacement of oat straw mulching was realized in all treatments (1 Mg ha^{-1}, 2 Mg ha^{-1}, 4 Mg ha^{-1}, 8 Mg ha^{-1}).

Figure 4. Plots to assess the soil losses by surface runoff, installed in the ridge of the peach's orchard row, covered by different amounts of oat (*Avena sativa*) straw mulching, which represent the treatments: 0 Mg ha^{-1} (**a**), 1 Mg ha^{-1} (**b**), 2 Mg ha^{-1} (**c**), 4 Mg ha^{-1} (**d**), 8 Mg ha^{-1} (**e**) dry biomass. Pictures dated 23 August 2015 (application of the treatments).

The oat straw used as mulching in the plots was collected in a peach orchard next to the studied area and forwarded to the laboratory to dry at a temperature of 65 °C (standard method to quantify dry biomass of plants) and, afterwards, was placed in the plots according to each treatment. In this same orchard, which was twelve years old, in its interrow, oat straw used as mulching was sampled in four random points, in an area of 1 m^2 each one, in its senescence period, to verify the oat straw yield in a management system where it has the objective to protect the soil. The average yield of oat straw was 3 Mg ha^{-1} (dry weight at a temperature of 65 °C). That twelve year-old orchard was chosen because it is next to the studied area and frequently uses oat straw in its interrows.

2.2. Soil Characterization of the Ridges of the Peach Orchard's Row

In order to characterize the soil in the ridges of peach orchard's row (i.e., where the plots for assessment of soil losses by surface runoff were placed), disturbed soil samples were collected within 0 to 0.10 m, 0.10 to 0.20 m and 0.20 to 0.40 m depth. Those samples were analyzed for particle size distribution, dispersible clay in water, particle density, and soil fertility (i.e., pH and soil nutrients). Undisturbed soil samples were also collected in the same depths in metal cylinders of 4.7 cm diameter and 3.0 cm height. These samples were used to evaluate soil porosity, bulk density and saturated hydraulic conductivity.

2.2.1. Soil Chemical Characterization

Disturbed soil samples were also analyzed for pH, organic matter (OM), phosphorus (P), potassium (K), sulfur (S), calcium (Ca), magnesium (Mg), iron (Fe), copper (Cu), zinc (Zn), manganese (Mn), sodium (Na), aluminum (Al) and potential acidity (H + Al), as described by Tedesco et al. [28]. Soil pH was measured using a 1:1 soil-to-water ratio. Through these determinations, the effective cation exchange capacity and that at pH 7.0 (respectively, CECeffective and CECpH7.0), base saturation (V) and aluminum saturation (m) were calculated. The H + Al was determined for the SMP index, while the extractant KCl 1 mol L^{-1} was used to determine Ca, Mg, Mn and Al, and the extractant Mehlich I was used to determine P, K, Na, Zn and Cu. To determine organic matter we used moist digestion.

2.2.2. Soil Physical Characterization

Particle size distribution analysis was performed by the pipette method [29]. The dispersion of the soil samples followed the method described by Suzuki et al. [30], i.e., 20 g of sample, 10 mL of 6% NaOH (chemical dispersant), 50 mL of distilled water and two nylon spheres (each one weighing 3.04 g, diameter of 1.71 cm and density of 1.11 g cm^{-3}) were put in 100 mL glass bottles, which were shacked horizontally at 120 rpm for four hours.

Afterwards, the soil particles were separated into sand (diameter between 2 and 0.053 mm) by sieving, and silt (diameter between 0.053 and 0.002 mm) by calculus between the difference of the sum of sand and clay (diameter < than 0.002 mm), which was determined by pipette. The sand fractions were separated by sieving in very coarse sand (2 to 1 mm), coarse sand (1 to 0.5 mm), medium sand (0.5 to 0.25 mm), fine sand (0.25 to 0.125 mm) and very fine sand (0.125 to 0.053 mm).

The results of the particle size distribution analysis were used for soil textural classification, using the soil texture triangle available from the National Resource Conservation Service/United States Department of Agriculture [31].

Dispersible clay in water was quantified following the same procedure used for total clay evaluation but without using the chemical dispersant.

The calculus of the degree of flocculation (DF, %) followed the Equation (1):

$$DF = [(total\ clay\text{-}dispersible\ clay\ in\ water)/total\ clay] \times 100 \tag{1}$$

The particle density was determined by the volumetric balloon method, according to Viana et al. [32].

2.2.3. Soil Porosity, Bulk Density, and Saturated Hydraulic Conductivity

The undisturbed soil samples were saturated through capillarity and balanced on a tension table (at 6 kPa tension) to determine macroporosity (pores > than 50 μm). After that, the samples were oven-dried (105 °C) to determine microporosity (pores < than 50 μm), total porosity and bulk density [33].

Samples with a preserved soil structure were also used to quantify the saturated hydraulic conductivity of the soil, using a permeameter of constant charge, as described by Klute and Dirksen [34]. Before beginning recording the measurements, the samples remained for some minutes with the water passing through the samples, to reach equilibrium and constancy. Three measurements for each sample were realized and we calculated the mean value.

Equation (2) below was used to calculate the saturated hydraulic conductivity:

$$KS = (V \times L)/[A \times t(h + L)] \tag{2}$$

where: KS = saturated hydraulic conductivity, mm h^{-1}; V = volume of water passed through the soil sample, mm^3; L = length of soil sample, mm; A = area of the transversal section of the soil sample, mm^2; t = time of lecture, hours; h = pressure potential (hydraulic charge) in the top of the soil sample, mm.

2.3. Soil and Nutrient Losses by Surface Runoff

In order to quantify the soil loss, the soil runoff plus water accumulated of rain in the PET bottle fixed in the lower edge of each plot was taken after each one of the ten rainfall events, sent to the laboratory and dried at 110 °C. Afterwards, the soil was broken manually and passed through a sieve with 2 mm mesh to separate particles larger and smaller than 2 mm. The total soil loss by surface runoff per hectare was quantified.

Soil runoff was collected 10 times between 23 August 2015 and 13 March 2016. A composite sample of soil with diameter < than 2 mm for each treatment (0 Mg ha^{-1}, 1 Mg ha^{-1}, 2 Mg ha^{-1}, 4 Mg ha^{-1}, 8 Mg ha^{-1} oat straw mulching dry biomass), collected in the ten events, was used to determine the particle size distribution and soil fertility indicators using the same procedures described above. Joining the soil of the 10 events of soil runoff to make a composite sample was necessary due to small amount of soil runoff in each event.

By multiplying the total amount of soil (<2 mm) runoff along the period of evaluation (data presented in Table 4, expressed in kg ha^{-1}), and the nutrient concentration in that soil (data presented in Table 6, expressed in kg dm^{-3}) the available P, K, Ca, and Mg runoff in one hectare (kg ha^{-1}), was calculated.

The cost of soil erosion (from 23 August 2015 to 13 March 2016) was calculated as the cost to replace the amount of soil nutrients (i.e., P, K, Ca and Mg) lost by runoff in one hectare using commercial mineral fertilizers. Specifically, superphosphate triple (41% P_2O_5) was considered for P, potassium chloride (50% K) for K, and dolomitic limestone (32% CaO + 6% MgO) for Ca and Mg replacement. The information about each nutrient concentration in the fertilizer was obtained in the Normative instruction number 39 of the Ministry of State of Agriculture, Livestock and Supply of Brazil [35], and the fertilizer cost was obtained using current market values (Pelotas City, Brazil).

The rainfall for the period of study was obtained from the monthly weather report available at Agrometeorology Laboratory of the Embrapa Temperate Climate (Embrapa/"Laboratório de Agrometeorologia" [36]), with data collected in an automatic weather meteorologic station installed in the Headquarters Weather Station of the Embrapa Temperate Climate/Pelotas City/"Rio Grande do Sul" State, around 14 km away of the experiment.

2.4. Data Analyses

The data were analyzed in terms of relative percentage; an analysis of variance and the Tukey test of means were performed considering 5% significance.

3. Results and Discussion

The ridges of the peach orchard's row where the plots to assess the soil losses by surface runoff were installed present a high soil fertility (Table 1). These results reflect the addition of high doses of chemical fertilizers before the orchard implementation, with high and very high nutrients content in the soil [37].

The soil runoff is basically the surface layer and the knowledge of its nutrient concentration is important to preview the possible environment impacts and economical losses due to soil runoff.

Tables 2 and 3 show the physical characterization of the soil ridges of peach orchard's row where the plots to assess the soil losses by surface runoff were installed. Overall, the soil had low average bulk density (1.12 Mg ha^{-1}), high macro (0.26 m^3 m^{-3}), micro (0.30 m^3 m^{-3}) and total (0.56 m^3 m^{-3}) porosity, and high saturated hydraulic conductivity (197 mm h^{-1}). These results were expected and reflect the short-term loose and disaggregated soil effect of tillage on those soil physical properties and processes. Within this area, deep tillage was performed around 1 month before soil sampling to incorporate fertilizer and to build the ridges where the peach plants were planted. In the short term, tillage can improve soil physical qualities for plant growth, however, this practice can also decrease aggregate stability [14] and promote soil surface crusting, which in turn can reduce the water infiltration rate and promote soil erosion [17]. Indeed, the tilled soil in the experimental area had a high content of dispersible clay in water and a low flocculation degree, suggesting a soil with high level of disaggregation (Table 2). Therefore, these results confirm the high potential of new planted peach orchards for nutrient losses and environmental degradation associated with runoff and interril soil erosion.

Table 1. Soil chemical characterization of the ridges of peach orchard's row where the plots to assess the soil losses by surface runoff were installed.

Soil		Depth, m			
Attribute	Unit	0–0.10	0.10–0.20	0.20–0.40	Mean
SOM	g kg^{-1}	27.6 (medium)	26.2 (medium)	27.6 (medium)	27.1 (medium)
P-Melich	mg dm^{-3}	108.1 (very high)	155.6 (very high)	202.1 (very high)	155.3 (very high)
Exch. K	mg dm^{-3}	95.0 (high)	128.0 (very high)	143.0 (very high)	122.0 (very high)
Ca	cmol$_c$ dm^{-3}	7.8 (high)	7.9 (high)	7.4 (high)	7.7 (high)
Mg	cmol$_c$ dm^{-3}	2.9 (high)	3.0 (high)	2.7 (high)	2.9 (high)
Na	mg dm^{-3}	11.0	12.0	13.0	12.0
Al	cmol$_c$ dm^{-3}	0.0	0.0	0.0	0.0
H + Al	cmol$_c$ dm^{-3}	2.0	1.6	2.5	2.0
CECeffective	cmol$_c$ dm^{-3}	11.0	11.3	10.5	10.9
CECpH7.0	cmol$_c$ dm^{-3}	13.0 (medium)	12.9 (medium)	13.0 (medium)	13.0 (medium)
pH water		6.4 (high)	6.4 (high)	6.0 (medium)	6.3 (high)
AlS	%	0.0 (very low)	0.0 (very low)	0.0 (very low)	0.0 (very low)
BS	%	85.0 (high)	87.0 (high)	81.0 (high)	84.0 (high)

SOM: soil organic matter; P: phosphorus; Exch. K: exchangeable potassium; Ca: calcium; Mg: magnesium; Na: sodium; Al: aluminum; H + Al: potential acidity; CEC: cation exchange capacity; AlS: aluminum saturation; BS: base saturation. In parentheses is the interpretation of the soil fertility [37].

Table 2. Soil physical and hydraulic characterization of the ridges of the peach orchard's row where the plots to assess the soil losses by surface runoff were installed.

Depth, m	BD, Mg m^{-3}	TP, m^3 m^{-3}	Macro, m^3 m^{-3}	Micro, m^3 m^{-3}	KS, mm h^{-1}	DCA, %	DF, %	PD Mg m^{-3}
0.00–0.10	1.05	0.593	0.259	0.333	142.71	8.94	31.65	2.56
0.10–0.20	1.12	0.538	0.276	0.262	300.58	8.42	31.54	2.52
0.20–0.40	1.20	0.548	0.231	0.318	148.86	8.92	29.04	2.54
Mean	1.12	0.560	0.255	0.304	197.38	8.76	30.74	2.54

BD: bulk density; TP: total porosity; Macro: macroporosity; Micro: microporosity; KS: saturated hydraulic conductivity; DCA: dispersible clay of soil in water; DF: degree of flocculation; PD: particle density.

Table 3. Particle size distribution and textural classification of the ridges of the peach orchard's row where the plots to assess the soil losses by surface runoff were installed.

Depth, m	Sand						Silt	Clay	Textural Classification [31]
	Total	Very Coarse	Coarse	Medium	Fine	Very Fine			
	%								
0–0.10	63.34	11.99	11.44	11.79	16.83	11.30	23.58	13.08	Sandy loam
0.10–0.20	64.10	13.16	11.74	11.31	17.58	10.31	23.60	12.30	Sandy loam
0.20–0.40	63.71	12.41	11.55	11.56	17.70	10.49	23.72	12.57	Sandy loam
Mean	63.72	12.52	11.58	11.55	17.37	10.70	23.63	12.65	

Total sand: particles with diameter between 2 and 0.05 mm; very coarse sand: diameter between 2 and 1 mm; coarse sand: diameter between 1 and 0.5 mm; medium sand: diameter between 0.5 and 0.25 mm; fine sand: diameter between 0.25 and 0.125 mm; very fine sand: diameter between 0.125 and 0.053 mm; silt: diameter between 0.053 and 0.002 mm; clay: diameter < than 0.002 mm.

The degree of flocculation is low (Table 2), and the sand content is a high sandy loam textural class (Table 3), reinforcing the necessity of soil and water conservation practices in that soil. Clay is an important agent of soil aggregation [38–44], therefore, the larger the clay dispersible in water, the larger the possibility to occur water erosion, especially in the topsoil that is more susceptible to the rainfall drop and runoff. The increment of clay dispersible in water decreases water infiltration or water conductivity [45,46] and favors the runoff probably because it closes the pores of soil [47].

It is important to know the relation between the particle size distribution and other soil physical attributes to understand the susceptibility of soil to erosion and sealing of its surface. According to Resende et al. [48], besides particle size distribution [16], other variables should be considered about water erosion: the depth [49], the slope [50–52] and its length [53], the porosity, and others, because they help us to preview the susceptibility to erosion, since water infiltration and storage are related to the variables cited. For example, Suzuki et al. [54] verified that the runoff correlated positively with coarse sand and saturated hydraulic conductivity of soil, while Keesstra et al. [10], using multivariate analysis, observed that vegetation cover, soil moisture and organic matter were negatively correlated with the bulk density, total runoff, runoff coefficient, sediment yield and soil erosion.

Regarding the particle sizes of soil and erosion, sand particles are difficult to be transported because of their size but they are easily detached from the soil mass; although silt soils generally are well aggregated, the aggregates break down easily when wetted, and the particles are easily detached and transported, and the clay particles are difficult to detach but they are transported across larger distances when separated from the soil [38].

The soil runoff in the plots with no straw mulching was larger and differed significantly from the other treatments (Figure 5); besides, according to the increase in the mulching, the soil runoff decreased at most times (Table 4 and Figure 5). This is because the soil is exposed to the rainfall according to the decrease of mulching, which is more susceptible to rainfall drops and splashes. Besides this, the topsoil was tilled to enhance the orchard's performance, breaking the soil aggregates and loosening it. Falling raindrops and running water are the two major agents in water erosion, and both are related to the energy necessary to detach and transport soil particles [38]. However, planting or mulching at the soil's surface intercepts raindrops and slows down runoff [38,49,51,55]. Other practices used together with straw mulching such as the disposal of branches from the yearly pruning on the interrows, harvesting manually, and opting for using a compact tractor contribute to avoiding soil compaction and probably soil erosion [56], as well as the use of terracing [57–60], such as dry-stone wall or earth bank terraces [53], and keyline arrangement [61,62].

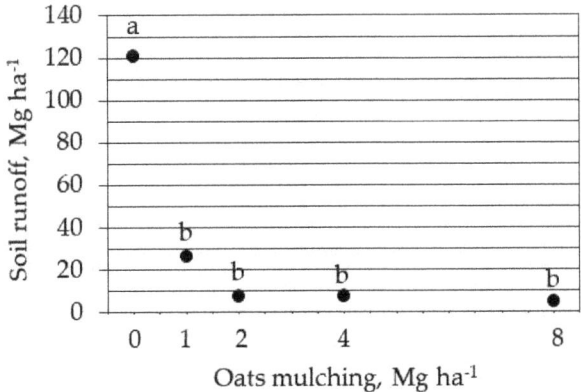

Figure 5. Tendency of total soil losses by surface runoff (particles with diameter < than 2 mm) in the period between 29 August 2015 to 13 March 2016, according to the amount of oat straw mulching in the plots. Total soil losses followed by same letters do not differ statistically from each other by Tukey's test at 5% significance.

Table 4. Soil losses by surface runoff (particles with diameter < than 0.002 mm) (Mg ha^{-1}) and percentage of losses by surface runoff in relation to the treatment without mulching, and rainfall accumulated up until the sampling date.

Sampling	[1] Rainfall	Treatment (Amount of Oat Straw Mulching, Mg ha^{-1})					
Date	Accumulated	0	1	2	4	8	Total
	mm			Mg ha^{-1}			
29 August 2015	110.5	0.86 (100%)	0.00 (0%)	0.00 (0%)	0.00 (0%)	0.00 (0%)	0.86
7 September 2015	28.9	0.27 (100%)	0.00 (0%)	0.00 (0%)	0.00 (0%)	0.00 (0%)	0.27
27 September 2015	284.0	11.53 (100%)	3.00 (26%)	1.35 (12%)	0.53 (5%)	1.03 (9%)	17.44
25 October 2015	299.2	18.73 (100%)	4.34 (23%)	2.29 (12%)	1.27 (7%)	0.94 (5%)	27.57
[2] 8 November 2015	42.4	0.00	0.00	0.00	0.00	0.00	0.00
14 November 2015	92.4	8.72 (100%)	2.19 (25%)	0.00 (0%)	0.00 (0%)	0.00 (0%)	10.91
8 December 2015	164.1	7.43 (100%)	2.97 (40%)	0.64 (9%)	0.89 (12%)	0.58 (8%)	12.51
[2] 13 January 2016	251.6	28.90 (100%)	7.12 (25%)	1.78 (6%)	4.86 (17%)	2.34 (8%)	45.00
16 February 2016	180.2	24.11 (100%)	4.81 (20%)	0.80 (3%)	0.00 (0%)	0.00 (0%)	29.72
13 March 2016	188.6	20.39 (100%)	1.83 (9%)	0.70 (3%)	0.00 (0%)	0.00 (0%)	22.92
Total		120.94	26.26	7.56	7.55	4.89	

[1] Source: "Laboratório de Agrometeorologia da Embrapa Clima Temperado" (Agrometeorology Laboratory of the Embrapa Temperate Climate) [36]. [2] On 8 November 2015, the oat straw mulching was replaced in the treatments 1 and 2 Mg ha^{-1} and on 13 January 2016 a new replacement of oat straw mulching was realized in all treatments (0 Mg ha^{-1}, 1 Mg ha^{-1}, 2 Mg ha^{-1}, 4 Mg ha^{-1}, 8 Mg ha^{-1} oat straw mulching dry biomass).

Those values of soil losses are high, because according to FAO [2], rates of tolerable soil loss calculated using soil production rates range from 0.2 to 2.2 Mg ha^{-1} year^{-1} and tolerable rates based on maintenance of crop production range from approximately 1 to 11 Mg ha^{-1} year^{-1}, and these ranges reinforce the need for site-specific studies to evaluate the different sensitivities of soils for the removal of surface soil through erosion.

These results agree with other studies, where the use of cover crops and/or mulching in orchards has the potential to reduce runoff and soil erosion [10,19,63].

In a rainfall simulation experiment using organic mulching in an urban forestry park, the runoff amount and runoff generation rate decreased by 28–83% and 21–83%, respectively, when using 0.25 kg m^{-2} and 0.50 kg m^{-2} of mulching, compared to bare soil [64]. Testing different mulching types of banana (*Musa* sp.) leaves, coconut (*Cocos nucifera*) leaves, and vetiver (*Vetiveria zizanoides*) and various amounts (0 Mg ha^{-1}, 10 Mg ha^{-1}, 20 Mg ha^{-1} and 40 Mg ha^{-1}) in farm fields with an 8% slope after seeding the plots with maize, the banana leaves at 10 Mg ha^{-1} and coconut leaves at 40 Mg ha^{-1} mitigated soil and nutrient erosion to, respectively, 28.9% and 57.3%, contributed to the mechanical barrier provided by the mulches, and also to the reduction of raindrops acting on the soil aggregates [65]. The author [65] verified that mulching also contributed to increasing the infiltration rate, lowering the temperature and, therefore, lowering evaporation.

On 8 November 2015 there was no soil runoff, even with rainfall before this date, corresponding to 14 days after the sampling in October. On this same date (8 November) the oat straw mulching was replaced in the treatments 1 and 2 Mg ha^{-1} because it was totally or almost totally decomposed in the plots, avoiding the zero-straw mulching. The larger soil runoff in the 4 Mg ha^{-1} mulching treatment compared to 2 Mg ha^{-1} mulching treatment, on 8 December and after, may be associated with this replacement of mulching, when the treatment with 4 Mg ha^{-1} mulching could be presenting less mulching than the treatment with 2 Mg ha^{-1} because of its decomposition since the installation of the experiment, considering the replacement of mulching on 8 November 2015 was realized only in the treatments with 1 and 2 Mg ha^{-1}.

On 13 January 2016 a new replacement of oat straw mulching was realized in all treatments. The straw mulching replacement along the experiment was necessary to maintain the same or almost the same cover density during the period, avoiding the zero-straw mulching, and to verify the biomass time of decomposition according to each treatment.

Considering the application of mulching in the treatments on 23 August 2015, the decomposition practically totaled 2 Mg ha^{-1} of oat straw mulching at around 80 days (23 August 2015 to 8 November 2015), while the larger amounts of straw mulching (4 and 8 Mg ha^{-1}) would take more than 140 days (23 August 2015 to 13 January 2016) to totally decompose, taking into account the conditions of the present study. The time of decomposition of the straw mulching is important because the longer it spends on the soil surface, the more soil protection against rainfall it provides. Besides, it was verified in the field that the plots with mulching presented a smaller incidence of spontaneous plants, especially at 4 and 8 Mg ha^{-1}, which was practically null. That is an important finding. In organic tree fruit fields, for example, the farmers have limited options for controlling weeds and furnish nutrients at the appropriate time and adequate amount [66].

Although the soil runoff was statistically the same with straw mulching (1 to 8 Mg ha^{-1}), from 2 Mg ha^{-1} straw mulching there was less soil runoff (Figure 5); it is possible to indicate this value as minimum amount of mulching in the peach orchard or any other condition of soil tilled to reduce soil erosion, but it is important to say that the time spent on decomposition and soil exposure will be greater than with larger amounts of mulching. This value (2 Mg ha^{-1}) is smaller than the 3 Mg ha^{-1} value, representing the average yield of oat straw mulching in its senescence (see Material and methods). In areas where the spontaneous weed is used, it would be interesting to evaluate its straw mulching yield if it is comparable to oat straw.

We verified soil runoff in all plots with different amounts of straw mulching, although with different amounts along the period of study, either because of rainfall intensity (not measured) or when the soil reached its capacity of infiltration. When the soil is exposed, it is more susceptible to rainfall. Then, when mulching or cover crops are used, the rainfall dropping onto the soil and topsoil compaction are decreased, and this reduces flooding speed [49,51,55,67,68]. According to some authors [51,67,68], the speed of the covering of plants is important, because the soil runoff is associated with the time of soil exposure, being susceptible to erosion. Water loss through runoff in Aquic Argiudoll (Luvic Phaeozem) soil was more related to the number of months in the year with the presence of crops than to the soil physical properties related to porosity and water flow [69].

According to Bertoni and Lombardi Neto [70], in Brazil, the soil runoff in agricultural areas is caused especially by water erosion, and this happens generally in the period of soil being tilled to crops' plantation, which is also the case in the present study; the tillage of the soil and implementation of the orchard changed its physical characteristics, and the soil was also exposed to rain and wind.

In general, there was an increase in accumulated rainfall that increased soil runoff (Figure 6). The total rainfall is not the most important variable when soil erosion is evaluated, the most relevant are the rainfall drop, the intensity (volume of rainfall during a certain period), speed and specially volume, duration and time to return the rainfall in the watershed [53,71]. Natural rainfalls larger than 70 mm resulted in similar runoff coefficients in an Aquic Argiudoll (Luvic Phaeozem) soil with a 3.5% slope, in natural plots under monocultures, rotation, pasture, and tilled soil without vegetation, while for intermediate and small rainfalls the runoff coefficients were different [69]. The rainfall in the "Rio Grande do Sul" State is well-distributed along the year, but its volume is different: the mean rainfall in the south is between 1299 mm and 1500 mm, while in the north it is between 1500 mm and 1800 mm [72]. The Pelotas mean annual rainfall is 1367 mm, according to the Pelotas Agroclimatology Station, in the period 1971–2000 [25], lower than the mean of the "Rio Grande do Sul" State.

According to Volk and Cogo [73], the main variables used to determine soil runoff are the rainfall intensity and flooding associated with it [52], the particle size distribution [16] and the degree of consolidation of the soil surface, the type of erosion (sheet, rill or gully), the soil cover [49,51,55], the microrelief or surface roughness resulting from soil tillage and the size and stability of soil aggregates [55].

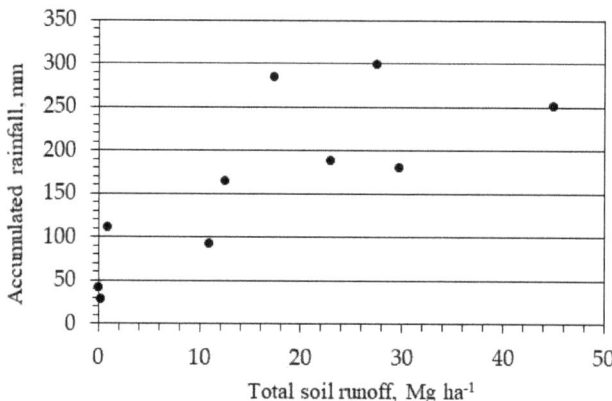

Figure 6. Total soil losses by surface runoff considering the sum of all treatments (amount of oat straw mulching) and accumulated rainfall until each sampling date.

Independent of rainfall intensity, the mulching prevented or reduced the runoff compared to bare soil, and the larger the amount of mulching, the greater the soil protection. In this sense, Suzuki et al. [54] verified less runoff under no tillage compared to conventional tillage.

The textural class of the soil runoff (Table 5) is the same one of the ridges of the peach orchard's row (Table 3). Comparing it with the soil depth 0–0.10 m of the ridges of peach orchard's row, the soil runoff has less clay, fine and very fine sand, and increases in the other particle sizes. Statistically, coarse, medium and fine sand did not differ significantly between treatments.

Table 5. [1] Particle size distribution of the soil runoff in the plots. [2] In parentheses is the percentage of increment or decrease of the particle compared to the soil depth 0–0.10 m (data available in Table 3).

Treatment	Total	Sand					Silt	Clay	Textural Classification
		Very Coarse	Coarse	Medium	Fine	Very Fine			
					%				
0 Mg ha^{-1}	64.07 c (+0.73)	10.98 b (−1.01)	14.78 a (+3.34)	12.90 a (+1.11)	14.70 a (−2.13)	10.70 a (−0.60)	28.29 a (+4.71)	7.65 ab (−5.43)	Sandy Loam
1 Mg ha^{-1}	67.32 bc (+3.98)	18.58 ab (+6.59)	15.97 a (+4.53)	12.30 a (+0.51)	12.77 a (−4.06)	7.70 b (−3.60)	24.04 bc (+0.46)	8.65 a (−4.43)	Sandy Loam
2 Mg ha^{-1}	70.13 ab (+6.79)	15.40 ab (+3.41)	15.53 a (+4.09)	13.33 a (+1.54)	15.47 a (−1.36)	10.40 ab (−0.90)	24.92 b (+1.34)	4.95 c (−8.13)	Sandy Loam
4 Mg ha^{-1}	73.73 a (+10.39)	22.33 a (+10.34)	16.20 a (+4.76)	13.13 a (+1.34)	13.80 a (−3.03)	8.27 ab (−3.03)	21.20 c (−2.38)	5.07 c (−8.01)	Sandy Loam
8 Mg ha^{-1}	67.53 bc (+4.19)	15.47 ab (+3.48)	14.87 a (+3.43)	12.40 a (+0.61)	14.60 a (−2.23)	10.20 ab (−1.10)	26.17 ab (+2.59)	6.30 bc (−6.78)	Sandy Loam

Total sand: particles with diameter between 2 and 0.05 mm; very coarse sand: diameter between 2 and 1 mm; coarse sand: diameter between 1 and 0.5 mm; medium sand: diameter between 0.5 and 0.25 mm; fine sand: diameter between 0.25 and 0.125 mm; very fine sand: diameter between 0.125 and 0.053 mm; silt: diameter between 0.053 and 0.002 mm; clay: diameter < than 0.002 mm. [1] Values obtained from a composite sample of soil runoff in each sampling date. [2] Calculation considering particle size of the soil runoff–particle size of the soil depth 0–0.10 m. Means followed by same letters in each column do not differ statistically from each other by the Tukey's test at 5% significance.

The soil runoff is basically composed of the topsoil of the ridges, generally with a larger amount of organic matter and nutrients (Table 6). In general, comparing with the soil depth 0–0.10 m of the ridges of peach orchard's row, the soil from runoff was more acid and, consequently, with slightly higher Al concentration and Al saturation. In addition, Na and K concentration was higher in the soil runoff than the 0–0.10 m depth.

Table 6. [1] Chemical characterization of the soil runoff in the plots. [2] In parentheses is the percentage of increase or decrease of the chemical element compared to the soil depth 0–0.10 m (data available in the Table 1) and the interpretation of the soil fertility [37].

Soil Attribute	Unit	Treatment (Amount of Oats Mulching)				
		0 Mg ha^{-1}	1 Mg ha^{-1}	2 Mg ha^{-1}	4 Mg ha^{-1}	8 Mg ha^{-1}
SOM	g kg^{-1}	27.6 (0.00/medium)	2.90 (+0.14/medium)	2.49 (−0.27/low)	2.76 (0.00/medium)	2.90 (+0.14/medium)
P-Melich	mg dm^{-3}	70.7 (−37.4/very high)	27.3 (−80.8/high)	146.5 (+38.4/very high)	122.3 (+14.2/very high)	97.0 (−11.1/very high)
Exch. K	mg dm^{-3}	103 (+8/high)	133 (+38/very high)	141 (+46/very high)	154 (+59/very high)	171 (+76/very high)
Ca	cmol$_c$ dm^{-3}	8.5 (+0.7/high)	7.8 (0.0/high)	7.8 (0.0/high)	8.1 (+0.3/high)	7.0 (−0.8/high)
Mg	cmol$_c$ dm^{-3}	2.9 (0.0/high)	2.7 (−0.2/high)	2.7 (−0.2/high)	2.7 (−0.2/high)	2.4 (−0.5/high)
Na	mg dm^{-3}	32 (+21)	32 (+21)	43 (+32)	35 (+24)	35 (+24)
Al	cmol$_c$ dm^{-3}	0.1 (+0.1)	0.1 (+0.1)	0.1 (+0.1)	0.1 (+0.1)	0.1 (+0.1)
H+Al	cmol$_c$ dm^{-3}	2.0 (0.0)	2.5 (+0.5)	2.0 (0.0)	2.0 (0.0)	2.2 (+0.2)
CECeffective	cmol$_c$ dm^{-3}	11.9 (+0.9)	11.1 (+0.1)	11.1 (+0.1)	11.4 (+0.4)	10.1 (−0.9)
CECpH7.0	cmol$_c$ dm^{-3}	13.8 (+0.8/medium)	13.5 (+0.5/medium)	13.0 (0.0/medium)	13.3 (+0.3/medium)	12.2 (−0.8/medium)
pH water 1:1		6.0 (−0.4/medium)	5.7 (−0.7/medium)	6.0 (−0.4/medium)	5.7 (−0.7/medium)	5.7 (−0.7/medium)
AlS	%	0.8 (+0.8/very low)	0.9 (+0.9/very low)	0.9 (+0.9/very low)	0.9 (+0.9/very low)	1.0 (+1.0/low)
BS	%	86 (+1/high)	81 (−4/high)	85 (0/high)	85 (0/high)	82 (−3/high)

SOM: soil organic matter; P: phosphorus; Exch. K: exchangeable potassium; Ca: calcium; Mg: magnesium; Na: sodium; Al: aluminum; H + Al: potential acidity; CEC: cation exchange capacity; AlS: aluminum saturation; BS: base saturation. [1] Values obtained from a composite sample of soil runoff in each sampling date. [2] Calculation considering chemical element of the soil runoff–chemical element of the soil depth 0–0.10 m.

The other variables, such as base saturation and Ca, Mg and organic matter levels (except 0 Mg ha^{-1}) (Table 6), did not present larger differences than 0–0.10 m depth, and may be associated with the lower clay content in the soil runoff (Table 5) compared to the 0–0.10 m depth (Table 3), since the reactivity and cation exchange capacity (CEC) of soil are derived from the clay. Troeh and Thompson [39] cite that the sequence of attractive forces between a cation and a micelle is the following one: $Al^{3+} > Ca^{2+} > Mg^{2+} > K^+ = NH_4^+ > Na^+$.

The soil surface has organic matter and nutrients, and in agricultural areas it has seeds, fertilizers and agrochemicals as well, and depending on soil runoff, this material may be carried to down in the relief, and may pollute and degrade soil and rivers, decrease the soil capacity of yield and increase costs of production, because it may be necessary for the addition of more fertilizers and interventions to stop soil erosion. Suzuki et al. [74] verified high concentrations of nutrients in the soil runoff, with the prevalence of silt and clay, in areas under annual crops. This has a strong relation with particle size due to CEC. The cations are adsorbed to the negative charges of the soil, and control the availability of Ca, Mg, K, Na, NH_4 and Al [75].

Along with mulching increments, the available nutrients P, K, Ca and Mg decreased in the soil runoff (Table 7). This was especially true for Ca; it presented expressive losses in surface runoff, followed by Mg, which was associated with the larger amount of these elements in the soil compared to P and K.

Table 7. Available nutrient losses in surface runoff according to their concentration in the soil runoff.

	P	K	Ca	Mg
Treatment	kg ha^{-1}			
0 Mg ha^{-1}	8.6	12.5	206.0	42.6
1 Mg ha^{-1}	0.7	3.5	41.0	8.6
2 Mg ha^{-1}	1.1	1.1	11.8	2.5
4 Mg ha^{-1}	0.9	1.2	12.3	2.5
8 Mg ha^{-1}	0.5	0.8	6.9	1.4

P: phosphorus; K: potassium; Ca: calcium; Mg: magnesium.

The losses for erosion are variable but the total number of bases lost in eroded soils may be almost the same number being exported by the harvested plants [39].

The cost to replace lost nutrients (Table 7) via mineral fertilizer (using respectively, superphosphate triple-41% P_2O_5, potassium chloride-50% K, dolomitic limestone-32% CaO + 6% MgO) would be US$ 75.4 per hectare, considering the larger losses for no-mulching (Table 8), and this cost would be reduced to US$ 2.70 per hectare for 8 Mg ha^{-1} oat straw mulching. It is important to highlight that other costs, such as transport and application of the fertilizer, fuel, depreciation, and others, were not considered in this cost, besides the impacts to the environment.

Table 8. Amount of mineral fertilizer necessary to replace the available nutrients lost in surface runoff and the cost of fertilizer, considering the treatments with larger (0 Mg ha^{-1} oat straw mulching) and smaller (8 Mg ha^{-1} oat straw mulching) losses.

Variables	Treatment	
	0 Mg ha^{-1}	**8 Mg ha^{-1}**
Superphosphate triple (41% P_2O_5), kg ha^{-1}	9	1
Potassium chloride (50% K), kg ha^{-1}	21	1
Dolomitic limestone (32% CaO + 6 % MgO), kg ha^{-1}	460	15
Cost of Superphosphate triple (US$ 240.00/ton), US$/ha	2.18	0.12
Cost of Potassium chloride (US$ 202.50/ton), US$/ha	4.19	0.28
Cost of Dolomitic limestone (US$ 150.00/ton), US$/ha	69.05	2.30
Total cost with mineral fertilizer, US$/ha	75.4	2.70

The cost of soil erosion varies according to its clay, organic matter, nutrient contents and other characteristics of soil but, due to concentrations in the topsoil layer, a ton of eroded soil may be more fertile and therefore more valuable than a ton of soil [38].

Other studies have showed the cost of soil erosion around the world. For example, Bucur et al. [76] verified mean annual losses of 10.24 kg ha^{-1} N, 0.62 kg ha^{-1} P_2O_5, 1.38 kg ha^{-1} K_2O, 0.66 kg ha^{-1} Ca^{2+}, 0.19 kg ha^{-1} Mg^{2+} and 195.95 kg ha^{-1} humus in a wheat–maize rotation, in a Cambic chernozem of Romania. Those values, however, decreased with the increase in crop rotation (i.e., the inclusion of pea, wheat, alfalfa, and perennial grasses into the cropping system), which protected the soil against erosion.

In vineyard fields in Spain, Martínez-Casasnovas and Ramos [77] verified that soil erosion exported 14.9 kg ha^{-1} of N and 11.5 kg ha^{-1} of total P, which represented 6 and 26.1% of the annual intakes and 2.4 and 1.2% of the annual income from the sale of the grapes, respectively. On the other hand, under the perennial crops of banana or banana-coffee, Onesimus et al. [78] observed a soil loss of, respectively, 38.5, 6.6 and 0.87 Mg ha^{-1} year^{-1}, with the replacement of NPK losses, caused by erosion, equaling a cumulative cost of, respectively, US$ 16,663, 4404 and 442 ha^{-1} year^{-1}, and the authors also verified that the total cost of replacing nutrients was higher, US$ 15,451 ha^{-1} year^{-1}, in areas without conservation practices (terraces), than in areas with terraces, equaling US$ 6,058 ha^{-1} year^{-1}.

Asfaw et al. [79] cite for their study that subsidizing fertilizers for the least productive farmers is a way to replace topsoil nutrients lost by soil erosion, but it does not provide cost-effective targeting criteria, being that erosion control practices are more effective in supporting this type of farmer.

The lack of information on erosion requires farmers to adopt soil conservation practices, and not adopting such practices affects farmers and society, since the society will bear the cost of repairing the off-site damage caused by soil erosion [80].

Our results come contribute information about water erosion and soil runoff using conservation practices such as mulching in peach orchards. Especially in the implementation of the orchard, when the soil is tilled, the use of mulching is efficient in reducing soil runoff by interrill erosion and consequently the costs of fertilizers exported by runoff.

4. Conclusions

The use of oat straw mulching was efficient to protect the soil from water erosion, especially the doses ≥ 2 Mg ha^{-1}, with considerably decreasing soil runoff by interril erosion from peach orchard.

The straw mulching decomposition time is important to protect soil against rainfall. Eighty days after its addition, 2 Mg ha^{-1} of oat straw mulching was totally decomposed. Meanwhile, the decomposition of the largest added amounts of oat straw mulching (4 and 8 Mg ha^{-1}) took more than 140 days. Furthermore, we visually verified in the field that the plots with straw mulching presented a smaller incidence of spontaneous plants, and was practically null at 4 and 8 Mg ha^{-1} straw mulching.

The textural class of the soil runoff is the same one of the ridges of peach orchard's row (sandy loam) but, with less clay and fine and very fine sand, and with increases in silt, and medium–large–very large sand compared with the topsoil of the ridges of peach orchard's row.

Compared with the topsoil of the ridges of peach orchard's row, the soil runoff is enriched with Na and K, but with more acid and with slightly larger Al concentrations and Al saturations.

With the incremental increase in straw mulching, the available nutrients P, K, Ca and Mg decreased in the soil runoff by interril erosion, and the cost to replace these nutrients via mineral fertilizer (using, respectively, superphosphate triple-41% P_2O_5, potassium chloride-50% K, dolomitic limestone-32% CaO + 6% MgO) is US$ 75.4 per hectare, considering the larger losses for no mulching, and this cost is reduced to US$ 2.70 per hectare for 8 Mg ha^{-1} oat straw mulching. We did not consider other costs such as transport and application of the fertilizer, fuel, and depreciation, nor did we assess the impacts on the environment.

Author Contributions: Conceptualization, L.E.A.S.S. and R.d.L.d.A.; methodology, L.E.A.S.S.; formal analysis, L.E.A.S.S. and R.d.L.d.A.; investigation, L.E.A.S.S., R.d.L.d.A. and W.R.d.S.A., M.F.R.; resources, L.E.A.S.S.; data curation, L.E.A.S.S.; writing—original draft preparation, L.E.A.S.S., R.d.L.d.A. and M.R.N.; writing—review and editing, L.E.A.S.S., R.d.L.d.A., W.R.d.S.A., M.F.R. and M.R.N.; funding acquisition, L.E.A.S.S. All authors have read and agreed to the published version of the manuscript.

Funding: This research was funded by CNPq–National Council for Scientific and Technological Development, funding number 484294/2013-0.

Data Availability Statement: Data available on request from the authors.

Acknowledgments: The authors would like to acknowledge the farmer for the support and kindness in authorizing our fieldwork in his orchard; the EMATER/RS-ASCAR of Pelotas City, "Rio Grande do Sul" State, Brazil, for support in choose the farm and appointments and discussions about the necessity and importance of the study to the farmers; and the CNPq–National Council for Scientific and Technological Development for financial support.

Conflicts of Interest: The authors declare no conflict of interest.

References

1. Karlen, D.L.; Rice, C.W. Soil degradation: Will humankind ever learn? *Sustainability* **2015**, *7*, 12490–12501. [CrossRef]
2. FAO—Food and Agriculture Organization of the United Nations, 2015. International Year of Soil Conference. 2015. Available online: http://www.fao.org/soils-2015/events/detail/en/c/338738/ (accessed on 30 August 2021).
3. IPCC—Intergovernmental Panel on Climate Change. Climate change and land. An IPCC Special Report on Climate Change, Desertification, Land Degradation, Sustainable Land Management, Food Security, and Greenhouse gas Fluxes in Terrestrial Ecosystems. Summary for Policymakers. 2019; 41p. Available online: https://www.ipcc.ch/srccl/ (accessed on 24 August 2021).
4. Yang, D.; Kanae, S.; Oki, T.; Koike, T.; Musiake, K. Global potential soil erosion with reference to land use and climate changes. *Hydrol. Process.* **2003**, *17*, 2913–2928. [CrossRef]

5.   Olsson, L.; Barbosa, H.; Bhadwal, S.; Cowie, A.; Delusca, K.; Flores-Renteria, D.; Hermans, K.; Jobbagy, E.; Kurz, W.; Li, D.; et al. Land Degradation. In *Climate Change and Land: An IPCC Special Report on Climate Change, Desertification, Land Degradation, Sustainable Land Management, Food Security, and Greenhouse Gas Fluxes in Terrestrial Ecosystems*; Shukla, P.R., Skea, J., Buendia, E.C., Masson-Delmotte, V., Pörtner, H.O., Roberts, D.C., Zhai, P., Slade, R., Connors, S., Van Diemen, R., et al., Eds.; Intergovernmental Panel on Climate Change: Geneva, Switzerland, 2019; pp. 345–436.
6.   IBGE—Instituto Brasileiro de Geografia e Estatística, Diretoria de Pesquisas, Coordenação de Agropecuária, Produção Agrícola Municipal. 2016. Áreas destinada à colheita e colhida, quantidade produzida, rendimento médio e valor da produção de pêssego segundo as grandes regiõ e unidades da federação produtora —Brasil 2016. Available online: https://www.ibge.gov. br/estatisticas/economicas/agricultura-e-pecuaria/9117-producao-agricola-municipal-culturas-temporarias-e-permanentes. html?utm_source=landing&utm_medium=explica&utm_campaign=producao_agropecuaria&t=downloads (accessed on 25 August 2021).
7.   Jie, Y.; Haijin, Z.; Xiaoan, C.; Le, S. Effects of tillage practices on nutrient loss and soybean growth in red-soil slope farmland. *Int. Soil Water Conserv. Res.* **2013**, *1*, 49–55. [CrossRef]
8.   Wolstenholme, B.N.; Moore-Gordon, C.; Ansermino, S.D. Some pros and cons of mulching avocado orchards. *South Afr. Avocado Grow. Assoc. Yearb.* **1996**, *19*, 87–91.
9.   Atucha, A.; Merwin, I.A.; Brown, M.G.; Gardiazabal, F.; Mena, F.; Adriazola, C.; Lehmann, J. Soil erosion, runoff and nutrient losses in an avocado (*Persea americana* Mill) hillside orchard under different groundcover management systems. *Plant Soil* **2013**, *368*, 393–406. [CrossRef]
10.   Keesstra, S.; Pereira, P.; Novara, A.; Brevik, E.C.; Azorin-Molina, C.; Parras-Alcántara, L.; Jordán, A.; Cerdà, A. Effects of soil management techniques on soil water erosion in apricot orchards. *Sci. Total Environ.* **2016**, *551–552*, 357–366. [CrossRef]
11.   Suzuki, L.E.A.S.; Reisser Júnior, C.; Miola, E.C.C.; Rostirolla, P.; Scherer, V.S.; Terra, V.S.S.; Pauletto, E.A. Efeito do manejo e da irrigação localizada sobre os atributos físicos e hídricos de um Argissolo cultivado com pessegueiro. *Pesqui. Agropecuária Gaúcha* **2021**, *27*, 127–147. [CrossRef]
12.   Stark, A.L.; Thorne, D.W. Peach orchard soil management studies. Bulletin No. 330, UAES Bulletins. Paper 291. 1948. Available online: https://digitalcommons.usu.edu/uaes_bulletins/291 (accessed on 28 January 2023).
13.   Youlton, C.; Espejo, P.; Biggs, J.; Norambuena, M.; Cisternas, M.; Neaman, A.; Salgado, E. Quantification and control of runoff and soil erosion on avocado orchards on ridges along steep-hillslopes. *Cien. Inv. Agr.* **2010**, *37*, 113–123. [CrossRef]
14.   Nunes, M.R.; Karlen, D.L.; Moorman, T.B. Tillage intensity effects on soil structure indicators—A US meta-analysis. *Sustainability* **2020**, *12*, 2071. [CrossRef]
15.   Nunes, M.R.; Karlen, D.L.; Veum, K.S.; Moorman, T.B.; Cambardella, C.A. Biological soil health indicators respond to tillage intensity: A US meta-analysis. *Geoderma* **2020**, *369*, 114335. [CrossRef]
16.   Castilho, S.C.P.; Cooper, M.; Silva, L.F.S. Micromorphometric analysis of porosity changes in the surface crusts of three soils in the Piracicaba region, São Paulo State, Brazil. *Acta Scientiarum. Agron.* **2015**, *37*, 385–395. [CrossRef]
17.   Baumhardt, R.L.; Stewart, B.A.; Sainju, U.M. North American soil degradation: Processes, practices, and mitigating strategies. *Sustainability* **2015**, *7*, 2936–2960. [CrossRef]
18.   FAO—Food and Agriculture Organization of the United Nations. *Soil Erosion: The Greatest Challenge to Sustainable Soil Management*; FAO: Rome, Italy, 2019; 100p, Available online: http://www.fao.org/3/ca4395en/ca4395en.pdf (accessed on 25 August 2021).
19.   Vicente-Vicente, J.L.; Gómez-Muñoz, B.; Hinojosa-Centeno, M.B.; Smith, P.; Garcia-Ruiz, R. Carbon saturation and assessment of soil organic carbon fractions in Mediterranean rainfed olive orchards under plant cover management. *Agric. Ecosyst. Environ.* **2017**, *245*, 135–146. [CrossRef]
20.   Walshl, B.D.; Salmins, S.; Buszard, D.J.; Mackenzie, A.F. Impact of soil management systems on organic dwarf apple orchards and soil aggregate stability, bulk density, temperature and water content. *Can. J. Soil Sci.* **1996**, *76*, 203–209. [CrossRef]
21.   Wade, M.K.; Sanchez, P.A. Mulching and green manure applications for continuous crop production in the Amazon basin. *Agron. J.* **1983**, *75*, 39–45. [CrossRef]
22.   Liu, Y.; Wang, J.; Liu, D.; Li, Z.; Zhang, G.; Tao, Y.; Xie, J.; Pan, J.; Chen, F. Straw mulching reduces the harmful effects of extreme hydrological and temperature conditions in citrus orchards. *PLoS ONE* **2014**, *9*, e87094. [CrossRef]
23.   Lordan, J.; Pascual, M.; Villar, J.M.; Fonseca, F.; Papió, J.; Montilla, V.; Rufat, J. Use of organic mulch to enhance water-use efficiency and peach production under limiting soil conditions in a three-year-old orchard. *Span. J. Agric. Res.* **2015**, *13*, e0904. [CrossRef]
24.   Bakshi, P.; Wali, V.K.; Iqbal, M.; Jasrotia, A.; Kour, K.; Ahmed, R.; Bakshi, M. Sustainable fruit production by soil moisture conservation with different mulches: A review. *Afr. J. Agric. Res.* **2015**, *10*, 4718–4729. [CrossRef]
25.   EMBRAPA—Empresa Brasileira de Pesquisa Agropecuária. *UFPEL—Universidade Federal de Pelotas. INMET—Instituo Nacional de Meteorologia. Normais Climatológicas Período: 1971/2000 (Mensal/Anual).* Available online: http://agromet.cpact.embrapa.br/ estacao/mensal.html (accessed on 2 June 2020).
26.   Soil Survey Staff. *Keys to Soil Taxonomy*, 12th ed.; USDA-Natural Resources Conservation Service: Washington, DC, USA, 2014; 142p. Available online: https://www.nrcs.usda.gov/wps/portal/nrcs/detail/soils/survey/class/taxonomy/?cid=nrcs142p2 _053580 (accessed on 8 December 2021).

27. Santos, H.G.; Jacomine, P.K.T.; Anjos, L.H.; Oliveira, V.A.; Lumbreras, J.F.; Coelho, M.R.; Almeida, J.A.; Araujo Filho, J.C.; Oliveira, J.B.; Cunha, T.J.F. *Sistema Brasileiro de Classificação de Solos*; Embrapa: Brasília, Brasil, 2018; Available online: https://ainfo.cnptia.embrapa.br/digital/bitstream/item/199517/1/SiBCS-2018-pdf (accessed on 29 November 2021)ISBN -9788570358004.

28. Tedesco, M.J.; Gianello, C.; Bissani, C.A.; Volkweiss, S.J. *Análises de Solo, Plantas e Outros Materiais*, 2nd ed.; Departamento de Solos, UFRGS: Porto Alegre, Brazil, 1995; 174p, (Boletim Técnico 5).

29. Gee, G.W.; Or, D. Particle-size analysis. In *Methods of Soil Analysis, Part 4: Physical Methods*, 5th ed.; Dane, J.H., Topp, C., Eds.; Soil Science Society of America: Madison, WI, USA, 2002; pp. 255–293.

30. Suzuki, L.E.A.S.; Reichert, J.M.; Albuquerque, J.A.; Reinert, D.J.; Kaiser, D.R. Dispersion and flocculation of Vertisols, Alfisols and Oxisols in Southern Brazil. *Geoderma Reg.* **2015**, *5*, 64–70. [CrossRef]

31. National Resource Conservation Service-NRCS/United States Department of Agriculture-USDA. Soil texture calculator. 2022. Available online: https://www.nrcs.usda.gov/wps/portal/nrcs/detail/?cid=nrcs142p2_054167 (accessed on 14 October 2022).

32. Viana, J.H.M.; Teixeira, W.G.; Donagemma, G.K. Densidade de partículas. In *Manual de Métodos de Análise de Solo, ver. ampl.*, 3rd ed.; Teixeira, P.C., Donagemma, G.K., Fontana, A., Teixeira, W.G., Eds.; Embrapa: Brasília, Brazil, 2017; pp. 76–81. Available online: http://www.infoteca.cnptia.embrapa.br/infoteca/handle/doc/1085209 (accessed on 15 April 2021).

33. Blake, G.R.; Hartge, K.H. Bulk density. In *Methods of Soil Analysis, Part 1: Physical and Mineralogical Methods*, 2nd ed.; Klute, A., Ed.; American Society of Agronomy, Soil Science Society of America: Madison, WI, USA, 1986; pp. 363–375. [CrossRef]

34. Klute, A.; Dirksen, C. Hydraulic conductivity and diffusivity: Laboratory methods. In *Methods of Soil Analysis, Part 1: Physical and Mineralogical Methods*; Klute, A., Ed.; American Society of Agronomy, Soil Science Society of America: Madison, WI, USA, 1986; pp. 687–734.

35. BRASIL. Ministério da Agricultura, Pecuária e Abastecimento. Instrução normativa nº 39, de 8 de agosto de 2018. Diário Oficial da União—Seção 1, n. 154, p. 19, 10 de agosto de 2018. Available online: https://www.in.gov.br/materia/-/asset_publisher/Kujrw0TZC2Mb/content/id/36278414/do1-2018-08-10-instrucao-normativa-n-39-de-8-de-agosto-de-2018-36278366 (accessed on 24 August 2021).

36. EMBRAPA—Empresa Brasileira de Pesquisa Agropecuária/Laboratório de Agrometeorologia. Boletim Climatológico Mensal. Available online: http://agromet.cpact.embrapa.br/online/Resumos_Mensais.htm (accessed on 2 July 2020).

37. SBCS—Sociedade Brasileira de Ciência do Solo. NRS—Núcleo Regional Sul. CQFS—Comissão de Química e Fertilidade do Solo. Manual de adubação e calagem para os estados do Rio Grande do Sul e Santa Catarina, 2016; 376p. Available online: http://www.sbcs-nrs.org.br/docs/Manual_de_Calagem_e_Adubacao_para_os_Estados_do_RS_e_de_SC-2016.pdf (accessed on 7 August 2022).

38. Troeh, F.R.; Hobbs, J.Á.; Donahue, R.L. *Soil and Water Conservation for Productivity and Environmental Protection*; Prentice-Hall: Hoboken, NJ, USA, 1980; 718p.

39. Troeh, F.R.; Thompson, L.M. *Solos e Fertilidade do Solo*; Dourado Neto, D.D.; Dourado, M.N., Translators; Organização Andrei Editora Ltda: São Paulo, Brazil, 2007.

40. Tombácz, E.; Libor, Z.; Illés, E.; Majzik, A.; Klumpp, E. The role of reactive surface sites and complexation by humic acids in the interaction of clay mineral and iron oxide particles. *Org. Geochem.* **2004**, *35*, 257–267. [CrossRef]

41. Bronick, C.J.; Lal, R. Soil structure and management: A review. *Geoderma* **2005**, *124*, 3–22. [CrossRef]

42. Wagner, S.; Cattle, S.R.; Scholten, T. Soil-aggregate formation as influenced by clay content and organic-matter amendment. *J. Plant Nutr. Soil Sci.* **2007**, *170*, 173–180. [CrossRef]

43. Reichert, J.M.; Norton, L.D.; Favaretto, N.; Huang, C.; Blume, E. Settling velocity, aggregate stability, and interrill erodibility of soils varying in clay mineralogy. *Soil Sci. Soc. Am. J.* **2009**, *73*, 1369–1377. [CrossRef]

44. Totsche, K.U.; Amelung, W.; Gerzabek, M.H.; Guggenberger, G.; Klumpp, E.; Knief, C.; Lehndorff, E.; Mikutta, R.; Peth, S.; Prechtel, A.; et al. Microaggregates in soils. *J. Plant Nutr. Soil Sci.* **2018**, *181*, 104–136. [CrossRef]

45. Igwe, C.A.; Agbatah, C. Clay and silt dispersion in relation to some physicochemical properties of derived savanna soils under two tillage management practices in southeastern Nigeria. *Acta Agric. Scand. Sect. B-Soil Plant Sci.* **2008**, *58*, 17–26. [CrossRef]

46. Parwada, C.; van Tol, J. Soil properties influencing erodibility of soils in the Ntabelanga area, Eastern Cape Province, South Africa. *Acta Agric. Scand. Sect. B-Soil Plant Sci.* **2017**, *67*, 67–76. [CrossRef]

47. Lunardi Neto, A.; Albuquerque, J.A.; De Almeida, J.A.; Mafra, Á.L.; Medeiros, J.C.; Alberton, A. Atributos físicos do solo em área de mineração de carvão influenciados pela correção da acidez, adubação orgânica e revegetação. *Revista Brasileira de Ciência do Solo* **2008**, *32*, 1379–1388. [CrossRef]

48. Resende, M.; Curi, N.; Rezende, S.B.; Corrêa, G.F. *Pedologia: Base Para Distinção de Ambientes*; UFLA: Lavras, Brazil, 2007; 322p.

49. Ramos, M.M.; Díaz, J.D.G.; Rivas, A.I.M.; Gómez, M.U.; Hernández, B.J.V.; García, P.R.; Asencio, C. Factors that influence soil hydric erosion in a temperate forest. *Rev. Mex. Cienc. For.* **2020**, *11*, 51–71. [CrossRef]

50. Farhan, Y.; Zregat, D.; Nawaiseh, S. Assessing the influence of physical factors on spatial soil erosion risk in Northern Jordan. *J. Am. Sci.* **2014**, *10*, 29–39.

51. Holz, D.J.; Williard, K.W.J.; Edwards, P.J.; Schoonover, J.E. Soil Erosion in Humid Regions: A Review. *J. Contemp. Water Res. Educ.* **2015**, *154*, 48–59. [CrossRef]

52. Shojaei, S.; Kalantari, Z.; Rodrigo-Comino, J. Prediction of factors afecting activation of soil erosion by mathematical modeling at pedon scale under laboratory conditions. *Sci. Rep.* **2020**, *10*, 20163. [CrossRef]

53. Pijl, A.; Reuter, L.E.H.; Quarella, E.; Vogel, T.A.; Tarolli, P. GIS-based soil erosion modelling under various steep-slope vineyard practices. *Catena* **2020**, *193*, 104604. [CrossRef]
54. Suzuki, L.E.A.S.; Matieski, T.; Strieder, G.; Pauletto, E.A.; Bordin, S.S.; Lima, L.S.C.; Collares, G.L.; Dai Prá, M. Perdas de solo por erosão hídrica e granulometria do material erodido em propriedades agrícolas. In *X ENES—Encontro Nacional de Engenharia de Sedimentos: Artigos selecionados, Capítulo X*; Poleto, C., Pletsch, A.L., Mello, E.L., Carvalho, N.O., Eds.; ABRH: Porto Alegre, Brazil, 2012; pp. 93–108.
55. Peng, L.; Tang, C.; Zhang, X.; Duan, J.; Yang, L.; Liu, S. Quantifying the effects of root and soil properties on soil detachment capacity in agricultural land use of Southern China. *Forests* **2022**, *13*, 1788. [CrossRef]
56. Ramos, M.F.; Almeida, W.R.S.; Amaral, R.L.; Suzuki, L.E.A.S. Degree of compactness and soil quality of peach orchards with different production ages. *Soil Tillage Res.* **2022**, *219*, 105324. [CrossRef]
57. Socci, P.; Errico, A.; Castelli, G.; Penna, D.; Preti, F. Terracing: From agriculture to multiple ecosystem services. *Oxf. Res. Encycl. Environ. Sci.* **2019**. [CrossRef]
58. Rutebuka, J.; Uwimanzi, A.M.; Nkundwakazi, O.; Kagabo, D.M.; Mbonigaba, J.J.M.; Vermeir, P.; Verdoodt, A. Effectiveness of terracing techniques for controlling soil erosion by water in Rwanda. *J. Environ. Manag.* **2021**, *277*, 111369. [CrossRef]
59. Pijl, A.; Wang, W.; Straffelini, E.; Tarolli, P. Soil and water conservation in terraced and non-terraced cultivations: An extensive comparison of 50 vineyards. *Land Degrad. Dev.* **2022**, *33*, 596–610. [CrossRef]
60. Rodrigo-Comino, J.; Seeger, M.; Iserloh, T.; González, J.M.S.; Ruiz-Sinoga, J.D.; Ries, J.B. Rainfall-simulated quantification of initial soil erosion processes in sloping and poorly maintained terraced vineyards—Key issues for sustainable management systems. *Sci. Total Environ.* **2019**, *660*, 1047–1057. [CrossRef]
61. Al- Siaede, R. Using Landscape analysis techniques to prevent silt accumulation in the reservoir of the Dwerige weir project and developing River basin, Missan, South Eastern IRAQ. *Iraqi J. Sci.* **2022**, *63*, 3031–3039. [CrossRef]
62. Giambastiani, Y.; Biancofiore, G.; Mancini, M.; Di Giorgio, A.; Riccardo Giusti, R.; Cecchi, S.; Gardin, L.; Errico, A. Modelling the effect of keyline practice on soil erosion control. *Land* **2023**, *12*, 100. [CrossRef]
63. González-Rosado, M.; Parras-Alcántara, L.; Aguilera-Huertas, J.; Lozano-García, B. Soil productivity degradation in a long-term eroded olive orchard under semiarid mediterranean conditions. *Agronomy* **2021**, *11*, 812. [CrossRef]
64. Wang, B.; Niu, J.; Berndtsson, R.; Zhang, L.; Chen, X.; Li, X.; Zhu, Z. Efficient organic mulch thickness for soil and water conservation in urban areas. *Sci. Rep.* **2021**, *11*, 6259. [CrossRef]
65. Lalljee, B. Mulching as a mitigation agricultural technology against land degradation in the wake of climate change. *Int. Soil Water Conserv. Res.* **2013**, *1*, 68–74. [CrossRef]
66. Granatstein, D.; Sánchez, E. Research knowledge and needs for orchard floor management in organic tree fruit systems. *Int. J. Fruit Sci.* **2009**, *9*, 257–281. [CrossRef]
67. Dechen, S.L.F.; Lombardi Neto, F.; Castro, O.M. Gramíneas e leguminosas e seus restos culturais no controle da erosão em Latossolo Roxo. *Revista Brasileira de Ciência do Solo* **1981**, *5*, 133–137.
68. Amado, T.J.C.; Matos, A.T.; Torres, L. Flutuação de temperatura e umidade do solo sob preparo convencional e em faixas na cultura da cebola. *Pesqui. Agropecuária Bras.* **1990**, *25*, 625–631.
69. Sasal, M.C.; Castiglioni, M.G.; Wilson, M.G. Effect of crop sequences on soil properties and runoff on natural-rainfall erosion plots under no tillage. *Soil Tillage Res.* **2010**, *108*, 24–29. [CrossRef]
70. Bertoni, J.; Lombardi Neto, F. *Conservação do Solo*; Ícone: São Paulo, Brazil, 1999; 355p.
71. Pruski, F.F. (Ed.) *Conservação de Solo e Água: Práticas Mecânicas para o Controle da Erosão Hídrica*, 2nd ed.; UFV: Viçosa, Brazil, 2009; 279p.
72. Rio Grande do Sul. Atlas Socioeconômico do Rio Grande do Sul. Clima, temperatura e precipitação. 7th ed., 2022. Available online: https://atlassocioeconomico.rs.gov.br/clima-temperatura-e-precipitacao (accessed on 26 December 2022).
73. Volk, L.B.S.; Cogo, N.P. Relações entre tamanho de sedimentos erodidos, velocidade da enxurrada, rugosidade superficial criada pelo preparo e tamanho de agregados em solo submetido a diferentes manejos. *Revista Brasileira Ciência Solo* **2009**, *33*, 1459–1471. [CrossRef]
74. Suzuki, L.E.A.S.; Bordin, S.S.; Matieski, T.; Rostirolla, P.; Strieder, G.; Nunes, M.R. Soil and nutrient losses by runoff from farmlands in Southern Brazil. *Rev. Ciências Agroambientais* **2021**, *19*, 1–15.
75. Ernani, P.R. *Química do Solo e Disponibilidade de Nutrientes*; O Autor: Lages, Brazil, 2008; 230p.
76. Bucur, D.; Jitareanu, G.; Ailincai, C.; Tsadilas, C.; Ailincai, D.; Mercus, A. Influence of soil erosion on water, soil, humus and nutrient losses in different crop systems in the Moldavian Plateau, Romania. *J. Food Agric. Environ.* **2007**, *5*, 261–264.
77. Martínez-Casasnovas, J.A.; Ramos, M.C. The cost of soil erosion in vineyard fields in the Penedès-Anoia Region (NE Spain). *Catena* **2006**, *68*, 194–199. [CrossRef]
78. Onesimus, S.; Kimaro, D.; Kasenge, V.; Isabirye, M.; Makhosi, P. Soil and nutrient losses in banana-based cropping systems of the Mount Elgon hillsides of Uganda: Economic implications. *Int. J. Agric. Sci.* **2012**, *2*, 256–262.

79. Asfawa, S.; Pallante, G.; Palma, A. Distributional impacts of soil erosion on agricultural productivity and welfare in Malawi. *Ecol. Econ.* **2020**, *177*, 106764. [CrossRef]

80. Telles, T.S.; Dechen, S.C.F.; De Souza, L.G.A.; Guimarães, M.F. Valuation and assessment of soil erosion costs. *Scientia Agricola* **2013**, *70*, 209–216. [CrossRef]

 *soil systems*

Article

Challenges in the Management of Environmentally Fragile Sandy Soils in Southern Brazil

Luis Eduardo Akiyoshi Sanches Suzuki [1,*], Fabrício de Araújo Pedron [2], Rodrigo Bomicieli de Oliveira [3] and Ana Paula Moreira Rovedder [2]

[1] Center of Technological Development, Federal University of Pelotas, Pelotas 96010-610, RS, Brazil
[2] Rural Sciences Center, Federal University of Santa Maria, Santa Maria 97105-9000, RS, Brazil
[3] EMATER/RS-ASCAR, Restinga Seca 97200-000, RS, Brazil
* Correspondence: luis.suzuki@ufpel.edu.br

Abstract: Quartzipsamments are environmentally fragile soils, being highly susceptible to water and wind erosion. Despite this, it seems that political and economic issues favor the advancement of agriculture in these soils. Therefore, studies are necessary for a better understanding of these soils and to minimize the impacts of land use. This work aims to characterize the morphological, physical–hydric, and chemical properties of Quartzipsamments under sandyzation in southwest Rio Grande do Sul State, Brazil. Soil morphology was evaluated in six profiles in areas under native field with the presence of gullies, and soil samples with preserved and non-preserved structures were collected to evaluate the physical–hydric and chemical properties. We verified that these soils have high macroporosity (0.253 to 0.373 m^3 m^{-3}) and saturated hydraulic conductivity (127.85 to 672.26 mm h^{-1}), and predominantly low organic matter (0.05 to 2.36%) and clay (23.03 to 126.29 g kg^{-1}) content, but correlation analysis showed that increasing pH and organic matter can improve the fertility of these soils. Quartzipsamments have a low volume of available water to plants (0.006 to 0.038 m^3 m^{-3}) and have a potential risk of leaching and aquifer contamination. The use of these soils demands the adoption of conservation practices.

Keywords: soil erosion; soil conservation; physical–hydric properties; soil morphology; soil fertility; Quartzipsamments; sandyzation

Citation: Suzuki, L.E.A.S.; Pedron, F.d.A.; Oliveira, R.B.d.; Rovedder, A.P.M. Challenges in the Management of Environmentally Fragile Sandy Soils in Southern Brazil. *Soil Syst.* 2023, 7, 9. https://doi.org/10.3390/soilsystems7010009

Academic Editor: Adriano Sofo

Received: 14 December 2022
Revised: 26 January 2023
Accepted: 30 January 2023
Published: 2 February 2023

1. Introduction

The southwestern region of Rio Grande do Sul State, Southern Brazil, is part of the Pampa Biome, a natural ecosystem rich in biological and pedological diversity with environmental, economic, and socio-cultural importance for Brazil, Uruguay, and Argentina [1], but the fragility of the soil, flora, and fauna makes the Pampa Biome vulnerable to agriculture conversion and degradation [2]. In this region, there are extensive areas in the process of sandyzation [3]. According to Bellanca and Suertegaray [4], such conditions have been interpreted in various ways, from natural origin and resulting from water processes acting on the lithology and specific soils, to anthropic causes associated with overgrazing and land use without conservation practices. In addition to these, Caneppele [5] also refers to the introduction of wheat and soybean crops and eucalyptus monocultures in the region as conditioning factors of erosion processes in these areas, as well as the use of heavy machinery and non-conservationist soil practices since 1970, causing soil compaction, the creation of preferential paths for drainage, and soil exposure through plowing or suppression of vegetation.

Quartzipsamments are predominantly in areas of the sandyzation process, and they have a sandy texture and a fragile structure, and are not very resistant to wind and water erosion [2], besides having low water availability, low natural fertility, and low cation exchange capacity [1,6,7].

Agricultural expansion in Quartzipsamments has been taking place [5], and few studies have been carried out on this soil and its use and management. In this sense, a better understanding of these environments is necessary so that the impacts of land use are minimized, since the intensive non-sustainable use of the land has taken negative consequences in this biome [2], at levels that are difficult to control.

In Brazil, Quartzipsamments together with other sandy and sandy loam soils represent 8% of the territory [8]. The authors state that in the past, these soils were of little agricultural relevance due to their limitations, even in areas favorable to mechanization, but currently, agriculture is establishing itself in these areas due to advances in production systems and agricultural practices. FAO [9] considers these soils as part of the group of Arenosols, covering about 900 million hectares or 7% of the earth's surface. In this sense, the study and understanding of the behavior and processes involved in sandy soils become relevant, either for its representative area in terms of Brazil and the world, or the advance of agricultural exploitation of these soils of low agricultural land suitability and high environmental fragility.

The fragility of the sandy areas, due to the morphogenetic soil characteristics where the ravine and gully processes are present, conditions a high risk of landscape degradation through the occupation of the soil by crops and forestation [10,11]. In a review of recent studies of sandy soils (considered by the authors those with sand > 50% and clay < 20%), Huang and Hartemink [12] consider these soils as more sensitive to climate change and anthropic activities when compared to others, and due to population growth and urbanization, they have been widely used in the supply of food and other products and services for society.

Although Quartzipsamments are environmentally fragile, they have been widely used for agricultural purposes. Despite their low agricultural land suitability, it seems that political and economic issues favor the advance of the agricultural use of Quartzipsamments in southern Brazil, but little information is available about these soils, especially their morphology, fertility, and physical–hydric properties, as well as practices for better agricultural use. Thus, this work aims to characterize the morphological, chemical, and physical–hydric properties of Quartzipsamments; point out some difficulties and challenges in the use and management of these soils; and propose strategies for better soil use. Our hypothesis is that Quartzipsamments are soils of low suitability for agricultural use due to their low fertility and available water to plants, besides being unstructured due to the low content of organic matter and clay, demanding conservation practices to improve their properties.

2. Materials and Methods

2.1. Sampling Sites

The study was conducted in areas under sandyzation in southwest Rio Grande do Sul State, southern Brazil, specifically in the cities of Quaraí, Manoel Viana, and São Francisco de Assis (Figure 1).

The annual average temperature and precipitation in the region are, respectively, around 17.8 °C and 1388 mm; torrential rains larger than 160 mm may occur in 24 h and frosts from April to November [13], and the mean monthly rainfall and temperature to the period 1981–2010 are presented in Figure 2.

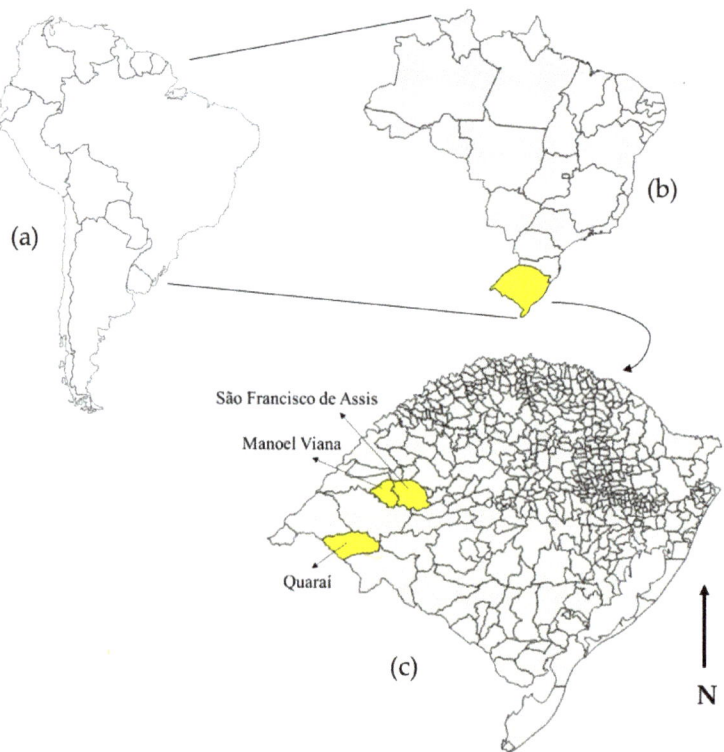

Figure 1. Map without scale from South America (**a**) Brazil and Rio Grande do Sul State highlighted (yellow) (**b**) and the cities of Quaraí (latitude 30°23′17″ S; longitude 56°29′56″ W; 112 m mean altitude; area ≅ 3238 km^2), Manoel Viana (latitude 29°35′07″ S; longitude 55°29′13″ W; 113 m mean altitude; area ≅ 1391 km^2), and São Francisco de Assis (latitude 29°33′01″ S; longitude 55°07′52″ W; 125 m mean altitude; area ≅ 2507 km^2), Rio Grande do Sul State (**c**).

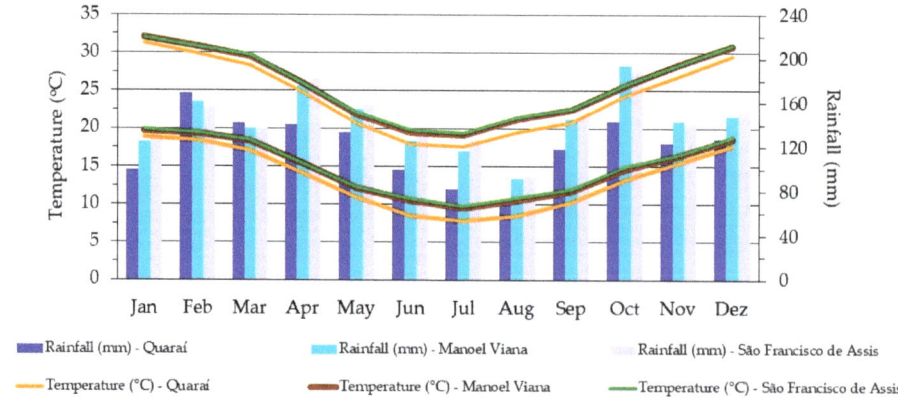

Figure 2. Mean monthly rainfall and temperature for the period 1981–2010. Climatological mean based on 30 years of data (1981–2010), using official INMET stations, and later interpolating for locations that do not have a meteorological data measurement station. Source: [14].

Six sites with Neossolos Quartzarênicos (NQ) (Brazilian Soil Classification System [15]), or Quartzipsamments, for "US Soil Taxonomy" [16], were chosen.

The sampling sites occur in undulating to slightly undulating relief, and the soils were sampled in areas under native field with the presence of gullies. Quartzipsamments NQ1 (Datum: UTM—WGS-84, Zone 21J, longitude 571,558 m E; latitude 6,629,425 m S; 146 m altitude) and NQ2 were sampled in Quaraí city, NQ3 (Datum: UTM—WGS-84, Zone 21J, longitude 657,414 m E; latitude 6,725,226 m S; 117 m altitude) and NQ4 (Datum: UTM—WGS-84, Zone 21J, longitude 655,845 m E; latitude 6,717,142 m S; 114 m altitude) in Manoel Viana city, and NQ5 (Datum: UTM—WGS-84, Zone 21J, longitude 678,086 m E; latitude 6,725,002 m S; 140 m altitude) and NQ6 (Datum: UTM—WGS-84, Zone 21J, longitude 682,571 m E; latitude 6,724,585 m S; 109 m altitude) in São Francisco de Assis city (Figure 3).

(a) (b)

(c) (d)

Figure 3. *Cont.*

(e)

(f)

(g)

(h)

(i)

(j)

Figure 3. *Cont.*

(k)   (l)

Figure 3. Quartzipsamments profiles NQ1 and NQ2 from Quaraí city (respectively, (**a,c**)), NQ3 and NQ4 from Manoel Viana city (respectively, (**e,g**)), NQ5 and NQ6 from São Francisco de Assis city (respectively, (**i,k**)), Rio Grande do Sul State, and their respective landscape (respectively, (**b,d,f,h,j,l**)). Each color on the measuring tape represents 10 cm in the profile pictures. Source: pictures taken by F.d.A. Pedron and L.E.A.S. Suzuki.

2.2. Morphological Analysis

The six soil profiles were described in the field according to Santos et al. [17] and Schoeneberger et al. [18], considering morphological procedures such as horizon sequence and depth, its boundary, texture, structure, consistency, and soil Munsell color.

2.3. Physical–Hydric Analysis

In each sampling site, the horizons of the profile were separated, and in the middle of the horizon, three samples with preserved structure by horizon were collected in cylinders with 0.047 m diameter and 0.030 m height, and one sample by horizon with a non-preserved structure.

The samples with a preserved structure were saturated for capillarity under 48 h. The saturated hydraulic conductivity of soil was determined after saturation in the laboratory using a permeameter of constant load [19].

Next, the samples were submitted and equilibrated in the tensions of 1 kPa and 6 kPa in the tension table and in the tensions of 10, 100, 500, and 1500 kPa in the Richards' pressure chamber [20]. Finally, the samples were oven-dried at a temperature of 105 °C. Using this information, the macroporosity (pores of diameter larger than 50 μm) to the tension of 6 kPa, the microporosity (pores of diameter smaller than 50 μm), the total porosity, the bulk density [21], and the volume of available water using the volumetric moisture between the field capacity (tension = 10 kPa) and the permanent wilt point (tension = 1500 kPa) were calculated.

The volumetric moisture was obtained by the ratio between the water retained in a determined tension and the volume of the cylinder used for sampling.

In the laboratory, the soil samples with a non-preserved structure were air-dried, broken individually and manually, and passed through a sieve of 2 mm mesh, the soil that passed through the sieve being used to determine the particle density by the volumetric balloon method [22], and the particle size distribution analysis using the pipette method [23]. The soil particles were separated in the fraction sand (2–0.053 mm) by sieving, silt (0.053–0.002 mm) by calculus between the difference of the sum of sand, and clay (<0.002 mm), which was determined by a pipette. The sand was sieved in very coarse sand

(2–1 mm), coarse sand (1–0.5 mm), medium sand (0.5–0.25 mm), fine sand (0.25–0.125 mm), and very fine sand (0.125–0.053 mm).

The results of the particle size distribution analysis were used to determine textural classification, using the soil texture triangle available from the USDA-NRCS [24] and according to Santos et al. [15].

Dispersible clay in water was quantified following the same procedure used for total clay evaluation but without using the chemical dispersant.

The degree of flocculation (DF, %) was calculated using the following equation:

$$DF = [(\text{total clay} - \text{clay disperse in water})/\text{total clay}] \times 100 \qquad (1)$$

2.4. Chemical Analysis

The soil samples with a non-preserved structure were also used for chemical characterization using the analytical procedures presented in Tedesco et al. [25] to determine: pH in water 1:1 (soil/water) (pH water) and KCl (pH KCl), calcium (Ca), magnesium (Mg), sodium (Na), potassium (K), aluminum (Al), potential acidity (cations H + Al), and organic carbon. Through these determinations were calculated the effective cation exchange capacity at pH 7.0 (respectively, CECeffective and CECpH7.0), base saturation, and aluminum saturation.

H + Al was determined for SMP index, while Ca, Mg, and Al were extracted with KCl 1 mol L^{-1} and measured with an atomic absorption spectrophotometer, and the extractant Mehlich I solution (0.05 mol L^{-1} HCl + 0.0125 mol L^{-1} H_2SO_4) was used for K and Na and measured by flame photometry.

Soil organic carbon was analyzed by the wet combustion method [25], and then soil organic matter was calculated according to Tedesco et al. [26]:

$$\text{Soil organic matter} = 1.724 \times \text{soil organic carbon} \qquad (2)$$

2.5. Data Analysis

The statistical analysis was realized by Pearson's correlation considering 10% significance, and coefficient of variation and mean values were determined using the statistical program SAS [27].

3. Results

The morphological data (Table 1) show that all Quartzipsamments profiles have a weak structure in subangular blocks that break down into single grains. They are very deep soils (>100 cm), with colors ranging from reddish (2.5 YR) to yellowish (10 YR), always with high values and chromas (≥4). The color indicates the good drainage of the analyzed profiles. The consistency verified was loose and non-plastic and non-sticky for virtually all profiles. The horizon boundary distinctness varied predominantly between clear and gradual.

Table 1. Morphological data of the Quartzipsamments profiles of the Southwest of Rio Grande do Sul State, southern Brazil.

Horizon	Depth (cm)	Moist Color	[1] Moist Consistency	Wet Consistency	Horizon Boundary	Structure
			NQ1			
Ap	0–20	10YR 6/4	L	NP/NS	I-C	SB-SG
A	20–65	10YR 4/4	VF	NP/NS	S-G	SB-SG
C1	65–94	10YR 5/6	L-VF	NP/NS	S-G	SB-SG
C2	94–150	10YR 6/6	L-VF	NP/NS	S-G	SB-SG

Table 1. *Cont.*

Horizon	Depth (cm)	Moist Color	[1] Moist Consistency	Wet Consistency	Horizon Boundary	Structure
			NQ2			
A	0–12	10YR 4.5/5	L	NP/NS	I-A	SB-SG
C1	12–42	10YR 4/5	L	NP/NS	S-C	SB-SG
C2	42–85	10YR 4/5	L	NP/NS	S-G	SB-SG
C3	85–145	10YR 5/8	L	NP/NS	S-G	SB-SG
			NQ3			
A	0–22	7.5YR 5/7	L	NP/NS	S-C	SB-SG
C1	22–53	7.5YR 5/6	L	NP/NS	S-C	SB-SG
C2	53–93	7.5YR 4.5/6	L	NP/NS	S-G	SB-SG
C3	93–150	7.5YR 7/4	L	NP/NS	S-G	SB-SG
			NQ4			
A	0–15	2.5YR 3.5/4	L	NP/NS	S-C	SB-SG
C1	15–40	2.5YR 4/4	L	NP/NS	S-C	SB-SG
C2	40–72	2.5YR 4/6	L	NP/NS	S-C	SB-SG
C3	72–140	2.5YR 4/7	L	NP/NS	S-C	SB-SG
			NQ5			
A	0–20	5YR 5/8	L	NP/NS	S-C	SB-SG
C1	20–66	5YR 5/8	L	NP/NS	S-G	SB-SG
C2	66–100	5YR 5/8	L	NP/NS	S-C	SB-SG
C3	100–170	5YR 5/8	L	NP/NS	S-C	SB-SG
			NQ6			
A1	0–18	5YR 5/4	L	NP/NS	S-C	SB-SG
A2	18–43	5YR 5/6	L	NP/NS	S-G	SB-SG
C1	43–60	5YR 5/8	L	NP/NS	S-G	SB-SG
C2	60–93	5YR 4/8	L	NP/NS	S-A	SB-SG
C3	93–125	5YR 4.5/8	L	NP/NS	S-C	SB-SG
C4	125–170	5YR 4/8	L	NP/NS	S-C	SB-SG

[1] Consistency: L—Loose, VF—Very Friable, NP—Non-plastic, NS—Non-sticky; Horizon boundary: A—Abrupt, C—Clear. G—Gradual, D—Diffuse, S—Smooth, I—Irregular; Structure: SB—Subangular Blocky, SG—Single Grain. NQ1 and NQ2: Quartzipsamments from Quaraí city; NQ3 and NQ4: Quartzipsamments from Manoel Viana city; NQ5 and NQ6: Quartzipsamments from São Francisco de Assis city.

The particle size distribution is shown in Tables 2 and 3. There is a predominance of medium and fine sand fractions, with mean values in the horizons ranging from, respectively, 218.13 to 756.38 g kg^{-1} and 178.66 to 577.04 g kg^{-1}. In the sand fraction, quartz mineralogical composition predominates. The silt and clay contents were extremely low, the mean values in the horizons ranging from, respectively, 4.53 to 57.35 g kg^{-1} and 23.03 to 126.29 g kg^{-1}. The textural class of horizons, according to Santos et al. [15], is sandy, and according to the USDA-NRCS [24], it is sand and loamy sand only for horizons with more than 100 g kg^{-1} of clay, such as C1 of NQ1 and NQ6, and C3 of NQ2.

Table 2. Sand size distribution of the Quartzipsamments profiles of the Southwest of Rio Grande do Sul State, southern Brazil.

		Sand					
Horizon	Depth	Total	Very Coarse	Coarse	Medium	Fine	Very Fine
	(cm)				(g kg^{-1})		
				NQ1			
Ap	0–20	954.94	1.54	12.94	633.33	287.51	19.62
A	20–65	947.57	0.33	12.42	477.15	393.57	64.10
C1	65–94	843.69	2.42	16.78	498.24	297.21	29.04
C2	94–150	904.97	3.26	16.65	414.45	422.12	48.49
Mean		912.79	1.89	14.70	505.79	350.10	40.31
CV, %		5.19	70.91	21.83	17.36	19.54	46.80

Table 2. *Cont.*

Horizon	Depth	Total	Very Coarse	Coarse	Medium	Fine	Very Fine
					Sand		
	(cm)				(g kg^{-1})		
				NQ2			
A	0–12	941.92	0.00	27.75	420.95	243.49	249.73
C1	12–42	947.26	0.00	30.51	431.11	419.48	66.16
C2	42–85	923.14	0.00	21.91	261.84	538.37	101.02
C3	85–145	846.04	4.77	27.41	543.40	248.72	21.74
Mean		914.59	1.19	26.90	414.33	362.52	109.66
CV, %		4.76	188.22	16.38	26.14	44.87	115.91
				NQ3			
A	0–22	948.51	0.00	29.29	402.02	445.93	71.27
C1	22–53	912.02	3.06	25.55	480.54	364.94	37.93
C2	53–93	897.23	2.72	11.71	348.15	458.68	75.97
C3	93–150	916.99	1.38	10.97	466.44	362.87	75.33
Mean		918.69	1.79	19.38	424.29	408.11	65.13
CV, %		2.18	76.36	45.81	13.38	12.14	26.52
				NQ4			
A	0–15	925.40	1.00	15.54	508.97	337.40	62.49
C1	15–40	874.73	1.76	14.72	473.65	349.89	34.71
C2	40–72	970.86	1.41	7.14	542.24	389.55	30.52
C3	72–140	957.40	2.04	16.06	577.46	333.79	28.05
Mean		932.10	1.55	13.37	525.58	352.66	38.94
CV, %		4.25	26.70	34.77	8.20	7.84	39.06
				NQ5			
A	0–20	949.14	3.48	29.33	615.37	278.91	22.05
C1	20–66	930.60	0.00	29.84	302.17	515.25	83.34
C2	66–100	914.61	0.07	33.18	350.22	484.30	46.84
C3	100–170	947.89	0.00	38.13	389.14	448.74	71.88
Mean		935.56	0.89	32.62	414.23	431.80	56.03
CV, %		1.62	185.16	15.40	31.03	22.59	45.43
				NQ6			
A1	0–18	914.35	2.73	18.22	443.06	406.10	44.24
A2	18–43	929.86	0.05	15.96	218.13	577.04	118.68
C1	43–60	834.04	4.54	22.59	541.46	243.44	22.01
C2	60–93	967.48	1.93	23.08	756.38	178.66	7.43
C3	93–125	955.04	3.31	14.53	505.03	393.05	39.12
C4	125–170	923.81	1.20	15.36	531.54	311.22	64.49
Mean		920.76	2.29	18.29	499.27	351.59	49.33
CV, %		4.86	69.79	23.87	33.73	38.84	76.22

Very coarse sand: 2–1 mm; Coarse sand: 1–0.5 mm; Medium sand: 0.5–0.25 mm; Fine sand: 0.5–0.25 mm; Very fine sand: 0.25–0.05 mm. CV: coefficient of variation. NQ1 and NQ 2: Quartzipsamments from Quaraí city; NQ3 and NQ4: Quartzipsamments from Manoel Viana city; NQ5 and NQ6: Quartzipsamments from São Francisco de Assis city.

Table 3. Silt and clay content, degree of flocculation (DF), and textural class [24] of the Quartzipsamments profiles of the Southwest of Rio Grande do Sul State, southern Brazil.

Horizon	Depth	Silt	Clay	DF	Textural
	(cm)	(g kg^{-1})		(%)	Class
NQ1					
Ap	0–20	18.32	26.73	73.49	Sand
A	20–65	14.93	37.50	78.70	Sand
C1	65–94	34.43	121.87	31.47	Loamy sand
C2	94–150	28.93	66.10	56.08	Sand
Mean		24.15	63.05	59.94	
CV, %		36.65	62.54	33.16	
NQ2					
A	0–12	28.05	30.03	38.46	Sand
C1	12–42	29.71	23.03	8.78	Sand
C2	42–85	31.29	45.57	28.59	Sand
C3	85–145	42.10	111.85	46.85	Loamy sand
Mean		32.79	52.62	30.67	
CV, %		21.53	71.43	49.66	
NQ3					
A	0–22	20.48	31.01	30.53	Sand
C1	22–53	35.26	52.72	32.44	Sand
C2	53–93	57.35	45.42	36.21	Sand
C3	93–150	28.42	54.59	47.74	Sand
Mean		35.38	45.94	36.73	
CV, %		41.89	21.63	22.34	
NQ4					
A	0–15	20.64	53.97	30.33	Sand
C1	15–40	31.86	93.42	54.18	Sand
C2	40–72	4.53	24.62	59.00	Sand
C3	72–140	9.44	33.17	72.69	Sand
Mean		16.62	51.30	54.05	
CV, %		68.84	55.36	31.71	
NQ5					
A	0–20	13.83	37.03	71.61	Sand
C1	20–66	38.79	30.62	39.46	Sand
C2	66–100	37.42	47.96	25.82	Sand
C3	100–170	21.60	30.50	77.03	Sand
Mean		27.91	36.53	53.48	
CV, %		40.78	20.88	43.27	
NQ6					
A1	0–18	37.55	48.10	45.88	Sand
A2	18–43	36.57	33.57	43.35	Sand
C1	43–60	39.66	126.29	56.84	Loamy sand
C2	60–93	7.51	25.01	53.99	Sand
C3	93–125	13.90	31.06	49.94	Sand
C4	125–170	16.27	59.92	80.79	Sand
Mean		25.24	53.99	55.13	
CV, %		53.90	66.45	8.24	

Silt: 0.05–0.002 mm; Clay: <0.002 mm. CV: coefficient of variation. NQ1 and NQ2: Quartzipsamments from Quaraí city; NQ3 and NQ4: Quartzipsamments from Manoel Viana city; NQ5 and NQ6: Quartzipsamments from São Francisco de Assis city.

The flocculation degree, which represents the resistance of the soil structure to disintegration, presented a wide range of values (8.78 to 80.79%) (Table 3).

The Quartzipsamments profiles showed high values of macroporosity and saturated hydraulic conductivity and low values of microporosity and available water (Table 4). These results are consistent with the characteristics of these sandy soils. The high conductivity was associated with macropores, the main responsible for the flow of air and water, and the low availability of water due to lower microporosity, pores responsible for retention and availability of water.

Table 4. Mean values of physical–hydric properties of Quartzipsamments profiles of the Southwest of Rio Grande do Sul State, southern Brazil.

Horizon	Depth (cm)	TP (m^3 m^{-3})	Macro (m^3 m^{-3})	Micro (m^3 m^{-3})	BD (Mg m^{-3})	PD (Mg m^{-3})	KS (mm h^{-1})	AW (m^3 m^{-3})
				NQ1				
Ap	0–20	0.522	0.373	0.150	1.32	2.63	223.51	0.013
A	20–65	0.443	0.282	0.161	1.48	2.67	170.37	0.028
C1	65–94	0.429	0.272	0.157	1.48	2.63	244.22	0.021
C2	94–150	0.432	0.279	0.153	1.47	2.67	293.76	0.027
Mean		0.457	0.302	0.155	1.44	2.65	233.83	0.022
CV, %		2.18	3.26	4.17	2.80	0.00	13.19	35.03
				NQ2				
A	0–12	0.458	0.323	0.135	1.44	2.70	398.19	0.015
C1	12–42	0.398	0.275	0.124	1.42	2.67	447.97	0.022
C2	42–85	0.385	0.253	0.132	1.41	2.70	430.58	0.019
C3	85–145	0.380	0.254	0.127	1.45	2.70	346.73	0.025
Mean		0.407	0.278	0.130	1.43	2.69	407.57	0.020
CV, %		4.93	7.50	6.80	4.70	0.00	8.88	32.81
				NQ3				
A	0–22	0.466	0.348	0.118	1.33	2.70	478.13	0.014
C1	22–53	0.419	0.300	0.120	1.43	2.67	367.27	0.021
C2	53–93	0.411	0.284	0.127	1.46	2.74	339.20	0.029
C3	93–150	0.401	0.260	0.141	1.46	2.70	342.98	0.022
Mean		0.424	0.300	0.127	1.42	2.70	377.15	0.022
CV, %		4.34	6.32	4.80	3.10	0.00	15.30	33.09
				NQ4				
A	0–15	0.480	0.308	0.171	1.43	2.67	220.83	0.023
C1	15–40	0.446	0.263	0.182	1.49	2.70	127.85	0.034
C2	40–72	0.449	0.294	0.155	1.42	2.70	262.80	0.032
C3	72–140	0.453	0.305	0.147	1.43	2.74	414.66	0.038
Mean		0.457	0.293	0.164	1.44	2.70	268.23	0.032
CV, %		4.99	7.99	4.49	3.60	0.00	19.71	40.12
				NQ5				
A	0–20	0.484	0.346	0.138	1.39	2.70	313.93	0.017
C1	20–66	0.452	0.316	0.136	1.39	2.74	347.01	0.027
C2	66–100	0.438	0.310	0.128	1.38	2.74	475.60	0.019
C3	100–170	0.436	0.313	0.123	1.47	2.74	458.28	0.022
Mean		0.452	0.322	0.131	1.41	2.73	399.49	0.021
CV, %		2.68	4.28	3.22	2.10	0.00	12.93	30.77
				NQ6				
A1	0–18	0.491	0.382	0.109	1.36	2.74	672.26	0.018
A2	18–43	0.464	0.359	0.105	1.39	2.70	671.87	0.022
C1	43–60	0.442	0.351	0.091	1.51	2.70	473.61	0.006
C2	60–93	0.409	0.295	0.114	1.42	2.70	402.51	0.017
C3	93–125	0.419	0.312	0.107	1.47	2.78	237.20	0.007
C4	125–170	0.461	0.378	0.083	1.64	2.74	645.44	0.007
Mean		0.448	0.346	0.102	1.47	2.73	513.80	0.013
CV, %		4.33	5.91	10.87	4.96	0.00	14.07	64.04

TP: total porosity; Macro: macroporosity; Micro: microporosity; BD: bulk density; PD: particle density; KS: saturated hydraulic conductivity; AW: available water (water volume between 10 kPa and 1500 kPa-tension). CV: coefficient of variation. NQ1 and NQ 2: Quartzipsamments from Quaraí city; NQ3 and NQ4: Quartzipsamments from Manoel Viana city; NQ5 and NQ6: Quartzipsamments from São Francisco de Assis city.

In general, the surface horizon showed higher total porosity and macroporosity (Table 4). Generally, the bulk density was higher with increasing depth.

The evaluated Quartzipsamments are acidic and, in general, with low fertility, with predominantly low Ca and Mg contents and very low K [28] (Table 5). The cation exchange capacity at pH 7.0 and the organic matter are predominantly low [28] (Table 6), reflecting a high risk of leaching and impact on the environment due to smaller adsorption sites and high hydraulic conductivity (Table 4) and, in terms of agricultural use, requiring an application in installments of fertilizers. These results corroborate the single-grain soil structure (Tables 1 and 4), since electrical charges and organic matter contribute to soil aggregation. The chemical results are consistent with the low clay content of these soils (Table 3) because, in this particle size, soil reactivity occurs [29].

Table 5. Mean values of chemical parameters and interpretation * of Quartzipsamments profiles of the Southwest of Rio Grande do Sul State, southern Brazil.

Horizon	Depth (cm)	pH Water	pH KCl	Ca	Mg	Na	K
				cmol$_c$ kg^{-1}			
				NQ1			
Ap	0–20	5.1	4.7	4.5 (High)	0.99 (Medium)	0.27	0.44 (Very low)
A	20–65	5.3	4.8	3.75 (Medium)	0.73 (Medium)	0.27	0.26 (Very low)
C1	65–94	5.4	4.9	4.00 (Medium)	0.82 (Medium)	0.27	0.20 (Very low)
C2	94–150	5.4	4.9	4.42 (High)	0.94 (Medium)	0.40	0.20 (Very low)
Mean		5.3	4.8	4.19 (High)	0.87 (Medium)	0.30	0.28 (Very low)
CV, %		1.82	1.65	12.92	16.23	39.42	40.39
				NQ2			
A	0–12	5.4	4.9	2.97 (Medium)	0.54 (Medium)	0.48	0.52 (Very low)
C1	12–42	5.4	4.9	0.84 (Low)	0.14 (Low)	0.35	0.29 (Very low)
C2	42–85	5.3	4.7	0.29 (Low)	0.08 (Low)	0.30	0.20 (Very low)
C3	85–145	5.3	4.8	1.36 (Low)	0.21 (Low)	0.27	0.17 (Very low)
Mean		5.4	4.8	1.37 (Low)	0.24 (Low)	0.35	0.30 (Very low)
CV, %		2.11	2.55	83.37	53.65	30.31	50.86
				NQ3			
A	0–22	5.0	4.5	2.16 (Medium)	0.35 (Low)	0.22	0.38 (Very low)
C1	22–53	5.1	4.5	0.90 (Low)	0.15 (Low)	0.14	0.13 (Very low)
C2	53–93	5.2	4.7	0.85 (Low)	0.11 (Low)	0.05	0.05 (Very low)
C3	93–150	5.2	4.7	0.13 (Low)	0.06 (Low)	0.11	0.04 (Very low)
Mean		5.1	4.6	1.01 (Low)	0.17 (Low)	0.13	0.15 (Very low)
CV, %		3.17	5.60	79.58	38.44	65.45	101.51
				NQ4			
A	0–15	5.3	4.8	10.16 (High)	3.71 (High)	0.19	0.32 (Very low)
C1	15–40	5.4	4.8	7.75 (High)	2.11 (High)	0.29	0.23 (Very low)
C2	40–72	5.2	4.8	3.08 (Medium)	0.55 (Medium)	0.15	0.14 (Very low)
C3	72–140	5.2	4.7	1.78 (Low)	0.28 (Low)	0.16	0.13 (Very low)
Mean		5.3	4.8	5.69 (High)	1.66 (High)	0.20	0.21 (Very low)
CV, %		0.87	0.67	64.41	82.58	31.41	40.48
				NQ5			
A	0–20	4.8	4.1	0.05 (Low)	0.05 (Low)	0.32	0.13 (Very low)
C1	20–66	5.0	4.3	0.31 (Low)	0.08 (Low)	0.22	0.08 (Very low)
C2	66–100	5.3	4.8	0.03 (Low)	0.03 (Low)	0.16	0.03 (Very low)
C3	100–170	5.2	4.8	1.05 (Low)	0.12 (Low)	0.18	0.05 (Very low)
Mean		5.1	4.5	0.36 (Low)	0.07 (Low)	0.22	0.07 (Very low)
CV, %		0.89	1.61	124.67	21.85	38.87	57.46
				NQ6			
A1	0–18	5.1	4.5	1.49 (Low)	0.23 (Low)	0.10	0.19 (Very low)
A2	18–43	5.0	4.5	0.04 (Low)	0.05 (Low)	0.35	0.11 (Very low)
C1	43–60	5.1	4.7	0.35 (Low)	0.08 (Low)	0.25	0.08 (Very low)
C2	60–93	5.1	4.5	0.05 (Low)	0.04 (Low)	0.08	0.07 (Very low)
C3	93–125	4.8	4.2	0.49 (Low)	0.07 (Low)	0.15	0.04 (Very low)
C4	125–170	4.9	4.2	0.81 (Low)	0.13 (Low)	0.40	0.07 (Very low)
Mean		5.0	4.4	0.54 (Low)	0.10 (Low)	0.22	0.09 (Very low)
CV, %		4.96	5.78	99.36	27.89	63.36	51.40

pH water: pH determined in water 1:1 (soil:water); pH KCl: pH determined in KCl; Ca: calcium; Mg: magnesium; Na: sodium; K: potassium; CV: coefficient of variation. * Interpretation: in parentheses is the interpretation of levels according to the "Manual de adubação e de calagem para os Estados do Rio Grande do Sul e Santa Catarina/Manual of fertilization and liming to the Rio Grande do Sul and Santa Catarina States" [28], while some chemical parameters do not have available the interpretation in the handbook. NQ1 and NQ 2: Quartzipsamments from Quaraí city; NQ3 and NQ4: Quartzipsamments from Manoel Viana city; NQ5 and NQ6: Quartzipsamments from São Francisco de Assis city.

Table 6. Mean values of chemical parameters and interpretation * of Quartzipsamments profiles of the Southwest of Rio Grande do Sul State, Brazil.

Horizon	Depth	Al^{3+}	H + Al	Effective	pH7.0	Bases	Al	OM
				CEC		**Saturation**		
	(cm)		(cmol$_c$ kg^{-1})				(%)	
				NQ1				
Ap	0–20	0.29	1.3	6.6	7.5$^{(Low)}$	83.3	4.36	1.58 $^{(Low)}$
A	20–65	0.29	1.1	5.3	6.1$^{(Low)}$	82.7	5.40	0.51 $^{(Low)}$
C1	65–94	0.17	1.5	5.5	6.7$^{(Low)}$	78.4	3.15	1.14 $^{(Low)}$
C2	94–150	0.23	1.1	6.2	7.0$^{(Low)}$	85.0	3.70	0.40 $^{(Low)}$
Mean		0.25	1.3	5.9	6.8$^{(Low)}$	82.4	4.15	0.91 $^{(Low)}$
CV, %		35.56	17.63	14.14	12.50	3.83	33.03	
				NQ2				
A	0–12	0.23	1.9	4.7	6.4$^{(Low)}$	70.7	4.83	2.02 $^{(Low)}$
C1	12–42	0.52	2.3	2.1	3.9$^{(Low)}$	41.5	24.17	2.18 $^{(Low)}$
C2	42–85	0.40	2.1	1.3	2.9$^{(Low)}$	29.3	31.87	0.77 $^{(Low)}$
C3	85–145	0.40	1.7	2.4	3.7$^{(Low)}$	54.8	16.60	0.42 $^{(Low)}$
Mean		0.39	2.0	2.6	4.2$^{(Low)}$	49.1	19.37	1.35 $^{(Low)}$
CV, %		29.67	14.70	52.71	33.78	30.48	53.73	
				NQ3				
A	0–22	0.17	1.9	3.3	5.0$^{(Low)}$	62.5	5.23	3.37 (Medium)
C1	22–53	0.40	1.9	1.7	3.2$^{(Low)}$	41.3	23.37	1.16 $^{(Low)}$
C2	53–93	0.52	2.3	1.6	3.3$^{(Low)}$	31.8	32.73	1.07 $^{(Low)}$
C3	93–150	0.52	2.5	0.9	2.8$^{(Low)}$	12.2	59.98	0.32 $^{(Low)}$
Mean		0.40	2.2	1.9	3.6$^{(Low)}$	37.0	30.33	1.48 $^{(Low)}$
CV, %		40.79	14.35	48.56	24.72	46.70	65.81	
				NQ4				
A	0–15	0.23	2.5	14.6	16.9 $^{(High)}$	85.3	1.57	4.38 (Medium)
C1	15–40	0.57	3.3	10.9	13.7 (Medium)	75.9	5.23	2.94 (Medium)
C2	40–72	0.69	3.5	4.6	7.4 $^{(Low)}$	52.8	14.92	1.93 $^{(Low)}$
C3	72–140	0.75	3.1	3.1	5.4 $^{(Low)}$	43.1	24.08	1.32 $^{(Low)}$
Mean		0.56	3.1	8.3	10.9 (Medium)	64.3	11.45	2.64 (Medium)
CV, %		39.54	14.03	59.76	45.20	27.56	82.95	
				NQ5				
A	0–20	0.52	2.7	1.1	3.2 $^{(Low)}$	17.2	48.05	1.56 $^{(Low)}$
C1	20–66	0.57	2.5	1.3	3.2 $^{(Low)}$	21.8	45.36	1.33 $^{(Low)}$
C2	66–100	0.40	2.7	0.6	2.9 $^{(Low)}$	8.3	62.20	0.84 $^{(Low)}$
C3	100–170	0.40	2.3	1.8	3.7 $^{(Low)}$	38.0	22.32	0.46 $^{(Low)}$
Mean		0.47	2.6	1.2	3.3 $^{(Low)}$	21.3	44.48	1.05 $^{(Low)}$
CV, %		26.19	8.37	35.15	10.06	43.42	31.00	
				NQ6				
A1	0–18	0.52	3.3	2.5	5.3 $^{(Low)}$	37.8	20.48	3.57 (Medium)
A2	18–43	0.52	3.3	1.1	3.9 $^{(Low)}$	14.4	48.08	2.36 $^{(Low)}$
C1	43–60	0.40	3.1	1.2	3.9 $^{(Low)}$	19.6	34.67	0.91 $^{(Low)}$
C2	60–93	0.34	2.3	0.6	2.5 $^{(Low)}$	9.4	59.25	0.53 $^{(Low)}$
C3	93–125	0.17	1.5	0.9	2.2 $^{(Low)}$	34.0	18.58	0.43 $^{(Low)}$
C4	125–170	0.23	1.3	1.6	2.7 $^{(Low)}$	53.1	13.89	0.05 $^{(Low)}$
Mean		0.36	2.5	1.3	3.4 $^{(Low)}$	28.1	32.49	1.31 $^{(Low)}$
CV, %		43.47	37.32	47.05	32.47	49.79	52.99	

Al^{3+}: aluminum; H + Al: potential acidity; CEC effective: effective cation exchange capacity; CECpH7.0: cation exchange capacity at pH 7.0; OM: organic matter; CV: coefficient of variation. * Interpretation: in parentheses is the interpretation of levels according to the "Manual de adubação e de calagem para os Estados do Rio Grande do Sul e Santa Catarina/Manual of fertilization and liming to the Rio Grande do Sul and Santa Catarina States" [28], while some chemical parameters do not have available the interpretation in the handbook. NQ1 and NQ 2: Quartzipsamments from Quaraí city; NQ3 and NQ4: Quartzipsamments from Manoel Viana city; NQ5 and NQ6: Quartzipsamments from São Francisco de Assis city.

According to Pearson's correlation analysis, the increment in macroporosity increases the total porosity ($r = 0.80$) and decreases the microporosity ($r = -0.50$), with a consequent increase in the saturated hydraulic conductivity ($r = 0.45$) and a decrease in available water ($r = -0.41$) (Table 7). The increase in bulk density decreases its total porosity ($r = -0.19$), although with no effect on macroporosity and microporosity, which suggests that there is a decrease in the size of the macropore with increasing bulk density, but no increase in microporosity. Because the vertical flow of water in the soil occurs mainly in macropores, while the water available to plants is in the micropores, increasing microporosity decreases hydraulic conductivity ($r = -0.75$) and increases available water ($r = 0.47$). Particle density showed an inversely proportional relationship with microporosity ($r = -0.45$) and a positive relationship with hydraulic conductivity ($r = 0.40$); that is, with an increase in particle density, there is a decrease in microporosity and an increase in hydraulic conductivity.

Table 7. Pearson's correlation between the soil physical–hydric properties of Quartzipsamments of the Southwest of Rio Grande do Sul State, southern Brazil.

	TP	Macro	Micro	BD	PD	KS
Macro	0.80 **					
Micro	ns	−0.50 **				
BD	−0.19 ***	ns	ns			
PD	ns	ns	−0.45 **	ns		
KS	ns	0.45 **	−0.75 **	ns	0.40 **	
AW	ns	−0.41 **	0.47 **	ns	ns	ns

TP: total porosity; Macro: macroporosity; Micro: microporosity; BD: bulk density; PD: particle density; KS: saturated hydraulic conductivity; AW: available water (water volume between 10 kPa and 1500 kPa-tension). ns: not significant; ** $p < 0.01$; *** $p < 0.10$.

Soil bulk density showed a positive and significant correlation with clay ($r = 0.45$) (Table 8), indicating that clay, even in small proportion in these soils, can occupy the spaces between larger particles (silt and sand), decreasing the size of the macropore, with a consequent decrease in total porosity and an increase in bulk density, but with no influence on microporosity, as seen in Table 7. At 10% probability, the degree of flocculation was positively correlated with total porosity ($r = 0.30$) and bulk density ($r = 0.33$).

Table 8. Pearson's correlation between the soil physical–hydric properties and the particle size and degree of flocculation of Quartzipsamments of the Southwest of Rio Grande do Sul State, southern Brazil.

	Sand					Silt	Clay	DF
	Very Coarse	Coarse	Medium	Fine	Very Fine			
TP	ns	ns	ns	ns	ns	ns	ns	0.30 ***
Macro	ns	ns	ns	ns	ns	ns	ns	ns
Micro	ns	ns	ns	ns	ns	ns	ns	ns
BD	ns	ns	ns	ns	ns	ns	0.45 *	0.33 ***
PD	ns	ns	ns	ns	ns	ns	ns	ns
KS	ns	ns	ns	ns	ns	ns	ns	ns
AW	ns	ns	ns	ns	ns	ns	ns	ns

TP: total porosity; Macro: macroporosity; Micro: microporosity; BD: bulk density; PD: particle density; KS: saturated hydraulic conductivity; AW: available water (water volume between 10 kPa and 1500 kPa-tension); DF: degree of flocculation. ns: not significant; * $p < 0.05$; *** $p < 0.10$.

Increasing pH water and pH KCl, there is an increase in the availability of Ca ($r = 0.47$ and 0.44, respectively), Mg ($r = 0.38$ and 0.35), and K ($r = 0.37$ and 0.37), and an increase in CECeffective ($r = 0.46$ and 0.43), CECpH7.0 ($r = 0.44$ and 0.42), and base saturation ($r = 0.45$ and 0.43) and a decrease in Al saturation for pH KCL ($r = -0.34$) (Table 9). Increasing organic matter, there is an increase in CECeffective ($r = 0.57$) and CECpH7.0 ($r = 0.66$), with a consequent increase in the availability of Ca, Mg, and K and base saturation, and a decrease in Al saturation, evidencing the effect of organic matter in the chemical improvement of these soils where the clay content is low, and the adjustment of pH improves soil fertility.

Table 9. Pearson's correlation between the soil chemical variables of Quartzipsamments of the Southwest of Rio Grande do Sul State, southern Brazil.

	pH Water	pH KCl	Ca	Mg	Na	K	Al^{3+}	H + Al	CEC Effective	CECpH7.0	Base Sat.	Al Sat.
pH KCl	0.94 **											
Ca	0.47 *	0.44 *										
Mg	0.38 ***	0.35 ***	0.96 **									
Na	ns	ns	ns	ns								
K	0.37 ***	0.37 ***	0.56 **	0.45 *	0.50 **							
Al^{3+}	ns	ns	ns	ns	ns	−0.35 ***						
H + Al	ns	ns	ns	ns	ns	ns	0.77 **					
CECeffective	0.46 *	0.43 *	0.99 **	0.97 **	ns	0.56 **	ns	ns				
CECpH7.0	0.44 *	0.42 *	0.97 **	0.97 **	ns	0.51 **	ns	ns	0.98 **			
Base sat.	0.45 *	0.44 *	0.82 **	0.67 **	−0.37 *	0.70 **	0.44 *	−0.48 *	0.79 **	0.70 **		
Al sat.	ns	−0.34 ***	−0.70 **	−0.55 **	−0.37 ***	−0.65 **	0.43 *	0.42 *	−0.67 **	−0.59 **	− 0.94 **	
OM	ns	ns	0.54 **	0.58 **	ns	0.53 **	ns	ns	0.45 *	0.57 **	0.66 **	ns	ns

pH water: pH determined in water 1:1 (soil:water); pH KCl: pH determined in KCl; Ca: calcium; Mg: magnesium; Na: sodium; K: potassium. Al^{3+}: aluminum; H+Al: potential acidity; CECeffective: effective cation exchange capacity; CECpH7.0: cation exchange capacity at pH 7.0; Base sat.: base saturation; Al sat.: aluminum saturation; OM: organic matter. ns: not significant; ** $p < 0.01$; * $p < 0.05$; *** $p < 0.10$.

The Na ($r = 0.47$) and K ($r = 0.42$) showed a positive and significant correlation with very fine sand, while Ca had a negative correlation with coarse sand, as well as CECeffective and CECpH7.0 (Table 10). The increase in pH water decreases the degree of flocculation of the soil ($r = −0.34$).

Table 10. Pearson's correlation between the soil chemical variables and the particle size and degree of flocculation of Quartzipsamments of the Southwest of Rio Grande do Sul State, southern Brazil.

	Sand					Silt	Clay	DF
	Very Coarse	Coarse	Medium	Fine	Very Fine			
pH water	ns	ns	ns	ns	ns	ns	ns	−0.34 ***
pH KCl	ns	ns	ns	ns	ns	ns	ns	ns
Ca	ns	−0.36 ***	ns	ns	ns	ns	ns	ns
Mg	ns	ns	ns	ns	ns	ns	ns	ns
Na	ns	ns	ns	ns	0.47 *	ns	ns	ns
K	ns	ns	ns	ns	0.42 *	ns	ns	ns
Al^{3+}	ns	ns	ns	ns	ns	ns	ns	ns
H + Al^{3+}	ns	ns	ns	ns	ns	ns	ns	ns
CECeffective	ns	−0.35 ***	ns	ns	ns	ns	ns	ns
CECpH7.0	ns	−0.33 ***	ns	ns	ns	ns	ns	ns
Base sat.	ns	ns	ns	ns	ns	ns	ns	ns
Al sat.	ns	ns	ns	ns	ns	ns	ns	ns
OM	ns	ns	ns	ns	ns	ns	ns	−0.36 ***

pH water: pH determined in water 1:1 (soil:water); pH KCl: pH determined in KCl; Ca: calcium; Mg: magnesium; Na: sodium; K: potassium; Al^{3+}: aluminum; H + Al: potential acidity; CECeffective: effective cation exchange capacity; CECpH7.0: cation exchange capacity at pH 7.0; Base sat.: base saturation; Al sat.: aluminum saturation; OM: organic matter; DF: degree of flocculation. ns: not significant; * $p < 0.05$; *** $p < 0.10$.

Increasing organic matter, there is an increase in total porosity ($r = 0.53$) and a decrease in bulk density ($r = −0.47$) (Table 11). The effect of some chemical variables on soil physics may be associated with electrical charges and aggregate formation. In these soils, the increase in microporosity has a positive effect on increasing the availability of Ca ($r = 0.74$), Mg ($r = 0.67$), K ($r = 0.38$), and on CECeffective ($r = 0.74$), CECpH7.0 ($r = 0.72$) and base saturation ($r = 0.58$), and decreasing Al saturation ($r = −0.37$).

Table 11. Pearson's correlation between the soil chemical and physical–hydric properties of Quartzipsamments of the Southwest of Rio Grande do Sul State, southern Brazil.

	TP	Macro	Micro	BD	PD	KS	AW
pH water	−0.33 ***	0.64 **	0.58 **	ns	−0.45 *	ns	0.50 **
pH KCl	ns	−0.53 **	0.51 **	ns	−0.47 *	ns	0.44 *
Ca	0.36 ***	ns	0.74 **	ns	−0.47 *	−0.58 **	ns
Mg	0.33 ***	ns	0.67 **	ns	−0.40 *	−0.51 **	ns
Na	ns	ns	ns	ns	−0.33 ***	ns	ns
K	0.39 *	ns	0.38 ***	−0.34 ***	−0.57 *	ns	ns
Al^{3+}	ns	ns	ns	ns	ns	ns	0.62 **
H+ Al^{3+}	ns	ns	ns	ns	ns	ns	ns
CECeffective	0.37 ***	ns	0.74 **	ns	−0.47 *	−0.56 **	ns
CECpH7.0	0.40 *	ns	0.72 **	ns	−0.41 *	−0.49 *	0.35 ***
Base sat.	ns	ns	0.58 **	ns	−0.54 **	−0.50 **	ns
Al sat.	ns	ns	−0.37 ***	ns	0.37 ***	0.36 ***	ns
OM	0.53 **	ns	ns	−0.47 *	ns	ns	ns

pH water: pH determined in water 1:1 (soil:water); pH KCl: pH determined in KCl; Ca: calcium; Mg: magnesium; Na: sodium; K: potassium. Al^{3+}: aluminum; H+Al: potential acidity; CECeffective: effective cation exchange capacity; CECpH7.0: cation exchange capacity at pH 7.0; Base sat.: base saturation; Al sat.: aluminum saturation; OM: organic matter; TP: total porosity; Macro: macroporosity; Micro: microporosity; BD: bulk density; PD: particle density; KS: saturated hydraulic conductivity; AW: available water (water volume between 10 kPa and 1500 kPa-tension). ns: not significant; ** $p < 0.01$; * $p < 0.05$; *** $p < 0.10$.

4. Discussion

The low clay contents in these soils are highlighted because practically all the reactivity of the soil, such as the cation exchange capacity, is available in this particle size [29]. The availability of nutrients to the soil in the medium and long term from weathering is not significant due to the predominantly quartz mineralogical composition. The flocculation degree resulted in a wide range of values (8.78 to 80.79%), and this soil property is especially important in terms of soil water erosion resistance because it represents the resistance of the soil structure to disintegration.

Although the bulk density values showed a wide range of variation (1.32 to 1.64 Mg m^{-3}), the mean value of the profiles did not show significant variation (1.41 to 1.47 Mg m^{-3}) (Table 4). According to FAO [9], Arenosols have high bulk density values, ranging from 1.5 to 1.7 Mg m^{-3}, and considering the quartz density of around 2.65 Mg m^{-3}, the total porosity is around 0.36 to 0.46 m^3 m^{-3}. Although the total porosity was obtained directly by weighing, the values (0.380 to 0.522 m^3 m^{-3}) are close to those indicated by FAO [9].

The range of macroporosity and microporosity values found are, respectively, 0.253 to 0.373 m^3 m^{-3} and 0.083 to 0.182 m^3 m^{-3}. Working with different uses of Quartzipsamment in Rio Grande do Sul State, Reichert et al. [30] found microporosity values similar to this study (0.073 to 0.169 m^3 m^{-3}) but higher to macroporosity (0.398 to 0.570 m^3 m^{-3}). The values found in the present study and by the aforementioned authors are completely opposite to what is usually found for clayey soils. For example, in comparative terms, Suzuki et al. [31], studying different land use and management systems, verified for an Oxisol (clay between 640 and 664 g kg^{-1}) macroporosity and microporosity values of, respectively, 0.219 to 0.017 m^3 m^{-3} and 0.476 to 0.354 m^3 m^{-3}, and total porosity of 0.573 to 0.460 m^3 m^{-3}, which refers to a lower agricultural and environmental suitability of Quartzipsamments when compared to other soils. Variations in bulk density according to its clay content are well documented [32–35]. The higher total porosity and macroporosity verified in most of the profiles in the superficial horizon can be associated with the action of the few roots in this superficial soil layer, while the greater values of bulk density with an increase in the depth can probably be related to the smaller action of roots and by the weight of the upper soil layers causing pressure on the subsurface layers.

The saturated hydraulic conductivity is extremely high in the Quartzipsamments evaluated, ranging from 127.85 to 672.26 mm h^{-1}, associated with the large volume of macropores (water-flow pores), as also observed by Reichert et al. [30]. According to

Mesquita and Moraes [36], the flow of water in saturated soil occurs preferentially in macropores (pores with diameter >50 μm). FAO [9] points out Arenosols as water-permeable soils, with the saturated hydraulic conductivity varying between 300 and 30,000 cm day^{-1} (125 and 12,500 mm h^{-1}), and depending on the particle size distribution and the organic matter content, the available water capacity can be less than 3 to 4% or greater than 15 to 17%. FAO [9] also mentions that because most pores are relatively large, much of the retained water is drained at a tension of only 100 kPa.

Suertegaray [7] even mentions that the water dynamics in Quartzipsamments from Southwest of Rio Grande do Sul State, with regard to erosion, are associated with concentrated superficial processes, which originate furrows, ravines, and gullies. According to the author, laminar flow is not characteristic of these areas due to the high infiltration capacity of these soils.

A fact that draws attention is the low coefficient of variation of the saturated hydraulic conductivity (8.88 to 19.71%) (Table 4). Generally, this property presents great spatial variability, generating a high coefficient of variation and requiring a greater number of samples to reduce this variability [36]. Coefficient of variation values can be greater than 200% for hydraulic conductivity [37]. According to the same authors, this variability is associated with types and land use, position in the landscape, depth, instruments and measurement methods, and experimental errors. The lower variability in Quartzipsamments allows the minimum number of soil samples in spatial variability studies. The low coefficient of variation of the saturated hydraulic conductivity in our study may be associated with the single-grain structure and high sand content in the soil profiles.

The volume of available water is extremely low (0.006 to 0.038 m^3 m^{-3}) (Table 4), associated with the sandy texture and low volume of micropores (pores responsible for water retention and availability). Reichert et al. [38] found for soils from Rio Grande do Sul State that field capacity and permanent wilting point increased in a similar proportion with increasing clay content of the soils. The authors also verified an average volume of available water to plants of 0.089 m^3 m^{-3} for sandy soils and 0.124 m^3 m^{-3} for very clayey soils, reaching 0.191 m^3 m^{-3} for silty clay soils.

Compaction, so harmful in soils with higher clay content, can increase microporosity in sandy soils and the volume of available water to plants. However, studies are needed to indicate the appropriate level of compaction to improve water retention and availability for plants without preventing their root growth by increasing soil bulk density.

Working with four profiles of sandy soils from the Brazilian semiarid region, Santos et al. [39] recommend lower irrigation rates and more frequent application, due to the lower water-holding capacity of these soils and greater risk of nutrient leaching.

In sandy soil (89% sand), the incorporation of clay together with organic matter increased aggregate stability, total soil porosity, and available water content and decreased soil bulk density. Moreover, increasing plant height and number of shoots of physic nut (*Jatropha curcas* L.) were observed [40]. While the biochar added to sandy soil (93.2% sand) increased water-holding capacity, decreased drainage, and increased available water for crop use [41].

The high acidity and low fertility of these soils require an adjustment of pH and mineral reserve, which would demand high costs with fertilizer. Due to high sand and low clay contents, these soils require split fertilization to reduce leaching rates and increase the efficiency of nutrient uptake by plants. According to the "Manual de adubação e de calagem para os Estados do Rio Grande do Sul e Santa Catarina/Manual of fertilization and liming to the Rio Grande do Sul and Santa Catarina States" [28], for these sandy soils (<20% clay) or with CEC < 7.5 cmolc dm^{-3}, it is recommended to avoid total corrective fertilization of K or P due to the possibility of leaching these nutrients or salinity problems, as well as the splitting of nitrogen fertilization.

Due to the smaller adsorption sites (low cation exchange capacity at pH 7.0 and organic matter) and the high hydraulic conductivity, there is a high risk of leaching and environmental damage, requiring an installment of fertilizer application in the agricultural use of these soils. In order to improve the chemical efficiency of these soils, it is recommended to increase their levels of organic matter, something that should occur gradually, with consequent improvement of the physical structure and biological activity of these soils. Meanwhile, due to the low levels of clay and organic matter, these soils need, at each crop, to be corrected through high doses of limestone and fertilizers, which can become financially unfeasible in the short and medium term.

Considering the very low K contents, for its total correction, the amount of 120 kg of K_2O per hectare is recommended [28]. To correct the pH, raising it to a value of 6.0, the equation indicated by the SBCS-NRS/CQFS [28] was used:

$$NC = -0.516 + 0.805OM + 2.435Al \qquad (3)$$

where NC is the limestone requirement in t ha^{-1} (with relative power of total neutralization—PRNT 100%), OM is the organic matter content (in %), and Al is the exchangeable aluminum content of the soil (in cmol$_c$ dm^{-3}).

From the calculations, considering the variation in Al and organic matter contents in the topsoil horizon, the limestone requirement ranged from 1.46 to 3.63 Mg ha^{-1}. This calculus of K_2O fertilizer and limestone reflects the high costs for fertility adjustment.

As reported by Donagemma et al. [8], a common practice adopted by technicians and farmers in sandy soils is the application of limestone doses above 6 Mg ha^{-1}, and may even exceed 10 Mg ha^{-1} because, according to farmers, when applying the dose of 2 Mg ha^{-1}, as recommended in corrective and fertilizer handbooks, crop yield will be very low. The authors explain this due to the low reactivity of limestone in these soils, associated with the low aluminum content and low buffering power of the soil, and possible losses of cations in depth, by leaching.

Hartemink and Huting [42], Bezabih et al. [43], Reichert et al. [30], and Olorunfemi et al. [44] also found in sandy soil a correlation between organic matter and CEC, and Bezabih et al. [43] also observed a correlation between pH and CEC, and porosity and bulk density with organic carbon and CEC. The results of Olorunfemi et al. [44] agree with the observations of this study on the negative correlation between base and aluminum saturation.

According to Reichert et al. [30], for most sandy soils, a large part of the CEC comes from organic matter. From the collection of published data from tropical sandy soils in western and eastern Africa, Blanchart et al. [45] showed that organic matter is the main determinant of soil fertility, nutrient storage, aggregate stability, and microbiological and enzymatic activities. According to the authors, although cultural practices or land use have a lower impact on the increase of organic matter when compared to clayey soils, this is the way to increase them and improve soil biofunction, which determines the agronomic and environmental potential of the sandy soils.

Reichert et al. [30] verified changes in the physical properties of a Quartzipsamment related to organic carbon, which lead to a resistance of this soil to degradation. According to the authors, the accumulation of organic matter in sandy soils is more important than in clayey soils due to their fragility and difficulty in increasing their organic matter content. In sandy soils, the soil aggregation is mainly controlled by carbon dynamics [46], and our study corroborates considering the low organic matter content and the single-grain soil structure.

The increase in organic matter can lead to a decrease in the use of mineral fertilizers [28] and, as a consequence, less leaching and risks of contamination of water resources (surface due to erosion and subsurface due to leaching). In addition, its sandy texture requires split fertilization to decrease nutrient leaching and increase plant uptake [28].

Alternatives are being studied to improve soil structure and fertility of sandy soils. For example, the use of clay and natural polymers increased the CEC and the number of available cations, and decreased the leached cations of coastal sandy soil (98% sand) [47]. While long-term application of organic amendments improved physical (bulk density was decreased, available water-holding capacity was increased), biological (enhanced the overall soil microbial activity, such as species number and diversity, especially of the desirable groups such as heterotrophic aerobes, actinomycetes, and pseudomonads), and chemical properties (increasing soil organic matter, carbon, pH, Mehlich 1-extractable P, K, Ca, Mg, Mn, Cu, Fe, and Zn concentrations, CEC) of a sandy soil [48].

The application of sewage sludge in sandy soil (87.69% sand) increased the organic matter content of soil; improved the infiltration rate, decreasing the water erosion under simulated high-intensity rainfall; decreased bulk density and increased the tendential air permeability of soil; and the soil compaction level was reduced in the first year after compost re-treatment. However, all the beneficial effects of sewage sludge last only for two years [49]. While the use of liming and catch crops alone did not influence cereal yield and straw and plant height, in the fourth year of study, all yield trait components significantly increased with the use of farmyard manure, liming, and catch crops together in a Podzol sandy soil (62.9% sand) [50].

It is known that current production techniques and technologies have evolved and improved and are accessible; however, it is worth highlighting the high cost of agricultural production of these soils, requiring a careful analysis in terms of production costs and sales value of the products.

Table 12 presents the difficulties and challenges specific to or associated with Quartzipsamments.

Table 12. Difficulties and challenges associated with the Quartzipsamments of the Southwest of Rio Grande do Sul State, southern Brazil.

Difficulties
— Sandy texture; — Low water-holding capacity; — Low volume of available water; — Low aggregation; — Low content of organic matter; — Low fertility; — High leaching; — High susceptibility to erosion; — Little vegetation cover; — Difficulty of plant species to develop; — Low agricultural land suitability; — Agricultural expansion; — High environmental fragility.
Challenges
— Identify plant species adapted to this soil and that produce a large volume of biomass to increase organic matter and physical structure of the soil; — Increase biological activity; — Due to agricultural expansion, identify the most appropriate uses and management for this soil; — From the use of this soil, what level of compaction it can reach, and what level is harmful to the development of plants; — The recommendation of fertilization (doses and splitting) that best suits this soil, considering the risks of contamination of surface water by erosion and subsurface water by leaching; — Soil and water conservation techniques best suited for these soils; — Quantity and frequency of irrigation; — For livestock use of these native pastures, what are the recommendations for better soil, animal, and pasture management; — Monitor the soil (physical, chemical, and biological variables), erosion processes, surface and subsurface water quality, and nutrient leaching and contaminant flow; — Evaluate the performance of growth and yield of crops managed in these soils; — Evaluate the economic viability of agricultural use of these soils.

Due to the sandy texture (Table 2) and low organic matter content of these soils (Table 6), reflected in poorly structured soils (Tables 1 and 4), they are very susceptible to erosion and the sandyzation process, requiring complex and permanent techniques and practices for the conservation of the soil, such as terracing and maintenance of permanent vegetation cover on the soil surface. Associated with this, there is still little vegetation cover, usually native pasture, although soybean cultivation is expanding in these soils and the use of the land with agriculture and livestock without grazing control and no adoption of conservation practices. Suertegaray et al. [10] highlight that the dynamics of land use, without prior recognition of its agricultural land suitability, is capable of intensifying the morphogenetic processes, weakening the landscape in a relatively shorter time than the dynamics of nature itself. Therefore, sustainable land use in Biome Pampa is just possible if the economical activities consider the soil suitability and the adaptations of its plant and animal communities [2].

In this sense, studying a toposequence with three sandy soils (sand between 89 and 95% and clay between 9 and 3%), Thomaz and Fidalski [51] observed that the soil position in the toposequence and the total sand content were the variables that best explained the erodibility interrill, in an experiment under simulated rainfall, and emphasized the need for differentiated management systems along the toposequence.

Furthermore, the low volume of available water (Table 4), considering these soils of low agricultural land suitability, and the high hydraulic conductivity and low water-holding capacity (Table 4), associated with the minimum contents of clay, silt (Table 3), and organic matter (Table 6), characterize these soils as very fragile environmentally [2], and should be used very carefully, especially due to the risks of leaching; transport of metals, pesticides, and other agrochemicals; and contamination of surface and subsurface water resources, besides the high susceptibility to erosion and sandyzation.

Suertegaray [7] reported that, due to the fragility of the soils where the sands occur, they are highly susceptible to erosion when their agricultural management occurs through heavy machinery, which forms rills that can evolve into the formation of ravines and gullies. According to the author, intensive pastoral activity, with animal overcrowding is indicated as a cause of erosion, linked to the formation of rills by the trampling of cattle through trails.

Figure 4 shows water and wind erosion and the extensive areas of bare and exposed Quartzipsamments to sandyzation.

Figure 4. Pictures showing the native field. Soil exposing the sandyzation and erosion in the Quartzipsamments of the Southwest of Rio Grande do Sul State, southern Brazil. Source: pictures taken by L.E.A.S. Suzuki.

The use of cover plants is efficient in reducing the transport of sand by the wind in these soils, being an effective alternative to contain the expansion of the sandyzation process [52].

In recent years, the return of soybean cultivation in Quartzipsamments has been observed; however, for their agricultural use, planning is essential to improve the structure of these soil, with the use of cover plants with a significant contribution of biomass to increase organic matter, soil physical structuring, and increase biological activity, and the use of soil and water conservation techniques. Moreover, cover crops can reduce evapotranspiration, a factor that can be limiting in these soils with low water-holding capacity. For example, Eltz and Rovedder [53] observed temperatures above 40 °C at 3 cm depth in exposed soil in a sandyzation area, but when this area was replanted with cover crops, the soil temperature was reduced by around 18.6%.

Among alternatives for cover crops, Rovedder [54] cites *Lupinus albescens* (lupine) as a potential specie for the recovery of sandy soils, while Reichert et al. (2016) found that eucalyptus was more efficient in increasing soil organic carbon after conversion from native field to conventional planting, eucalyptus and unvegetated area.

Silva [55] cites the dwarf butia tree (*Butia lalemanti*), a common palm in the sandy top relief and deep soil, mainly in the cities of São Francisco de Assis and Manoel Viana, as a contributor to the minimization of laminar erosion caused by floods, besides being a specie adapted to water stress.

Gass et al. [56] cite that the advance of monocrops on the sands of Rio Grande do Sul State have occurred without the knowledge of the potential use of native plant species endemic to the region, most of which are unknown in their food, medicinal, ornamental, aromatic properties, condiments, and for use in projects for the recovery of degraded areas.

From our study, we verify the low agricultural suitability and the high environmental fragility of Quartzipsamments in Rio Grande do Sul State. The agricultural use of these soils will be intensified, especially due to political and economic issues; therefore, studies are needed to guide farmers, extensionists, and stakeholders on the best way to use and manage them. Any incentive to use these sandy soils and the availability of technologies (e.g., irrigation, fertilizers, and machinery, among others) requires technical monitoring, specific recommendations for these soils, and monitoring of the soil and the environment. Specific recommendations include lower irrigation rates and more frequent application, due to the lower water-holding capacity of these soils and greater risk of nutrient leaching; split fertilization to reduce leaching and increase the efficiency of nutrient uptake by plants; include plant species adapted to this soil and climatic conditions in a rotation crop system, to produce enough volume of biomass to increase organic matter content, with consequent improvement of the soil physical structure and biological activity; maintenance of permanent vegetation cover on the soil surface to reduce water evaporation and erosion; and use of techniques and practices for soil and water conservation, such as terracing and keyline arrangement.

5. Conclusions

Quartzipsamments are soils with high macroporosity and saturated hydraulic conductivity, and these soils play an important role in aquifer recharge. On the other hand, their low content of clay and organic matter hinders the soil's physical structuring and biological activity, making these soils susceptible to water and wind erosion and the sandyzation process, besides a high possibility of leaching.

Their sandy texture, extremely low volume of available water to plants, low fertility, and cation exchange capacity, besides wind erosion, hinder the development of vegetation in these soils and, given these characteristics, make these soils of low suitability for agricultural use and with a high risk of leaching and aquifer contamination. However, we verified that the increase in pH and organic matter can improve the fertility of these soils (especially Ca, Mg, K, cation exchange activity), and for organic matter, an improvement in the physical structure of the soil (increase in total porosity and decrease in bulk density) can occur also.

Some difficulties and challenges in the use and management of Quartzipsamments were pointed out in our study, and specific recommendations for better use and management of these soils were indicated.

Author Contributions: Conceptualization, L.E.A.S.S. and F.d.A.P.; methodology, L.E.A.S.S., F.d.A.P., R.B.d.O. and A.P.M.R.; formal analysis, L.E.A.S.S., F.d.A.P. and R.B.d.O.; investigation, L.E.A.S.S., F.d.A.P. and R.B.d.O.; resources, F.d.A.P.; data curation, L.E.A.S.S. and F.d.A.P.; writing—original draft preparation, L.E.A.S.S. and F.d.A.P.; writing—review and editing, L.E.A.S.S., F.d.A.P., R.B.d.O. and A.P.M.R. All authors have read and agreed to the published version of the manuscript.

Funding: This research received no external funding.

Institutional Review Board Statement: Not applicable.

Informed Consent Statement: Not applicable.

Data Availability Statement: Data available on request from the authors.

Conflicts of Interest: The authors declare no conflict of interest.

References

1. Pedron, F.A.; Dalmolin, R.S.D. *Solos Arenosos do Bioma Pampa Brasileiro*; UFSM: Santa Maria, Brazil, 2019; 280p.
2. Roesch, L.F.W.; Vieira, F.C.B.; Pereira, V.A.; Schünemann, A.L.; Teixeira, I.F.; Senna, A.J.T.; Stefenon, V.M. The Brazilian Pampa: A fragile biome. *Diversity* **2009**, *1*, 182–198. [CrossRef]
3. Suertegaray, D.M.A.; Guasseli, L.A.; Verdum, R.; Basso, L.A.; Medeiros, R.M.V.; Bellanca, E.T.; Bertê, A.M.A. Projeto Arenização no Rio Grande do Sul, Brasil: Gênese, Dinâmica e Espacialização. In *Anais do X Simpósio Brasileiro de Sensoriamento Remoto*; INPE: Foz do Iguaçu, Brazil, 2001; pp. 349–356. Available online: http://marte.sid.inpe.br/col/dpi.inpe.br/lise/2001/09.14.12.00/doc/0349.356.234.pdf (accessed on 7 August 2022).
4. Bellanca, E.T.; Suertegaray, D.M.A. Sítios Arqueológicos e Areias no Sudoeste do Rio Grande do Sul. *Mercator–Rev. De Geogr. Da UFC* **2003**, *4*, 99–114. Available online: http://www.mercator.ufc.br/mercator/article/view/155 (accessed on 7 August 2022).
5. Caneppele, J.C.G. Espacialização da Arenização a Partir da Ecodinâmica e da Cartografia Ambiental. Master's Thesis, Universidade Federal do Rio Grande do Sul, Instituto de Geociências, Porto Alegre, Brazil, 2017; 129p. Available online: http://hdl.handle.net/10183/171442 (accessed on 7 August 2022).
6. Azevedo, A.C.; Kaminski, J. Considerações sobre os solos dos campos de areia no Rio Grande do Sul. *Ciência Ambiente* **1995**, *11*, 65–70.
7. Suertegaray, D.M.A. Erosão nos campos sulinos: Arenização no sudoeste do Rio Grande do Sul. *Rev. Bras. Geomorfologia* **2011**, *12*, 61–74. [CrossRef]
8. Donagemma, G.K.; de Freitas, P.L.; Balieiro, F.C.; Fontana, A.; Spera, S.T.; Lumbreras, J.F.; Viana, J.H.M.; Araújo Filho, J.C.; Santos, F.C.; de Albuquerque, M.R.; et al. Characterization, agricultural potential, and perspectives for the management of light soils in Brazil. *Pesqui. Agropecu. Bras.* **2016**, *51*, 1003–1020. [CrossRef]
9. FAO-Food and Agriculture Organization of the United Nations. *Lecture Notes on the Major Soils of the World*; FAO-Food and Agriculture Organization of the United Nations: Rome, Italy, 2001; 300p. Available online: https://www.fao.org/publications/card/es/c/3aa8d214-f539-5c44-a603-9e4031c7fadd/ (accessed on 7 August 2022).
10. Suertegaray, D.M.A.; Verdum, R.; Medeiros, R.M.; Guasselli, L.; Frank, M.W. Caracterização Hidrogeomorfológica e Uso do Solo em Áreas de Ocorrência de Areais: São Francisco de Assis/Manuel Viana. In *Anais do VIII Simpósio Brasileiro de Sensoriamento Remoto*; MCT/INPE: Salvador, Brazil, 1996; pp. 663–669. Available online: https://docplayer.com.br/41367129-Caracterizacao-hidrogeomorfologica-e-uso-do-solo-em-areas-de-ocorrencia-de-areais-sao-francisco-de-assis-manuel-viana.html (accessed on 7 August 2022).
11. Da Costa, B.S.C.; Sluter, C.R.; Lescheck, A.L.; Rodrigues, E.L.S. Large-scale cartographic representation of relief features from sandyzation process. *Bull. Geod. Sci.* **2020**, *26*, e2020019. [CrossRef]
12. Huang, J.; Hartemink, A.E. Soil and environmental issues in sandy soils. *Earth Sci. Rev.* **2020**, *208*, 103295. [CrossRef]
13. Maluf, J.R.T. Nova classificação climática do Estado do Rio Grande do Sul. *Rev. Bras. Agrometeorol.* **2000**, *8*, 141–150.
14. Instituto Rio Grandense do Arroz–IRGA. *Médias Climatológicas*; Instituto Rio Grandense do Arroz–IRGA: Porto Alegre, Brazil. Available online: https://irga.rs.gov.br/medias-climatologicas (accessed on 23 January 2023).
15. Santos, H.G.; Jacomine, P.K.T.; Anjos, L.H.; Oliveira, V.A.; Lumbreras, J.F.; Coelho, M.R.; Almeida, J.A.; Araujo Filho, J.C.; Oliveira, J.B.; Cunha, T.J.F. *Sistema Brasileiro de Classificação de Solos*; Embrapa: Brasília, Brasil, 2018. Available online: https://ainfo.cnptia.embrapa.br/digital/bitstream/item/199517/1/SiBCS-2018-ISBN-9788570358004.pdf (accessed on 29 November 2021).
16. Soil Survey Staff. *Keys to Soil Taxonomy*, 12th ed.; USDA-Natural Resources Conservation Service: Washington, DC, USA, 2014. Available online: https://www.nrcs.usda.gov/wps/portal/nrcs/detail/soils/survey/class/taxonomy/?cid=nrcs142p2_053580 (accessed on 8 December 2021).
17. Santos, R.D.; Santos, H.G.; Ker, J.C.; Anjos, L.H.C.; Shimizu, S.H. *Manual de Descrição e Coleta de Solo no Campo*, 7th ed.; Revisada e Ampliada; Sociedade Brasileira de Ciência do Solo: Viçosa, Brazil, 2015; 102p.
18. Schoeneberger, P.J.; Wysocki, D.A.; Benham, E.C.; Soil Survey Staff. *Field Book for Describing and Sampling Soils*; Natural Resources Conservation Service, National Soil Survey Center: Lincoln, NE, USA, 2012.
19. Klute, A.; Dirksen, C. Hydraulic conductivity and diffusivity: Laboratory methods. In *Methods of Soil Analysis, Part 1: Physical and Mineralogical Methods*; Klute, A., Ed.; American Society of Agronomy, Soil Science Society of America: Madison, WI, USA, 1986; pp. 687–734.
20. Klute, A. Water retention: Laboratory methods. In *Methods of Soil Analysis: Physical and Mineralogical Methods*; Klute, A., Ed.; American Society of Agronomy, Soil Science Society of America: Madison, WI, USA, 1986; pp. 635–660.
21. Blake, G.R.; Hartge, K.H. Bulk density. In *Methods of Soil Analysis, Part 1: Physical and Mineralogical Methods*, 2nd ed.; Klute, A., Ed.; American Society of Agronomy, Soil Science Society of America: Madison, WI, USA, 1986; pp. 363–375. [CrossRef]

22. Viana, J.H.M.; Teixeira, W.G.; Donagemma, G.K. Densidade de partículas. In *Manual de Métodos de Análise de Solo*, 3rd ed.; Teixeira, P.C., Donagemma, G.K., Fontana, A., Teixeira, W.G., Eds.; Embrapa: Brasília, Brazil, 2017; pp. 76–81. Available online: https://www.embrapa.br/busca-de-publicacoes/-/publicacao/1085209/manual-de-metodos-de-analise-de-solo (accessed on 1 February 2023).
23. Gee, G.W.; Or, D. Particle-size analysis. In *Methods of Soil Analysis, Part 4: Physical Methods*, 5th ed.; Dane, J.H., Topp, C., Eds.; Soil Science Society of America: Madison, WI, USA, 2002; pp. 255–293.
24. United States Department of Agriculture (USDA)-National Resource Conservation Service (NRCS). Soil Texture Calculator. 2022. Available online: https://www.nrcs.usda.gov/wps/portal/nrcs/detail/?cid=nrcs142p2_054167 (accessed on 14 October 2022).
25. Tedesco, M.J.; Gianello, C.; Bissani, C.A.; Volkweiss, S.J. *Análises de Solo, Plantas e Outros Materiais*, 2nd ed.; Departamento de Solos; UFRGS: Porto Alegre, Brazil, 1995; 174p.
26. Nelson, E.W.; Sommers, L.E. Total carbon, organic carbon and organic matter. In *Methods of Soil Analysis: Chemical and Microbiological Properties, Part 2*; Page, A.L., Miller, R.H., Keeney, D.R., Eds.; Agronomy Monograph 9; ASA and Soil Science Society of America: Madison, WI, USA, 1982; pp. 539–579.
27. SAS Institute Inc. *SAS/STAT User's Guide*; SAS Institute Inc.: Cary, NC, USA, 1999.
28. Sociedade Brasileira de Ciência do Solo-SBCS. *Núcleo Regional Sul-NRS. Comissão de Química e Fertilidade do Solo. Manual de Adubação e Calagem para os Estados do Rio Grande do Sul e Santa Catarina*; Sociedade Brasileira de Ciência do Solo-SBCS. Núcleo Regional Sul-NRS. Comissão de Química e Fertilidade do Solo: Viçosa, Brazil, 2016; 376p. Available online: http://www.sbcs-nrs.org.br/docs/Manual_de_Calagem_e_Adubacao_para_os_Estados_do_RS_e_de_SC-2016.pdf (accessed on 7 August 2022).
29. Ernani, P.R. *Química do Solo e Disponibilidade de Nutrientes*; O Autor: Lages, Brazil, 2008; 230p.
30. Reichert, J.M.; Amado, T.J.C.; Reinert, D.J.; Rodrigues, M.F.; Suzuki, L.E.A.S. Land use effects on subtropical, sandy soil under sandyzation/desertification processes. *Agric. Ecosyst. Environ.* **2016**, *233*, 370–380. [CrossRef]
31. Suzuki, L.E.A.S.; Reinert, D.J.; Alves, M.C.; Reichert, J.M. Medium-term no-tillage, additional compaction, and chiseling as affecting clayey subtropical soil physical properties and yield of corn, soybean and wheat crops. *Sustainability* **2022**, *14*, 9717. [CrossRef]
32. Reichert, J.M.; Suzuki, L.E.A.S.; Reinert, D.J.; Horn, R.; Håkansson, I. Reference bulk density and critical degree-of-compactness for no-till crop production in subtropical highly weathered soils. *Soil Tillage Res.* **2009**, *102*, 242–254. [CrossRef]
33. Suzuki, L.E.A.S.; Reichert, J.M.; Reinert, D.J. Degree of compactness, soil physical properties and yield of soybean in six soils under no-tillage. *Soil Res.* **2013**, *51*, 311–321. [CrossRef]
34. Suzuki, L.E.A.S.; Reinert, D.J.; Reinert, D.J.; de Lima, C.L.R. Degree of compactness and mechanical properties of a subtropical Alfisol with eucalyptus, native forest, and grazed pasture. *For. Sci.* **2015**, *61*, 716–722. [CrossRef]
35. Suzuki, L.E.A.S.; Reinert, D.J.; Alves, M.C.; Reichert, J.M. Critical limits for soybean and black bean root growth, based on macroporosity and penetrability, for soils with distinct texture and management systems. *Sustainability* **2022**, *14*, 2958. [CrossRef]
36. Mesquita, M.G.B.F.; Moraes, S.O. A dependência entre a condutividade hidráulica saturada e atributos físicos do solo. *Ciênc. Rural* **2004**, *34*, 963–969. [CrossRef]
37. Deb, S.K.; Shukla, M.K. Variability of hydraulic conductivity due to multiple factors. *Am. J. Environ. Sci.* **2012**, *8*, 489–502. [CrossRef]
38. Reichert, J.M.; Albuquerque, J.A.; Kaiser, D.R.; Reinert, D.J.; Urach, F.L.; Carlesso, R. Estimation of water retention and availability in soils of Rio Grande do Sul. *Rev. Bras. Ciênc. Solo* **2009**, *33*, 1547–1560. [CrossRef]
39. Dos Santos, N.G.N.; Olszevski, N.; Salviano, A.M.; Cunha, T.J.F.; Giongo, V.; Pereira, J.S. Granulometric fractions and physical-hydric behavior of sandy soils. *Rev. Agrar.* **2019**, *12*, 318–327. [CrossRef]
40. Djajadi, D.; Heliyanto, B.; Hidayah, N. Changes of physical properties of sandy soil and growth of physic nut (*Jatropha curcas* L.) due to addition of clay and organic matter. *Agrivita* **2011**, *33*, 245–250. [CrossRef]
41. Radwan, N.M.; Marzouk, E.R.; El-Melegy, A.M.; Hassan, M.A. Improving sandy soil properties by using biochar. *SINAI J. Appl. Sci.* **2020**, *9*, 157–168. [CrossRef]
42. Hartemink, A.E.; Huting, J. Sandy soils in Southern and Eastern Africa: Extent, properties and management. In Proceedings of the Management of Tropical Sandy Soils for Sustainable Agriculture, Khon Kaen, Thailand, 27 November–2 December 2005; pp. 54–59. Available online: https://www.fao.org/3/ag125e/AG125E08.htm (accessed on 7 August 2022).
43. Bezabih, B.; Aticho, A.; Mossisa, T.; Dume, B. The effect of land management practices on soil physical and chemical properties in Gojeb Sub-river Basin of Dedo District, Southwest Ethiopia. *J. Soil Sci. Environ. Manag.* **2016**, *7*, 154–165. [CrossRef]
44. Olorunfemi, I.E.; Fasinmirin, J.T.; Akinola, F.F. Soil physico-chemical properties and fertility status of longterm land use and cover changes: A case study in forest vegetative zone of Nigeria. *Eurasian J. Soil Sci.* **2018**, *7*, 133–150. [CrossRef]
45. Blanchart, E.; Albrecht, A.; Bernoux, M.; Brauman, A.; Chotte, J.L.; Feller, C.; Ganry, F.; Hien, E.; Manlay, R.; Masse, D.; et al. Organic matter and biofunctioning in tropical sandy soils and implications for its management. In Proceedings of the Management of Tropical Sandy Soils for Sustainable Agriculture, Khon Kaen, Thailand, 27 November–2 December 2005; pp. 224–241. Available online: https://cupdf.com/document/organic-matter-and-biofunctioning-in-tropical-sandy-soils-of-organic-matter.html?page=1 (accessed on 7 August 2022).
46. Tisdall, J.M.; Oades, J.M. Organic matter and water stable aggregates in soils. *J. Soil Sci.* **1982**, *33*, 141–163. [CrossRef]
47. Minhal, F.; Ma'as, A.; Hanudin, E.; Sudira, P. Improvement of the chemical properties and buffering capacity of coastal sandy soil as affected by clay and organic by-product application. *Soil Water Res.* **2020**, *15*, 93–100. [CrossRef]

48. Ozores-Hampton, M.; Stansly, P.A.; Salame, T.P. Soil chemical, physical, and biological properties of a sandy soil subjected to long-term organic amendments. *J. Sustain. Agric.* **2011**, *35*, 243–259. [CrossRef]
49. Aranyos, J.T.; Tomócsik, A.; Makádi, M.; Mészáros, J.; Blaskó, L. Changes in physical properties of sandy soil after long-term compost treatment. *Int. Agrophys.* **2016**, *30*, 269–274. [CrossRef]
50. Lipiec, J.; Usowicz, B. Quantifying cereal productivity on sandy soil in response to some soil-improving cropping systems. *Land* **2021**, *10*, 1199. [CrossRef]
51. Thomaz, E.L.; Fidalski, J. Interrill erodibility of different sandy soils increases along a catena in the Caiuá Sandstone Formation. *Rev. Bras. Ciênc. Solo* **2020**, *44*, e0190064. [CrossRef]
52. Rovedder, A.P.M.; Eltz, F.L.F. Revegetação com plantas de cobertura em solos arenizados sob erosão eólica no Rio Grande do Sul. *Rev. Bras. Ciênc. Solo* **2008**, *32*, 315–321. [CrossRef]
53. Eltz, F.L.F.; Rovedder, A.P.M. Revegetação e temperatura do solo em áreas degradadas no sudoeste do Rio Grande do Sul. *Rev. Bras. Agrociência* **2005**, *11*, 193–200. Available online: https://periodicos.ufpel.edu.br/ojs2/index.php/CAST/article/view/1211 (accessed on 1 February 2023).
54. Rovedder, A.P.M. Potencial do Lupinus Albescens Hook. & Arn. Para Recuperação de Solos Arenizados do Bioma Pampa. Ph.D. Thesis, Universidade Federal de Santa Maria, Santa Maria, Brazil, 2007; 126p. Available online: https://repositorio.ufsm.br/handle/1/3384 (accessed on 7 August 2022).
55. da Silva, L.A.P. Narrativas das Percepções e Conectividade de Caminhantes nas Paisagens dos Areais Pampeanos: Perspectivas Ambientais para Geração de Ambiências. Master's Thesis, Universidade Federal do Rio Grande do Sul, Instituto de Geociências, Porto Alegre, Brazil, 2008; 154p. Available online: https://lume.ufrgs.br/handle/10183/15719 (accessed on 7 August 2022).
56. Gass, S.L.B.; Verdum, R.; Vieira, L.F.S.; Caneppele, J.C.G.; Laurent, F. Os areais do sudoeste do Rio Grande do Sul, Brasil, como patrimônio geomorfológico. *Physis Terrae Rev. Ibero-Afro-Am. Geogr. Física Ambiente* **2019**, *1*, 101–119. [CrossRef]

soil systems

Review

Conservation Agriculture and Soil Organic Carbon: Principles, Processes, Practices and Policy Options

Rosa Francaviglia [1,*], María Almagro [2] and José Luis Vicente-Vicente [3]

[1] CREA, Research Centre for Agriculture and Environment, 00184 Rome, Italy
[2] IFAPA, Andalusian Institute of Agricultural and Fisheries Research and Training, Camino de Purchil s/n, 18004 Granada, Spain
[3] ZALF, Leibniz Centre for Agricultural Landscape Research, 15374 Müncheberg, Germany
* Correspondence: r.francaviglia@gmail.com

Abstract: Intensive agriculture causes land degradation and other environmental problems, such as pollution, soil erosion, fertility loss, biodiversity decline, and greenhouse gas (GHG) emissions, which exacerbate climate change. Sustainable agricultural practices, such as reduced tillage, growing cover crops, and implementing crop residue retention measures, have been proposed as cost-effective solutions that can address land degradation, food security, and climate change mitigation and adaptation by enhancing soil organic carbon (SOC) sequestration in soils and its associated co-benefits. In this regard, extensive research has demonstrated that conservation agriculture (CA) improves soil physical, chemical, and biological properties that are crucial for maintaining soil health and increasing agroecosystem resilience to global change. However, despite the research that has been undertaken to implement the three principles of CA (minimum mechanical soil disturbance, permanent soil organic cover with crop residues and/or cover crops, and crop diversification) worldwide, there are still many technical and socio-economic barriers that restrict their adoption. In this review, we gather current knowledge on the potential agronomic, environmental, and socio-economic benefits and drawbacks of implementing CA principles and present the current agro-environmental policy frameworks. Research needs are identified, and more stringent policy measures are urgently encouraged to achieve climate change mitigation targets.

Keywords: reduced tillage; permanent soil cover; crop diversification; soil and water conservation; ecosystem services; carbon sequestration; climate change mitigation and adaptation; adoption barriers; economic incentives; agro-environmental policies

Citation: Francaviglia, R.; Almagro, M.; Vicente-Vicente, J.L. Conservation Agriculture and Soil Organic Carbon: Principles, Processes, Practices and Policy Options. *Soil Syst.* **2023**, *7*, 17. https://doi.org/10.3390/soilsystems7010017

Academic Editor: Luis Eduardo Akiyoshi Sanches Suzuki

Received: 16 January 2023
Revised: 15 February 2023
Accepted: 20 February 2023
Published: 22 February 2023

1. Background and Rationale

The concept of conservation agriculture (CA) was born in the 1930s when Edward Faulkner first questioned the utility of ploughing in a manuscript called *Ploughman's Folly*, and it gained popularity during the 1960s in the mid-western United States as a means of preventing soil degradation after the Dust Bowl ecological disaster that occurred in the 1930s. Since then, research on adapting CA practices to cropping systems has been undertaken worldwide. In addition to reducing tillage intensity, CA also implies the application of organic amendments, such as manure, compost, and by-products from agro-industry [1], and the improvement of N management if mineral fertilizers are adopted to decrease N_2O emissions [2].

The exploitation of agricultural soils based on crop monocultures and deep tillage with inversion of the layers has resulted in progressive soil structure degradation and compaction and reductions in soil organic matter content. These detrimental developments have triggered negative cascade effects on the soil biota and fertility, increasing soil water and wind erosion and CO_2 emissions [3,4]. Among alternative management systems to conventional agriculture that aim at the sustainability of crop systems, CA represents one of the most advanced models.

2. Adoption of Conservation Agriculture (CA)

CA is defined by the Food and Agriculture Organization [5,6] as "a sustainable agricultural production system for the protection of water and agricultural soil that integrates agronomic, environmental and economic aspects". CA is based on three principles (Figure 1): minimum mechanical soil disturbance through conservation tillage (i.e., no tillage, minimum tillage), permanent soil organic cover with crop residues and/or cover crops, and crop diversification through rotations and associations involving at least three different crops (including a legume crop). The benefits of CA are shown in Table 1.

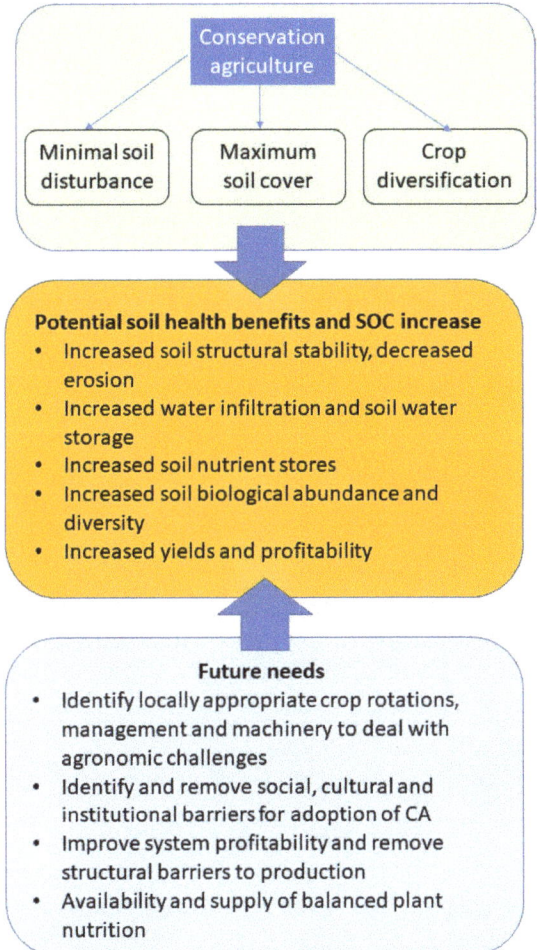

Figure 1. Principles of conservation agriculture, benefits of increasing SOC, and future needs. Modified from [7].

Kassam et al. [8] analysed the spread of the adoption of CA in 2015–2016 in different countries based on data available from government statistics, no-till farmer organizations, ministries of agriculture, non-governmental organizations, and research and development organizations.

The highest cropland areas were in South and North America (Table 2), with 69.9 and 63.2 M ha of cropland areas employed for CA, representing 38.7 and 35.0% of the total cropland employed for CA, respectively. However, CA represented 63.2% of the cropland area in South America and 28.1% in North America. The corresponding values for Australia/New Zealand and Asia were 22.7 and 13.9 M ha (12.6 and 7.7% of total cropland), representing 45.4 and 4.1% of croplands in the respective regions. Cropland areas employed for CA decreased in the order Russia/Ukraine > Europe > Africa from 5.7 to 1.5 M ha; i.e., from 3.6 to 1.1% of these regions' total cropland areas, respectively. Globally, the total cropland area employed for CA was 180.4 M ha, equivalent to 12.5% of total cropland.

Table 1. Benefits of conservation agriculture [9].

Target	Soil Cover	Minimal or No Soil Disturbance	Legumes in the Rotation	Crop Diversification
Simulate "forest floor" conditions	X	X		
Reduce evaporative loss of moisture from soil surface	X			
Reduce evaporative loss from upper soil layers	X	X		
Minimize oxidation of SOM and CO_2 loss		X		
Minimize compaction due to intense rainfall and the passage of machinery	X	X		
Minimize temperature fluctuations at the soil surface	X			
Maintain supply of OM as substrate for soil biota	X			
Increase and maintain nitrogen levels in the root zone	X	X	X	X
Increase CEC of the root zone	X	X	X	X
Maximize rain infiltration and minimize runoff	X	X		
Minimize soil loss in runoff	X	X		
Maintain natural layering of soil horizons through actions of soil biota	X	X		
Minimize weeds	X	X		X
Increase rate of biomass production	X	X	X	X
Speed up recuperation of soil porosity by soil biota	X	X	X	X
Reduce labour input		X		
Reduce fuel-energy input		X		
Recycle nutrients	X	X	X	X
Reduce pests and diseases				X
Rebuild damaged soil conditions and dynamics	X	X	X	X

Table 2. Cropland areas employed for CA by region in 2015–16, CA area as percentage of global total cropland, and CA area as percentage of cropland of each region [8].

Region	CA Cropland Area (M ha)	Total Cropland CA Area (%)	CA Area Cropland in the Region (%)
South America	69.90	38.7	63.2
North America	63.18	35.0	28.1
Australia/New Zealand	22.67	12.6	45.5
Asia	13.93	7.7	4.1
Russia/Ukraine	5.70	3.2	3.6
Europe	3.56	2.0	5.0
Africa	1.51	0.8	1.1
Global Total	180.44	100	12.5

3. Principles: Conservation Tillage, Permanent Plant Cover, and Crop Diversification

3.1. Conservation Tillage (CT)

Tillage is needed for different agricultural processes (e.g., seedbed preparation, weed control, crop residue management, improving soil aeration and avoiding soil compaction, optimizing soil temperature and moisture regimes). However, as a consequence, soil physical and chemical properties (structure, bulk density, pore size distribution, and fertility condition) are also altered, ultimately leading to good or poor crop performance [10]. Appropriate tillage practices, such as CT, aim to avoid soil degradation without compromising crop yields and while maintaining agroecosystem stability [11].

CT, as defined by the Conservation Tillage Information Center (CTIC, West Lafayette, Indiana, USA), excludes those tillage operations that invert the soil and bury crop residues. It consists of reducing the ploughing depth occasionally or continuously, applying shallower tillage with other implements, and/or reducing the intensity of seedbed preparation. Thus, it minimizes soil disturbance and reduces losses in soil and water, for which at least

30% of the soil surface must be covered by crop residues. Therefore, CT is a general term that includes specific operations, such as no-tillage, minimum tillage, reduced tillage, and mulch tillage practices [12–14]. Interest in CT systems increased globally after the 1930s following the Dust Bowl events, as they were seen as a way to halt soil erosion and promote water conservation [15]. However, extensive research has further demonstrated the multiple environmental benefits of adopting CT, such as enhancement of soil organic carbon (SOC) content, maintenance of agricultural productivity, and savings in the costs—in terms of time, fuel, and machinery—of seedbed preparation [13,14]. Moreover, it has been demonstrated that leaving crop residues on the soil surface also reduces evapotranspiration, improves infiltration, and suppresses weed growth [12,16]. According to the CTIC, there are five types of CT systems.

(1)   No tillage (NT)

The NT system is a specialized type of CT consisting of a one-pass planting and fertilizer operation in which the soil and the surface residues are minimally disturbed [17]. NT systems eliminate all pre-planting mechanical seedbed preparation except for the opening of a narrow (2–3 cm wide) strip or small hole in the ground for seed placement that ensures adequate seed–soil contact [11]. Retaining crop residues and leaving them on the soil surface is pivotal for soil and water conservation. Weed control can be managed using herbicides, a brush cutter, or biological control methods, such as crop rotation, intercropping, or vegetation strips. However, the use of herbicides may have detrimental effects on the soil system and its functions; thus, they should be applied with caution. Indeed, the new European agro-environmental policy framework discourages the use of herbicides; thus, use of mechanical or biological control methods should be boosted. Among the potential benefits of NT compared to other tillage systems are that it is more effective in controlling soil erosion, it improves soil water storage capacity, and it results in lower energy costs per unit of production and higher grain yields, especially in low-slope areas. However, as already stated, major disadvantages of NT are the heavy use of herbicides for weed control and the risk of soil compaction and nutrient stratification [18,19] in intensive agricultural systems (e.g., low residue input, machine traffic).

(2)   Mulch tillage

Mulch tillage is based on the principles of causing the least disturbance to the soil and leaving the maximum percentage of crop residue on the soil surface. For this purpose, in addition to in situ crop residues, the use of live mulch derived from cover crop residues is becoming a common practice. This practice can be adopted in herbaceous and woody crop systems by either allowing spontaneous plant cover to become established or by growing cover crops in the fallow period (in the case of herbaceous crop systems) or in the inter-tree rows (in the case of woody crop systems). Regardless of the type of plant cover used, this practice consists of maintaining plant cover that can protect the soil for as long as possible without causing the problem of competition for water and nutrients with the main crop. To do so, in accordance with the crop type and climate conditions, the spontaneous or seeded plant covers are mowed before the water-limiting period starts, and their residues are left on the soil surface as mulch.

(3)   Strip or zonal tillage

Strip tillage is a practice in which soil disturbance is limited to the crop rows while the rest of the soil is left undisturbed [20]. This tillage practice emerged as an alternative soil management practice in attempts to solve and mitigate the problems derived from conventional tillage or direct seeding methods [21]. The seedbed is divided into a seedling zone (5–10 cm wide), which is mechanically tilled to optimize the soil and micro-climate environment for germination and establishment of seedlings, and an inter-row zone, which is left undisturbed and protected by mulch or managed using chiselling to improve water infiltration and root development [22]. Today, strip tillage can benefit from the use of global positioning system (GPS) guidance equipment [23].

(4)   Ridge till

Ridge tillage consists of leaving the soil undisturbed before planting and then tilling about one third of the soil surface when planting with sweeps or row cleaners. Crops are planted in rows on cultivated ridges, while weeds are controlled with herbicides. This tillage practice gained popularity as a conservation agriculture practice for maize and soybean production in the USA [17].

(5)   Reduced and minimum tillage or occasional tillage

Reduced tillage (RT) is a soil management practice that consists of reducing the total number of tillage passes per year needed before seed planting (in both annual and perennial crops) or for soil aeration and decompaction (particularly in perennial crops). RT is also called minimum tillage and shallow tillage since, in some cases, it refers to reducing the depth at which the soil is tilled and/or using a cultivator or chisel plough to avoid soil inversion. Occasional tillage refers to the practice of one-time tillage, where tillage is conducted once every 5 or 10 years—depending on the soil, climate, and crop type—in an otherwise continuous NT system. This tillage practice is generally applied to mitigate the potential negative effects that tillage cessation may cause in some cases, such as soil compaction and nutrient stratification, particularly in rainfed perennial cropping systems [24,25].

3.1.1. Context of Application

CT is presently applied worldwide under a wide variety of climate conditions and with a wide variety of soil types and crops. However, the potential benefits and drawbacks of CT vary with climate (dry vs. moist), soil type (clayey vs. sandy), crop type (arable vs. perennial), and management (rainfed vs. irrigated); therefore, CT must be locally adapted or combined with other practices to become more cost-effective. This practice has been widely adopted in humid, sub-humid, and tropical regions, particularly for arable crops. However, the adoption of CT can be a challenge in dry regions because of (1) low biomass production and (2) the fact that crop residues are needed as fuel or animal feed. Nevertheless, and despite the fact that CT has been proven to have several environmental benefits, there are still some limitations and barriers to overcome, as discussed in the following sections.

3.1.2. Potential for SOC Sequestration

The enhancement of SOC content when shifting from conventional (intensive) tillage to CT has been demonstrated worldwide [2,26]. However, variations in SOC sequestration rates can be found among studies depending on the climate conditions, soil characteristics, initial SOC levels, crop type (arable vs. woody cropping systems), management (rainfed vs. irrigated), and the duration of the experiments. Results from various meta-analyses and modelling studies indicate SOC sequestration rates ranging from 0.27 to 1.1 t ha^{-1} yr^{-1} when CT is adopted in Mediterranean woody cropping systems [1,27,28]. Under Mediterranean conditions, average values about five times higher were reported for woody compared to arable crops for SOC sequestration rates [29]. Under tropical conditions, SOC sequestration rates oscillated between 0.12 and 1.56 t SOC ha^{-1} yr^{-1} depending on the crop type and climate regime [2,26]. As would be expected, higher SOC sequestration rates were estimated for moist compared to dry conditions regardless of the crop type [26]. However, SOC sequestration rates were generally higher for arable compared to woody crops under tropical conditions. Under boreal conditions, a local study estimated SOC sequestration rates of between 0.28 and 0.39 t ha^{-1} yr^{-1} across different soil types [30].

3.1.3. Co-Benefits

The enhancement of SOC content when shifting from conventional (intensive) tillage to CT has multiple beneficial effects, as has been demonstrated worldwide [2,27]. Several studies have demonstrated that CT improves soil physical, chemical, and biological properties crucial for maintaining soil condition and health. Indeed, conservation agriculture is indicated by the Intergovernmental Panel on Climate Change (IPCC) as one of the frameworks aimed at addressing land degradation, food security, and greenhouse gas (GHG) emissions [31]. For example, CT prevents soil sealing [32]. It is also well-known that increasing soil carbon sequestration by reducing tillage intensity (frequency and depth) improves soil biodiversity [33]. The presence of a vegetation cover due to CT increases soil biodiversity and can provide a habitat for arthropod predators and parasitoids, promoting biological control of pests and pathogens [16,34,35]. In addition, the build-up of soil organic matter derived from below-ground plant biomass inputs provides food and energy sources for microorganisms, favouring microbial growth and activity. Microorganisms decompose organic matter and increase nutrient availability for crops [33,36]. The presence of plant cover improves soil structure, porosity, aggregate stability, and water infiltration compared to bare soil. Therefore, CT also influences water regulation through the increase in soil water infiltration, which in turn fosters groundwater storage and lessens surface runoff, improving the availability of water for crops [16,37]. Moreover, it has been demonstrated that CT reduces soil erosion by water and wind due to the development of a vegetation cover. Reductions in runoffs of between 30% and 65% and in erosion of between 63 and 80% with decreasing tillage intensity have been observed worldwide [38–40], which ultimately lead to reductions in the nutrient losses resulting from erosion. Additionally, CT has been proven to improve nitrogen availability [35,41]. Other beneficial effects of CT are that the presence of the plant cover enhances soil aggregation, thus improving the protection of SOC against erosion, tillage operations, and abrupt soil temperature and moisture fluctuations [42].

Generally, CT positively impacts crop yields because the enhancement of the organic matter inputs into the system improves water infiltration and storage capacities and the availability of nutrients in soils [43]. In any case, the benefits are sometimes observed a few years after adoption, and the magnitude of the impacts of CT on crop yields depends on pedoclimatic conditions, crop types (arable or woody), and management practices (e.g., rotations, irrigation, and fertilization) [44,45].

In relation to climate change mitigation and adaptation, CT generally reduces GHG emissions compared to conventional (intensive) tillage systems. First, the reduction in the number of passes per year not only mitigates direct CO_2 emissions from the machinery but also prevents the peaks of CO_2 emissions from soils that typically occur after tillage operations [42]. Second, since CA usually includes improvements in N management, N_2O emissions from soils are reduced or, at least, a decrease in yield-scaled N_2O emissions is achieved [46,47]. Moreover, the presence of plant cover all year round not only protects the soil against erosion while improving its water retention capacity but also increases its buffer capacity against temperature extremes, making soils more resilient to extreme rainfall events, droughts, and warming [33,39,42].

CT has also been proven to have other socio-economic benefits, such as: (i) fuel, fertilizer, and pesticide savings; (ii) reducing erosion and flood risks and associated damage to infrastructure; (iii) sustainable preservation of cultural landscapes; and (iv) maintenance of crop yields, agricultural activity, and long-term employment, contributing to maintaining the local population in rural areas [16,48]. The impact of SOC sequestration goes beyond improving soil properties and synergizes with other biophysical ecosystem services, positively affecting further non-material ecosystem services—or nature's contribution to people—by providing learning opportunities and inspiration, as well as physical and physiological experiences, and supporting identities [49,50].

3.1.4. Possible Drawbacks and Recommendations

Various biophysical, technical, social, economic, cultural, and political barriers can restrict the adoption of CT worldwide. For any given location, the success or failure of CT will depend on one or more of the following factors.

(a)   Biophysical barriers

Local pedoclimatic conditions (soil type—particularly its organic matter content and texture—rainfall amount and distribution, and temperature), together with slope, crop type (arable vs. woody crops, water requirements, growing period, rooting characteristics), and management (rainfed vs. irrigated, conventional vs. organic), determine the viability of field operations, as well as whether crops will be established and their yields. For example, in arid and semiarid regions, the adoption of CT may be hampered because of competition for water and nutrients between the plant cover and the main crop [51,52], this effect being more visible when aridity and temperature increase [43]. On the other hand, in water-logged and heavy-clay soils (e.g., rice fields), reduced tillage is hampered. Moreover, on-site and off-site soil and water contamination problems may arise if pesticides and inorganic fertilizers are applied in high doses [16]. It is also important to note that the positive impacts of reducing tillage operations (i.e., reducing direct emissions from the activities and fostering sinks via SOC sequestration) can be counterbalanced by the increase in soil N_2O emissions in cases where higher doses of inorganic fertilizer are needed, as has been pointed out in a recent meta-analysis [53]. However, the results vary depending on: (1) the duration of the experiment and (2) the management type (e.g., fertilizer type and application rate, use and type of spontaneous and planted cover crops, and crop residue management); therefore, no general conclusions can be drawn. In this regard, CT needs to be accompanied by wise management of nitrogen and weeds.

(b)   Technical barriers

One technical barrier that hinders the wider adoption of NT practices in Europe is the unavailability of proper machinery, such as direct drilling machines or machinery to manage crop residues or cover crops [54]. For instance, the adoption of direct sowing is still a challenge for many crops—in particular, small-seed crops—in silty soils prone to crust and heavy compaction in the topsoil [55]. Due to their weight, direct drilling machines may cause soil compaction, hampering the germination of seeds and the effective establishment of seedlings [56]. Although machines can be adapted to specific soil conditions to reduce soil compaction, the increase in purchase costs makes them unaffordable for farmers. In addition to these technical constraints, one important limitation is the heavy dependence on herbicides and pesticides, which can lead to severe pollution of soil, water, and biodiversity resources. To overcome this problem, the development of cheap alternative methods for weed control is pivotal. In this regard, wise management of ground cover and cover crops (i.e., selection of species and varieties to combat weeds, promotion of mechanical instead of chemical termination, leaving plant residues on the soil surface, combining CT with other practices to control weeds, etc.) is recommended.

(c)   Economic barriers

The absence of financial incentives or subsidies to motivate farmers or compensate them for possible yield losses restricts the adoption of CT practices in many regions. Generally, yields are reduced in the short term, but this trend can be reverted in the long term, especially if CT is adopted in combination with other practices (e.g., addition of organic or green manure [57]). As already mentioned, CT normally encompasses the use of agrochemicals (pesticides and mineral fertilizers), resulting in increased costs that farmers cannot afford [58]. In many cases, investments in adapted machinery are necessary but not affordable by farmers because of limited finance and access to capital for implementation. Uncertainty about the development of policies and market fluctuations, together with internal farm factors (such as farm size, debt, tenure, and family status), are other important barriers to overcome.

(d)   Social, cultural, and political barriers

Lack of access to appropriate technologies, practices, and equipment is a major barrier in many countries [48]. Moreover, there are other many factors that hinder the adoption of CT, such as farm size and type, the availability of a power source, family structure and composition, the labour situation, access to cash and credit facilities, peer pressure, the degree of autonomy in choosing and implementing results, and community support [59,60]. The main cultural factors that hamper the adoption of CT are lack of awareness among farmers, lack of innovativeness, lack of motivation, and lack of understanding of the agroecosystem [61,62]. In some regions, CT is in conflict with an important cultural symbol of hard work, as tillage is generally believed to symbolize a hard worker, and with the social recognition that a field properly ploughed is "clean" [56,62].

The objectives and priorities of each government will determine how agriculture is managed at the regional and national levels. Lack of economic incentives and support from governments, including subsidies [60]—and, in particular, the lack of strictness in legislation and standards [63]—are the main reasons why the adoption of CA, despite its well-known agro-environmental benefits in the long term, continues to fail in many countries. In this regard, carbon schemes and other political initiatives are urgently needed (see Section 6).

3.2. Permanent Plant Cover

Permanent plant cover refers to those practices involving the growth of a permanent spontaneous or seeded plant cover within the crop system (intercropping systems) or between periods of normal crop production for soil protection and improvement. In the case of spontaneous plant covers, weeds grow in accordance with the pedoclimatic conditions of the area, and species are typically wild species. When the plant cover is seeded by employing what is known as a cover crop, species are selected for which the products can be harvested for food or feed. They may be leguminous (e.g., vetch) so that the cover crop can help to improve the N content, the crops may be used for forage or human consumption (e.g., rye, rapeseed), or mixtures of two or more species may be employed. Spontaneous plant cover can either be removed with a reduced tillage operation so that the plant residues are quickly incorporated into the soil or left on the soil surface; thus, the incorporation of the C and other nutrients will be slower. When seeded cover crops are harvested, their residues are usually left on the soil surface.

3.2.1. Context of Application

Growing a permanent plant cover in intercropping systems is more commonly found with woody crops, since competition for water and nutrients between the woody crop and the plant cover is lower than in the case of arable crops, for which plant covers are usually adopted between normal crop production periods.

Permanent plant covers can be adopted worldwide. However, in rainfed agriculture, they are highly dependent on the precipitation regime. Thus, in arid climates, water availability conditions can place strong limitations on the growth of a permanent plant cover. Regarding the species, a wide variety of wild or seeded species can be grown; therefore, the species composition of the plant cover should be adapted to the specific pedoclimatic conditions and management practices [64]. In the case of seeded cover crops, economic viability plays an important role [65].

3.2.2. Potential for SOC Sequestration

The immediate effect of protecting soil and improving soil conditions is an increase in SOC content. However, the extent of this increase varies with the type of crop, the pedoclimatic conditions, and the specific management practices. Thus, it can range from 0.27 to 1.03 t C ha^{-1} yr^{-1} (Table 3).

Table 3. Summary of meta-analyses assessing SOC sequestration rates in different locations and climatic zones. Authors' elaboration published in [66].

Location	Climate Zone	Additional C Storage Potential (t C ha^{-1} yr^{-1})	Duration (Years)	Cropping System	Reference
Regional	Warm temperate dry	0.27	10.6	A + W	[27]
Global	Arid, temperate, and tropical	0.56	8.5	A	[67]
Regional	Warm temperate dry	0.43	5.6	AC + W	[43]
Regional	Warm temperate dry	1.01	6.7	PC + W	[43]
Global	Temperate and tropical	0.32	11.9	A	[68]
Regional	Warm temperate dry	1.03	7.7	W	[1]

A—arable crops, W—woody crops, AC—annual cover crops, PC—permanent cover crops.

The highest values were achieved for woody crops under warm temperate conditions at around 1.0 t C ha^{-1} yr^{-1}, whereas this figure was between 0.3 and 0.6 t C ha^{-1} yr^{-1} for arable crops. This was mainly due to the lower soil disturbance when the plant cover is grown in the inter-row area of woody crops than in the case of arable crops. Average SOC sequestration rates are typically higher in low-duration experiments, when SOC levels are closer to the equilibrium, than in longer ones [69].

3.2.3. Co-Benefits

Cover crops and spontaneous plant covers have been reported to not only increase SOC content but also improve other physical (e.g., aggregate stability, water infiltration, and bulk density) [70], chemical (e.g., N, P, and K contents) [71,72], and biological (e.g., microbial diversity, abundance, and activity) properties [73,74], leading to a decrease in the effects of wind and water erosion [70,71] and to higher and more stable yields [75]. SOC increases and improvements in other chemical properties are especially visible with spontaneous plant covers, where the biomass is left on the soil surface or incorporated into the soil with reduced tillage. Cover crops are harvested and eventually used for animal feed or biofuel production [76].

Another benefit of planting cover crops is the weed control resulting from the competition for light, water, and nutrients or the release of allelopathic exudates [75,77]. Cover crops reduce weed density and biomass during the growth of the subsequent cash crop by 10% and 5%, respectively [69]. Weed competition among winter and early-season weeds has been found to have an important role during cover crop growth. On the other hand, growth of spontaneous plant cover in orchards can improve biodiversity and, thus, pollination services and pest control [78].

For cover crops, the diversity of species is also a key driver for the delivery of ecosystem services. However, it becomes a problem when it comes to selecting which species should be planted and how they should be mixed. Indeed, in addition to species diversity, the functional complementary between species is of high importance when mixing plant species. Thus, objective criteria for the selection of species with functional complementary and, thus, the maximization of the delivered ecosystem services have been established [79].

However, it is not only biophysical (provisioning, regulating, and supporting) services that are improved but also cultural and economic ones. The improvement in the soil properties and the competition with weeds in the case of cover crops can lead to reductions in inorganic fertilization and pesticide application, thus leading to lower dependence on external inputs and positive effects on human safety [71]. On the other hand, these better soil conditions might lead to higher and more stable yields [80] because of the increase in the soil resilience. Moreover, in the case of spontaneous plant covers in orchards, improvements to landscape quality (e.g., rural aesthetics) must be considered, as well as other derived socio-economic benefits (e.g., ecotourism and recreational activities) [81].

3.2.4. Possible Drawbacks and Recommendations

The main trade-off from cover crops might be the GHG emissions associated with the decomposition of the organic matter [71], which can be mitigated by using cover crops with low C:N ratios and minimizing the tillage intensity [82,83]. However, these extra emissions might take place only after removing the plant cover. In this context, it has been found that, in olive orchards with spontaneous plant cover and conventional management (weeds controlled with herbicides), the overall amount of CO_2 emissions was negative in both treatments (i.e., sinks) but, in the case of the spontaneous plant cover management, the CO_2 uptake was double that for the conventional management (-140 and -70 g C m^{-2} yr^{-1}, respectively) [84]. This was mainly due to the increase in the photosynthesis of the plant cover during the growing season, which offset the CO_2 emissions after the removal of the plant cover.

In addition to the direct effects from the different management practices on ecosystem services, trade-offs between the different ecosystem services affected by the management practices should be considered [71]. For instance, two positive impacts of cover cropping are the increase in seed production and the increase in faunal activity. However, increases in granivorous faunal activity increase seed predation and, therefore, the subsequent cover crop growth may be negatively affected. These authors suggested applying various tillage activities that could help to control the populations of these insects. Another trade-off is the increase in the soil N content from leguminous cover crops that improve nutrient cycling but, at the same time, may stimulate nematode populations and weed abundance [85,86]. Therefore, determining the right mixture between legume and non-legume species and the right tillage activities could be a way to mitigate these trade-offs between ecosystem services and, at the same time, maintain yields [71]. Cover crop mixtures represent an optimal way to overcome some of these trade-offs [87–89]. Nevertheless, these trade-offs are less common with woody crops under spontaneous plant covers where different and adapted species appear and, therefore, greater self-regulation is achieved.

3.3. Crop Diversification

Crop diversification (CD) is a farming system that encourages the cultivation of different plant species in the same field as opposed to monoculture farming [90,91]. There are different options for implementing CD, such as crop rotations (at least two crops in different years), multiple cropping (different crops grown in succession during the same year), and intercropping (crops grown together on the same field). In intercropping, crops can be planted in alternate rows and harvested together (row intercropping) or in wide rows and mechanically harvested separately (strip intercropping), or they can be sown together (mixed intercropping); i.e., with no separation between rows or strips.

In addition to allowing a higher number of crops to be grown and alternated on a field, CD has several objectives [7]:

(a) Covering and protecting the soil from climatic agents in a continuous and effective way;
(b) Maintaining and improving soil structure through the action of the root systems of the plants;
(c) Stimulating biological activity in the soil and eliminating periods with no crop cover;
(d) Limiting environmental risks due to nitrate leaching, erosion and surface runoff, and loss of biodiversity.

3.3.1. Context of Application

CD can potentially be applied worldwide, but barriers to its adoption can emerge from biophysical constraints and cultural and socio-economic factors. In arid and semiarid environments, the climate is warm, and low rainfall limits the cultivation of summer crops if irrigation cannot be supplied; thus, cropping systems are mainly based on winter crops, such as cereals and pulses. Conversely, in cold and wet environments, cropping systems are mainly based on spring–summer crops, since low temperatures, snow accumulation, and the surplus of water during the autumn–winter months can restrict crop growth.

3.3.2. Potential for SOC Sequestration

In a global data analysis of 97 paired treatments from long-term experiments (Figure 2), the results indicated that enhancing rotation complexity (i.e., changing from monoculture farming to continuous rotation cropping or from crop–fallow to continuous monoculture or rotation cropping or increasing the number of crops in a rotation system) increased SOC by 0.15 ± 0.11 t C ha^{-1} on average [92]. However, changing from continuous corn to corn–soybean rotation did not help sequester C (-0.19 ± 0.19 t C ha^{-1}) due to the lower residue return and C input in the rotation compared to the corn monoculture. This result was consistent with findings from the Midwestern USA [93] reporting a SOC loss of 0.15 t C ha^{-1} for corn–soybean rotations with NT and residue incorporation. Not considering corn–soybean rotation, the average SOC sequestration rates were 0.20 ± 0.12 t C ha^{-1} and 0.16 ± 0.14 t C ha^{-1} under conventional tillage and 0.26 ± 0.56 t C ha^{-1} with NT rotations. Rotations with grass, hay, or pasture increased SOC by 0.19 ± 0.08 t C ha^{-1} on average. Decreasing the fallow period in wheat experiments (e.g., changing from a wheat–fallow rotation to a wheat–wheat–fallow rotation) and rotating wheat with one or more different crops (e.g., wheat–sunflower or wheat–legume rotations) increased SOC by 0.51 ± 0.47 t C ha^{-1} and were more effective than changing from wheat–fallow to continuous wheat farming (0.06 ± 0.08 t C ha^{-1}).

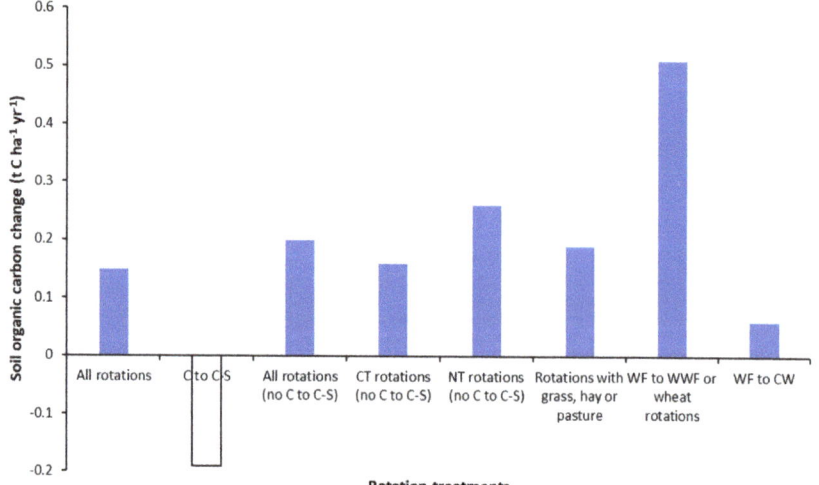

Figure 2. Effects of rotation complexity on SOC change. C to C-S—continuous corn to corn–soybean, CT—conventional tillage, NT—no tillage, WF to WWF—wheat–fallow to wheat–wheat–fallow, WF to CW—wheat–fallow to continuous wheat. Authors' elaboration based on [92].

The effects of the number of crops included in a rotation were investigated in a meta-analysis including 122 studies with 454 observations [94]. The results indicated that total soil C (TC) increased by 3.6% on average with the addition of one or more crops in the rotation compared to a monoculture. TC increased by 1.9% with two crops in the rotation, 7.5% with three crops, and 3.7% with four crops. The highest TC responses to rotation were found for soybean (11%), sorghum (7.9%), and wheat (2.9%) monocultures, but rotations did not increase soil C compared to corn monocultures. The introduction of a cover crop in the rotation increased TC by 7.8%, but no significant effect was found in rotations without cover crops. Mean annual temperature and rainfall were correlated positively with rotation effects on TC.

A recent data analysis (304 paired samples) assessed SOC content as affected by CD in different European regions [90]. SOC increased by 18% compared to the control treatment (no rotation/no legumes) when adopting more complex rotations and introducing legume crops. In contrast, SOC decreased in long rotations without legumes (6%) and in short rotations with legumes (3–5%). Furthermore, SOC increases were greater in semiarid climates (11%) compared to humid and sub-humid conditions. The results also indicated greater SOC increases (28%) 2–10 years after adopting CD; in contrast, SOC changes were showed a decreasing trend after 11–20 years (6%) and became definitely negative (−6%) in sites where CD had been adopted for very long time periods (>20 years), showing that a steady-state condition was reached.

Table 4 illustrates the Spearman rank correlation analysis for changes in SOC and several pedoclimatic and predictive variables of CD (e.g., rotation, tillage, fertilization, and residue management). The significant negative coefficient found for the duration of the experiment in years (−0.45) indicates that SOC changes were greater when CD had been established more recently, while significant positive coefficients for SOC changes were found for crop rotations ≥ 3 years (0.61), legumes in the rotations (0.60), conventional tillage (0.22), and the removal of crop residues (0.58). Negative coefficients were also found for no tillage (−0.32), residue incorporation and mulching (−0.31 and −0.33, respectively), mixed fertilization (−0.16), autumn–winter cereals in the Southern Mediterranean region (−0.45), and clay and loam textures types (−0.28 and −0.41, respectively). Positive coefficients were found for semiarid climates (0.16), autumn–winter cereals of the Northern Mediterranean region (0.17), and sandy clay loam soil textures (0.61).

Table 4. Spearman rank correlation coefficients (rs) for changes in SOC content and the predictive variables for crop diversification. Authors' data based on [88].

Variable	Coefficient
SOC control *	0.20
Rotation every 3 years *	0.61
Years *	−0.45
Legumes *	0.60
Cover crop	−0.04
Conventional tillage *	0.22
No tillage *	−0.32
Mineral fertilization	0.08
Mixed fertilization *	−0.16
Organic fertilization	0.11
Residue incorporated *	−0.31
Residue mulched *	−0.33
Residue removed *	0.58
Semiarid *	0.16
Subhumid	−0.14
MedNCerAw *	0.17
MedSCerAW *	−0.45
BorFodMix	0.11
Clay *	−0.28
Loam *	−0.41
Sandy clay loam *	0.61

The asterisks (*) indicate both positive and negative correlations with significant coefficients at $p < 0.05$ above rs = |0.15|.

3.3.3. Co-Benefits

Adopting CD provides further benefits for soil properties. It can lead to an overall improvement in soil structure resulting from the aggregation of mineral particles and organic materials. Germination and rooting of crops are facilitated by the higher resistance of the soil aggregates to physical stress [95,96], and better soil aggregation also improves carbon storage due to the physical protection of organic materials [97]. Furthermore, soil crusting and erosion are avoided [98].

CD also improves soil biological properties because different crop species have different C:N ratios (residue qualities), which enhance the activities of different types of soil microorganisms. Furthermore, adopting CD is more effective if coupled with other CA practices (e.g., RT or NT). For example, in the temperate conditions of the northeastern USA [99], adopting NT and cover crops in a crop rotation (maize and perennial grass) system improved active C, respiration, and protein content. Similarly, a rotation system with wheat and forage crops enhanced the microbial biomass carbon under both rainfed and irrigated conditions (by 0.4 and 14.9%, respectively) in comparison to continuous wheat cropping [100]. Furthermore, soil microbial richness and diversity were increased by C (15.11 and 3.36%, respectively) [36].

Crop rotations also spread out the need for labour, reduce equipment costs and peak labour demand, smooth out price fluctuations in markets, and increase local community interaction for labour [101]. However, the possible lack of a market for the alternative crops adopted for CD can represent an economic barrier [102].

Crop yields are generally higher if crops are cultivated after unrelated species, which is known as the break-crop effect. Cultivating a break crop increased wheat yield from 0.5 to 1.2 t ha^{-1}, particularly when wheat was cultivated after legumes (e.g., faba beans and chickpeas) [103]. It has also been reported that longer and more complex crop rotations increased yields by 12% compared to monocropping systems, and the increase was lower (5%) for the shortest rotations (2 years) [104].

Nitrogen fertilization is the main agricultural contributor of soil nitrous oxide (N_2O) emissions to the atmosphere [105]. Since legumes fix the atmospheric N that is available for plant nutrition, their adoption in CD implies the supply of lower amounts of N fertilizers, thus reducing the N_2O emissions and mitigating their global warming potential. Some examples have been reported from the Northern Great Plains of North America [106], France [107], and Australia [108].

3.3.4. Possible Drawbacks and Recommendations

Local pedoclimatic conditions (e.g., rainfall and soil texture) can limit the cultivation of some crops in specific environments, and the market opportunities for each crop included in the diversification scheme must be considered. As already mentioned, cereals (e.g., wheat) have high nitrogen requirements; thus, including N_2-fixing legumes in the planned rotations can both increase cereal yields and limit nitrogen losses through N_2O emissions and leaching [109].

Farmers often perceive CD negatively because they fear possible decreases in yields and economic benefits. However, the crops adopted in diversification should be those already grown locally as monocultures, since they have been proven to be suitable for the soil and climatic conditions and provide good yields. Therefore, farmers only need to learn how to use them in rotations, multiple cropping, or intercropping systems. However, not all farmers are skilled in CD. Therefore, providing adequate training to agricultural technical advisors is crucial to successfully disseminate diversified cropping systems among farmers [102,104].

4. Processes

4.1. The Soil Carbon Balance and Different Processes of SOC Loss in Agroecosystems

Soil carbon storage in agricultural systems is governed by the difference that exists between the carbon inputs from crop biomass (roots plus aboveground crop residues after harvesting and pruning) and any endogenous (e.g., ground covers) and/or exogenous (e.g., manure, compost, sludge, and/or cover crops) organic matter added to the soil, on the one hand, and the carbon outputs, as affected by erosion, leaching, and the decomposition and mineralization of plant material and organic matter at both short- and long-term scales, on the other hand (Figure 3).

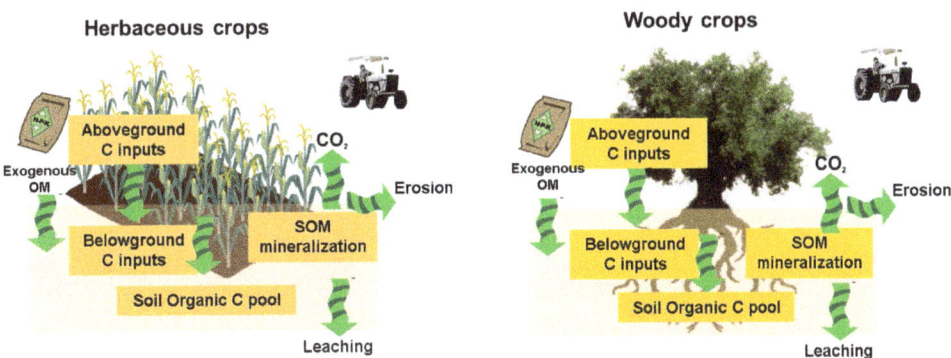

Figure 3. The soil carbon balance in herbaceous (**left**) and woody (**right**) cropping systems depends on the difference between carbon inputs and outputs throughout the agroecosystem, as explained in the text. © María Almagro.

To date, most studies focused on the assessment of the effects of CA practices on the soil carbon balance have been based on empirical data derived from field plots and laboratory assays, and scaling up the potential for carbon sequestration from the farm to the global scale still remains a challenge for the scientific community. An excellent indicator of the effectiveness of a certain CA practice is undoubtedly the increase in SOC content, given its well-known agro-environmental benefits and its potential for climate change mitigation [16,110,111]. However, further research and a robust monitoring, verification, and reporting framework are still needed to increase carbon gains and address the limitations of SOC sequestration [112]. In this regard—and given the huge uncertainty associated with SOC estimations at the farm level, particularly in the short term—it is recommended that long-term monitoring programs assess SOC changes a decade after the implementation of the CA practice. However, shorter-term assessments may be needed to guide policy debates and decisions. To address this, estimations of the carbon gains and losses occurring throughout the agroecosystem when a certain management practice is adopted can be used to anticipate decisions concerning agriculture management based on early assessments of SOC net balances. In other words, if the annual amounts of carbon entering the soil due to the addition of organic amendments, the implementation of cover crops, and crop residue retention exceed the carbon losses through erosion, leaching, and decomposition, the SOC balance will be positive and SOC sequestration will be achieved in the short term. However, if the opposite occurs, organic carbon will be lost from the soil system. Notably, each of the described process causing carbon gains or losses at the agroecosystem level contributes differently depending on the specific site (i.e., local climate, soil and crop type, slope, etc.) and management (i.e., rainfed vs. irrigated regime, low- vs. high-input systems) conditions, which, in turn, drive the direction and magnitude of the impact on the net soil carbon balances.

4.2. Erosion by Water

Soil water erosion refers to lateral movement of soil downhill caused by significant rainfall events. Soils in natural ecosystems are considered to be under steady-state conditions, as the loss of soil material due to erosion from a given area is approximately balanced by the formation of soil as a result of weathering [113]. However, manmade actions, such as intensive agriculture, deforestation, and soil sealing, have increased soil erosion rates by 10–40 times globally, causing on-site and off-site negative environmental impacts [114]. On-site negative impacts include the loss of carbon and nutrients, such as nitrogen and phosphorus, from the topsoil, reducing soil fertility and crop productivity and causing land degradation and desertification, as recognized by the Soil Thematic Strategy [115] of the European Commission. Off-site effects include sedimentation of reservoirs, eutrophication of water bodies, and damage to infrastructure [116]. Although normally overlooked, this process is relevant for the soil carbon and nutrient budgets of agroecosystems. Moreover, soil erosion interacts with other relevant processes, such as organic matter decomposition, SOC sequestration, and net primary productivity [117–119].

A global review of empirical data indicated that soil erosion rates from conventionally ploughed agricultural fields (~1 mm yr^{-1}) are, on average, one to two orders of magnitude greater than the rates of soil formation, erosion under native vegetation, and long-term geological erosion [113]. These results prove that conventional tillage is unsustainable, particularly in Mediterranean regions where extreme high-rainfall events can cause great soil losses through erosion in a few hours [120]. The same study also indicated that NT systems produce erosion rates much closer to soil formation rates (~3 mm century^{-1}, as reported in [121]), highlighting their contributions to soil conservation, mitigation of climate change through the retention of carbon in soils, and sustainable agriculture. However, the success of the various CT practices that can be adopted depends on the local environmental conditions (e.g., soil type, climate, and management practices), as well as the socio-economic, cultural, and political contexts [48,54]. Moreover, the combination of two or more CA practices is generally more effective than the adoption of a single agricultural practice. A comprehensive overview of different soil and water conservation practices in Europe and in the Mediterranean Basin indicated that annual runoff and soil loss rates can be reduced by 20–74% if CA practices are adopted [122]. However, this review also concluded that vegetation management practices (such as the adoption of cover crops and mulching) were the most effective in reducing annual runoff and soil loss rates, followed by mechanical techniques (such as terraces, contour bounds, and geotextiles) and soil management practices (such as NT, RT, contour tillage, and soil amendment), which were the least effective in controlling runoff and erosion. These results highlight the importance of ensuring permanent soil cover in order to reduce soil erosion rates globally. Nevertheless, the more erosion-prone conditions are (i.e., erodible soils, steeper slopes, areas with low-frequency occurrence of high-intensity rainfall events), the more effective these CA practices will be in reducing runoff and soil erosion rates.

4.3. Decomposition

Decomposition refers to the physical, biological, and chemical breakdown and leaching of soluble compounds of plant biomass residues (leaves, shoots, and roots) and soil organic matter (SOM), along with the subsequent mineralisation and humification of organic compounds [123]. Decomposition is one of the most important processes in terrestrial ecosystems because it controls SOM formation and the release of organic nutrients and energy for plant growth and soil microorganisms [124,125]. Moreover, it is a major component of carbon and nutrient cycling in ecosystems and a key driver of soil fluxes of carbon dioxide (CO_2), methane (CH_4), and nitrous oxide (N_2O) into the atmosphere. It is estimated that 60 Pg of CO_2 is emitted annually by the decomposition of plant litter and SOM [126].

Among the main environmental drivers of plant litter decomposition are temperature, moisture availability, the chemical composition of plant litter, and the soil biotic community structure and activity, which altogether control carbon and nutrient sequestration efficiency in agricultural soils [127]. However, despite the importance of this process, it is still unclear which environmental factors control it and how we can ensure that a significant proportion of the decomposed plant material is returned to the soil instead of released into the atmosphere in the form of CO_2 and N_2O. This is particularly important in arid and semiarid environments, such as in many Mediterranean regions, where solar ultraviolet (UV) radiation has been identified as a significant driver of plant litter decomposition [128,129]. In this process—known as photodegradation—solar radiation directly breaks down organic matter components, releasing CO_2 and other gases and, thus, promoting the direct loss of carbon and nutrients from ecosystems into the atmosphere without incorporation into the SOM pool [128]. Photodegradation is a complex process in which several abiotic (e.g., ambient temperature and moisture, plant residue chemical composition) and biotic (e.g., local microbial community response to solar UV radiation) factors interact; as a result, its net effect on plant litter decomposition can be positive, negative, or neutral [130,131]. A recent meta-analysis showed that exposure of plant residues to solar radiation sped up decomposition by 23% [129]. Therefore, photodegradation can negatively impact the SOC content and fertility level in semiarid agricultural soils if crop residues are not wisely managed in these environments. In other words, if photodegradation dominates the decomposition process under certain environmental conditions, facilitating the direct loss of carbon and nutrients from plant residues into the atmosphere, then it may be desirable to incorporate them into the soil through RT (rather than leaving them on the soil surface as mulching) to promote SOC sequestration and fertility.

On the other hand, under arid and semiarid conditions, the decomposition of plant residues mediates soil inorganic carbon (SIC) dynamics and can, therefore, change the net carbon balances of agricultural systems, converting them into sources or sinks depending on their management (rainfed vs. irrigated, the chemical composition of crop residues, and crop residue incorporation into the soil vs. mulching) and local conditions (mainly mean annual precipitation and soil pH). Specifically, the fate of the released CO_2 during the decomposition and mineralization of plant residues can lead to formation or dissolution of pedogenic carbonate, leading to its sequestration or to its direct release into the atmosphere, depending on the aridity conditions and soil pH [132].

4.4. Leaching

Soil leaching is the downward movement of nutrients (i.e., nitrate, phosphorus, and base cations) and other constituents in the soil profile, such as dissolved organic carbon (DOC) and dissolved inorganic carbon (DIC), following the percolation of rain or irrigation water. This process occurs when the soil pores become filled with water and water moves downward in the soil, hampering the availability of soil nutrients for plants and, therefore, reducing soil fertility and plant yield [133]. Moreover, leaching may cause environmental problems, such as eutrophication, when large amounts of certain nutrients move into ground- and surface water.

Natural ecosystems normally have a point of equilibrium between demand and supply for nutrients, with a closed loop recycling essential nutrients. However, in agricultural cropping systems, the supply of nutrients normally exceeds the demand; therefore, leaching occurs. Global change drivers, such as climate change, land-use change, and agriculture intensification and contamination, affect soil leaching trends.

Leaching of DOC and DIC represents a relatively small but continuous loss of carbon from terrestrial ecosystems. However, only a few studies have estimated carbon losses through leaching in different land-use systems; thus, their contribution to the net ecosystem carbon balance is uncertain [134]. Additionally, climate change may increase the frequency of extreme precipitation events in arid and semiarid regions, leading to increases in SOC losses through both leaching and respiration [135].

For instance, the levels of SOC leached across Europe from forests, grasslands, and croplands have been estimated to be 15.1, 32.4, and 20.5 g C m^{-2} yr^{-1}, respectively, which represent 4, 14, and 8% of net ecosystem exchanges, respectively [134]. On the other hand, leaching of biogenic DIC in the same land-use types accounted for lower losses (8.3, 24.1, and 14.6 g m^{-2} yr^{-1} for forests, grasslands, and croplands, respectively) [134].

Additionally, leaching of carbon stored in surface litter and soil layers is considered a main source of DIC and DOC in inland waters [136]. In particular, SIC is more prone to leaching in arid and semiarid regions than SOC via sporadic high precipitation events [135]. This is of great relevance, since SIC stocks and sequestration rates are between two and ten times higher than those for SOC in these areas [137].

5. Practices

5.1. Conventional Tillage

As mentioned above, SOC sequestration rates vary among studies depending on the local climate conditions, soil characteristics, initial SOC levels, crop type (arable vs. woody cropping systems), previous and current management (rainfed vs. irrigated; low vs. high input systems), and the duration of the experiments. Nevertheless, the increase in SOC in soils is limited in time by the carbon saturation level, and, after a certain point, the rate of accumulation slows down towards a plateau, depending on the soil type, the length of the growing period, and the climatic conditions [26].

It was demonstrated that reducing tillage improved soil aggregation and the protection of organic carbon within the aggregates against erosion or ploughing in two organic rainfed almond orchards under semiarid Mediterranean conditions [42]. The promising results from reducing tillage intensity and frequency were further confirmed by Martínez-Mena et al. [41], who demonstrated that passing from conventional moldboard ploughing at a 40 cm depth (5–7 passes yr^{-1}) to minimum tillage at a 20 cm depth (2 passes yr^{-1}) in a rainfed cereal field and an organic almond field reduced soil erosion by 65% and 85%, respectively, preventing the carbon losses associated with this process. As a result, SOC stocks at a 30 cm depth increased by 37% and 25%, respectively, in the cereal field and the almond field after six years. On the other hand, however, it was also found that shifting from minimum tillage at a 15 cm depth (twice per year) to NT did not significantly reduce soil CO_2 emissions from the soil and negligibly improved SOC stocks (by 1%) after four years in an organic rainfed almond orchard under the same semiarid Mediterranean conditions [42]. Furthermore, crop yields decreased abruptly from the beginning of the cessation of tillage, making this practice unsustainable for local farmers. The failure of NT in this particular case study can be explained by the fact that no fertilization was applied [39], highlighting the importance of adopting NT in combination with other practices, such as addition of organic or green manure, in order to improve N management in semiarid rainfed woody crop systems [138]. Indeed, in an irrigated woody cropping system (i.e., *Citrus limon*) where drip ferti-irrigation was applied together with the addition of pruning residues as mulching, NT was proved to be successful in enhancing SOC stocks, soil aggregation, and OC physicochemical protection at 0–5 cm soil depths after 20 years, thus improving soil structure and halting carbon losses [139]. Nevertheless, given the high spatial variability observed when measuring SOC in agricultural fields, long-term studies are encouraged to assess SOC stock trends over time and thereby estimate average SOC sequestration rates more accurately. For example, SOC was sequestered at a rate of 1 t C ha^{-1} yr^{-1} when shifting from conventional to RT at a 20 cm depth after 10 years in an organic rainfed woody crop system, while a rate of 0.33 t C ha^{-1} yr^{-1} was obtained when shifting from RT to NT at a 15 cm depth under the same conditions [140].

For traditional cereal–fallow rotation and a continuous cropping system with barley under semiarid Mediterranean conditions in northeastern Spain, the results indicated that the adoption of RT (with chisel ploughing at 25–30 cm depths) and NT in formerly conventionally tilled (with mouldboard ploughing at 30–40 cm depths) fields improved soil aggregate formation and stability, as well as the OC content associated with them, after 15 years, particularly in the NT system under continuous cropping [141].

In the north of France, the effects of changing from conventional full-inversion tillage to NT and shallow tillage in combination with different crop management systems (i.e., crop types, residue removal, rotation, and use of catch crops) on SOC stocks were compared after 41 years [142]. The authors demonstrated that tillage and crop residue management had no significant effects on SOC stocks after 41 years at either the formerly ploughed layer (i.e., 0–28 cm) or in whole soil profile (0–58 cm). In the shallow and NT systems, SOC content increased in the surface layer (0–10 cm), reaching a plateau after 24 years, but declined continuously in the subsurface layer (10–28 cm) at rates of 0.42–0.44% yr^{-1}. In both the RT and NT systems, SOC sequestration rates increased rapidly during the first four years and then remained more or less constant at average rates of 2.17 and

1.31 t C ha^{-1} yr^{-1}, respectively, for the next 24 years, after which they started to decrease. The authors attributed these drops to the water balance in those years, stating that the studied cropping systems sequestered less SOC in wet compared to dry periods, which is the opposite of what occurs under semiarid conditions.

In Lithuania, impacts on SOC sequestration were assessed when shifting from conventional tillage to RT and NT in combination with different fertilization levels in a crop rotation system including winter wheat, spring oilseed rape (*Brassica napus* L.), spring wheat, spring barley (*Hordeum vulgare* L.), and pea (*Pisum sativum* L.) in which crop residue retention was implemented [143]. The SOC sequestration rates were estimated in two long-term (11 years) experiments set up on loam and sandy-loam textured soil. In this study case, NT enhanced SOC sequestration by 5–35% compared to the conventional and RT systems when fertilizer was applied. Specifically, the adoption of NT increased the SOC stocks in the loam soil by 27 and 7% and the SOC stocks in the sandy-loam soil by 29 and 33% compared to the conventional and RT systems, respectively.

The abovementioned contrasting findings highlight the importance of understanding the effects of tillage on SOC sequestration and its interaction with environmental and management factors before drawing conclusions on the potential of CT itself for SOC sequestration.

5.2. Cover Cropping

In a meta-analysis of 51 studies and 144 datasets, an average value for the SOC sequestration rate of about 1 t C ha^{-1} yr^{-1} for spontaneous plant covers and cover crops was estimated for Mediterranean woody crops [1]. However, it has been shown that, under Mediterranean climatic conditions, the proportion of non-protected SOC (i.e., available for decomposition and, therefore, not really sequestered SOC) might be between 10% and 50% of the total SOC [144]. Nevertheless, these authors also found that the amount of total SOC in the spontaneous plant cover would be two times higher than that found for conventional management, and statistically significant differences were found for all SOC fractions and the two considered depths (0–5 cm and 5–15 cm), suggesting that the consequences of vegetation cover for SOC extend beyond particulate organic matter and might affect all protected SOC fractions in the first 15 cm.

Similar results were found for grass and legume cover crops in vineyards in Australia [145], where significantly higher concentrations of total, coarse, and fine organic C for the grass–legume mixture and grass-only cover crops were found. However, for the legume-only cover crops, significantly higher values were achieved only for coarse SOC (Table 5). In this study, it was also found that, for mixed cover crops, the total N was generally higher, and extractable N was 75% higher than for the control; furthermore, importantly, plant-available N was 17% greater than with legumes alone. Therefore, a combination of grass and legumes had a positive effect not only on total SOC, including fine particles, but also on the total and plant-available N.

However, even though it is not defined as really sequestered SOM, easily mineralizable organic carbon might play an important role in microbial activity. In an integrated crop–livestock (ICL) system in the USA that included livestock grazing on cover crops and crop residues in agricultural systems, it was found that easily mineralizable SOC and labile C might play important roles in shifting the bacterial community structure and composition in the soil [146]. In particular, these authors found that, in ICL systems, compared to the control, cold-water- and hot-water-soluble carbon levels were increased by 88% and 185%, respectively. These increases in easily mineralizable organic C were associated with significant increases in microbial enzymatic activities (dehydrogenase, fluorescein diacetate, urease, and β-glucosidase activities).

Table 5. Means ($\pm$ standard errors) for dependent variables, bulk soil, and coarse and fine SOC fractions by treatment obtained from linear mixed effects models examining the effects of cover crop type on dependent variables at four sites (n = 16). Different lowercase letters represent significant differences between treatment groups (α = 0.05). Authors' elaboration based on [145].

Cover Crop	SOC Bulk Soil (mg g^{-1})	SOC Coarse Fraction (mg g^{-1})	SOC Fine Fraction (mg g^{-1})
Grass only	14.22 $\pm$ 1.22 b	7.42 $\pm$ 1.43 b	35.50 $\pm$ 4.74 b
Legume only	13.62 $\pm$ 1.10 a	6.96 $\pm$ 1.15 b	32.77 $\pm$ 4.23 a
Mixture	14.64 $\pm$ 1.29 b	9.36 $\pm$ 1.86 b	34.56 $\pm$ 4.55 b
Control	11.41 $\pm$ 1.02 a	5.36 $\pm$ 1.01 a	30.57 $\pm$ 3.91 a
p-value	<0.01	<0.01	0.02

A clear relationship between cover crops, SOC, and soil biological parameters was also found for Andisols in Japan with arable crops [147]. Results showed that a combination of NT and the use of rye as a cover crop could enhance SOC and soil health parameters (total N, available P, exchangeable K-Mg, CEC, bulk density, soil penetration resistance, and substrate-induced respiration) in soybean crops After a Z-score assessment, these authors found a positive effect from the use of rye as a cover crop, especially for soil biological and chemical features, and it significantly increased the cover crop biomass input

Finally, regarding the SOC dynamics over time and long-term SOC sequestration, a simulation study of SOC dynamics was performed for NT with cover crops (winter cereal) and conventional tillage in a continuous maize system in the USA for the period 1970–2099. The results showed that, in 1970–2018, the SOC gains were 0.22 t C ha^{-1} yr^{-1}. However, sequestration rates under climate change were much lower, with gains equal to 0.031 t C ha^{-1} yr^{-1} with NT compared to conventional tillage in the IPCC RCP 8.5 scenarios and lower SOC losses in the case of the RCP 2.6 scenarios of -0.002 vs. -0.017 t C ha^{-1} yr^{-1} for NT with cover crops and conventional tillage, respectively [148].

5.3. Crop Diversification

A recent meta-analysis [149] demonstrated that CD generally improves pollination and pest control, water regulation, carbon sequestration, nutrient cycling, and crop yields, although exceptions to this general trend were also observed. In this context, long-term field experiments (LTEs) can be used as robust research instruments for the study of ecosystem productivity and sustainability because they capture the changes in and relationship between cropping systems, agricultural management, and the fluctuating environment at different time points over long periods. LTEs also make it possible to quantify the effects of CD through crop rotations on SOC storage.

Two monocultures (continuous corn and continuous soybean) and three rotations (soybean–corn, soybean–winter wheat, and soybean–winter wheat–corn) were evaluated in an LTE using conventional tillage (CT) and NT in Canada (Ontario) [150]. After 11 years (Figure 4), for the soybean–corn rotation compared to continuous corn farming, SOC was higher by 9.6 t C ha^{-1} (18.8%) using CT, lower by 18.4 t C ha^{-1} (26%) using NT, and lower by 3.8 t C ha^{-1} (6.3%) on average. For the soybean–winter wheat–corn rotation compared to continuous corn farming, SOC was slightly higher in using (0.4 t C ha^{-1}, 0.8%) and lower using NT (1.3 t C ha^{-1}, 1.8%) and on average (0.3 t C ha^{-1}, 0.5%). For the soybean–winter wheat rotation compared to continuous soybean farming, SOC was higher by 33.8 t C ha^{-1} (74.8%) using CT, 17.5 t C ha^{-1} (28.1%) using NT, and 26.3 t C ha^{-1} (49.5%) on average. For the soybean–winter wheat–corn rotation compared to continuous soybean farming, SOC was higher by 6.3 t C ha^{-1} (13.9%) using CT, 7.3 t C ha^{-1} (11.7%) using NT, and 6.8 t C ha^{-1} (12.8%) on average. The overall results suggest the efficacy of the incorporation of winter wheat in the rotations, adopting soybean–wheat and soybean–winter wheat–corn rotations rather than monocultures based on corn and soybean or soybean–corn rotations.

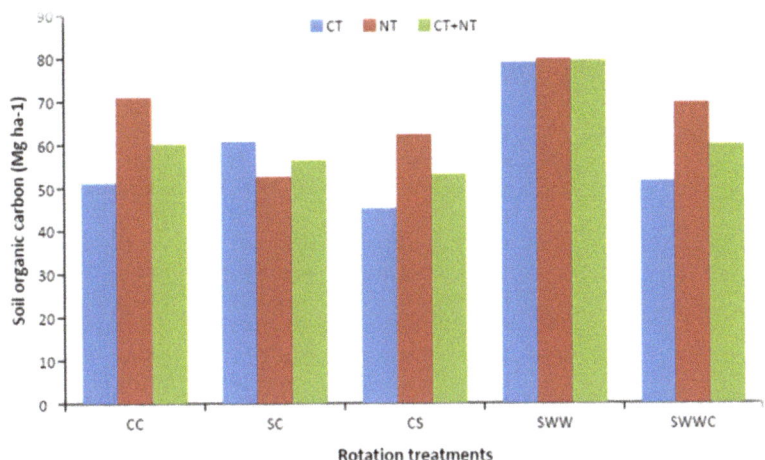

Figure 4. Effects of rotation treatments on SOC storage. CC—continuous corn, SC—soybean–corn, CS—continuous soybean, SWW—soybean–winter wheat, SWWC—soybean–winter wheat–corn, CT—conventional tillage, NT—no tillage. Authors' elaboration based on [150].

A study performed in the semiarid Pampean region of Argentina (Buenos Aires Province) over 15 years examined three treatments with and without fertilizer inputs: continuous wheat (WW), 1 year of wheat followed by 1 year of grazing of natural grasses (WG), and 2 years of wheat followed by 2 years of legume (clover, vetch) and grass (barley, oat, triticale) mixtures (WL) [151]. The results demonstrated the positive influence of the inclusion of legumes (WL) on SOC, as well as that of alternate cattle grazing (WG), while continuous wheat showed the lowest SOC storage. Compared to continuous wheat with no fertilization, SOC increased by 3.1 t C ha^{-1} (7.7%) and 3.8 t C ha^{-1} (9.5%) in the WG and WL treatments, respectively. With fertilization (64 kg N ha^{-1} and 16 kg P ha^{-1}), SOC increased by 1.8 t C ha^{-1} (4.1%) and 7.6 t C ha^{-1} (17.4%) in the WG and WL treatments, respectively (Figure 5).

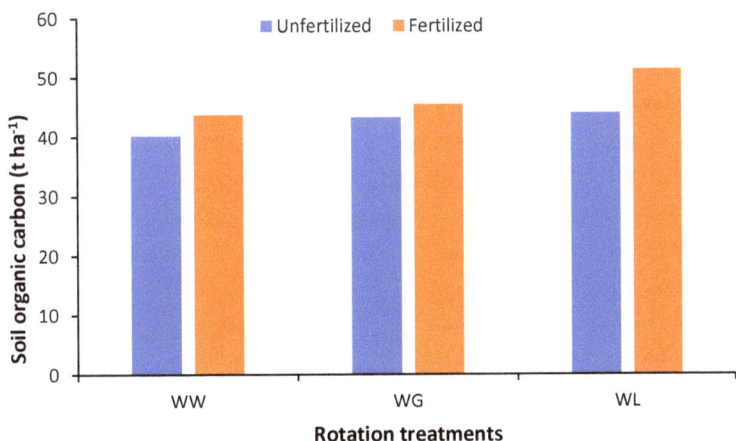

Figure 5. Effects of rotation treatments on SOC storage. WW—continuous wheat, WG—1 year of wheat followed by 1 year of grazing of natural grasses, WL—2 years of wheat followed by 2 years of legume and grass mixtures. Authors' elaboration based on [151].

An LTE (18 years) performed in the Western Corn Belt (NE, USA) evaluated three monocultures (continuous corn, continuous soybean, and continuous grain sorghum), two 2 year rotations (corn–soybean and grain sorghum–soybean), and two 4 year rotations (oat+clover–grain sorghum–soybean–corn and soybean–grain sorghum–oat+clover–corn) using three nitrogen fertilization levels (0, low, and high—i.e., 0–90–180 kg N ha^{-1}—for corn and sorghum and 0–34–68 kg N ha^{-1} for soybean and oat+clover) [152]. Compared to the corn monoculture (Table 6), SOC increased in the no-fertilizer treatment by 5.6 t C ha^{-1} (1.8%), 9.9 t C ha^{-1} (21.0%), and 7.8 t C ha^{-1} (16.5%) with the corn–soybean, oat+clover–grain sorghum–soybean–corn, and soybean–grain sorghum–oat+clover–corn rotations, respectively. In relation to soybean monoculture, SOC increased in the oat+clover–grain sorghum–soybean–corn rotation by 3.7 t C ha^{-1} (7.0%) and 4.0 t C ha^{-1} (7.5%) with low and high fertilization rates, respectively. Compared to sorghum monoculture, SOC increased in the oat+clover–grain sorghum–soybean–corn rotation by 3.7 t C ha^{-1} (7.0%) and 4.0 t C ha^{-1} (7.5%) with low and high fertilization rates, respectively. SOC increased in the oat+clover–grain sorghum–soybean–corn rotation by 3.3 t C ha^{-1} (6.2%) and 3.0 t C ha^{-1} (5.5%) with low and high fertilization rates, respectively. The overall results indicated that the 4 year rotations with oat+clover crops represented the best option compared to the corn, soybean, and grain sorghum monocultures or 2 year rotations.

Table 6. SOC comparisons from 2002 for each rotation and N level.

N Fertilization	0 N		Low N		High N	
Rotation	Delta SOC	%	Delta SOC	%	Delta SOC	%
Corn						
C-SB vs. CC	5.6	11.8	3.4	6.9	1.5	2.9
OCL-SG-SB-C vs. CC	9.9	21.0	7.0	14.2	6.2	12.1
SB-SG-OCL-C vs. CC	7.8	16.5	4.2	8.6	3.9	7.6
Soybean						
C-SB vs. CSB	−1.4	−2.5	0.1	0.1	−0.7	−1.3
SG-SB vs. CSB	−5.3	−9.8	−3.1	−6.0	−4.3	−8.0
OCL-SG-SB-C vs. CSB	3.0	5.5	3.7	7.0	4.0	7.5
SB-SG-OCL-C vs. CSB	0.9	1.6	0.9	1.7	1.7	3.2
Sorghum						
SG-SB vs. CSG	−6.1	−11.1	−3.5	−6.6	−5.3	−9.7
OCL-SG-SB-C vs. CSG	2.2	4.0	3.3	6.2	3.0	5.5
SB-SG-OCL-C vs. CSG	0.1	0.1	0.5	0.9	0.7	1.3

CC—continuous corn, CSB—continuous soybean, CSG—continuous grain sorghum, C-SB—corn–soybean, SG-SB—grain sorghum–soybean, OCL-SG-SB-C—oat+clover–grain sorghum–soybean–corn, SB-SG-OCL-C—soybean–grain sorgum–oat+clover–corn. Authors' elaboration based on [152].

6. Policy Options

6.1. European Union Policy Options

6.1.1. The Soil Thematic Strategy

The Thematic Strategy for Soil Protection is a Communication from the European Commission to the other European institutions [115] involving a 10 year work program for the European Commission. The strategy aims at protecting soil and preserving its capacity to perform its functions in environmental, economic, social, and cultural terms. The strategy includes a legislative framework with four goals: (1) protecting and sustainably using soil, (2) integrating soil protection into national and EU policies, (3) improving knowledge in this area, and (4) increasing public awareness. The proposal for a Directive represents a key component of the strategy, enabling Member States to adopt context-specific measures (e.g., identification of areas at risk of erosion, organic matter depletion, soil compaction, or salinisation) as part of the obligation to adopt programmes of measures addressing causes and impacts.

The European Environment Agency [153] indicated that the lack of a comprehensive and coherent policy framework to protect land and soil is a key gap that may limit the EU's ability to meet future goals. A new policy framework is, therefore, needed, as the 2006 EU Soil Thematic Strategy [115] is no longer adapted to the current policy context and the scientific evidence. This impasse seems close to an end, since the EU Biodiversity Strategy for 2030 [154] provides an update to the 2006 EU Soil Thematic Strategy, aiming to achieve land degradation neutrality by 2030. It highlights the importance of increasing efforts to protect soil fertility, reduce erosion, and increase soil organic matter. Thus, the EU has put soil and land at the core of most of the Sustainable Development Goals (SDGs) of the UN Agenda 2030, particularly SDG 15.3: "combat desertification, restore degraded land and soil, including land affected by desertification, drought and floods, and strive to achieve a land degradation-neutral world by 2030".

6.1.2. The Common Agricultural Policy (CAP) (2023–2027)

In this review, it was shown that increasing SOC with different sustainable management practices results in potential synergies with other ecosystem services. The main European Union instrument used to address sustainability issues in agriculture is the Common Agricultural Policy (CAP). The so-called "Green Architecture" of the new CAP has three specific objectives relating to environmental and climate issues:

1. Contribute to climate change mitigation and adaptation;
2. Foster sustainable development and efficient management of natural resources;
3. Contribute to the protection of biodiversity, thus enhancing ecosystem services and preserving habitats and landscapes.

The new architecture is based on three components, retaining two pillars of the previous architecture (Pillar 1 relating to direct payments) [155]:

(1) Eco-schemes (voluntary, Pillar 1): direct payments to farmers for the implementation of sustainable management. This is a novel feature of the new Green Architecture, and such schemes can be adapted to the specific needs of the different Member States at the national and/or regional levels. Eco-schemes are intended to play an important role in the new CAP, since 100% of the funding comes directly from the EU and, therefore, no extra funding from Member States is needed;

(2) Agri-environment–climate measures (AECM) (voluntary, Pillar 2): these measures aim to address environmental and climate challenges using Rural Development Programmes;

(3) Enhanced conditionality (mandatory, Pillar 1): this component sets out the basic and mandatory requirements that farmers and managers must fulfil in order to receive payments. The requirements refer to the implementation of good agricultural and environmental conditions (GAECs); e.g., maintenance of permanent grasslands, banning of burning arable stubble, implementation of buffer strips in water courses, use of tools for nutrient management, adoption of reduced tillage, avoidance of bare soils in sensitive periods, crop rotation, preservation of a share of the total agricultural area for landscape measures, and banning of the conversion of permanent grasslands in Natura 2000 sites.

The different sustainable management practices addressed in this review relate to the three different CAP components. Importantly, the avoidance of bare soils in most sensitive periods, the use of crop rotations, and the maintenance of a certain ratio of permanent grassland to agricultural areas are practices included in the conditionality, and they involve some of the management techniques previously assessed (e.g., reduced tillage, crop diversification, and cover crops). Fulfilling these requirements would enable farmers to receive the area- and animal-based payments under both Pillars 1 and 2.

6.1.3. The European Green Deal

The European Green Deal [156] was set out by the European Commission in December 2019 with two overall objectives:

- "Transform the EU into a fair and prosperous society, with a modern, resource-efficient and competitive economy where there are no net emissions of greenhouse gases in 2050 and where economic growth is decoupled from resource use";
- "Protect, conserve and enhance the EU's natural capital, and protect the health and well-being of citizens from environment-related risks and impacts".

In order to achieve these goals, the European Commission has established a set of transformative policies addressing different environmental and socio-economic challenges. Most of them are directly or indirectly related to SOC sequestration and preserving or improving soil-supporting functions. They are briefly described below.

Climate Initiatives

The aim of the climate initiatives is to achieve climate neutrality by 2050 [157], enshrining this objective in legislation. To do so, the first European "Climate Law" will be launched, and the targets of reducing GHG emissions by at least 50% compared to 1990 levels by 2030 and towards 55% have been proposed. More specifically, and beyond the creation of a trading system in industry, the aim is to include GHG emissions and removals from land use, land use changes, and forestry [158]. Finally, and in addition to the future efforts in mitigation, the Commission will adopt a new and more ambitious [158] EU strategy on adaptation to climate change, including nature-based solutions [159], where SOC sequestration will play a central role.

From "Farm to Fork": Designing a Fair, Healthy, and Environmentally Friendly Food System

Within the frame of the Green Deal, the EU has developed the "from farm to fork" concept [160] and adapted it to EU biophysical and socio-economic conditions. Thus, the goals of this initiative are "to reduce the environmental and climate footprint of the EU food system and strengthen its resilience, ensure food security in the face of climate change and biodiversity loss and lead a global transition towards competitive sustainability from farm to fork and tapping into new opportunities".

The aim is, therefore, threefold. First, this initiative aims to ensure that the food chain has a neutral or positive environmental impact (preserving and restoring the land-, freshwater-, and sea-based resources on which the food system depends), helping to mitigate climate change and facilitate adaption to its impacts, protect resources (land, soil, water, air) and animal health and welfare, and reverse the loss of biodiversity. Second, the aim is to ensure food security, nutrition, and public health. Third, the initiative aims to preserve the affordability of food while generating fairer economic returns in the supply chain.

Again, SOC sequestration and sustainable management practices in agriculture will play important roles in achieving a more sustainable and fairer European food system. For instance, organic farming is supposed to be promoted through the implementation of the Action Plan on Organic Farming [161], which aims to achieve organic farming on 25% of the total agricultural land in the EU by 2030.

Preserving and Restoring Ecosystems and Biodiversity

The EU has launched the Biodiversity Strategy for 2030 [154] in order to address the biodiversity loss that is threatening food systems and the implementation of healthy and nutritious diets while preserving rural livelihoods and agricultural production in the face of reductions in pollination. The strategy is framed as part of the ambition to "ensure that by 2050 all of the world's ecosystems are restored, resilient, and adequately protected". The strategy links agricultural land management and biodiversity preservation by:

- "Bringing nature back to agricultural land" through the promotion of eco-schemes and results-based payment schemes and by ensuring that the CAP strategic plans include realistic and robust climate and environmental criteria and targets. These plans should include practices such as organic farming, agro-ecology, and agro-forestry. Furthermore, as also suggested by the EU Pollinators Initiative [162], the overall use of chemical pesticides should be reduced by 50% by 2030. The strategy also aims to restore at least 10% of agricultural areas occupied by high-diversity landscape features (inter alia, buffer strips, rotational or non-rotational fallow land, hedges, non-productive trees, terrace walls, and ponds) in order to enhance SOC sequestration and prevent soil erosion and depletion. Finally, the decline in genetic diversity will be addressed by modifying the marketing rules for traditional crop varieties in order to promote their conservation and sustainable use;

- "Addressing land take and restoring ecosystems" in order to protect soil fertility, reduce soil erosion, and increase SOC through the adoption of sustainable management practices. To promote these practices, the Commission updated the EU Soil Thematic Strategy in 2021. Soil sealing and rehabilitation of contaminated brownfields will be part of the Strategy for a Sustainable Built Environment;
- "Bringing nature back to cities" by calling on European cities of at least 20,000 inhabitants to develop ambitious Urban Greening Plans by the end of 2021 incorporating nature-based solutions;
- "Reducing pollution" through the implementation of the EU Chemicals Strategy for Sustainability [163], the Zero Pollution Action Plan for Air, Water and Soil [164], and the Nutrient Management Action Plan in 2022 [165], which aim to reduce the use of fertilizers by at least 20% and the risks related to and use of pesticides.

The Need for the Integration of CAP Reform and the Green Deal

At this point, it should be clear that there are many links between the new CAP and the Green Deal. However, the new CAP does not directly consider these links, but they are implicitly included in the conditionality and its indicators. However, in the case of eco-schemes, Member States will be in charge of expanding the sustainability of the agricultural sector beyond the requirements of the conditionality throughout the development of CAP Strategic Plans.

The working paper of the European Commission concludes that the CAP reform proposal is compatible with the Green Deal and the associated strategies and initiatives. Nevertheless, to consider these linkages, among others, realistically, it proposes:

- an adequate "no backsliding" principle obliging Member States to be more ambitious in their CAP Strategic Plans than at present regarding environmental and climate-related goals;
- an ambitious system of conditionality to maintain key standards (in particular, for crop rotation, soil cover, and maintenance of permanent grassland and agricultural land devoted to non-productive areas or features);
- mandatory eco-schemes.

The eco-schemes are of critical importance in implementing the climate, air, water, soil, and biodiversity EU goals with regard to country-based and regional specificities. Thus, they should cover those management practices not included in the conditionality. Although some attempts to link soil management with different EU policies have been developed (e.g., Healthy Soils, the EU Soil Observatory, the European Soil Data Centre) [166,167], the reality is that the eco-schemes proposed by many EU countries—and, especially, those related to conservation agriculture—are not ambitious enough, being at best reformative and addressing some specific issues.

Therefore, the challenges of the new CAP for the future are: (1) to integrate the CAP with the Green Deal and its policies and instruments; (2) to increase the ambition of the CAP, particularly the eco-schemes and the enhanced conditionality; and (3) to adopt a systemic view so that the new architecture of the CAP can contribute to the systemic transformation of the agri-food system (Figure 6).

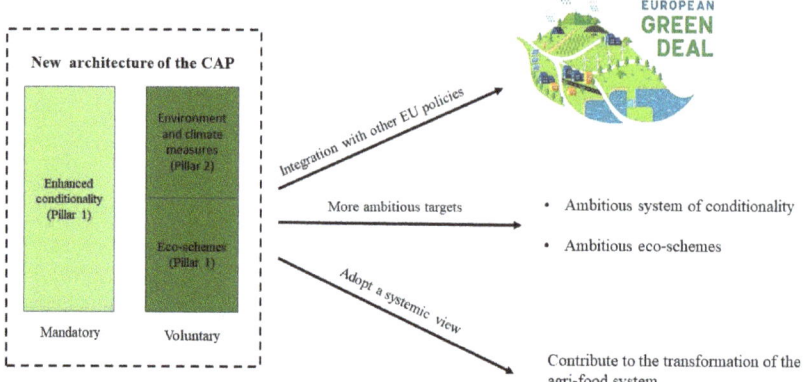

Figure 6. The three challenges related to the new architecture of the European Common Agricultural Policy (CAP). Authors' elaboration based on the cited literature.

6.2. Other International Relevant Policies

6.2.1. The 4 per 1000 Initiative

Climate change is expected to have relevant impacts on SOC dynamics, since the rising atmospheric CO_2 concentration could increase biomass production and the crop residues returned to soils. However, increasing temperatures could reduce SOC by accelerating microbial decomposition. The 4 per 1000 initiative Soils for Food Security and Climate, launched by the French Government in 2015 during the 21st Session of the Conference of the Parties of the United Nations Framework Convention on Climate Change in Paris (http://4p1000.org/, accessed on 14 February 2023), is a voluntary action plan aiming at better management of SOC in agricultural soils. The objective is to achieve a 4‰ annual growth rate for SOC stocks in the top 40 cm of soils (i.e., 0.4 per cent per year) as a compensation for the global emissions of greenhouse gases (GHGs) from anthropogenic sources, thus limiting global warming to 2 °C.

Sequestration rates differ between countries and climatic conditions, but a general trend for the relationships between different management practices and SOC accumulation rates has been observed [168]: afforestation—~0.6 t C ha^{-1} yr^{-1}, conversion to pasture—~0.5 t C ha^{-1} yr^{-1}, organic amendments—~0.5 t C ha^{-1} yr^{-1}, residue incorporation—~0.35 t C ha^{-1} yr^{-1}, no or reduced tillage—~0.3 t C ha^{-1} yr^{-1}, and crop rotation—~0.2 t C ha^{-1} yr^{-1}. However, there is a tendency to find higher C sequestration potential (10–30 per 1000) in croplands with low initial SOC stock ($\leq$30 t C ha^{-1}) (i.e., high C saturation deficit). In addition, sequestration rates can reach up to 20 per 1000 within the first 5 years after the adoption of sustainable management practices and up to 10 per 1000 after 20 years, then becoming limited to 4 per 1000 after 40 years. However, despite these data, there are still some scientific and policy challenges for the implementation of this initiative [169]:

- The scarcity of scientific data. Research data on rates of SOC sequestration resulting from the implementation of recommended management practices (RMPs) for land use and agricultural management combinations are not widely available;
- The finite capacity of soil carbon sinks. The potential for SOC sequestration in global croplands is finite (0.4 to 1.2 Gt). Thus, SOC sequestration by itself cannot offset all emissions but must be part of a wider set of actions, including the adoption of RMPs that reduce C emissions and enhance C sinks;
- Resource-poor farmers and small landholders who are unable to adopt RMPs because of weak institutional support and poor access to essential inputs. These farmers' degraded and depleted soils need urgent restoration through SOC sequestration and the adoption of RMPs;

- Financial commitments. The adoption of RMPs would require economic resources;
- Permanence. Incentivising the continuous use of RMPs and restorative land uses, as has been undertaken by some successful programs in the EU and USA, is of crucial importance and must be addressed;
- Implementation of the Paris Agreement's 4 per 1000 program. Even though the limitation of global warming to 1.5 °C is required, the word "soil" is never mentioned. Therefore, this is a new challenge for soil scientists and agronomists.

6.2.2. Sustainable Development Goals (SDGs)

During the United Nations General Assembly in 2015 [170], the 2030 Agenda for Sustainable Development was adopted. The 2030 Agenda indicates a set of 17 Sustainable Development Goals (SDGs) aiming to "end hunger and poverty, to protect the planet, and to ensure peace and prosperity for all". Each SDG includes specific Targets to be achieved between 2015 and 2030 and to be implemented at the national scale.

Soil sustainable management is directly related to half of the SDGs and indirectly relevant for the other SDGs [171]. The 2030 Agenda adopted specific Targets aiming to restore degraded soils, achieve land degradation neutrality worldwide, and implement agricultural practices to improve soil quality and reduce soil contamination [170].

The stock of SOC has strong interactions with all environmental compartments (e.g., water and air) and supports many soil-derived ecosystem services [172]. Thus, increasing SOC stocks is related to many SDGs and Targets, such as Target 2.4 (improving land and soil quality), Target 15.3 (achieving a land degradation-neutral world), and Goal 13 on climate action, which evaluates climate change and its impacts, aiming to regulate C storage and GHGs [173] and use soil as a C pool [115].

7. Concluding Remarks, Future Research Needs, and Policy Recommendations

The widespread adoption of conservation agriculture (CA) principles (i.e., conservation tillage (CT), permanent plant cover, and crop diversification (CD)) could contribute to the mitigation of climate change without compromising food security from local to global scales. However, there are still many scientific knowledge gaps to be filled, as well as biophysical, technical, socio-economic, cultural, and political barriers to overcome, before its adoption can be enabled among farmers worldwide. In this regard, the success or failure of the adoption of any CA practice will depend on the environmental–socio-economic context; therefore, institutional guidance should be planned and created from local to regional scales. Likewise, providing adequate training for farmers to help them implement CA—particularly the adoption of cover crops and CD—and opening up market opportunities for new products are necessary steps in the transition to more sustainable and diversified cropping systems.

CA improves the physical, chemical, and biological properties of soil that are crucial for maintaining soil condition and health, and adopting any of the three principles of CA has beneficial effects on soil organic carbon (SOC), as has been demonstrated worldwide. However, SOC sequestration rates and their co-benefits vary among studies depending on the local pedoclimatic and management conditions, and further research is needed to determine the optimal agricultural management practices within each environmental, socio-economic, and legal context. Attention must also be paid to trade-offs. Furthermore, each of the three CA principles needs to be accompanied by wise and integrated nitrogen and weed control management to ensure sustainable crop yields.

The challenge will be monitoring and verifying that the different sustainable management practices are being applied correctly and assessing how they impact the different ecosystem services. For these purposes, suitable and feasible indicators must be clearly defined. In this regard, the increase in SOC content is an excellent indicator of the effectiveness of a certain CA practice, given its well-known agro-environmental benefits and its potential for climate change mitigation. However, further research and a robust monitoring, verification, and reporting framework are still needed to accurately assess

SOC gains and address the limitations of SOC sequestration. Given the huge uncertainty associated with SOC estimations at the farm level, particularly in the short term, long-term monitoring programs are recommended to accurately assess the SOC gains associated with CA practices. However, short-term monitoring is also needed to guide policy decisions on agriculture management using early assessments of net SOC balances. Likewise, a better understanding of the major processes involved in SOC losses—i.e., erosion, abiotic decomposition, and leaching—and how to curb them is necessary to guarantee the success of CA practices.

Many different strategies, initiatives, and regulations relating to soil ecosystem services have been developed at the regional, national, and international levels, and more will arise in the upcoming years. However, since soil ecosystem services are closely interlinked with other biophysical and socio-economic services, strong coherence between the different initiatives (i.e., the Common Agricultural Policy (CAP), the Farm to Fork Strategy, the Biodiversity Strategy for 2030, the 4 per 1000 initiative, and the Climate Law) is highly recommended. The Sustainable Development Goals (SDGs) could be a suitable framework to achieve this coherence. In addition, considering the new CAP and the recommendations already made by the staff of the European Commission, we encourage Member States to propose ambitious and mandatory eco-schemes in their CAP Strategic Plans involving management practices that aim to increase SOC content and improve and protect soil conditions (e.g., CT, cover crops, diversify cropping systems, etc.) by setting up specific indicators and targets for SOC accumulation in the upcoming years. Nevertheless, trade-offs between increasing SOC storage and GHG emissions should be included in the assessments.

Author Contributions: All authors contributed equally in the different phases of the development of the manuscript: Conceptualization, investigation, literature review, R.F., M.A., J.L.V.-V.; Writing—original draft preparation, R.F., M.A., J.L.V.-V.; Writing—review and editing, and visualization, R.F., M.A., J.L.V.-V. All authors have read and agreed to the published version of the manuscript.

Funding: This research received no external funding.

Data Availability Statement: Not applicable.

Acknowledgments: María Almagro acknowledges the financial support from the Spanish Ministry of Science, Innovation and Universities through the "Ramón y Cajal" Program (RYC2020-029181-I).

Conflicts of Interest: The authors declare no conflict of interest.

References

1. Vicente-Vicente, J.L.; García-Ruiz, R.; Francaviglia, R.; Aguilera, E.; Smith, P. Soil carbon sequestration rates under Mediterranean woody crops using recommended management practices: A meta-analysis. *Agric. Ecosyst. Environ.* **2016**, *235*, 204–214. [CrossRef]
2. Powlson, D.S.; Stirling, C.M.; Thierfelder, C.; White, R.P.; Jat, M.L. Does conservation agriculture deliver climate change mitigation through soil carbon sequestration in tropical agro-ecosystems? *Agric. Ecosyst. Environ.* **2016**, *220*, 164–174. [CrossRef]
3. Kirschenmann, F. Alternative agriculture in an energy- and resource-depleting future. Renew. *Agric. Food Syst.* **2010**, *25*, 85–89. [CrossRef]
4. Wezel, A.; Goris, M.; Bruil, J.; Félix, G.; Peeters, A.; Bàrberi, P.; Bellon, S.; Migliorini, P. Challenges and Action Points to Amplify Agroecology in Europe. *Sustainability* **2018**, *10*, 1598. [CrossRef]
5. Conservation Agriculture. Available online: https://www.fao.org/conservation-agriculture/overview/what-is-conservation-agriculture/en/ (accessed on 13 February 2023).
6. Kassam, A.; Friedrich, T.; Shaxson, F.; Pretty, J. The spread of Conservation Agriculture: Justification, sustainability and uptake. *Int. J. Agric. Sustain.* **2009**, *7*, 292–320. [CrossRef]
7. Page, K.L.; Dang, Y.P.; Dalal, R.C. The Ability of Conservation Agriculture to Conserve Soil Organic Carbon and the Subsequent Impact on Soil Physical, Chemical, and Biological Properties and Yield. *Front. Sustain. Food Syst.* **2020**, *4*, 31. [CrossRef]
8. Kassam, A.; Friedrich, T.; Derpsch, R. Global spread of Conservation Agriculture. *Int. J. Environ. Stud.* **2018**, *76*, 29–51. [CrossRef]
9. Friedrich, T.; Kassam, A.H.; Shaxson, F. Conservation Agriculture. In *Agriculture for Developing Countries*; Science and Technology Options Assessment (STOA) Project; European Technology Assessment Group: Karlsruhe, Germany, 2009.
10. Ohiri, A.C.; Ezumah, H.C. Tillage effects on cassava (Manihot esculenta) production and some soil properties. *Soil Tillage Res.* **1990**, *17*, 221–229. [CrossRef]
11. Lal, R. Tillage effects on soil degradation, soil resilience, soil quality, and sustainability. *Soil Tillage Res.* **1993**, *27*, 1–8. [CrossRef]

12. Antapa, P.L.; Angen, T.V. Tillage practices and residue management in Tanzania. In *Organic-Matter Management and Tillage in Humid and Sub-Humid Africa*; Food and Agriculture Organization of the United Nations: Rome, Italy, 1990; pp. 49–57.
13. Greenland, D.J. Soil management and soil degradation. *J. Soil Sci.* **1981**, *32*, 301–322. [CrossRef]
14. Claassen, R.; Bowman, M.; McFadden, J.; Smith, D.; Wallander, S. *Tillage Intensity and Conservation Cropping in the United States*; EIB-197; U.S. Department of Agriculture, Economic Research Service: Washington, DC, USA, 2018; p. 27.
15. Unger, P.W.; Langdale, G.W.; Papendick, R.I. Role of crop residues—Improving water conservation and use. In *Cropping Strategies for Efficient Use of Water and Nitrogen*; Hargrove, W.L., Ed.; American Society of Agronomy: Madison, WI, USA, 1988; Volume 51, pp. 69–100.
16. Stavi, I.; Bel, G.; Zaady, E. Soil functions and ecosystem services in conventional, conservation, and integrated agricultural systems. A review. *Agron. Sustain. Dev.* **2016**, *36*, 32. [CrossRef]
17. Parr, J.F.; Papendick, R.I.; Hornick, S.B.; Meyer, R.E. The use of cover crops, mulches and tillage for soil water conservation and weed control. In *Organic-Matter Management and Tillage in Humid and Sub-Humid Africa*; Food and Agriculture Organization of the United Nations: Rome, Italy, 1990; pp. 246–261.
18. Blanco-Canqui, H.; Ruis, S.J. No-tillage and soil physical environment. *Geoderma* **2018**, *326*, 164–200. [CrossRef]
19. Peixoto, D.S.; Silva, B.M.; de Oliveira, G.C.; Moreira, S.G.; da Silva, F.; Curi, N. A soil compaction diagnosis method for occasional tillage recommendation under continuous no tillage system in Brazil. *Soil Tillage Res.* **2019**, *194*, 104307. [CrossRef]
20. Haramoto, E.R.; Brainard, D.C. Strip Tillage and Oat Cover Crops Increase Soil Moisture and Influence N Mineralization Patterns in Cabbage. *Hortscience* **2012**, *47*, 1596–1602. [CrossRef]
21. Lahmar, R. Adoption of conservation agriculture in Europe: Lessons of the KASSA project. *Land Use Policy* **2010**, *27*, 4–10. [CrossRef]
22. Garcia-Franco, N.; Almagro, M. Strip, precision, zone tillage. In *Recarbonizing Global Soils: A Technical Manual of Best Management Practices*; FAO, ITPS, Eds.; Food and Agriculture Organization of the United Nations: Rome, Italy, 2021; Volume 3, pp. 85–95.
23. Nowatzki, J.; Endres, G.; Dejong-Hughes, J.; Aakre, D. *Strip Till for Field Crop Protection*; University of Minnesota Extension and NDSU Extension Service: St. Paul, MN, USA; Fargo, ND, USA, 2011; p. 10.
24. Blanco-Canqui, H.; Wortmann, C.S. Does occasional tillage undo the ecosystem services gained with no-till? A review. *Soil Tillage Res.* **2020**, *198*, 104534. [CrossRef]
25. Almagro, M.; Díaz-Pereira, E.; Boix-Fayos, C.; Zornoza, R.; Sánchez-Navarro, V.; Re, P.; Fernández, C.; Martínez-Mena, M. The combination of crop diversification and no tillage enhances key soil quality parameters related to soil functioning without compromising crop yields in a low-input rainfed almond orchard under semiarid Mediterranean conditions. *Agric. Ecosyst. Environ.* **2023**, *345*, 108320. [CrossRef]
26. Gonzalez-Sanchez, E.J.; Veroz-Gonzalez, O.; Conway, G.; Moreno-Garcia, M.; Kassam, A.; Mkomwa, S.; Ordoñez-Fernandez, R.; Triviño-Tarradas, P.; Carbonell-Bojollo, R. Meta-analysis on carbon sequestration through Conservation Agriculture in Africa. *Soil Tillage Res.* **2019**, *190*, 22–30. [CrossRef]
27. Aguilera, E.; Lassaletta, L.; Gattinger, A.; Gimeno, B.S. Managing soil carbon for climate change mitigation and adaptation in Mediterranean cropping systems: A meta-analysis. *Agric. Ecosyst. Environ.* **2013**, *168*, 25–36. [CrossRef]
28. Pardo, G.; del Prado, A.; Martínez-Mena, M.; Bustamante, M.A.; Martín, J.R.; Álvaro-Fuentes, J.; Moral, R. Orchard and horticulture systems in Spanish Mediterranean coastal areas: Is there a real possibility to contribute to C sequestration? *Agric. Ecosyst. Environ.* **2017**, *238*, 153–167. [CrossRef]
29. Francaviglia, R.; Di Bene, C.; Farina, R.; Salvati, L.; Vicente-Vicente, J.L. Assessing "4 per 1000" soil organic carbon storage rates under Mediterranean climate: A comprehensive data analysis. *Mitig. Adapt. Strat. Glob. Chang.* **2019**, *24*, 795–818. [CrossRef]
30. Sheehy, J.; Regina, K.; Alakukku, L.; Six, J. Impact of no-till and reduced tillage on aggregation and aggregate-associated carbon in Northern European agroecosystems. *Soil Tillage Res.* **2015**, *150*, 107–113. [CrossRef]
31. Special Report on Climate Change, Desertification, Land Degradation, Sustainable Land Management, Food Security, and GHG Fluxes in Terrestrial Ecosystems. Available online: https://www.ipcc.ch/site/assets/uploads/2019/08/Fullreport-1.pdf) (accessed on 12 December 2022).
32. Alliaume, F.W.A.H.; Rossing, W.A.H.; Tittonell, P.; Jorge, G.; Dogliotti, S. Reduced tillage and cover crops improve water capture and reduce erosion of fine textured soils in raised bed tomato systems. *Agric. Ecosyst. Environ.* **2014**, *183*, 127–137. [CrossRef]
33. Blanco-Canqui, H.; Francis, C.A. Building resilient soils through agroecosystem redesign under fluctuating climatic regimes. *J. Soil Water Conserv.* **2016**, *71*, 127A–133A. [CrossRef]
34. Li, Y.; Li, Z.; Cui, S.; Jagadamma, S.; Zhang, Q.P. Residue retention and minimum tillage improve physical environment of the soil in croplands: A global meta-analysis. *Soil Tillage Res.* **2019**, *194*, 104292. [CrossRef]
35. Paredes, D.; Cayuela, L.; Gurr, G.M.; Campos, M. Is ground cover vegetation an effective biological control enhancement strategy against olive pests? *PLoS ONE* **2015**, *10*, e0117265. [CrossRef]
36. Venter, Z.S.; Jacobs, K.; Hawkins, H.-J. The impact of crop rotation on soil microbial diversity: A meta-analysis. *Pedobiologia* **2016**, *59*, 215–223. [CrossRef]
37. Almagro, M.; de Vente, J.; Boix-Fayos, C.; García-Franco, N.; Melgares de Aguilar, J.; González, D.; Solé-Benet, A.; Martínez-Mena, M. Sustainable land management practices as providers of several ecosystem services under rainfed Mediterranean agroecosystems. *Mitig. Adapt. Strat. Glob. Chang.* **2016**, *21*, 1029–1043. [CrossRef]

38. Biddoccu, M.; Ferraris, S.; Pitacco, A.; Cavallo, E. Temporal variability of soil management effects on soil hydrological properties, runoff and erosion at the field scale in a hillslope vineyard, North-West Italy. *Soil Tillage Res.* **2017**, *165*, 46–58. [CrossRef]
39. Martínez-Mena, M.; Carrillo-López, E.; Boix-Fayos, C.; Almagro, M.; Franco, N.G.; Díaz-Pereira, E.; Montoya, I.; de Vente, J. Long-term effectiveness of sustainable land management practices to control runoff, soil erosion, and nutrient loss and the role of rainfall intensity in Mediterranean rainfed agroecosystems. *Catena* **2020**, *187*, 104352. [CrossRef]
40. Preiti, G.; Romeo, M.; Bacchi, M.; Monti, M. Soil loss measure from Mediterranean arable cropping systems: Effects of rotation and tillage system on C-factor. *Soil Tillage Res.* **2017**, *170*, 85–93. [CrossRef]
41. Martínez-Mena, M.; Perez, M.; Almagro, M.; Garcia-Franco, N.; Díaz-Pereira, E. Long-term impacts of sustainable management practices on soil properties and crop yields in rainfed Mediterranean almond agroecosystems. *Eur. J. Agron.* **2021**, *123*, 126207. [CrossRef]
42. Almagro, M.; Garcia-Franco, N.; Martínez-Mena, M. The potential of reducing tillage frequency and incorporating plant residues as a strategy for climate change mitigation in semiarid Mediterranean agroecosystems. *Agric. Ecosyst. Environ.* **2017**, *246*, 210–220. [CrossRef]
43. Morugán-Coronado, A.; Linares, C.; Gómez-López, M.D.; Faz, Á.; Zornoza, R. The impact of intercropping, tillage and fertilizer type on soil and crop yield in fruit orchards under Mediterranean conditions: A meta-analysis of field studies. *Agric. Syst.* **2020**, *178*, 102736. [CrossRef]
44. Alvarez, R.; Steinbach, H.S. A review of the effects of tillage systems on some soil physical properties, water content, nitrate availability and crops yield in the Argentine Pampas. *Soil Tillage Res.* **2009**, *104*, 1–15. [CrossRef]
45. Van den Putte, A.; Govers, G.; Diels, J.; Gillijns, K.; Demuzere, M. Assessing the effect of soil tillage on crop growth: A meta-regression analysis on European crop yields under conservation agriculture. *Eur. J. Agron.* **2010**, *33*, 231–241. [CrossRef]
46. Bhatia, A.; Pathak, H.; Jain, N.; Singh, P.K.; Tomer, R. Greenhouse gas mitigation in rice–wheat system with leaf color chart-based urea application. *Environ. Monit. Assess.* **2012**, *184*, 3095–3107. [CrossRef]
47. Chauhan, B.S.; Mahajan, G.; Sardana, V.; Timsina, J.; Jat, M.L. Productivity and sustainability of the rice–wheat cropping system in the Indo-Gangetic Plains of the Indian subcontinent: Problems, opportunities, and strategies. *Adv. Agron.* **2012**, *117*, 315–369.
48. Sanz, M.J.; De Vente, J.L.; Chotte, J.-L.; Bernoux, M.; Kust, G.; Ruiz, I.; Almagro, M.; Alloza, J.A.; Vallejo, R.; Castillo, V.; et al. *Sustainable Land Management Contribution to Successful Land-Based Climate Change Adaptation and Mitigation: A Report of the Science-Policy Interface*; United Nations Convention to Combat Desertification (UNCCD): Bonn, Germany, 2017; p. 178.
49. Smith, P.; Adams, J.; Beerling, D.J.; Beringer, T.; Calvin, K.V.; Fuss, S.; Griscom, B.; Hagemann, N.; Kammann, C.; Kraxner, F.; et al. Land-Management Options for Greenhouse Gas Removal and Their Impacts on Ecosystem Services and the Sustainable Development Goals. *Annu. Rev. Environ. Resour.* **2019**, *44*, 255–286. [CrossRef]
50. Smith, P.; Nkem, J.; Calvin, K.; Campbell, D.; Cherubini, F.; Grassi, G.; Korotkov, V.; Hoang, A.L.; Lwasa, S.; McElwee, P.; et al. Interlinkages between Desertification, Land Degradation, Food Security and Greenhouse Gas Fluxes: Synergies, Trade-offs and Integrated Response Options. In *Climate Change and Land: An IPCC Special Report on Climate Change, Desertification, Land Degradation, Sustainable land Management, Food Security, and Greenhouse Gas Fluxes in Terrestrial Ecosystems*; Shukla, P.R., Skea, J., Calvo Buendia, E., Masson-Delmotte, V., Portner, H.-O., Roberts, D.C., Zhai, P., Slade, R., Connors, S., van Diemen, R., et al., Eds.; Intergovernmental Panel on Climate Change: Geneva, Switzerland, 2019; p. 122.
51. Cooper, J.; Baranski, M.; Stewart, G.; Nobel-de Lange, M.; Bàrberi, P.; Fließbach, A.; Peigné, J.; Berner, A.; Brock, C.; Casa-grande, M.; et al. Shallow non-inversion tillage in organic farming maintains crop yields and increases soil C stocks: A meta-analysis. *Agron. Sustain. Dev.* **2016**, *36*, 22. [CrossRef]
52. Martínez-Mena, M.; Garcia-Franco, N.; Almagro, M.; Ruiz-Navarro, A.; Albaladejo, J.; de Aguilar, J.M.; Gonzalez, D.; Querejeta, J.I. Decreased foliar nitrogen and crop yield in organic rainfed almond trees during transition from reduced tillage to no-tillage in a dryland farming system. *Eur. J. Agron.* **2013**, *49*, 149–157. [CrossRef]
53. Guenet, B.; Gabrielle, B.; Chenu, C.; Arrouays, D.; Balesdent, J.; Bernoux, M.; Bruni, E.; Caliman, J.-P.; Cardinael, R.; Chen, S.; et al. Can N2O emissions offset the benefits from soil organic carbon storage? *Glob. Chang. Biol.* **2020**, *27*, 237–256. [CrossRef]
54. Costantini, E.A.C.; Antichi, D.; Almagro, M.; Hedlund, K.; Sarno, G.; Virto, I. Local adaptation strategies to increase or maintain soil organic carbon content under arable farming in Europe: Inspirational ideas for setting operational groups within the European innovation partnership. *J. Rural. Stud.* **2020**, *79*, 102–115. [CrossRef]
55. Sasal, M.C.; Léonard, J.; Andriulo, A.; Boizard, H. A contribution to understanding the origin of platy structure in silty soils under no tillage. *Soil Tillage Res.* **2017**, *173*, 42–48. [CrossRef]
56. Chinseu, E.; Dougill, A.; Stringer, L. Why do smallholder farmers dis-adopt conservation agriculture? Insights from Malawi. *Land Degrad. Dev.* **2019**, *30*, 533–543. [CrossRef]
57. Pittelkow, C.M.; Liang, X.; Linquist, B.A.; van Groenigen, K.J.; Lee, J.; Lundy, M.E.; van Gestel, N.; Six, J.; Venterea, R.T.; van Kessel, C. Productivity limits and potentials of the principles of conservation agriculture. *Nature* **2015**, *517*, 365–368. [CrossRef]
58. Ingram, J.; Mills, J.; Frelih-Larsen, A.; Davis, M.; Merante, P.; Ringrose, S.; Molnar, A.; Sánchez, B.; Ghaley, B.B.; Karaczun, Z. Managing Soil Organic Carbon: A Farm Perspective. *Eurochoices* **2014**, *13*, 12–19. [CrossRef]
59. Borgström, S.; Zachrisson, A.; Eckerberg, K. Funding ecological restoration policy in practice—Patterns of short-termism and regional biases. *Land Use Policy* **2016**, *52*, 439–453. [CrossRef]
60. Runhaar, H. Governing the transformation towards 'nature-inclusive' agriculture: Insights from the Netherlands. *Int. J. Agric. Sustain.* **2017**, *15*, 340–349. [CrossRef]

61. Ferwerda, W. *4 Returns, 3 Zones, 20 Years: A Holistic Framework for Ecological Restoration by People and Business for Next Generations*; Rotterdam School of Management, Erasmus University: Rotterdam, The Netherlands; International Union of Conservation of Nature, Commission on Ecosystem Management and Commonland: Gland, Switzerland, 2015.
62. Schoonhoven, Y.; Runhaar, H. Conditions for the adoption of agro-ecological farming practices: A holistic framework illustrated with the case of almond farming in Andalusia. *Int. J. Agric. Sustain.* **2018**, *16*, 442–454. [CrossRef]
63. Ahnström, J.; Höckert, J.; Bergeå, H.L.; Francis, C.A.; Skelton, P.; Hallgren, L. Farmers and nature conservation: What is known about attitudes, context factors and actions affecting conservation? *Renew. Agric. Food Syst.* **2009**, *24*, 38–47. [CrossRef]
64. Schomberg, H.H.; Wietholter, S.; Griffin, T.S.; Reeves, D.W.; Cabrera, M.L.; Fisher, D.S.; Endale, D.M.; Novak, J.M.; Balkcom, K.S.; Raper, R.L.; et al. Assessing Indices for Predicting Potential Nitrogen Mineralization in Soils under Different Management Systems. *Soil Sci. Soc. Am. J.* **2009**, *73*, 1575–1586. [CrossRef]
65. Reeves, D. Cover crops and rotations. In *Crop Residue Management*; Hatfield, J., Stewart, B.A., Eds.; Lewis Publishers: Boca Raton, FL, USA, 1994; pp. 124–172.
66. Francaviglia, R.; Vicente-Vicente, J.L. Cover cropping. In *Recarbonizing Global Soils: A Technical Manual of Best Management Practices*; FAO, ITPS, Eds.; Food and Agriculture Organization of the United Nations: Rome, Italy, 2021; Volume 3, pp. 3–13.
67. Jian, J.; Du, X.; Reiter, M.S.; Stewart, R.D. A meta-analysis of global cropland soil carbon changes due to cover cropping. *Soil Biol. Biochem.* **2020**, *143*, 107735. [CrossRef]
68. Poeplau, C.; Don, A. Carbon sequestration in agricultural soils via cultivation of cover crops—A meta-analysis. *Agric. Ecosyst. Environ.* **2015**, *200*, 33–41. [CrossRef]
69. Smith, P. An overview of the permanence of soil organic carbon stocks: Influence of direct human-induced, indirect and natural effects. *Eur. J. Soil Sci.* **2005**, *56*, 673–680. [CrossRef]
70. Blanco-Canqui, H.; Shaver, T.M.; Lindquist, J.L.; Shapiro, C.A.; Elmore, R.W.; Francis, C.A.; Hergert, G.W. Cover Crops and Ecosystem Services: Insights from Studies in Temperate Soils. *Agron. J.* **2015**, *107*, 2449–2474. [CrossRef]
71. Daryanto, S.; Fu, B.; Wang, L.; Jacinthe, P.-A.; Zhao, W. Quantitative synthesis on the ecosystem services of cover crops. *Earth-Sci. Rev.* **2018**, *185*, 357–373. [CrossRef]
72. Kaspar, T.; Singer, J. The Use of Cover Crops to Manage Soil. In *Soil Management: Building a Stable Base for Agriculture*; Hatfield, J., Sauer, T., Eds.; American Society of Agronomy and Soil Science Society of America: Madison, WI, USA, 2011; pp. 321–337.
73. Muhammad, I.; Wang, J.; Sainju, U.M.; Zhang, S.; Zhao, F.; Khan, A. Cover cropping enhances soil microbial biomass and affects microbial community structure: A meta-analysis. *Geoderma* **2021**, *381*, 114696. [CrossRef]
74. Vukicevich, E.; Lowery, T.; Bowen, P.; Úrbez-Torres, J.R.; Hart, M. Cover crops to increase soil microbial diversity and mitigate decline in perennial agriculture. A review. *Agron. Sustain. Dev.* **2016**, *36*, 48. [CrossRef]
75. Adetunji, A.T.; Ncube, B.; Mulidzi, R.; Lewu, F.B. Management impact and benefit of cover crops on soil quality: A review. *Soil Tillage Res.* **2020**, *204*, 104717. [CrossRef]
76. Blanco-Canqui, H.; Ruis, S.J.; Proctor, C.A.; Creech, C.F.; Drewnoski, M.E.; Redfearn, D.D. Harvesting cover crops for biofuel and livestock production: Another ecosystem service? *Agron. J.* **2020**, *112*, 2373–2400. [CrossRef]
77. Masilionyte, L.; Maiksteniene, S.; Kriauciuniene, Z.; Jablonskyte-Rasce, D.; Zou, L.; Sarauskis, E. Effect of cover crops in smothering weeds and volunteer plants in alternative farming systems. *Crop Prot.* **2017**, *91*, 74–81. [CrossRef]
78. Demestihas, C.; Plénet, D.; Génard, M.; Raynal, C.; Lescourret, F. Ecosystem services in orchards. A review. *Agron. Sustain. Dev.* **2017**, *37*, 12. [CrossRef]
79. Chapagain, T.; Lee, E.A.; Raizada, M.N. The Potential of Multi-Species Mixtures to Diversify Cover Crop Benefits. *Sustainability* **2020**, *12*, 2058. [CrossRef]
80. Lee, H.; Lautenbach, S.; Nieto, A.P.G.; Bondeau, A.; Cramer, W.; Geijzendorffer, I.R. The impact of conservation farming practices on Mediterranean agro-ecosystem services provisioning—A meta-analysis. *Reg. Environ. Chang.* **2019**, *19*, 2187–2202. [CrossRef]
81. Wratten, S.D.; Gillespie, M.; Decourtye, A.; Mader, E.; Desneux, N. Pollinator habitat enhancement: Benefits to other ecosystem services. *Agric. Ecosyst. Environ.* **2012**, *159*, 112–122. [CrossRef]
82. Basche, A.D.; Miguez, F.E.; Kaspar, T.C.; Castellano, M.J. Do cover crops increase or decrease nitrous oxide emissions? A meta-analysis. *J. Soil Water Conserv.* **2014**, *69*, 471–482. [CrossRef]
83. Kaye, J.P.; Quemada, M. Using cover crops to mitigate and adapt to climate change. A review. *Agron. Sustain. Dev.* **2017**, *37*, 4. [CrossRef]
84. Chamizo, S.; Serrano-Ortiz, P.; López-Ballesteros, A.; Sánchez-Cañete, E.P.; Vicente-Vicente, J.L.; Kowalski, A.S. Net ecosystem CO_2 exchange in an irrigated olive orchard of SE Spain: Influence of weed cover. *Agric. Ecosyst. Environ.* **2017**, *239*, 51–64. [CrossRef]
85. DuPont, S.T.; Ferris, H.; Van Horn, M. Effects of cover crop quality and quantity on nematode-based soil food webs and nutrient cycling. *Appl. Soil Ecol.* **2009**, *41*, 157–167. [CrossRef]
86. Hill, E.C.; Renner, K.A.; Sprague, C.L.; Davis, A.S. Cover Crop Impact on Weed Dynamics in an Organic Dry Bean System. *Weed Sci.* **2016**, *64*, 261–275. [CrossRef]
87. Baraibar, B.; Hunter, M.C.; Schipanski, M.E.; Hamilton, A.; Mortensen, D.A. Weed Suppression in Cover Crop Monocultures and Mixtures. *Weed Sci.* **2018**, *66*, 121–133. [CrossRef]
88. Blesh, J. Functional traits in cover crop mixtures: Biological nitrogen fixation and multifunctionality. *J. Appl. Ecol.* **2018**, *55*, 38–48. [CrossRef]

89. Finney, D.M.; Murrell, E.G.; White, C.M.; Baraibar, B.; Barbercheck, M.E.; Bradley, B.A.; Cornelisse, S.; Hunter, M.C.; Kaye, J.P.; Mortensen, D.A.; et al. Ecosystem Services and Disservices Are Bundled in Simple and Diverse Cover Cropping Systems. *Agric. Environ. Lett.* **2017**, *2*, 170033. [CrossRef]
90. Francaviglia, R.; Álvaro-Fuentes, J.; Di Bene, C.; Gai, L.; Regina, K.; Turtola, E. Diversified Arable Cropping Systems and Management Schemes in Selected European Regions Have Positive Effects on Soil Organic Carbon Content. *Agriculture* **2019**, *9*, 261. [CrossRef]
91. Kremen, C.; Iles, A.; Bacon, C.M. Diversified Farming Systems: An agroecological, systems-based alternative to modern industrial agriculture. *Ecol. Soc.* **2012**, *17*, 44. [CrossRef]
92. West, T.O.; Post, W.M. Soil organic carbon sequestration rates by tillage and crop rotation: A global data analysis. *Soil Sci. Soc. Am. J.* **2002**, *66*, 1930–1946. [CrossRef]
93. Christopher, S.F.; Lal, R.; Mishra, U. Regional Study of No-Till Effects on Carbon Sequestration in the Midwestern United States. *Soil Sci. Soc. Am. J.* **2009**, *73*, 207–216. [CrossRef]
94. McDaniel, M.D.; Tiemann, L.K.; Grandy, A.S. Does agricultural crop diversity enhance soil microbial biomass and organic matter dynamics? A meta-analysis. *Ecol. Appl.* **2014**, *24*, 560–570. [CrossRef]
95. Angers, D.A.; Carson, J. Plant-induced changes in soil structure: Processes and feedbacks. *Biogeochemistry* **1998**, *42*, 55–72. [CrossRef]
96. Lynch, J.M.; Bragg, E. Microorganisms and Soil Aggregate Stability. *Adv. Soil Sci.* **1985**, *2*, 133–171. [CrossRef]
97. Jastrow, J.D.; Miller, R.M. Soil aggregate stabilisation and carbon sequestration: Feedbacks through organomineral associations. In *Soil Processes and the Carbon Cycle*; Lal, R., Kimble, J.M., Follett, R.F., Stewart, B.A., Eds.; CRC Press: Boca Raton, FL, USA, 1997; pp. 207–223.
98. LE Bissonnais, Y. Aggregate stability and assessment of soil crustability and erodibility: I. Theory and methodology. *Eur. J. Soil Sci.* **1996**, *47*, 425–437. [CrossRef]
99. Nunes, M.R.; van Es, H.M.; Schindelbeck, R.; Ristow, A.J.; Ryan, M. No-till and cropping system diversification improve soil health and crop yield. *Geoderma* **2018**, *328*, 30–43. [CrossRef]
100. Martiniello, P. Biochemical parameters in a Mediterranean soil as affected by wheat–forage rotation and irrigation. *Eur. J. Agron.* **2007**, *26*, 198–208. [CrossRef]
101. Francis, C.A. Crop rotations. In *Encyclopedia of Soils in the Environment*; Hillel, D., Rosenzweig, C., Powlson, D., Scow, K., Singer, M., Sparks, D., Eds.; Academic Press: Cambridge, MA, USA, 2005; pp. 318–322.
102. Di Bene, C.; Gómez-López, M.; Francaviglia, R.; Farina, R.; Blasi, E.; Martínez-Granados, D.; Calatrava, J. Barriers and Opportunities for Sustainable Farming Practices and Crop Diversification Strategies in Mediterranean Cereal-Based Systems. *Front. Environ. Sci.* **2022**, *10*, 861225. [CrossRef]
103. Angus, J.F.; Kirkegaard, J.A.; Hunt, J.R.; Ryan, M.H.; Ohlander, L.; Peoples, M.B. Break crops and rotations for wheat. *Crop Pasture Sci.* **2015**, *66*, 523–552. [CrossRef]
104. Francaviglia, R.; Álvaro-Fuentes, J.; Di Bene, C.; Gai, L.; Regina, K.; Turtola, E. Diversification and Management Practices in Selected European Regions. A Data Analysis of Arable Crops Production. *Agronomy* **2020**, *10*, 297. [CrossRef]
105. Bouwman, A.F.; Beusen, A.H.W.; Griffioen, J.; Van Groenigen, J.W.; Hefting, M.M.; Oenema, O.; Van Puijenbroek, P.J.T.M.; Seitzinger, S.; Slomp, C.P.; Stehfest, E. Global trends and uncertainties in terrestrial denitrification and N_2O emissions. *Philos. Trans. R. Soc. B Biol. Sci.* **2013**, *368*, 20130112. [CrossRef]
106. Lemke, R.L.; Zhong, Z.; Campbell, C.A.; Zentner, R. Can pulse crops play a role in mitigating greenhouse gases from North American agriculture? *Agron. J.* **2007**, *99*, 1719–1725. [CrossRef]
107. Jeuffroy, M.H.; Baranger, E.; Carrouée, B.; de Chezelles, E.; Gosme, M.; Hénault, C.; Schneider, A.; Cellier, P. Nitrous oxide emissions from crop rotations including wheat, oilseed rape and dry peas. *Biogeosciences* **2013**, *10*, 1787–1797. [CrossRef]
108. Schwenke, G.D.; Herridge, D.F.; Scheer, C.; Rowlings, D.W.; Haigh, B.M.; McMullen, K.G. Soil N2O emissions under N2-fixing legumes and N-fertilised canola: A reappraisal of emissions factor calculations. *Agric. Ecosyst. Environ.* **2015**, *202*, 232–242. [CrossRef]
109. Plaza-Bonilla, D.; Nolot, J.-M.; Raffaillac, D.; Justes, E. Innovative cropping systems to reduce N inputs and maintain wheat yields by inserting grain legumes and cover crops in southwestern France. *Eur. J. Agron.* **2017**, *82*, 331–341. [CrossRef]
110. Bossio, D.A.; Cook-Patton, S.C.; Ellis, P.W.; Fargione, J.; Sanderman, J.; Smith, P.; Wood, S.; Zomer, R.J.; von Unger, M.; Emmer, I.M.; et al. The role of soil carbon in natural climate solutions. *Nat. Sustain.* **2020**, *3*, 391–398. [CrossRef]
111. Paustian, K.; Lehmann, J.; Ogle, S.; Reay, D.; Robertson, G.P.; Smith, P. Climate-smart soils. *Nature* **2016**, *532*, 49–57. [CrossRef]
112. Smith, P.; Soussana, J.-F.; Angers, D.; Schipper, L.; Chenu, C.; Rasse, D.P.; Batjes, N.H.; van Egmond, F.; McNeill, S.; Kuhnert, M.; et al. How to measure, report and verify soil carbon change to realize the potential of soil carbon sequestration for atmospheric greenhouse gas removal. *Glob. Chang. Biol.* **2020**, *26*, 219–241. [CrossRef]
113. Montgomery, D.R. Soil erosion and agricultural sustainability. *Proc. Natl. Acad. Sci. USA* **2007**, *104*, 13268–13272. [CrossRef]
114. Verheijen, F.G.; Jones, R.J.; Rickson, R.J.; Smith, C.J. Tolerable versus actual soil erosion rates in Europe. *Earth-Sci. Rev.* **2009**, *94*, 23–38. [CrossRef]
115. European Commission. *Communication from the Commission to the Council, the European Parliament, the European Economic and Social Committee and the Committee of the Regions, Thematic Strategy for Soil Protection, COM 231 Final*; European Commission: Brussels, Belgium, 2006.

116. Cantón, Y.; Solé-Benet, A.; de Vente, J.; Boix-Fayos, C.; Calvo-Cases, A.; Asensio, C.; Puigdefábregas, J. A review of runoff generation and soil erosion across scales in semiarid south-eastern Spain. *J. Arid Environ.* **2011**, *75*, 1254–1261. [CrossRef]
117. Almagro, M.; Martínez-Mena, M. Litter decomposition rates of green manure as affected by soil erosion, transport and deposition processes, and the implications for the soil carbon balance of a rainfed olive grove under a dry Mediterranean climate. *Agric. Ecosyst. Environ.* **2014**, *196*, 167–177. [CrossRef]
118. Lal, R. Forest soils and carbon sequestration. *For. Ecol. Manag.* **2005**, *220*, 242–258. [CrossRef]
119. Throop, H.L.; Archer, S.R. Resolving the Dryland Decomposition Conundrum: Some New Perspectives on Potential Drivers. *Prog. Bot.* **2009**, *70*, 171–194. [CrossRef]
120. Martínez-Mena, M.; López, J.; Almagro, M.; Albaladejo, J.; Castillo, V.; Ortiz, R.; Boix-Fayos, C. Organic carbon enrichment in sediments: Effects of rainfall characteristics under different land uses in a Mediterranean area. *Catena* **2012**, *94*, 36–42. [CrossRef]
121. The Mission for Soil Health and Food: Opportunities for Earth Observation. Available online: https://www.copernicus. eu/system/files/2020-12/NT%2025nov%20-%20The%20Horizon%20Europe%20Missions%20-%20Challenges%20and%20 opportunities%20for%20Copernicus_0.pdf (accessed on 12 December 2022).
122. Maetens, W.; Poesen, J.; Vanmaercke, M. How effective are soil conservation techniques in reducing plot runoff and soil loss in Europe and the Mediterranean? *Earth-Sci. Rev.* **2012**, *115*, 21–36. [CrossRef]
123. Halbritter, A.H.; De Boeck, H.J.; Eycott, A.E.; Reinsch, S.; Robinson, D.A.; Vicca, S.; Berauer, B.; Christiansen, C.T.; Estiarte, M.; Grünzweig, J.M.; et al. The handbook for standardized field and laboratory measurements in terrestrial climate change experiments and observational studies (ClimEx). *Methods Ecol. Evol.* **2019**, *11*, 22–37. [CrossRef]
124. Berg, B.; McClaugherty, C. *Plant Litter: Decomposition, Humus Formation, Carbon Sequestration*; Springer: Berlin/Heidelberg, Germany, 2008; p. 286.
125. Currie, W.S. Relationships between carbon turnover and bioavailable energy fluxes in two temperate forest soils. *Glob. Chang. Biol.* **2003**, *9*, 919–929. [CrossRef]
126. Houghton, R.A. Balancing the Global Carbon Budget. *Annu. Rev. Earth Planet. Sci.* **2007**, *35*, 313–347. [CrossRef]
127. Chenu, C.; Angers, D.A.; Barré, P.; Derrien, D.; Arrouays, D.; Balesdent, J. Increasing organic stocks in agricultural soils: Knowledge gaps and potential innovations. *Soil Tillage Res.* **2019**, *188*, 41–52. [CrossRef]
128. Austin, A.T.; Vivanco, L. Plant litter decomposition in a semi-arid ecosystem controlled by photodegradation. *Nature* **2006**, *442*, 555–558. [CrossRef]
129. King, J.Y.; Brandt, L.A.; Adair, E.C. Shedding light on plant litter decomposition: Advances, implications and new directions in understanding the role of photodegradation. *Biogeochemistry* **2012**, *111*, 57–81. [CrossRef]
130. Almagro, M.; Maestre, F.T.; Martínez-López, J.; Valencia, E.; Rey, A. Climate change may reduce litter decomposition while enhancing the contribution of photodegradation in dry perennial Mediterranean grasslands. *Soil Biol. Biochem.* **2015**, *90*, 214–223. [CrossRef]
131. Wang, J.; Liu, L.; Wang, X.; Chen, Y. The interaction between abiotic photodegradation and microbial decomposition under ultraviolet radiation. *Glob. Chang. Biol.* **2015**, *21*, 2095–2104. [CrossRef]
132. Sanderman, J. Can management induced changes in the carbonate system drive soil carbon sequestration? A review with particular focus on Australia. *Agric. Ecosyst. Environ.* **2012**, *155*, 70–77. [CrossRef]
133. Cameron, K.C.; Di, H.J.; Moir, J.L. Nitrogen losses from the soil/plant system: A review. *Ann. Appl. Biol.* **2013**, *162*, 145–173. [CrossRef]
134. Kindler, R.; Siemens, J.A.N.; Kaiser, K.; Walmsley, D.C.; Bernhofer, C.; Buchmann, N.; Cellier, P.; Eugster, W.; Gleixner, G.; Grünwald, T.; et al. Dissolved carbon leaching from soil is a crucial component of the net ecosystem carbon balance. *Glob. Chang. Biol.* **2011**, *17*, 1167–1185. [CrossRef]
135. Liu, T.; Wang, L.; Feng, X.; Zhang, J.; Ma, T.; Wang, X.; Liu, Z. Comparing soil carbon loss through respiration and leaching under extreme precipitation events in arid and semiarid grasslands. *Biogeosciences* **2018**, *15*, 1627–1641. [CrossRef]
136. Spencer, R.G.; Aiken, G.R.; Wickland, K.P.; Striegl, R.G.; Hernes, P.J. Seasonal and spatial variability in dissolved organic matter quantity and composition from the Yukon River basin, Alaska. *Glob. Biogeochem. Cycles* **2008**, *22*, GB4002. [CrossRef]
137. Lal, R.; Kimble, J.M. Pedogenic carbonates and the global carbon cycle. In *Global Change and Pedogenic Carbonate*; Lal, R., Kimble, J.M., Eswaran, H., Stewart, B.A., Eds.; CRC Press: Boca Raton, FL, USA, 2000; pp. 1–14.
138. Soto, R.L.; Martínez-Mena, M.; Padilla, M.C.; de Vente, J. Restoring soil quality of woody agroecosystems in Mediterranean drylands through regenerative agriculture. *Agric. Ecosyst. Environ.* **2021**, *306*, 107191. [CrossRef]
139. Garcia-Franco, N.; Wiesmeier, M.; Colocho Hurtarte, L.C.; Fella, F.; Martínez-Mena, M.; Almagro, M.; García Martínez, E.; Kögel-Knabner, I. Pruning debris incorporation and reduced tillage improve soil organic matter stabilization and structure of salt-affected soils in a semi-arid Mediterranean Citrus tree orchard. *Soil Tillage Res.* **2021**, *213*, 105129. [CrossRef]
140. Almagro, M.; Martínez-Mena, M. Conservation, reduced and superficial tillage. In *Recarbonizing Global Soils: A Technical Manual of Best Management Practices*; FAO, ITPS, Eds.; Food and Agriculture Organization of the United Nations: Rome, Italy, 2021; Volume 3, pp. 72–84. [CrossRef]
141. Álvaro-Fuentes, J.; Arrúe, J.L.; Gracia, R.; López, M.V. Tillage and cropping intensification effects on soil aggregation: Temporal dynamics and controlling factors under semiarid conditions. *Geoderma* **2008**, *145*, 390–396. [CrossRef]
142. Dimassi, B.; Mary, B.; Wylleman, R.; Labreuche, J.; Couture, D.; Piraux, F.; Cohan, J.-P. Long-term effect of contrasted tillage and crop management on soil carbon dynamics during 41 years. *Agric. Ecosyst. Environ.* **2014**, *188*, 134–146. [CrossRef]

143. Feiziene, D.; Feiza, V.; Slepetiene, A.; Liaudanskiene, I.; Kadziene, G.; Deveikyte, I.; Vaideliene, A. Long-Term Influence of Tillage and Fertilization on Net Carbon Dioxide Exchange Rate on Two Soils with Different Textures. *J. Environ. Qual.* **2011**, *40*, 1787–1796. [CrossRef] [PubMed]

144. Vicente-Vicente, J.L.; Gómez-Muñoz, B.; Hinojosa-Centeno, M.B.; Smith, P.; Garcia-Ruiz, R. Carbon saturation and assessment of soil organic carbon fractions in Mediterranean rainfed olive orchards under plant cover management. *Agric. Ecosyst. Environ.* **2017**, *245*, 135–146. [CrossRef]

145. Ball, K.R.; Baldock, J.A.; Penfold, C.; Power, S.A.; Woodin, S.J.; Smith, P.; Pendall, E. Soil organic carbon and nitrogen pools are increased by mixed grass and legume cover crops in vineyard agroecosystems: Detecting short-term management effects using infrared spectroscopy. *Geoderma* **2020**, *379*, 114619. [CrossRef]

146. Sekaran, U.; Kumar, S.; Gonzalez-Hernandez, J.L. Integration of crop and livestock enhanced soil biochemical properties and microbial community structure. *Geoderma* **2021**, *381*, 114686. [CrossRef]

147. Wulanningtyas, H.S.; Gong, Y.; Li, P.; Sakagami, N.; Nishiwaki, J.; Komatsuzaki, M. A cover crop and no-tillage system for enhancing soil health by increasing soil organic matter in soybean cultivation. *Soil Tillage Res.* **2021**, *205*, 104749. [CrossRef]

148. Huang, Y.; Ren, W.; Grove, J.; Poffenbarger, H.; Jacobsen, K.; Tao, B.; Zhu, X.; McNear, D. Assessing synergistic effects of no-tillage and cover crops on soil carbon dynamics in a long-term maize cropping system under climate change. *Agric. For. Meteorol.* **2020**, *291*, 108090. [CrossRef]

149. Tamburini, G.; Bommarco, R.; Wanger, T.C.; Kremen, C.; van der Heijden, M.G.A.; Liebman, M.; Hallin, S. Agricultural diversification promotes multiple ecosystem services without compromising yield. *Sci. Adv.* **2020**, *6*, eaba1715. [CrossRef]

150. Van Eerd, L.L.; Congreves, K.A.; Verhallen, A.; Hayes, A.; Hooker, D.C. Long-term tillage and crop rotation effects on soil quality, organic carbon, and total nitrogen. *Can. J. Soil Sci.* **2014**, *94*, 303–315. [CrossRef]

151. Miglierina, A.M.; Iglesias, J.O.; Landriscini, M.R.; Galantini, J.A.; Rosell, R.A. The effects of crop rotation and fertilization on wheat productivity in the Pampean semiarid region of Argentina. 1. Soil physical and chemical properties. *Soil Tillage Res.* **2000**, *53*, 129–135. [CrossRef]

152. Varvel, G.E. Soil Organic Carbon Changes in Diversified Rotations of the Western Corn Belt. *Soil Sci. Soc. Am. J.* **2006**, *70*, 426–433. [CrossRef]

153. European Environment Agency. *The European Environment: State and Outlook 2020*; Publications Office of the European Union: Luxembourg, 2020; Volume 60.

154. European Commission. Communication from the Commission to the European Parliament, the Council, The European Economic and Social Committee and the Committee of the Regions. In *EU Biodiversity Strategy for 2030: Bringing Nature Back into Our Lives COM/2020/380 Final*; European Commission: Brussel, Belgium, 2020.

155. Lampkin, N.; Stolze, M.; Meredith, S.; de Porras, M.; Haller, L.; Mészáros, D. *Eco-Schemes in the New CAP: A Guide for Managing Authorities*; IFOAM EU, FIBL, IEEP: Brussels, Belgium, 2020; p. 76.

156. European Commission. Communication from the Commission to the European Parliament, the European Council, the Council, The European Economic and Social Committee and the Committee of the Regions. In *The European Green Deal COM/2019/640 Final*; European Commission: Brussel, Belgium, 2019.

157. European Commission. *A Clean Planet for All—A European Strategic Long-Term Vision for a Prosperous, Modern, Competitive and Climate Neutral Economy*; COM/2018/773; European Commission: Brussel, Belgium, 2018.

158. European Commission. *Report from the Commission to the European Parliament and the Council on the Implementation of the EU Strategy on Adaptation to Climate Change COM/2018/738 Final*; European Commission: Brussel, Belgium, 2018.

159. European Commission. Nature-Based Solutions. 2017. Available online: https://ec.europa.eu/info/research-and-innovation/research-area/environment/nature-based-solutions_en (accessed on 12 December 2022).

160. European Commission. *From Farm to Fork Strategy: For a Fair, Healthy and Environmentally-Friendly Food System*; European Union: Brussel, Belgium, 2020.

161. European Commission. *Organic Farming—Action Plan for the Development of EU Organic Production*; European Commission: Brussel, Belgium, 2021.

162. European Commission. Communication from the Commission to the European Parliament, the Council, the European Economic and Social Committee and the Committee of the Regions. In *EU Pollinators Initiative COM (2018) 395 Final*; European Commission: Brussel, Belgium, 2018.

163. European Commission. Communication from the Commission to the European Parliament, the Council, the European and Social Committee and the Committee of the Regions. In *Chemicals Strategy for Sustainability Towards a Toxic-Free Environment COM (2020) 667 Final*; European Commission: Brussel, Belgium, 2020.

164. Towards a Zero Pollution Ambition for Air, Water and Soil. Available online: https://ec.europa.eu/info/law/better-regulation/have-your-say/initiatives/12588-EU-Action-Plan (accessed on 12 December 2022).

165. European Union. *Regulation (EU) 2018/841 on the Inclusion of Greenhouse Gas Emissions and Removals from Land Use, Land Use Change and Forestry in the 2030 Climate and Energy Framework, and Amending Regulation (EU) No 525/2013 and Decision No 529/2013/EU*; European Parliament and European Council: Brussels, Belgium, 2018.

166. Panagos, P.; Montanarella, L.; Barbero, M.; Schneegans, A.; Aguglia, L.; Jones, A. Soil priorities in the European Union. *Geoderma Reg.* **2022**, *29*, e00510. [CrossRef]

167. Panagos, P.; Van Liedekerke, M.; Borrelli, P.; Köninger, J.; Ballabio, C.; Orgiazzi, A.; Lugato, E.; Liakos, L.; Hervas, J.; Jones, A.; et al. European Soil Data Centre 2.0: Soil data and knowledge in support of the EU policies. *Eur. J. Soil Sci.* **2022**, *73*, e13315. [CrossRef]
168. Minasny, B.; Malone, B.P.; McBratney, A.B.; Angers, D.A.; Arrouays, D.; Chambers, A.; Chaplot, V.; Chen, Z.-S.; Cheng, K.; Das, B.S.; et al. Soil carbon 4 per mille. *Geoderma* **2017**, *292*, 59–86. [CrossRef]
169. Lal, R. Beyond COP 21: Potential and challenges of the "4 per Thousand" initiative. *J. Soil Water Conserv.* **2016**, *71*, 20A–25A. [CrossRef]
170. UNGA Resolution 70/1. Transforming our world: The 2030 Agenda for Sustainable Development. In *Resolution Adopted by the General Assembly on 25 September 2015*; United Nations: New York, NY, USA, 2015.
171. Jónsson, J.Ö.G.; Davíðsdóttir, B.; Jónsdóttir, E.M.; Kristinsdóttir, S.M.; Ragnarsdóttir, K.V. Soil indicators for sustainable development: A transdisciplinary approach for indicator development using expert stakeholders. *Agric. Ecosyst. Environ.* **2016**, *232*, 179–189. [CrossRef]
172. Lorenz, K.; Lal, R.; Ehlers, K. Soil organic carbon stock as an indicator for monitoring land and soil degradation in relation to United Nations' Sustainable Development Goals. *Land Degrad. Dev.* **2019**, *30*, 824–838. [CrossRef]
173. Lal, R. Soil health and carbon management. *Food Energy Secur.* **2016**, *5*, 212–222. [CrossRef]

soil systems

Review

Soil Health Assessment and Management Framework for Water-Limited Environments: Examples from the Great Plains of the USA

Rajan Ghimire [1,2,*], Vesh R. Thapa [2], Veronica Acosta-Martinez [3], Meagan Schipanski [4], Lindsey C. Slaughter [5], Steven J. Fonte [4], Manoj K. Shukla [2], Prakriti Bista [2], Sangamesh V. Angadi [1,2], Maysoon M. Mikha [6], Olufemi Adebayo [2] and Tess Noble Strohm [4]

[1] Agricultural Science Center Clovis, New Mexico State University, 2346 State Road 288, Clovis, NM 88101, USA
[2] Department of Plant and Environmental Sciences Las Cruces, New Mexico State University, Las Cruces, NM 88003, USA
[3] US Department of Agriculture, Agricultural Research Services, Cropping Systems Research Laboratory, Wind Erosion and Water Conservation Unit, Lubbock, TX 79415, USA
[4] Department of Soil and Crop Sciences, Colorado State University, Fort Collins, CO 80523, USA
[5] Department of Plant and Soil Science, Texas Tech University, Lubbock, TX 79409, USA
[6] USDA–ARS Central Plains Resources Management Research, Akron, CO 80720, USA
* Correspondence: rghimire@nmsu.edu

Citation: Ghimire, R.; Thapa, V.R.; Acosta-Martinez, V.; Schipanski, M.; Slaughter, L.C.; Fonte, S.J.; Shukla, M.K.; Bista, P.; Angadi, S.V.; Mikha, M.M.; et al. Soil Health Assessment and Management Framework for Water-Limited Environments: Examples from the Great Plains of the USA. *Soil Syst.* **2023**, 7, 22. https://doi.org/10.3390/soilsystems7010022

Academic Editor: Luis Eduardo Akiyoshi Sanches Suzuki

Received: 2 February 2023
Revised: 24 February 2023
Accepted: 27 February 2023
Published: 2 March 2023

Abstract: Healthy soils provide the foundation for sustainable agriculture. However, soil health degradation has been a significant challenge for agricultural sustainability and environmental quality in water-limited environments, such as arid and semi-arid regions. Soils in these regions is often characterized by low soil organic matter (SOM), poor fertility, and low overall productivity, thus limiting the ability to build SOM. Soil health assessment frameworks developed for more productive, humid, temperate environments typically emphasize building SOM as a key to soil health and have identified the best management practices that are often difficult to implement in regions with water limitations. This study reviewed existing soil health assessment frameworks to assess their potential relevance for water-limited environments and highlights the need to develop a framework that links soil health with key ecosystem functions in dry climates. It also discusses management strategies for improving soil health, including tillage and residue management, organic amendments, and cropping system diversification and intensification. The assessment of indicators sensitive to water management practices could provide valuable information in designing soil health assessment frameworks for arid and semi-arid regions. The responses of soil health indicators are generally greater when multiple complementary soil health management practices are integrated, leading to the resilience and sustainability of agriculture in water-limited environments.

Keywords: conservation agriculture; cover crops; semi-arid region; soil carbon; soil functions

1. Introduction

Farm productivity and economic profitability have been linked to effective soil health management [1–3]. Since the widespread adoption of the concept of healthy soil a few decades ago, there has been a consensus that soil health indicates the capacity of soils to function within an ecosystem and land-use boundaries, such as sustaining productivity, maintaining environmental quality, and promoting plant and animal health [4]. Soil health depends on complex biophysical and biochemical interactions in time and space, leading to the creation of a suitable environment for plant growth. It emphasizes soil as a living, dynamic system that provides multiple ecosystem services such as carbon (C) sequestration, nutrient cycling and storage, soil water retention and availability, erosion control, and crop productivity [5,6].

Soil health management in water-limited environments could benefit from an improved understanding of the linkages between soil health indicators and water conservation. However, such information is lacking, in part due to the relatively low adoption of soil health management practices, such as cover crops, improved crop rotations, conservation tillage, etc., in water-limited environments compared to more mesic or humid environments, or due to challenges in implementing soil health-promoting practices in semi-arid row crop systems because of the short-term losses in profitability. In addition, soil organic matter (SOM) has been the central component of soil health assessment due to its perceived impacts on soil's physical, chemical, and biological properties. While numerous studies have shown the critical role of SOM content in soil biological activity and diversity, nutrient cycling, cation exchange capacity (CEC), soil bulk density, aggregate stability and structure, and water storage and infiltration [7–10], the response of SOM to management changes in water-limited environments is typically very slow. Measurable changes in SOM accumulation can take decades in arid and semi-arid regions because precipitation limits plant biomass production and soil C inputs [11–13]. Producers and landowners in dry areas are looking for indicators that are more responsive to management changes while being inexpensive, reproducible, accessible through commercial laboratories or at-home testing, and able to provide management guidance [8].

Measuring responsive parameters such as microbial communities (specifically, fungal communities), enzyme activities, and labile SOM components could be valuable for water-limited regions. Saprophytic fungi and arbuscular mycorrhizal fungi (AMF) have survived and functioned better than most bacterial groups in semi-arid areas [12,14]. These fungal groups can also respond faster to sudden increases in soil moisture than bacterial communities in semi-arid regions. Similarly, labile soil organic C (SOC) and nitrogen (N) components can respond to management changes within 2–4 years [11,15,16]. The SOM components that serve as early indicators of soil health improvements include mineralizable C (soil respiration) and N, permanganate oxidizable C (POXC), particulate organic matter (POM), microbial biomass C (MBC) and N (MBN), dissolved organic C and N, and available soil nutrients [17,18]. Soil physical indicators such as aggregate stability, infiltration rates, and saturated conductivity (Ksat) can also respond rapidly to management changes [19]. Specifically, soil aggregate stability could be a valuable physical indicator of soil health in arid and semi-arid regions due to its rapid response to management changes, its relationship with many soil functions, and its sensitivity to changes in management [20]. Well-aggregated soils increase infiltration rates, thus improving water capture and storage compared to poorly aggregated soils. Studies demonstrated a rapid increase in soil aggregate fractions with cover cropping in the limited irrigation and dryland conditions of the central and southern Great Plains [7,21]. Small proportional changes in surface soil C (<20% increase) were positively associated with much larger changes in soil aggregation (>200% increase) and microbial biomass (>300% increase) in intensified, continuous dryland cropping systems relative to traditional wheat (*Triticum aestivum* L.)-fallow rotations [22]. Similarly, the higher soil water infiltration in continuous wheat was attributed to the greater aggregate stability compared to that in the wheat-sorghum (*Sorghum bicolor* L. Moench)-fallow rotation in semi-arid Texas High Plains [23].

No specific set or number of indicators or threshold scores define healthy soil. Soil health varies within soil types, climates, environmental conditions, and agricultural management practices [24]. The selection of the appropriate indicators will help producers and landowners identify the right management strategies to improve soil health. In arid and semi-arid regions, these indicators should be low-cost, sensitive to management changes, and responsive to soil water dynamics. This review discusses approaches for soil health assessment, examines the linkages between different soil health indicators and soil functions in water-limited environments, and ultimately discusses alternative management practices with the potential to improve soil health and agricultural sustainability.

2. Approaches for Soil Health Assessment

Soil health assessment indicates how well soil contributes to ecosystem services and can predict the ability of soils to provide those services if an adopted management scenario continues. Soil health is often evaluated by measuring various indicators within three main categories: physical, chemical, and biological properties of soil, which provide insight into key soil functions. Soil physical indicators primarily reflect limitations to seedling emergence, root growth, and soil water infiltration or the movement and storage of water in the soil profile. Examples of physical soil health indicators include the topsoil depth, bulk density, porosity, aggregate stability, infiltration rate, texture, crusting, and compaction [19]. Soil chemical indicators often relate to soil nutrient availability and the ability of soils to support plant nutrient uptake. Soil pH, electrical conductivity (EC), SOM, cation-exchange capacity, nutrient concentrations, and elements that may be potential contaminants (heavy metals, radioactive compounds, etc.) are important chemical indicators, while soil biological communities (macro- and microorganisms) of different sizes, diversity, and activities serve as the biological indicators of soil health [15–18]. Soil microbial communities are central to multiple ecosystem services, and they both drive and are constrained by many physical and chemical soil processes.

The relevance of different soil functional indicators changes from site to site. The relative importance of indicators related to soil water functions, such as water movement and retention, would be greater for arid and semi-arid regions. In contrast, nutrient provisioning and availability may be prioritized in areas with plenty of water. The indicators selected for assessing soil health must be: (a) responsive to changes in climate and soil management practices so that growers can use them as a basis for prioritizing management practices, (b) easy to sample, measure, and interpret for growers, (c) cheap and relatively accessible to many growers and applicable to field conditions, and (d) able to represent critical agronomic and soil ecological processes [6,20]. Soil health indicators developed for more productive, humid regions may not be responsive to management changes in water-limited environments due to the differences in the soil type, climate, crops and cropping intensity, and agricultural management practices.

Current soil health assessment frameworks do not account for regional differences in climate, soil conditions, and management. Different government agencies, non-government organizations, and universities have developed metrics for soil health assessment that may have broader relevance. For instance, Cornell University's comprehensive soil health assessment (CASH) identified 39 potential indicators [20] and narrowed them down to 12–13 parameters to make the evaluation simple, cost-effective, and universal. These indicators are aggregate stability, penetration resistance, available water capacity, bulk density, soil pH, phosphorus (P), potassium (K), nitrate-N, organic matter content, soil proteins, soil respiration, and soil pathogen population. The Soil Health Institute (SHI) has also endorsed 18 primary indicators as "Tier 1" and 12 secondary indicators as "Tier 2" [25]. The "Tier 1" list mostly included physical and chemical components rather than biological ones. The United States Department of Agriculture's Natural Resources Conservation Service (USDA-NRCS) proposed a similar set of physical (aggregate stability, available water capacity, bulk density, infiltration, slaking, soil crust, soil structure, and macropores), chemical (reactive carbon, soil EC, soil nitrate, and soil pH), and biological (earthworm count, POM, potentially mineralizable N, soil enzymes, soil respiration, and total organic C) indicators [26]. In these various soil health assessment matrices, soil properties identified as major indicators are a group of soil properties that have defined thresholds (i.e., rankings of poor to good) or have been benchmarked nationally [19]. The Soil Management Assessment Framework (SMAF) [27] and the Haney Soil Health Test (HSHT) [28] have also proposed a suite of indicators for monitoring soil health, but they are not as comprehensive as the CASH, SHI, or USDA-NRCS frameworks for soil health assessment. The Haney test does not even provide region-specific soil health information. More recently, the Soil Health Assessment Protocol and Evaluation (SHAPE) tool has been proposed to help overcome the geographical limitations of SMAF and CASH by leveraging a nationally distributed dataset and incorporating more edaphic and climatic factors [29]. Similarly, Zvomuya et al. [30] emphasized certain indicators such as soil salinity (EC), cation exchange capacity, and calcium carbonate content for arid and semi-arid regions. However, these regionally relevant parameters are not emphasized in the major soil health assessment frameworks.

In all soil health assessment frameworks, high emphasis is given to indicator selection. While indicator selection is critical, soil health assessment goes beyond identifying indicators. Typical steps in the development of the assessment framework involve quantifying the response of selected soil indicators, providing an appropriate score for each indicator based on the criteria set for defining the weight of each parameter, creating assessment metrics, and, finally, assigning soil health scores (Figure 1). There are multiple ways to integrate data into a final soil health score. Some approaches for integrating the measured indicators and developing the soil health index include: (i) weighted additive scores for individual indicators and (ii) the use of statistical tools such as multiple regression, principal component analysis, or factor analysis [21,31,32]. Expert opinion can also be used for scoring soil health [33]. However, these steps are not regionally tailored to address soil health issues specific to a particular region or specific soil functions and have a regionally tailored assessment matrix. Therefore, developing a regionally tailored scoring matrix that emphasizes water-sensitive indicators could provide a more representative soil health assessment framework for water-limited environments. More research on region-specific minimum data development and alternative scoring functions based on the relative response of indicators is needed for effective soil health assessment in regions varying in soils, climate, and agricultural systems.

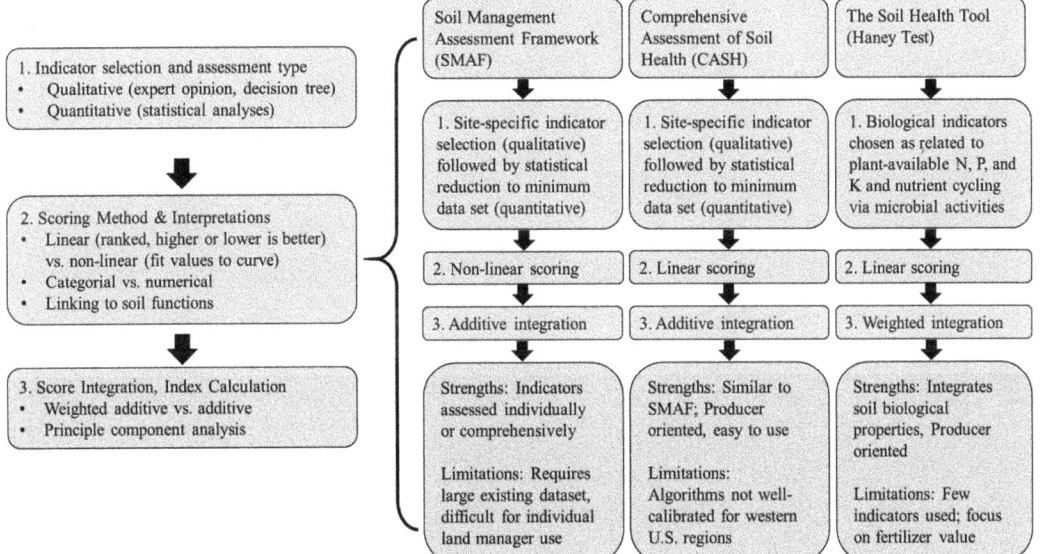

Figure 1. Sequential steps in soil health assessment, with examples shown for commonly used soil health index tools in the U.S. under various frameworks.

There has been a widespread interest among researchers, policymakers, and agricultural stakeholders in soil health assessment and management. The Soil Health Institute's North American Soil Health assessment project evaluated 31 different soil health indicators on soil samples collected from 125 long-term agricultural research sites across North America. This project aimed to give farmers, ranchers, and others science-based measurements for evaluating the health of their soils. This project can provide information on region-specific as well as universal indicators for assessing soil health by engaging farmers and agricultural stakeholders in identifying and prioritizing soil health indicators, developing assessment metrics, and interpreting soil health results. Given that soil has enormous heterogeneity, soil management is site-specific, and its ecosystem services vary with the soil and climatic condition. In addition, soil health management is linked with agricultural sustainability and environmental quality. Most indicators are developed based on research conducted in experimental farms with replicated plots or a small field section and are not validated on working farms, where direct replication is generally not possible. This means that the conditions under which they are developed vs. used may not be comparable (e.g., [27]), which adds to the complexity of employing them for on-farm soil health assessment. The broader validation of soil health indicators through on-farm testing and the engagement of stakeholders in the process (Figure 2) will establish the soil health assessment framework with broader acceptance.

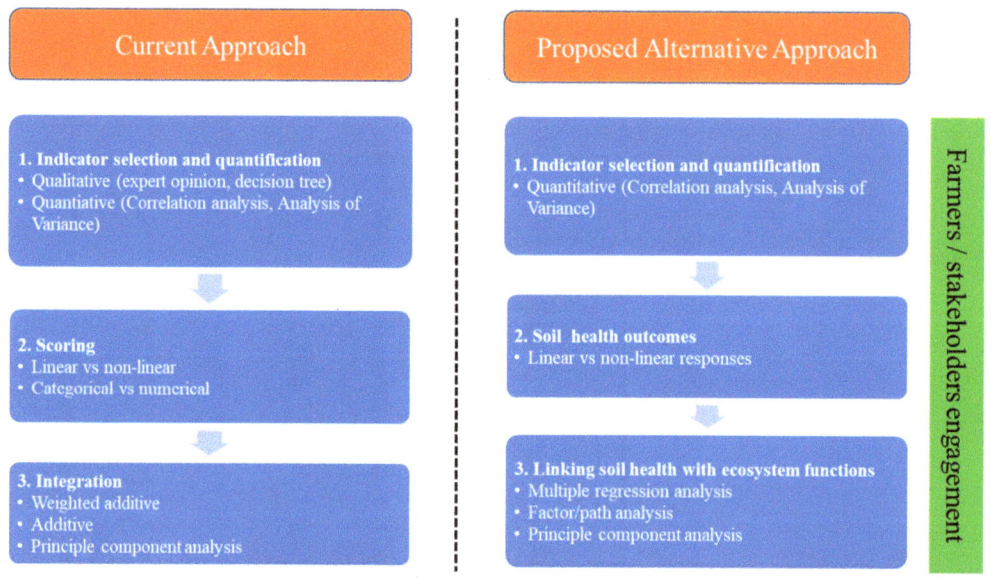

Figure 2. Current and proposed framework for soil health assessment in water-limited environments.

Current soil health assessment does not consider inputs from stakeholders. Accounting for region-specific differences in the response of various indicators, engaging stakeholders in soil health assessment, and linking soil health assessment with key ecosystem services could benefit farmers and landowners in arid and semi-arid regions. Different indicators should be used depending on the soil health goals or targeted soil use. Engaging stakeholders in selecting the most representative indicators, on-farm trials, and goal-based indicator identification could enhance the adoption of soil health practices. The adoption rate of soil health practices is often higher when farmers experience changes in their observations.

3. Linking Soil Health with Essential Water Functions

Developing an effective and reliable soil health assessment framework for water-limited environments requires an improved understanding of the linkages between soil health indicators and essential water functions. The ability of soils to infiltrate and retain precipitation or irrigation water is a function of soil physical properties such as aggregation, porosity, compaction, and soil texture, as well as site factors such as slope, residue cover, and surface roughness. Soil aggregate formation is strongly influenced by soil biology, particularly soil fungi [34], rooting activity, and soil macrofauna [35]. Soil chemical properties such as pH, EC, SOM content, and nutrients determine the diversity and abundance of microbial communities. Fungal communities respond rapidly to management changes; they respond even in sandy soils (55% sand content) before detectable changes in SOM were observed in the Great Plains semi-arid region [36]. Various physical, chemical, and biological soil properties influence soil functions that have implications for soil water conservation (Table 1). However, their direct and indirect relationships with essential water functions have not been studied well in arid and semi-arid regions. Measuring the response of soil properties, including more sensitive biological communities or processes, along with water storage and movement, is likely essential for comprehensive soil health assessment in water-limited environments.

Table 1. Soil health indicators, their functions, and their implications for soil water conservation.

Indicator/Method	Soil Function	Implications for Water Conservation and Related Soil Functions
Physical		
Bulk density	Porosity	Higher water infiltration with less compaction
Soil texture	Porosity	A direct baseline measure of soil water storage capacity
Soil aggregates (%)	Soil structure	Soil structure, higher water storage in well-aggregated soil
Wet aggregate stability	Soil structure	Capacity of soil to resist crusting and water erosion and to facilitate infiltration
Water infiltration	Soil water dynamics	Soil water capture, water use efficiency, and heat transfer
Soil water retention	Soil water dynamics	Soil water storage and plant available water
Soil depth	Soil water dynamics	Soil water storage and availability for crops
Chemical		
Soil pH	Soil acidity/alkalinity	Nutrient availability, creating a suitable environment for plant and microbial growth
Electrical conductivity	Salinity	Nutrient availability, plant and microbial growth, soil structure, and water-holding capacity
Soil organic C	Microbial substrate availability, nutrient provision, buffering	Direct measure of SOM status (58% of SOM) and baseline potential of water storage
Plant available nutrients	Nutrient provision	Nutrient availability for crop and microbial growth
Biological		
Microbial biomass C	Microbial community size	Soil processes such as decomposition, N fixation, C sequestration, nutrient availability
FAME profiling Fungal: AMF, saprophytic; Bacteria: G+, G−, Actinobacteria	Microbial community size and diversity of microbial groups	Mediate key soil processes such as decomposition, nutrient cycling, and water uptake, especially depending on the microbial groups (e.g., higher fungal populations can provide greater decomposition, cementing agents for aggregate stability, and a higher diversity of enzymes in soils to decompose a wide variety of substrates). AMF can provide an additional benefit to drought resilience.
Three-day CO_2 mineralization	Microbial activity	Indicate decomposition vs. sequestration of carbon, SOM storage, nutrient/water cycling
Particulate Organic Matter (POM)	Fresh residue C	Early indication of C sequestration and water conservation
Permanganate oxidizable carbon (POXC)	Diversity of C sources	This C pool can represent simple C sources available for microbial decomposition, substrates from root exudates, and microbial biomass C.
N mineralization	Crop N supply	Integrative indicator of labile N and microbial activity for increasing N availability
Enzyme activity assays: β-glucosidase, β-glucosaminidase, acid/alkaline phosphatase, arylsulfatase	Nutrient cycling	Indicator of potential enhancement in SOM and nutrient cycling and availability with a direct linkage to water changes in soil
Soil macrofauna	Residue/nutrient turnover	Soil aggregation and water dynamics, decomposition and nutrient cycling, pest control

The soil health literature often cites the general claim that increasing SOM by 1% enhances the water-holding capacity by >250,000 L ha^{-1} (25 mm) [37]. While there is generally a positive correlation between SOM and water-holding capacity [38], this relationship is influenced by multiple factors [39,40]. In addition, SOM changes may be relatively small and take decades to detect in arid and semi-arid environments. Therefore, increasing SOM enough to have a meaningful impact on soil water holding capacity is challenging in environments where an increase in SOM by 1% often represents a doubling of baseline SOM stocks. Since building SOM in water-limited environments can be quite difficult and it can take many years to generate detectable changes, there are approaches that could provide some insights into compositional changes in SOM that may affect soil water dynamics. The evaluation of SOM components such as mineralizable C (soil respiration) and N, POXC, POM, MBC, MBN, dissolved organic C and N, and available soil nutrients could serve as an early indicator of soil health improvements [17,18]. Similarly, isotopic methodologies can be explored to better characterize compositional changes in SOM dynamics and how not only soil management but also more frequent droughts and climate variability in semi-arid regions interfere with soil health and productivity.

The selection of indicators in soil health assessment should reflect soil water functions and beyond, e.g., soil erosion, biodiversity conservation, dust prevention, SOC sequestration, etc., to become more effective for water-limited environments. Risks associated with the implementation of soil health management systems in water-limited environments vary with the evapotranspiration gradient, with considerably higher risks in the hot, dry areas than in temperate drylands, where the majority of dryland cropping systems still include summer fallow [41]. However, there is a significant knowledge gap in soil health management in water-limited environments due to the lack of research-based information in understanding the relationship between various soil health indicators and essential soil water functions. The more rapid response of parameters such as microbial communities (specifically, fungal communities), enzyme activities, and labile SOM components could provide valuable insight into soil health management in water-limited environments. Similarly, soil aggregate stability could be a valuable physical indicator of soil health in arid and semi-arid regions. However, quantifying the relationship of these soil properties with soil water functions is critical for a reliable estimate of soil health in water-limited regions.

4. Implementing Soil Health Management to Improve Water Functions

Linkages between soil health and water functions in dry environments can be established by an improved understanding of the interaction between soil management, soil health indicators, and water functions related to these dynamic soil properties (Figure 3). Management selection in arid and semi-arid regions is affected by low rainfall, high climatic variability (specifically, heat and drought), and low inherent soil fertility statuses. Options for soil health improvement are limited, and the relative response of selected soil health management systems is small. Understanding the complex interactions between climatic factors, inherent and dynamic soil properties, and associated soil functions can help in designing the best management practices. Therefore, management selection should emphasize practices adapted to arid and semi-arid regions. Alternative soil management practices, their soil health response, and their linkages to soil water functions are discussed in the following sections.

4.1. Minimizing Soil Disturbance

Soil disturbance disrupts soil's physical structure, impacting soil's biological communities and associated microbially mediated processes. Tillage is the major disturbance activity in cultivated soils, which is typically practiced for seedbed preparation, weed control, crop residue mixing and incorporation, and fertilizer and amendment application. Producers in semi-arid row crop environments may also employ tillage to increase soil aeration and disrupt soil surface crusts formed after rainfall events. However, these benefits are short-term; a poor soil structure can cause several soil issues. Intensive tillage exposes soil to direct raindrop impact at the surface, thereby increasing the susceptibility of aggregates to disruption [42]. It reduces water and air-filled pore spaces between aggregates, thus restricting infiltration, increasing surface crusting, and leading to wind and water erosion. In addition, it disrupts roots and fungal hyphal networks, reducing the enmeshing action of soil particles in those hyphal networks, and decreases aggregate stability [34]. Increased soil temperature and soil aeration from tillage are expected to increase soil microbial activity, thereby increasing SOM mineralization, in which SOC is converted to carbon dioxide (CO_2)

and lost to the atmosphere [43]. Additionally, while tillage can stimulate microbial activity, it typically has negative impacts on soil macrofauna communities, so all of the functions they provide are diminished, especially those related to soil structure [44]. Earthworms are probably the most important soil engineers, at least where there is adequate moisture, and they are also the most susceptible organisms to tillage [35]. Therefore, intensive tillage often leads to poor soil health and inefficient water capture and use because of the poor soil structure and low SOM content.

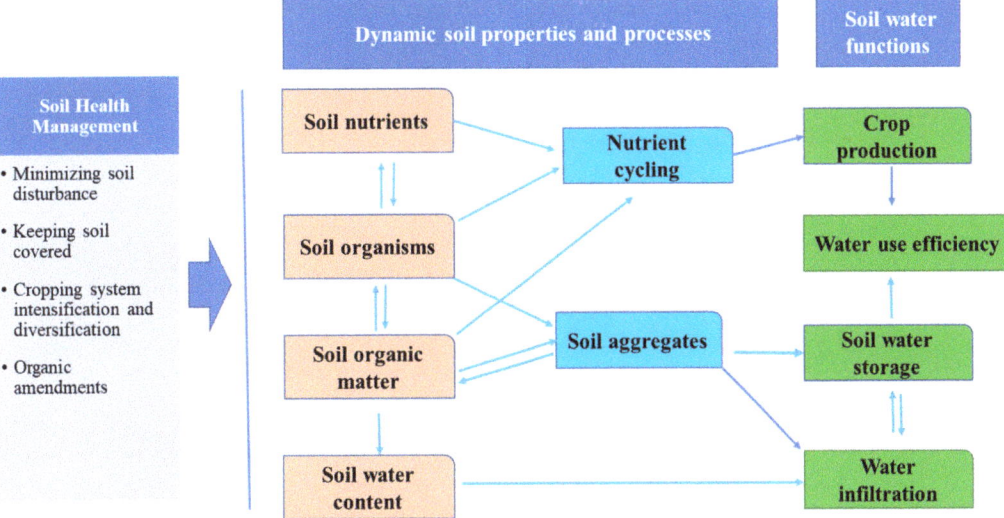

Figure 3. Interactions among soil management, soil health status, and soil functions in a water-limited environment.

Reducing tillage frequency and intensity can increase fungal growth, aggregate formation and stability, SOM accumulation, and soil health improvements by reversing the negative effects of intensive tillage. A study in eastern Montana and western North Dakota showed that conventional tillage increased the CO_2 flux by 62–118% as compared to no-tillage [45], while a study in eastern New Mexico reported a 26% greater wet aggregate stability and 9–15% greater permanganate oxidizable C under a 0–20 cm depth of no-tillage and strip-tillage compared to conventional tillage [46]. Research from Akron, Colorado, revealed that no-tillage and reduced-tillage resulted in 21% more SOC at the 0–30 cm depth than conventional tillage and moldboard plow [47]. Although minimizing soil disturbance through conservation tillage has also been linked to increased water infiltration, erosion resistance, soil aeration, and soil C stabilization [42,48], a quantitative relationship between soil aggregation, SOC storage, and soil water functions has not been established in these studies. Similarly, reduced- and no-till systems have been shown to support soil macrofauna communities in irrigated systems of eastern Colorado, along with associated improvements in soil aggregation and water infiltration [9,44]. Comprehensive research on the linkages between soil health improvements with reduced- and no-tillage management and soil water functions, including infiltration and water storage, will help design the best soil management practices for arid and semi-arid environments.

4.2. Keeping the Soil Covered

Maintaining soil cover with living or dead crop residues provides another mechanism for enhancing soil health and water functions in water-limited regions. Residue cover protects soil from wind and water erosion, while crop residue removal can reduce SOM by reducing C inputs and increasing susceptibility to soil loss. Soil cover increases precipitation storage efficiency and reduces the soil evaporation rates, making more soil moisture available for plant use [49]. Crop residue accumulation after 12 years of no-tillage management in three sites in eastern Colorado increased water sorptivity via improving soil aggregation, bulk density, and porosity and favored greater water infiltration and precipitation use efficiency [50]. Moreover, the surface cover reduces soil compaction by dissipating the raindrop energy, suppresses weed growth by limiting the amount of sunlight available to weed seedlings, and provides a protective habitat for soil organisms, positively affecting soil health. Carbon and other essential elements in the plant residues become a source of nutrition for soil flora and fauna, including bacteria and fungi, which mediate 90% of the soil ecosystem functions [51]. In semi-arid western Kansas, fields with spring triticale ($\times$ *Triticosecale* Wittm.) and spring lentil (*Lens culinaris* Medik.) residues had greater soil aggregates than bare soil, with spring lentils reducing the wind erodible fraction by 160% [52]. Dryland studies in Nebraska have demonstrated that wheat stubble increased the non-growing season soil water storage by 2–2.5 inches compared to bare soil [53]. After four crop seasons, SOC and total N, light fraction organic matter-C, and N were greater in soils with straw retention than in those with straw removal in semi-arid Canada [54]. Govaerts et al. [55] also reported that the SOC and total N were 1.15 and 1.17 times greater with straw retained than with straw removed, respectively, in semi-arid Mexico.

Keeping the soil covered by cover crops provides vegetative cover, controls soil erosion, enhances soil aggregation, adds organic matter, and increases soil biological activity [15,52], which could significantly improve soil water functions. Besides providing ground cover, cover cropping can maximize cropping intensity and diversity, thereby contributing to increased microbial substrate diversity, the proliferation of diverse soil organisms, and improved nutrient cycling. Dryland cropping systems with a history of winter cover crops (rye (*Secale cereale* L.)) improved soil microbial biomass and enzyme activities compared to cropping systems without winter cover crops [12]. Replacing fallow with hairy vetch (*Vicia villosa* Roth), rye, and mustard (*Brassica juncea* L.) as winter cover crops increased the mean weight diameter of dry soil aggregates and the wet aggregate stability in sweet corn (*Zea mays* L.)-fallow rotation in central and southern New Mexico [56]. However, growing cover crops in arid and semi-arid regions can deplete soil water storage, affecting the subsequent crop yield. A study on soil water storage with cover cropping in irrigated corn demonstrated a depletion in soil water at cover crop termination but greater soil water storage at the main crop harvest, suggesting overall positive effects of cover cropping on soil water storage [57]. While this study suggests that a careful selection of cover crop species and planting and termination timing could benefit cropping systems, it also discusses a complex relationship between soil health and water dynamics.

4.3. Cropping Systems Intensification and Diversification

Farmers have been attempting the intensification and diversification of cropping across arid and semi-arid regions. However, their response to soil health and water dynamics is inconsistent across the regions. For example, cover crops could be a promising option for increasing the complexity of rotations and extending the duration of photosynthetic capture in annual crop rotations, thus increasing organic C inputs to the soil. Increasing crop diversity with cover cropping can also diversify microbial substrates and support long-term improvements in soil health. Legume species in cover crop mixtures can fix atmospheric N in their root nodules and increase the soil N content, while grass cover crops have a dense fibrous root system and produce more root biomass, contributing to greater root-derived C in the soil [58]. The greater root biomass and length density of grass cover crops increase root channels and improve soil aggregation through enmeshing action. The rhizosphere of living brassica species, i.e., canola (*Brassica napus* L.) roots, releases a fumigant-like compound (2-phenylethyl isothiocyantae) that helps suppress pest populations and soil-borne diseases [59]. Diversified cropping systems improve the retention and cycling of nutrients and maintain soil biodiversity [60]. Research from semi-arid western Kansas comparing winter triticale, winter lentil, spring lentil, spring pea

(*Pisum sativum* L.), and spring triticale cover crops revealed an up to 12.2–17.4% increase in SOC (0–10 cm depth) with spring pea than with continuous winter wheat and fallow [52]. Mixtures of legumes, grasses, and oilseed cover crops produced greater belowground biomass, root C and N, and soil biodiversity than either species alone [61,62]. In eastern New Mexico, diverse cover crops that included cereals, legumes, and brassicas had 31% and 41% greater microbial community sizes and fungal fatty acid methyl ester (FAME) markers, respectively, at a 0–15 cm depth compared to fallow [15]. A six-species mixture of legumes, grasses, and brassicas in the same study plots increased the combined enzyme activity of acid phosphatase, β-glucosidase, and β-glucosaminidase by 44% and that of potentially mineralizable C (PMC) by 39% at termination time compared to fallow [11,15].

Crop rotations, which include growing a variety of crop species, can benefit the soil food web, improve nutrient cycling, and reduce soil-borne diseases and pests [15,19,21]. Crop rotation and intensification using a variety of crops, including low-water users, taproots, fibrous roots, high-C crops, legumes, and non-legumes, increase soil cover, contributing to key functions such as rainfall infiltration, SOM formation, and stabilization [21]. Several on-farm and research station experiments across the Central Great Plains have demonstrated that crop diversification and reducing the frequency of summer fallow periods through cropping intensification can improve the chemical, biological, and physical metrics of soil health, supporting improved profitability [22]. Cotton (*Gossypium hirsutum* L.), when rotated with peanut (*Arachis hypogaea* L.), sorghum, rye, or wheat, increased enzyme activities in comparison to continuous cotton in semi-arid soils from west Texas [63]. Another study from Akron, Colorado, reported greater soil fungal markers in rotations that reduced fallow and increased crop diversity from a typical winter wheat-fallow to a corn-proso millet (*Panicum miliaceum* L.)-winter wheat or corn-fallow-winter wheat for 15 years [64]. These shifts in the microbial community composition led to an increase in C and P cycling enzyme activities in both diversified rotations. Both SOC and total N were higher for sorghum-wheat-soybean (*Glycine max* L.) than for continuous sorghum from 0–55 cm in central Texas [65]. Intensive cropping (wheat-soybean double-crop and sorghum-wheat-soybean) increased SOC by 15–21% and total N by 19% at depths of 0–55 compared to continuous soybean, regardless of the tillage regime [66].

Improved knowledge of the relationship between soil health and soil water dynamics could help develop a soil health framework for water-limited regions because the potential longer-term benefits of cropping system intensification regarding soil health may have variable effects on soil water functions. For example, changes in the soil water content, infiltration, and water conservation in intensified rotations can have short-term trade-offs with crop productivity in water-limited regions [67,68]. A study in eastern NM revealed that, although the cropping system scale water balance was positive for the cover crop-corn rotation, the cover crops depleted 47–91 mm of soil water during their growth [57]. If rainfall or irrigation water is not available during the early growth of the main crop, the cash crop yield might be significantly reduced. Soil health indicators that capture system-level responses may not represent the seasonal and inter-annual dynamics of soil water storage and depletion. Therefore, careful planning and selecting the right crop or cover crops in the rotation are essential for agricultural sustainability and soil health in water-limited regions.

4.4. Role of Organic Amendments in Soil Health and Water Dynamics

Because increasing C inputs in arid and semi-arid areas are limited by water availability, organic amendments, such as manure and compost, could be a low-cost alternative for rapidly improving soil health and sustaining crop production in water-limited environments, especially in areas close to feedlots, dairies, or similar operations. Dairy enterprises are concentrated in eastern New Mexico, where about 329,000 milking cows produce more than 1.2 million metric tons of dry manure annually [69,70]. Similarly, beef cattle produce an additional 1.2 million metric tons of dry manure annually in eastern Colorado [71]. Since eastern New Mexico and Colorado have surplus manure from dairy and beef cattle, this represents a potentially convenient and inexpensive option for improving soil health, farm productivity, and profitability. Composted manure can increase nutrient availability, SOM storage, soil biological activity, aggregation, water-holding capacity, and aeration, ultimately supporting crop production and promoting the rural economy [72,73]. Moreover, heat generated during composting reduces weed seeds and pathogens [73]. Manure and composts often release nutrients slowly and can increase nutrient-use efficiency compared to chemical fertilizers. Recent studies in the central and southern Great Plains regions show that selected soil health indicators, including aggregation, enzyme activity, and particulate organic matter, respond positively to compost applications [74–76]. However, eastern New Mexico/West Texas farmers typically apply composted dairy manure based on crops' N needs, which may adversely affect soil health by creating an imbalance in other soil nutrients and an accumulation of salts [77,78]. The global warming potential of N in the form of increased N_2O emissions following compost application could also be a concern. However, associated increases in SOC can help offset the global warming potential of N in compost in semi-arid soils [79,80].

While improved soil microbial and biochemical functioning with poultry litter application has been reported [81], organic fecal materials such as manure derived from livestock contain organically bound N, P, K, calcium (Ca), and micronutrients, which might not be readily available for crops [81]. A high rate of manure and compost can also lead to salt accumulation [74]. The high salt concentration in soils decreases the microbial population and soil water potential, creating a water deficit condition and subsequently reducing the water use efficiency of crops. Additionally, a high accumulation of P and K in the soil increases the runoff and leaching of these nutrients, increasing environmental risks such as eutrophication. Therefore, salt accumulation should be carefully considered in nutrient management plans that integrate organic fecal materials in cropping systems. In addition, the challenges semi-arid and arid regions may experience are the hot temperatures associated with the faster decomposition of the compost and the lower long-lasting effects of the organic compost substrates in the soil. Organic amendment application should be integrated with other soil health practices that provide ground cover and lower soil temperature, thereby reducing the rate of organic matter decomposition and loss [43,74].

5. Challenges and Opportunities in Soil Health Assessment and Management in Water-Limited Regions

The soil health assessment and management framework for arid and semi-arid regions should be cost-effective, feasible, and linked to soil water conservation. Although there is no consensus on soil health indicators for water-limited environments, the importance of adopting alternative management to improve soil health is well-established. There are many challenges in identifying a minimum set of indicators for soil health assessment and using certain soil management practices in water-limited regions (Figure 4). Limited data on soil health responses to alternative management practices and their relationship with soil water functions are available, and the available data show highly variable responses to management alternatives. For example, cover crops may deplete soil water and nutrients and negatively impact the subsequent cash crop yield if careful planning and management for planting, species selection, and termination are not adopted. The early termination of cover crops in hot and dry regions could maintain soil moisture for the following cash crop but may not accumulate as much biomass carbon as needed to increase SOM. Inter-seeding cover crops into main crops should be carried out carefully to avoid competition between cover crops and the cash crop for water and nutrients. While inter-seeding before the main crop canopy closure or prior to harvest, when the crop canopy begins to re-open, would minimize the competition for water and nutrients, overcoming challenges in adopting soil health practices, such as selecting cover crop species that are drought- and shade-tolerant and relatively easy to establish would increase the possibility of improving soil health in water-limited areas.

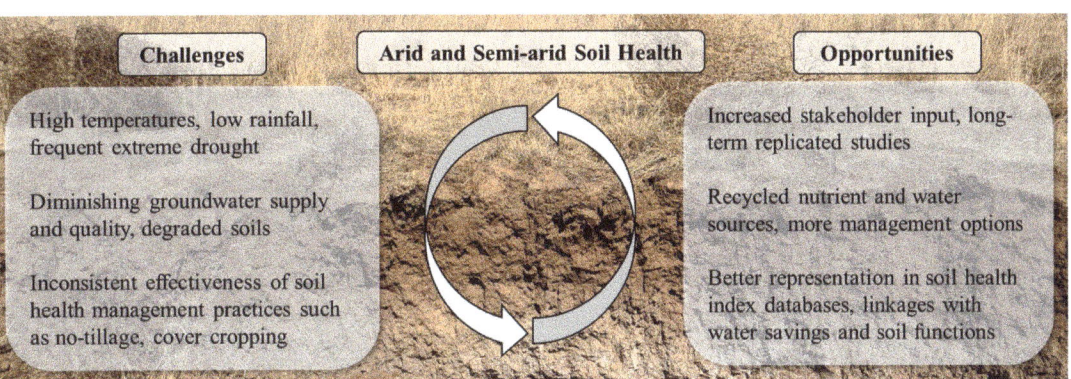

Figure 4. Challenges and opportunities of soil health assessment and management in arid and semi-arid (water-limited) environments of the western United States.

Multiple soil health management practices are often practiced together to enhance soil health benefits. Although the relative response of management strategies is variable, adopting multiple practices often provides synergistic, additive effects on soil health. A recent study that examined crop water use in a semi-arid cotton system after decades of implementing no-till practices with a rye or mixed species cover showed that, even though water use by cover crops depleted soil water prior to termination, which persisted in the early stages of cotton growth, the addition of cover crops resulted in greater water infiltration and storage throughout the growing season compared to conventionally tilled continuous cotton [82]. Similarly, cover crops and irrigation enhanced soil enzyme activities and promoted soil microbial community development [15,83]. Fields covered with cover crops and conservation tillage practices had 5 to 6 °C lower soil temperature and 3.5 to 4.9 °C lower soil surface air temperature and stored more soil water than conventional tillage without cover cropping in eastern New Mexico [84]. Similarly, rye, as a no-tilled cover crop, increased SOC, reduced penetration resistance, and increased infiltration by 34% compared to conventional tillage without cover crops in a 0–10 cm depth of cotton cropping systems in Lubbock, Texas [85]. A recent study across 96 dryland no-till fields in eastern Colorado and western Nebraska found 17% more SOC at 0–10 cm, twice as much aggregate stability, and three times greater fungal biomass in continuous rotations (no summer fallow) than in wheat-fallow [22]. Another study comparing diverse long-term cropping systems across the Great Plains revealed greater microbial biomass and mineralizable N under reduced-till diversified crop rotations than under conventional crop-fallow systems [56]. Planting winter and summer annual crops (corn, proso millet, and sunflower (*Helianthus annuus* L.) in sequence with winter wheat and fallow improved pest management in Akron, Colorado [86]. Another study in eastern New Mexico showed that cover crops significantly reduced the soil volumetric water content at cover crop termination and up to 30 days after sorghum planting [57]. However, the total soil water extraction at sorghum harvest was 8–89% higher under fallow than under cover crops, leading to an 18–23% greater forage sorghum yield after cover crops than after fallow. Therefore, cropping systems representing maximum biomass production and eventually returning to the soil are crucial for enhancing SOC, nutrient cycling, and aggregate stability, thereby improving soil health.

Climate change has added complexity to soil management and agricultural sustainability in dry environments. Increased temperature and a decreased amount and increased variability in the amount, intensity, and frequency of precipitation added challenges to agricultural production in water-limited environments of the Great Plains [8]. As the water supply is projected to decrease in the region, the importance of management practices that improve soil health is even greater in helping producers adapt to a changing climate.

Therefore, the development of a soil health assessment and management framework that is cost-effective, feasible, and promotes water conservation must be identified through multi-state collaborative research and implemented in the entire dry regions for thriving agriculture in the context of climate change. However, immense opportunities to improve soil health due to increasing research and growing interest in sustainable cropping practices in arid and semi-arid regions cannot be overlooked (Figure 4). New and innovative farming practices are rapidly evolving, a new source of water is identified, and recycling technology is being developed, demonstrating more opportunities to improve soil health. A robust soil health assessment and management framework based on stakeholder engagement is needed to mitigate challenges and maximize benefits from soil and water conservation practices in water-limited environments.

6. Conclusions

Our review evaluates existing soil health assessment frameworks and highlights the need to develop region-specific, stakeholder-driven approaches for a more reliable estimate of soil health in water-limited environments. Soil health assessment for water-limited environments likely cannot rely on the same primary indicators as more humid regions, or, at a minimum, the weighting of the different indicators will differ by the climate context. This lack of attention to soil health indicators and practices relevant to improved water dynamics constrains the adoption of soil health management practices in arid and semi-arid regions of the USA. For example, improving soil's physical and biological functions are likely more relevant in water-limited regions than emphasizing increasing the total SOM to increase the soil water holding capacity. It is also important to explore new approaches that can address changes in the SOM chemical composition, as it is common for these soils to have low SOM content, and it is unlikely that changes in the SOM quantity may be observed within a decade or longer and that it could take drastic management changes. Therefore, more responsive soil health indicators such as fungal biomass, labile organic matter fractions, and soil aggregates may indicate changes in key soil functions in arid and semi-arid regions. Developing a minimum dataset based on regional multi-location research is needed. In addition, designing cropping systems based on soil health goals and adopting no-tillage or reduced-tillage, cover cropping, diverse crop rotations, residue management, and organic amendments such as manure or compost could improve soil health and agricultural sustainability in dry regions. Since soil health management in arid and semi-arid environments is often challenged by soil water availability for biomass production, water management should be a primary consideration. Challenges and opportunities unique to water-limited regions lie in the proper management of crop residues, cover crop planting and termination timing, seeding rate or species selection, tillage practices, and organic amendments such as manure and compost, which affect the soil water and nutrient dynamics. Combinations of multiple soil health management practices may rapidly improve soil water functions and enhance the resilience and sustainability of agriculture in water-limited environments.

Author Contributions: Conceived the manuscript and wrote the original draft: R.G.; Literature collection and contribution to the initial draft: V.R.T., V.A.-M. and M.S.; Writing, revision, and discussion: S.J.F., M.K.S., S.V.A., P.B., M.M.M., L.C.S., O.A. and T.N.S. All authors have read and agreed to the published version of the manuscript.

Funding: This research was funded by project No. 2022-67019-36106 of the USDA National Institute for Food and Agriculture's Agriculture and Food Research Initiative, and additional support was provided by project No. GR0006511 of USDA Natural Resources Conservation Service.

Institutional Review Board Statement: Not applicable.

Informed Consent Statement: Not applicable.

Data Availability Statement: Not applicable.

Acknowledgments: We thank funding agencies for providing support to complete this study. We thank funding agencies for providing support to complete this study. Trade names and company names are included for the benefit of the reader and do not infer any en-dorsement or preferential treatment of the product by researchers or USDA-ARS. USDA-ARS, New Mexico State University, and Colorado State University are equal opportunity providers and em-ployers.

Conflicts of Interest: The authors declare no conflict of interest. The funders had no role in the design of the study; in the collection, analyses, or interpretation of the data; in the writing of the manuscript; or in the decision to publish the results.

References

1. Paustian, K.; Lehmann, J.; Ogle, S.; Reay, D.; Robertson, G.P.; Smith, P. Climate-smart soils. *Nature* **2016**, *532*, 49–57. [CrossRef] [PubMed]
2. Hendrickson, J.R.; Liebig, M.A.; Sassenrath, G.F. Environment and integrated agricultural systems. *Renew. Agric. Food Syst.* **2008**, *23*, 304–313. [CrossRef]
3. Liebig, M.A.; Tanaka, D.L.; Krupinsky, J.M.; Merrill, S.D.; Hanson, J.D. Dynamic cropping systems: Contributions to improve agroecosystem sustainability. *Agron. J.* **2007**, *99*, 899–903. [CrossRef]
4. Doran, J.W.; Parkin, T.B. Defining and assessing soil quality. In *Defining and Assessing Soil Quality*; John Wiley & Sons, Ltd.: Hoboken, NJ, USA, 1994; pp. 1–21. [CrossRef]
5. Romig, D.E.; Garlynd, M.J.; Harris, R.F.; McSweeney, K. How farmers assess soil health and quality. *J. Soil Water Conserv.* **1995**, *50*, 229–236.
6. Doran, J.W.; Zeiss, M.R. Soil health and sustainability: Managing the biotic component of soil quality. *Appl. Soil Ecol.* **2000**, *15*, 3–11. [CrossRef]
7. Kelly, C.; Schipanski, M.; Kondratieff, B.; Sherrod, L.; Schneekloth, J.; Fonte, S.J. The effects of dryland cropping system intensity on soil function and associated changes in macrofauna communities. *Soil Sci. Soc. Am. J.* **2020**, *84*, 1854–1870. [CrossRef]
8. Cano, A.; Núñez, A.; Acosta-Martinez, V.; Schipanski, M.; Ghimire, R.; Rice, C.; West, C. Current knowledge and future research directions to link soil health and water conservation in the Ogallala Aquifer region. *Geoderma* **2018**, *328*, 109–118. [CrossRef]
9. Melman, D.A.; Kelly, C.; Schneekloth, J.; Calderón, F.; Fonte, S.J. Tillage and residue management drive rapid changes in soil macrofauna communities and soil properties in a semi-arid cropping system of Eastern Colorado. *Appl. Soil Ecol.* **2019**, *143*, 98–106. [CrossRef]
10. Shukla, M.K.; Lal, R.; Ebinger, M. Determining soil quality indicators by factor analysis. *Soil Tillage Res.* **2006**, *87*, 194–204. [CrossRef]
11. Ghimire, R.; Ghimire, B.; Mesbah, A.O.; Sainju, U.M.; Idowu, O.J. Soil health response of cover crops in winter wheat–fallow system. *Agron. J.* **2019**, *111*, 2108–2115. [CrossRef]
12. Acosta-Martínez, V.; Lascano, R.; Calderón, F.; Booker, J.D.; Zobeck, T.M.; Upchurch, D.R. Dryland cropping systems influence the microbial biomass and enzyme activities in a semi-arid sandy soil. *Biol. Fertil. Soils* **2011**, *47*, 655–667. [CrossRef]
13. Jacinthe, P.-A.; Shukla, M.K.; Ikemura, Y. Carbon pools and soil biochemical properties in manure-based organic farming systems of semi-arid New Mexico. *Soil Use Manag.* **2011**, *27*, 453–463. [CrossRef]
14. Acosta-Martínez, V.; Cotton, J.; Gardner, T.; Moore-Kucera, J.; Zak, J.; Wester, D.; Cox, S. Predominant bacterial and fungal assemblages in agricultural soils during a record drought/heat wave and linkages to enzyme activities of biogeochemical cycling. *Appl. Soil Ecol.* **2014**, *84*, 69–82. [CrossRef]
15. Thapa, V.R.; Ghimire, R.; Acosta-Martínez, V.; Marsalis, M.A.; Schipanski, M.E. Cover crop biomass and species composition affect soil microbial community structure and enzyme activities in semi-arid cropping systems. *Appl. Soil Ecol.* **2021**, *157*, 103735. [CrossRef]
16. Ghimire, R.; Norton, J.B.; Stahl, P.D.; Norton, U. Soil microbial substrate properties and microbial community responses under irrigated organic and reduced-tillage crop and forage production systems. *PLoS ONE* **2014**, *9*, e103901. [CrossRef] [PubMed]
17. Franzluebbers, A.J.; Stuedemann, J.A.; Schomberg, H.H.; Wilkinson, S.R. Soil organic C and N pools under long-term pasture management in the Southern, Piedmont USA. *Soil Biol. Biochem.* **2000**, *32*, 469–478. [CrossRef]
18. Weil, R.R.; Islam, K.R.; Stine, M.A.; Gruver, J.B.; Samson-Liebig, S.E. Estimating active carbon for soil quality assessment: A simplified method for laboratory and field use. *Am. J. Altern. Agric.* **2003**, *18*, 3–17. [CrossRef]
19. Stewart, R.D.; Jian, J.; Gyawali, A.J.; Thomason, W.E.; Badgley, B.D.; Reiter, M.S.; Strickland, M.S. What we talk about when we talk about soil health. *Agric. Environ. Lett.* **2018**, *3*, 180033. [CrossRef]
20. Moebius-Clune, B.N. *Comprehensive Assessment of Soil Health: The Cornell Framework Manual*; Cornell University: Geneva, NY, USA, 2016.
21. Thapa, V.R.; Ghimire, R.; VanLeeuwen, D.; Acosta-Martínez, V.; Shukla, M. Response of soil organic matter to cover cropping in water-limited environments. *Geoderma* **2022**, *406*, 115497. [CrossRef]
22. Rosenzweig, S.T.; Fonte, S.J.; Schipanski, M.E. Intensifying rotations increases soil carbon, fungi, and aggregation in semi-arid agroecosystems. *Agric. Ecosyst. Environ.* **2018**, *258*, 14–22. [CrossRef]

23. Baumhardt, R.L.; Johnson, G.L.; Schwartz, R.C. Residue and long-term tillage and crop rotation effects on simulated rain infiltration and sediment transport. *Soil Sci. Soc. Am. J.* **2012**, *76*, 1370–1378. [CrossRef]
24. Lehman, R.M.; Acosta-Martinez, V.; Buyer, J.S.; Cambardella, C.A.; Collins, H.P.; Ducey, T.F.; Halvorson, J.J.; Jin, V.L.; Johnson, J.M.; Kremer, R.J.; et al. Soil biology for resilient, healthy soil. *J. Soil Water Conserv.* **2015**, *70*, 12A–18A. [CrossRef]
25. Soil, Health Institute—Enriching Soil, Enhancing Life n.d. Available online: https://soilhealthinstitute.org/ (accessed on 5 July 2022).
26. Soil, Health | NRCS Soils n.d. Available online: https://www.nrcs.usda.gov/wps/portal/nrcs/main/soils/health/ (accessed on 5 July 2022).
27. Andrews, S.S.; Karlen, D.L.; Cambardella, C.A. The Soil Management, Assessment Framework. *Soil Sci. Soc. Am. J.* **2004**, *68*, 1945–1962. [CrossRef]
28. Haney, R.L.; Haney, E.B.; Smith, D.R.; Harmel, R.D.; White, M.J. The soil health tool—Theory and initial broad-scale application. *Appl. Soil Ecol.* **2018**, *125*, 162–168. [CrossRef]
29. Nunes, M.R.; Veum, K.S.; Parker, P.A.; Holan, S.H.; Karlen, D.L.; Amsili, J.P.; van Es, H.M.; Wills, S.A.; Seybold, C.A.; Moorman, T.B. The soil health assessment protocol and evaluation applied to soil organic carbon. *Soil Sci. Soc. Am. J.* **2021**, *85*, 1196–1213. [CrossRef]
30. Zvomuya, F.; Janzen, H.H.; Larney, F.J.; Olson, B.M. A long-term field bioassay of soil quality indicators in a semi-arid environment. *Soil Sci. Soc. Am. J.* **2008**, *72*, 683–692. [CrossRef]
31. Andrews, S.S.; Karlen, D.L.; Mitchell, J.P. A comparison of soil quality indexing methods for vegetable production systems in Northern, California. *Agric. Ecosyst. Environ.* **2002**, *90*, 25–45. [CrossRef]
32. Ikemura, Y.; Shukla, M.K. Soil quality in organic and conventional farms for an arid ecosystem of New Mexico. *J. Org. Syst.* **2009**, *4*, 34–47.
33. Cherubin, M.R.; Karlen, D.L.; Cerri, C.E.P.; Franco, A.L.C.; Tormena, C.A.; Davies, C.A.; Cerri, C.C. Soil, Quality Indexing, Strategies for Evaluating, Sugarcane Expansion in Brazil. *PLoS ONE* **2016**, *11*, e0150860. [CrossRef]
34. Rillig, M.C.; Aguilar-Trigueros, C.A.; Bergmann, J.; Verbruggen, E.; Veresoglou, S.D.; Lehmann, A. Plant root and mycorrhizal fungal traits for understanding soil aggregation. *New Phytol.* **2015**, *205*, 1385–1388. [CrossRef]
35. Lavelle, P.; Spain, A.; Fonte, S.; Bedano, J.C.; Blanchart, E.; Galindo, V.; Grimaldi, M.; Jimenez, J.J.; Velasquez, E.; Zangerlé, A. Soil aggregation, ecosystem engineers and the C cycle. *Acta Oecologica* **2020**, *105*, 103561. [CrossRef]
36. Acosta-Martínez, V.; Burow, G.; Zobeck, T.M.; Allen, V.G. Soil microbial communities and function in alternative systems to continuous cotton. *Soil Sci. Soc. Am. J.* **2010**, *74*, 1181–1192. [CrossRef]
37. NRCS. Soil, Health Key, Points. Available online: https://www.nrcs.usda.gov/Internet/FSE_DOCUMENTS/stelprdb1082147.pdf. (accessed on 5 July 2022).
38. Rawls, W.J.; Pachepsky, Y.A.; Ritchie, J.C.; Sobecki, T.M.; Bloodworth, H. Effect of soil organic carbon on soil water retention. *Geoderma* **2003**, *116*, 61–76. [CrossRef]
39. Minasny, B.; McBratney, A.B. Limited effect of organic matter on soil available water capacity. *Eur. J. Soil Sci.* **2018**, *69*, 39–47. [CrossRef]
40. Bagnall, D.K.; Morgan, C.L.S.; Cope, M.; Bean, G.M.; Cappellazzi, S.; Greub, K.; Liptzin, D.; Norris, C.L.; Rieke, E.; Tracy, P.; et al. Carbon-sensitive pedotransfer functions for plant available water. *Soil Sci. Soc. Am. J.* **2022**, *86*, 612–629. [CrossRef]
41. Rosenzweig, S.T.; Carolan, M.S.; Schipanski, M.E. A dryland cropping revolution? Linking an emerging soil health paradigm with shifting social fields among wheat growers of the High Plains. *Rural Sociol.* **2020**, *85*, 545–574. [CrossRef]
42. Six, J.; Conant, R.T.; Paul, E.A.; Paustian, K. Stabilization mechanisms of soil organic matter: Implications for C-saturation of soils. *Plant Soil* **2002**, *241*, 155–176. [CrossRef]
43. Nilahyane, A.; Ghimire, R.; Thapa, V.R.; Sainju, U.M. Cover crop effects on soil carbon dioxide emissions in a semi-arid cropping system. *Agrosyst. Geosci. Environ.* **2020**, *3*, e20012. [CrossRef]
44. Deleon, E.; Bauder, T.A.; Wardle, E.; Fonte, S.J. Conservation tillage supports soil macrofauna communities, infiltration, and farm profits in an irrigated maize-based cropping system of Colorado. *Soil Sci. Soc. Am. J.* **2020**, *84*, 1943–1956. [CrossRef]
45. Sainju, U.M.; Jabro, J.D.; Stevens, W.B. Soil carbon dioxide emission and carbon content as affected by irrigation, tillage, cropping system, and nitrogen fertilization. *J. Environ. Qual.* **2008**, *37*, 98–106. [CrossRef]
46. Thapa, V.R.; Ghimire, R.; Mikha, M.M.; Idowu, O.J.; Marsalis, M.A. Land use effects on soil health in semi-arid drylands. *Agric. Environ. Lett.* **2018**, *3*, 180022. [CrossRef]
47. Mikha, M.M.; Vigil, M.F.; Benjamin, J.G. Long-term tillage impacts on soil aggregation and carbon dynamics under wheat-fallow in the central Great Plains. *Soil Sci. Soc. Am. J.* **2013**, *77*, 594–605. [CrossRef]
48. Shaver, T.M.; Peterson, G.A.; Ahuja, L.R.; Westfall, D.G.; Sherrod, L.A.; Dunn, G. Surface soil physical properties after twelve years of dryland no-till management. *Soil Sci. Soc. Am. J.* **2002**, *66*, 1296–1303. [CrossRef]
49. Shaver, T.M.; Peterson, G.A.; Ahuja, L.R.; Westfall, D.G. Soil sorptivity enhancement with crop residue accumulation in semi-arid dryland no-till agroecosystems. *Geoderma* **2013**, *192*, 254–258. [CrossRef]
50. Schneekloth, J.; Calderón, F.; Nielsen, D.; Fonte, S.J. Tillage and residue management effects on irrigated maize performance and water cycling in a semi-arid cropping system of Eastern Colorado. *Irrig. Sci.* **2020**, *38*, 547–557. [CrossRef]
51. Nannipieri, P.; Ascher, J.; Ceccherini, M.T.; Landi, L.; Pietramellara, G.; Renella, G. Microbial diversity and soil functions. *Eur. J. Soil. Sci.* **2003**, *54*, 655–670. [CrossRef]

52. Blanco-Canqui, H.; Holman, J.D.; Schlegel, A.J.; Tatarko, J.; Shaver, T.M. Replacing fallow with cover crops in a semi-arid soil: Effects on soil properties. *Soil Sci. Soc. Am. J.* **2013**, *77*, 1026–1034. [CrossRef]
53. Klein, R.N. *Improving Your Success in No-Till. Cover Your Acres Proceedings*; Kansas State University: Manhattan, KS, USA, 2008; pp. 22–26.
54. Malhi, S.S.; Lemke, R.; Wang, Z.H.; Chhabra, B.S. Tillage, nitrogen and crop residue effects on crop yield, nutrient uptake, soil quality, and greenhouse gas emissions. *Soil Tillage Res.* **2006**, *90*, 171–183. [CrossRef]
55. Govaerts, B.; Sayre, K.D.; Ceballos-Ramirez, J.M.; Luna-Guido, M.L.; Limon-Ortega, A.; Deckers, J.; Dendooven, L. Conventionally tilled and permanent raised beds with different crop residue management: Effects on soil C and N dynamics. *Plant Soil* **2006**, *280*, 143–155. [CrossRef]
56. Antosh, E.; Idowu, J.; Schutte, B.; Lehnhoff, E. Winter cover crops effects on soil properties and sweet corn yield in semi-arid irrigated systems. *Agron. J.* **2020**, *112*, 92–106. [CrossRef]
57. Paye, W.S.; Acharya, P.; Ghimire, R. Water productivity of forage sorghum in response to winter cover crops in semi-arid irrigated conditions. *Field Crops Res.* **2022**, *283*, 108552. [CrossRef]
58. Amsili, J.P.; Kaye, J.P. Root traits of cover crops and carbon inputs in an organic grain rotation. *Renew. Agric. Food Syst.* **2021**, *36*, 182–191. [CrossRef]
59. Rumberger, A.; Marschner, P. 2-Phenylethylisothiocyanate concentration and microbial community composition in the rhizosphere of canola. *Soil Biol. Biochem.* **2003**, *35*, 445–452. [CrossRef]
60. Liebig, M.; Carpenter-Boggs, L.; Johnson, J.M.F.; Wright, S.; Barbour, N. Cropping system effects on soil biological characteristics in the Great Plains. *Renew Agric. Food Syst.* **2006**, *21*, 36–48. [CrossRef]
61. Sainju, U.M.; Singh, B.P.; Whitehead, W.F. Tillage, cover crops, and nitrogen fertilization effects on cotton and sorghum root biomass, carbon, and nitrogen. *Agron. J.* **2005**, *97*, 1279–1290. [CrossRef]
62. Brozović, B.; Jug, D.; Đurđević, B.; Vukadinović, V.; Tadić, V.; Stipešević, B. Influence of winter cover crops incorporation on weed infestation in popcorn maize (*Zea mays* everta Sturt.) organic production. *Agric. Conspec. Sci.* **2018**, *83*, 77–81.
63. Acosta-Martínez, V.; Zobeck, T.M.; Gill, T.E.; Kennedy, A.C. Enzyme activities and microbial community structure in semi-arid agricultural soils. *Biol. Fertil. Soils* **2003**, *38*, 216–227. [CrossRef]
64. Acosta-Martínez, V.; Mikha, M.M.; Vigil, M.F. Microbial communities and enzyme activities in soils under alternative crop rotations compared to wheat–fallow for the central Great Plains. *Appl. Soil Ecol.* **2007**, *37*, 41–52. [CrossRef]
65. Dou, F.; Wright, A.L.; Hons, F.M. Dissolved and soil organic carbon after long-term conventional and no-tillage sorghum cropping. *Commun. Soil Sci. Plant Anal.* **2008**, *39*, 667–679. [CrossRef]
66. Dou, F.; Wright, A.L.; Hons, F.M. Depth distribution of soil organic C and N after long-term soybean cropping in Texas. *Soil Tillage Res.* **2007**, *94*, 530–536. [CrossRef]
67. Allen, B.L.; Pikul, J.L., Jr.; Waddell, J.T.; Cochran, V.L. Long-term lentil green-manure replacement for fallow in the semi-arid northern Great, Plains. *Agron. J.* **2011**, *103*, 1292–1298. [CrossRef]
68. Miller, P.R.; Lighthiser, E.J.; Jones, C.A.; Holmes, J.A.; Rick, T.L.; Wraith, J.M. Pea green manure management affects organic winter wheat yield and quality in semi-arid Montana. *Can. J. Plant Sci.* **2011**, *91*, 497–508. [CrossRef]
69. Nennich, T.D.; Harrison, J.H.; VanWieringen, L.M.; Meyer, D.; Heinrichs, A.J.; Weiss, W.P.; St-Pierre, N.R.; Kincaid, R.L.; Davidson, D.L.; Block, E. Prediction of manure and nutrient excretion from dairy cattle. *J. Dairy Sci.* **2005**, *88*, 3721–3733. [CrossRef]
70. USDA ERS—Organic Production, n.d. Available online: https://www.ers.usda.gov/data-products/organic-production.aspx (accessed on 5 July 2022).
71. Sims, J.T.; Maguie, R.O. Manure management. In *Encyclopedia of Soils in the Environment*; Hillel, D., Ed.; Elsevier: Amsterdam, The Netherlands, 2004; Volume 4.
72. Delgado, J.A.; Follett, R.F. Carbon and nutrient cycles. *J. Soil Water Conserv.* **2002**, *57*, 455–464.
73. Butler, T.J.; Muir, J.P. Dairy manure compost improves soil and increases tall wheatgrass yield. *Agron. J.* **2006**, *98*, 1090–1096. [CrossRef]
74. Acharya, P.; Ghimire, R.; Cho, Y. Linking soil health to sustainable crop production: Dairy compost effects on soil properties and sorghum biomass. *Sustainability* **2019**, *11*, 3552. [CrossRef]
75. Calderón, F.J.; Vigil, M.F.; Benjamin, J. Compost input effect on dryland wheat and forage yields and soil quality. *Pedosphere* **2018**, *28*, 451–462. [CrossRef]
76. Liu, J.; Calderón, F.J.; Fonte, S.J. Compost inputs, cropping system, and rotation phase drive aggregate-associated carbon. *Soil Sci. Soc. Am. J.* **2021**, *85*, 829–846. [CrossRef]
77. Maltais-Landry, G.; Scow, K.; Brennan, E.; Vitousek, P. Long-term effects of compost and cover crops on soil phosphorus in two California agroecosystems. *Soil Sci. Soc. Am. J.* **2015**, *79*, 688–697. [CrossRef]
78. Helton, T.J.; Butler, T.J.; McFarland, M.L.; Hons, F.M.; Mukhtar, S.; Muir, J.P. Effects of dairy manure compost and supplemental inorganic fertilizer on coastal Bermudagrass. *Agron. J.* **2008**, *100*, 924–930. [CrossRef]
79. Dungan, R.S.; Leytem, A.B.; Tarkalson, D.D. Greenhouse gas emissions from an irrigated cropping rotation with dairy manure utilization in a semi-arid climate. *Agron. J.* **2021**, *113*, 1222–1237. [CrossRef]
80. Brempong, M.B.; Norton, U.; Norton, J.B. Compost and soil moisture effects on seasonal carbon and nitrogen dynamics, greenhouse gas fluxes and global warming potential of semi-arid soils. *Int. J. Recycl. Org. Waste Agric.* **2019**, *8*, 367–376. [CrossRef]

81. Acosta-Martínez, V.; Harmel, R.D. Soil microbial communities and enzyme activities under various poultry litter application rates. *J. Environ. Qual.* **2006**, *35*, 1309–1318. [CrossRef] [PubMed]
82. Burke, J.A.; Lewis, K.L.; DeLaune, P.B.; Cobos, C.J.; Keeling, J.W. Soil water dynamics and cotton production following cover crop use in a semi-arid ecoregion. *Agronomy* **2022**, *12*, 1306. [CrossRef]
83. Calderón, F.J.; Nielsen, D.; Acosta-martínez, V.; Vigil, M.F.; Lyon, D. Cover crop and irrigation effects on soil microbial communities and enzymes in semi-arid agroecosystems of the central Great Plains of North America. *Pedosphere* **2016**, *26*, 192–205. [CrossRef]
84. Thapa, V.R.; Ghimire, R.; Duval, B.D.; Marsalis, M.A. Conservation systems for positive net ecosystem carbon balance in semi-arid drylands. *Agrosyst. Geosci. Environ.* **2019**, *2*, 190022. [CrossRef]
85. DeLaune, P.B.; Mubvumba, P.; Lewis, K.L.; Keeling, J.W. Rye cover crop impacts soil properties in a long-term cotton system. *Soil Sci. Soc. Am. J.* **2019**, *83*, 1451–1458. [CrossRef]
86. Anderson, R.L. Improving sustainability of cropping systems in the central Great Plains. *J. Sustain. Agric.* **2005**, *26*, 97–114. [CrossRef]

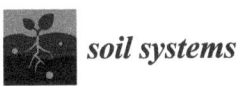

soil systems

MDPI

Article

Silica and Biochar Amendments Improve Cucumber Growth under Saline Conditions

Manar Al-Toobi [1,2], Rhonda R. Janke [1], Muhammad Mumtaz Khan [1], Mushtaque Ahmed [3], Waleed M. Al-Busaidi [1] and Abdul Rehman [1,4,*]

1   Department of Plant Sciences, College of Agricultural and Marine Sciences, Sultan Qaboos University, Al-Khoud 123, Oman; manar.altobi@ea.gov.om (M.A.-T.)
2   Azzan Bin Qais International School, Muscat 102, Oman
3   Soil, Water & Agricultural Engineering, College of Agricultural and Marine Sciences, Sultan Qaboos University, Al-Khoud 123, Oman
4   Department of Agronomy, Faculty of Agriculture and Environment, The Islamia University of Bahawalpur, Bahawalpur 63100, Pakistan
*   Correspondence: abdurehmanuaf@gmail.com or a.rehman@iub.edu.pk

Abstract: Rapidly increasing salinization of arable land is a major threat to crop production globally, and the soil of regions with arid environments, such as Oman, are more prone to this menace. In this work, two complementary studies were carried out to evaluate the effect of soil amendments on soil physicochemical properties and growth of cucumber seedlings. In the first study, high- and low-saline soils were used with or without perlite. The amendments tested included mango wood biochar, silica, and biochar + silica, while no amendment was taken as the control. The second study included two cucumber cultivars and irrigation water with two salinity treatments, along with the same four soil amendments. The results showed that soil amendment with biochar alone or with silica enhanced the soil organic matter and NO_3, P, and K concentration, while silica amendment substantially enhanced the soil Si level in both studies. Saline soil and irrigation water inhibited seedling emergence and plant growth in both experiments. However, the addition of biochar and silica alone or in combination increased the cucumber seedling dry weight from 39.5 to 77.3% under salt stress compared to the control. Likewise, silica and biochar + silica reduced the sap Na accumulation by 29–31.1% under high salinity. Application of biochar under high salinity resulted in 87.2% increase in sap K. Soil amendments with biochar and silica or their combination have the potential to reduce the adverse effect of salt stress on cucumber.

Keywords: salinization; biomass production; nutrient; irrigation water salinity; mango wood biochar

Citation: Al-Toobi, M.; Janke, R.R.; Khan, M.M.; Ahmed, M.; Al-Busaidi, W.M.; Rehman, A. Silica and Biochar Amendments Improve Cucumber Growth under Saline Conditions. *Soil Syst.* **2023**, *7*, 26. https://doi.org/10.3390/soilsystems7010026

Academic Editor: Luis Eduardo Akiyoshi Sanches Suzuki

Received: 1 February 2023
Revised: 7 March 2023
Accepted: 9 March 2023
Published: 12 March 2023

1. Introduction

Soil salinity is one of the leading environmental constraints limiting agricultural productivity in many regions of the world. Globally, about 0.8 billion ha of land is considered as saline [1]. Secondary salinization, which is human-induced salinization, affects more than 75 million ha [2]. Every single year, up to 1.5–2 million ha of land around the world is lost due to salinity [2], and the monetary loss in the agriculture sector due to soil salinity is more than USD 27.3 billion [3]. The majority of irrigated land worldwide is saline compared to nonirrigated land [2,4,5]. The accumulation of salt in the soil produces a high soil solution osmotic pressure, which reduces water availability to plants, leading to wilting and poor growth. Saline-stress-induced deleterious effects on plant growth include photosynthesis reduction due to limited chlorophyll biosynthesis [6], impaired photosynthesis machinery [7], reduced osmotic potential, nutrient imbalance, specific ion toxicity, or a combination of all these factors [8–10]. Salt stress also enhances the accumulation of reactive oxygen species, leading to loss of membrane integrity and electrolyte leakage [11].

Biochar (BC) is a carbon-rich material produced by pyrolysis of biomass under low oxygen supply. Its application can improve plant growth under salt stress conditions as it reduces the soil bulk density, electrical conductivity, and exchangeable Na^+ and Cl^- ions in saline soil [12]. For instance, application of maple (*Acer pseudoplatanus* L.) residue BC (50 and 100 g kg^{-1} soil) in saline soils was found to reduce Na^+ uptake and reactive oxygen species generation in root and leaves and increase the cation exchange capacity of soil, chlorophyll content index, leaf area, and nutrient uptake in mung bean (*Vigna radiata* (L.) R. Wilczek) [13]. Although silicon (Si) is a nonessential plant nutrient, it still plays a positive role in the growth and development of plants. Increased Si uptake can improve the ability of plants to grow under suboptimal growth environments [14]. Adequate Si supply can help ameliorate the adverse effects of NaCl on plants through activation of signaling molecules and regulation of phytohormones under salt stress conditions. For instance, jasmonic acid signaling was found to upregulate the genes involved in Si uptake under salt stress conditions, which accelerated the antioxidant defense system and osmolyte production under salt stress conditions [15]. In addition to the role of Si in the activation of signaling molecules and regulation of phytohormones, it has been shown to effectively reduce degradation of photosynthetic pigments, improve gas exchange traits, inhibit lipid peroxidation and leaf electrolyte leakage, and enhance osmotic adjustment of plants. The application of Si has been found to increase K^+ concentration and reduce Na^+ absorption, transport, and accumulation in plants [16].

The Sultanate of Oman is located in an arid region where the annual rainfall is less than 100 mm, resulting in the majority of land being unsuitable for agriculture without irrigation [17]. The main factor contributing to the issue of salinity is the high-salt content of the groundwater in Oman, especially in the Al-Batinah coastal region, where seawater intrusion into the aquifers has increased salinity levels [18]. This has resulted in serious consequences for agriculture, with limited water resources exacerbating the problem [19].

There are several reports regarding the role of BC and Si in abiotic stress tolerance in plants. However, to the best of our knowledge, very little information is available regarding the individual and interactive effects of BC and Si application in cucumber. Cucumber is a widely grown commercial greenhouse crop in Oman. The present study was carried out to evaluate the individual and interactive effects of soil BC and Si amendments on soil physicochemical properties, growth, and nutrient dynamics of cucumber seedlings under conditions of high and low soil electrical conductivity (EC) and high and low salt levels in irrigation water.

2. Materials and Methods

2.1. Experiment 1

Two types of soils were used. One was from a greenhouse at a local farm with a relatively high EC level due to past fertilization practices (EC ~5.0 dS m^{-1}, Na^+ ~500 ppm), and the other was from a nonfarmed deposit of soil from Sohar farm with low salt and nutrient content (EC ~1.4 dS m^{-1}, Na^+ ~160 ppm). The pH, EC, and water-soluble Na^+ were measured in a 1:2 (30 g:60 mL) mixture of soil and distilled water according to standard methods [20] using a hand-held calibrated meter (EUTECH, OAKTON, 35425-10). Each of these two soils were used as 100% soil or mixed with commercial-grade perlite in a 1:1 ratio (*v/v*) to improve drainage. Then, in each of the four soil mixtures, four basic treatments were compared—control, BC, Si, and a mixture of BC and Si—that were added at a rate of 10% by volume to each pot and mixed well. All treatments were replicated 3 times for a total of 48 pots. Plastic pots of 8 cm diameter and a volume of 470 mL were used.

In Experiment 1, four cucumber seeds (*Cucumis sativus*) of the variety Jabbar F1 were planted in each pot and watered with tap water (EC = 1.19 dS m^{-1}; pH = 7.2) to saturation. Hand watering was performed twice per week as needed, and care was taken to only apply the amount that could be absorbed by the soil to minimize leaching.

Biochar was produced at the AES (Agricultural Experiment Station) at SQU using local dried mango wood. The wood was burned using prototype BC apparatus, and the final product was crushed and sieved to 2.0 mm as detailed in our previous study [21]. The Si used in this experiment was manufactured by Agripower®, a company based in Australia (https://agripower.com.au/agrisilica-granular-fertiliser/, accessed on 1 March 2022), and was added as off-white granules, which dissolved easily when water was added.

The experiment was conducted in a growth chamber under controlled environmental conditions. After planting, pots were placed on four clear plastic trays in a completely randomized design on four shelves of a growth chamber/incubator with vertical fluorescent lighting on the front, back, and sides. To avoid the effect of distance to light, the pots were frequently rotated within each tray, and the trays were rotated from shelf to shelf. No effect of position within the chamber on plant growth was observed in this or subsequent experiments. The growth chamber (SANYO, MLR-350HT) was set at a continuous temperature of 25/20 ± 2 °C (day/night) and 50% humidity. Photosynthetic active photon flux of 380 μmol s^{-1} m^{-2} was measured using a Hydrofarm LGBQM Quantum PAR meter at the surface of the pots, and a light/dark photoperiod (16/8 h) was maintained during the experiment.

2.2. Experiment 2

The nonfarmed low-fertility, low EC soil from Experiment 1 was used for all treatments in this experiment, mixed at a ratio of 3:1 with commercial-grade perlite. Two salt treatments were applied to the water in order to simulate a farm with saline irrigation water. Two cucumber genotypes (SV8975CB and Jabbar F1) were compared, and the same soil amendments were compared (no amendment, BC, Si, and BC + Si) at the same rate of 10% by volume in the same sized pots with 3 replications for a total of 48 pots.

Two seeds were planted in each pot, which were thinned to one plant per pot and watered with tap water for a week before exposing them to two salt levels: high-salt water (EC ~3 dS m^{-1}) and low-salt water (EC ~0.5 dS m^{-1}). Plants were watered with NaCl salt solution every 2 days, and the EC level of 3.0 dS m^{-1} was chosen to stress the cucumber seedlings but not kill them [22]. A balanced soluble nutrient solution was used once a week to improve fertility. After each watering session, the plant trays were rotated on growth chamber shelves to obtain more even light effects. Both experiments were conducted for approximately 5.5 to 6 weeks in the spring of 2018.

2.3. Plant Observations

In Experiment 1, the number of emerged seedlings were recorded twice a week, and the emergence percentage was calculated. A vigor rating of 1 (poor growth, weak stem, and few leaves) to 5 (excellent growth, strong stem, and many leaves) was used as visual observation of plant health, and an index was created combining germination × vigor. In Experiment 2, plant height was recorded before harvesting, and the number of leaves was counted on the fifth week after emergence. The chlorophyll density was measured in the fifth week using a SPAD meter (SPAD-502Plus). In both experiments, plants were harvested in week 5.5 or week 6. The root and shoots were separated and weighed immediately for fresh weight using an electric balance. The harvested plant samples were then oven dried at 60 °C for 48 h to dry weight.

After the fresh weight measurements, sap from the shoots was extracted using a garlic press and collected on a small tissue paper to determine the sap nutrient concentration. The concentration of Na$^+$, K$^+$, and NO$_3^-$ in plant sap was analyzed using different meters (HORIBA B-722 for Na$^+$, HORIBA B-731 for K$^+$, and HORIBA B-743 for NO$_3^-$) that were calibrated with nutrient solutions provided by the manufacturer.

2.4. Soil Analysis

Phosphorus levels were determined using the "Olsen method" of phosphorus extraction using 5 g of soil and NaHCO$_3$ solution according to [23]. Water-soluble nutrient concentrations were determined in a 1:2 solution of soil and distilled water (*v/v*) and included tests for pH, EC, and water soluble Na^+, NO_3, and K^+ using the hand-held meters previously described [24]

The water-holding capacity of each soil was estimated by saturating 30 mL of soil in a filter paper in a funnel. After six hours, the soil was weighed and the saturated soil samples were oven dried for 48 h at 80 °C to a constant value [25]. Then, the water content was calculated as water held in the soil divided by soil dry weight. The organic matter was estimated in soil via loss on ignition (LOI) [26] in a muffle furnace for 2 h as 450 °C.

The concentration of soluble Si in soil was measured according to the methods described in ICARDA [23]. The most abundant form of Si is monomeric silica acid (H_3SiO_4), which can be extracted with 0.01 M $CaCl_2$ solution, even in soils with a high level of $CaCO_3$. The absorbance of blank, standards, and samples were recorded on a SPECTRONIC 200E spectrophotometer (Thermo Scientific) at 660 nm. The calibration curve for the standards was prepared, and the absorbance was plotted against the respective Si concentration to convert absorbance to soil concentration.

2.5. Data Analysis

The experimental data were analyzed using analysis of variance at $p < 0.05$ using Minitab (Minitab® 17.3.1) and the PROC GLM procedure. Significant treatment means were separated using Tukey's honestly significant difference (HSD) test. Microsoft Excel program was used to develop the figures and calculate standard errors.

3. Results

3.1. Experiment 1

3.1.1. Soil Physicochemical Properties

The soils used in this experiment (S) significantly ($p < 0.001$) differed for all studied soil physicochemical traits. However, the soil amendment treatments (T) were significantly different for soil organic matter (OM), pH, NO_3 ($p < 0.05$), P ($p < 0.001$), K^+ ($p < 0.05$), and Si ($p < 0.001$) concentration in soil (Table 1). The interaction S × T was significant ($p < 0.001$) for pH, EC, P, Na, and Si (Table 1). The water-holding capacity (WHC) was higher in the high EC soil obtained from the local greenhouse compared to the never-cropped low EC soil. Likewise, OM was higher in the high-saline soils and also increased by BC application alone or in combination with Si (Table 1). The high EC greenhouse soils also had the highest concentrations of NO_3, P, K, Na, and Si compared to the uncropped soil (Table 1).

Table 1. Soil nutrient values and significance levels in Experiment 1.

Treatments	WHC	OM	pH	EC	NO₃	P	K	Na	Si
Soil Salinity (S)	(%)	(%)		(dS m⁻¹)	(ppm)	(ppm)	(ppm)	(ppm)	(ppm)
High-salt soil	49.4 B	12.8 A	6.9 B	3.6 A	253 A	29 A	435 A	415 A	47 A
High-salt soil + perlite	58.4 A	11.6 A	6.8 B	2.0 B	167 B	27 B	208 B	230 B	41 B
Low-salt soil	21.1 D	3.4 B	7.5 A	0.4 C	92 C	4 C	10 C	59 C	14 D
Low-salt soil + perlite	29.6 C	4.0 B	7.5 A	0.3 C	99 C	4 C	13 C	68 C	17 C
$p < 0.05$	**	**	**	**	**	**	**	**	**
Treatments (T)									
Control	36.3	6.0 B	7.2 AB	1.5	125 B	14 C	158 AB	179	24 B
BC	40.3	8.8 A	7.3 A	1.6	155 AB	17 B	193 A	186	25 B
Si	38.9	7.2 B	7.2 AB	1.5	150 AB	13 C	133 B	196	36 A
BC + Si	42.8	9.7 A	7.1 B	1.7	181 A	21 A	182 AB	211	35 A
$p < 0.05$	ns	**	*	ns	*	**	*	ns	**
S × T	ns	Ns	**	**	ns	**	ns	**	**

Means sharing the same letters in the column do not differ significantly at $p < 0.05$. * = $p < 0.05$; ** = $p < 0.01$. WHC = water-holding capacity; OM = organic matter; EC = electrical conductivity; BC = biochar; Si = silica; BC + Si = biochar + silica.

The BC treatments resulted in significantly higher levels of P, and the Si treatments resulted in higher levels of soil Si. Nitrate was highest in the BC plus Si treatment but was only significantly different from the control.

Looking more closely at the interaction effects, the saline soils had lower pH compared to low-saline soils, probably as a result of the fertilization history. In this regard, the lowest soil pH (6.47) was observed for high-saline soil amended with Si + BC, while low-saline soils amended with BC had the highest soil pH (Figure 1a). The soil EC was largely influenced by the soil salinity level, and perlite addition reduced the soil EC. The highest soil EC was recorded for saline soil without perlite (3.75 dS m^{-1}), while low-saline soil exhibited the lowest value of EC (0.34 dS m^{-1}) irrespective of soil amendment (Figure 1b). Phosphorus level was higher in high-saline soils. BC + Si application in high-saline soil with perlite had the highest soil P concentration (44.1 ppm), while the lowest P concentration (2.66 ppm) was recorded in soils with low salt irrespective of soil amendment (Figure 1c). Soil Si levels were highest overall in the high EC greenhouse soils, probably due to prior application of compost and/or peat moss, which also contains high levels of soluble Si. Application of Si amendment further increased the Si concentration (53.9 ppm) for both Si application alone or in combination with BC. The lowest soil Si concentration was measured for the control (9.8 ppm) and BC amendment (10.2 ppm) in low-saline soils (Figure 1d). The Na concentration was highest in the high EC soils without perlite irrespective of soil amendment and lower when perlite was added (Figure 1e).

3.1.2. Germination, Growth, and Sap Nutrient Concentrations

The analysis of variance revealed that soil salinity significantly ($p < 0.001$) affected all the studied traits of cucumber in Experiment 1 (Table 2). However, the four treatments were not significant, except for sap K concentration. The soil salinity by treatment (S × T) interaction was only significant for the sap NO$_3$ concentration (Table 2). All soil types exhibited similar germination rates except the high-saline soil without perlite. Higher vigor ratings and index were noted for all soils except for the high-saline soil without perlite (Table 2). The plants grown in high-saline soils could not survive after germination, and the BC and Si treatments did not reduce the effect of salt enough to increase germination or survival. The highest root and top fresh weight were noted for low saline + perlite and high saline + perlite soils, respectively. Again, there was no significant effect of BC or Si treatments on root or shoot fresh weight (FW) (Table 2). The sap Na and K concentration was highest in plants grown on high saline + perlite soil, while low salinity reduced sap Na and K accumulation. Application of BC alone and BC + Si enhanced sap K ac-cumulation in cucumber (Table 2). In the case of sap NO$_3$, application of Si substantially enhanced NO$_3$ accumulation (690 ppm) in the high salt + perlite soil, while none of the soil amendments enhanced NO$_3$ uptake in the low-salt stress condition (Figure 1f).

Table 2. Germination, plant vigor, FW, sap nutrient concentration, and significance levels in Experiment 1.

Treatments	Final Germination Count	Average Vigor (0–5 Rating)	Index	Root FW (g Plant^{-1})	Tops FW (g Plant^{-1})	Na (ppm)	NO$_3$ (ppm)	K (ppm)
Soil Salinity (S)								
High-salt soil	0.5 B	0.125 B	0.125 B	0	0	0	0	0
High-salt soil + perlite	3.17 A	0.770 A	2.562 A	0.44 B	1.85 A	1609 A	475 A	2787 A
Low-salt soil	2.75 A	0.687 A	2.354 A	0.54 B	0.99 B	626 B	192 B	1829 B
Low-salt soil + perlite	3.416 A	0.854 A	2.979 A	1.15 A	1.11 B	905 B	173 B	1715 B
$p < 0.05$	***	***	***	***	***	***	***	***
Treatments (T)								
Control	2.583	0.541	1.916	0.53	1.15	951	228	1446 B
BC	2.583	0.625	1.958	0.84	1.33	991	283	2877 A
Si	2.50	0.625	2.00	0.74	1.48	1176	346	1517 B
BC + Si	2.166	0.645	2.145	0.73	1.31	1070	261	2600 A
$p < 0.05$	Ns	ns	ns	ns	ns	ns	ns	***
S × T	Ns	ns	ns	ns	ns	ns	*	ns

Means sharing the same letters in the column do not differ significantly at $p < 0.05$. * = $p < 0.05$; *** = $p < 0.001$. FW = fresh weight; index = germination × vigor; BC = biochar; Si = silica; BC + Si = biochar + silica.

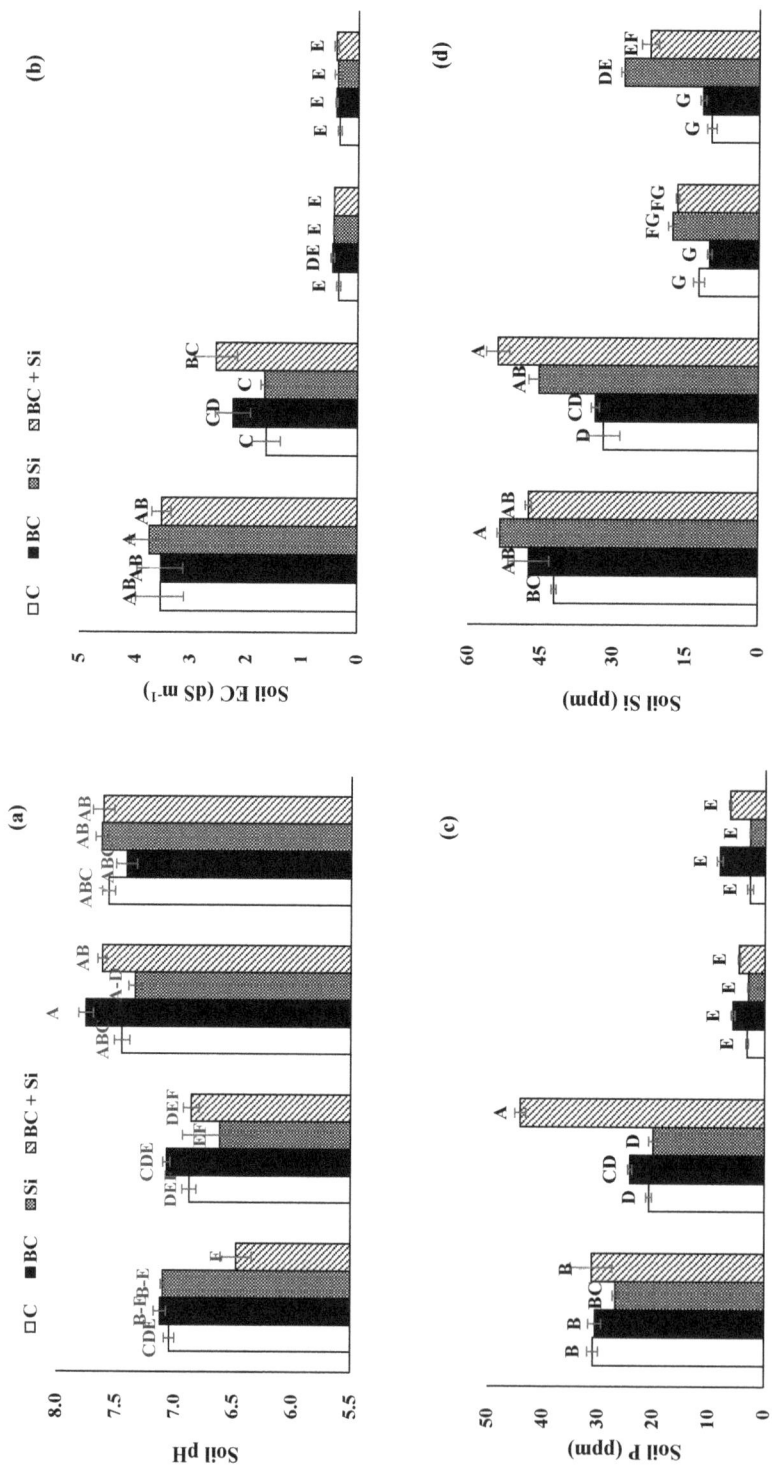

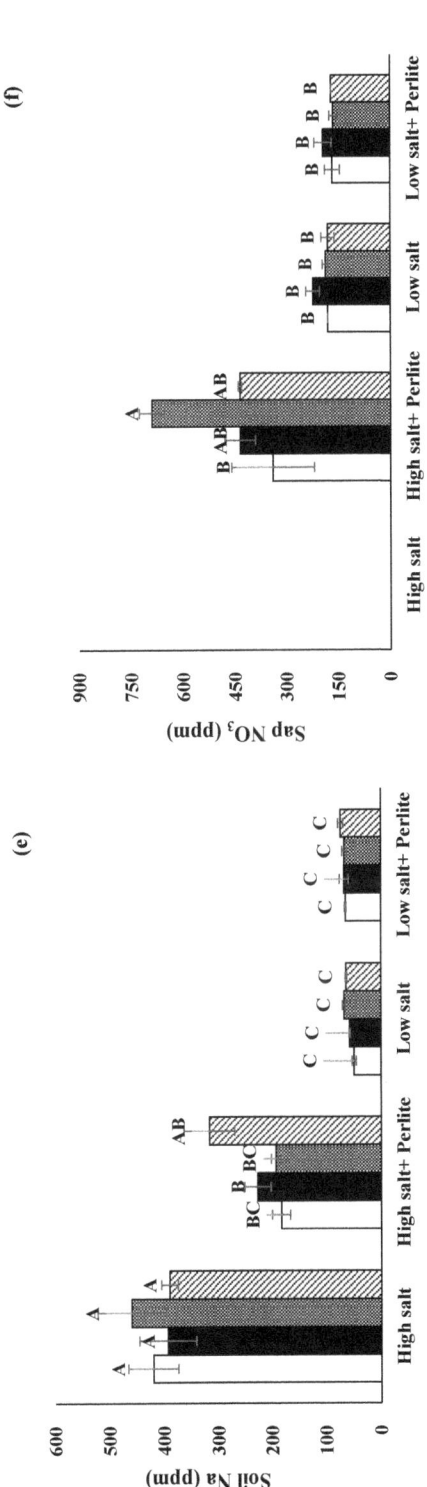

Figure 1. Effect of soil type and soil amendment on soil (**a**) pH, (**b**) EC, (**c**) phosphorus, (**d**) Si, (**e**) Na, and (**f**) plant sap NO_3 concentrations of cucumber ± S.E. C = control; BC = biochar; Si = silica; BC + Si = biochar + silica. Means sharing the same letters do not differ significantly at $p < 0.05$.

3.2. Experiment 2

3.2.1. Soil Physicochemical Properties

The analysis of variance revealed that irrigation water salinity significantly influenced all soil physicochemical properties except Si concentration (Table 3). Soil amendment treatments also influenced all traits except EC. The interactions $S \times T$ were only significant for WHC, pH, and soil K and Na concentrations (Table 3).

Table 3. Soil nutrient values and significance levels for Experiment 2.

Treatments	WHC	OM	pH	EC	Si	K	Na	NO$_3$
Irrigation Water Salinity (S)	(%)	(%)		(dS m^{-1})	(ppm)	(ppm)	(ppm)	(ppm)
High salt	38.9 B	3.18 A	8.37 B	2.26 A	19.1	35.3 A	395 A	140.7 B
Low salt	36.9 A	2.91 B	8.56 A	0.85 B	20.4	25.7 B	128 B	208.8 A
$p < 0.05$	*	*	**	**	ns	**	**	**
Treatments (T)								
Control	33.8 C	2.24 B	8.30 B	1.43	14.1 B	12.9 C	260 AB	165.9 B
BC	39.3 AB	3.58 A	8.64 A	1.50	16.5 B	58.8 A	240 B	191.7 A
Si	37.1 B	2.54 B	8.32 B	1.65	23.3 A	9.2 C	292 A	162.4 B
BC + Si	41.5 A	3.82 A	8.60 A	1.65	25.0 A	41.0 B	253 B	179.0 AB
$p < 0.05$	**	**	**	ns	**	**	*	*
$S \times T$	*	ns	*	ns	ns	*	*	ns

Means sharing the same letters in the column do not differ significantly at $p < 0.05$. * = $p < 0.05$; ** = $p < 0.01$; WHC = water-holding capacity; OM = organic matter; EC = electrical conductivity; BC = biochar; Si = silica; BC + Si = biochar + silica.

Soil amendments improved the WHC, with the highest WHC (43.2%) in the BC + Si treatment, followed by BC alone and then Si alone. The lowest WHC was in the control treatment (33.4%). The higher salinity irrigation treatment seemed to have higher WHC in both BC soil treatments (Figure 2a). Biochar application alone or in combination with Si increased the soil pH (8.70) irrespective of soil salinity level. However, the lowest soil pH was noted for the control (8.26) and Si treatment (8.10) for high-salinity irrigation treatment (Figure 2b). High-salinity irrigation treatment exhibited higher OM accumulation, while BC addition alone or in combination with Si enhanced (3.95%) the soil OM level (Table 3). In the case of soil EC, high-salinity irrigation had the highest EC, while none of the soil amendment treatments significantly influenced the soil EC (Table 3). Application of Si alone or in combination with BC substantially increased (26.4 ppm) the soil Si concentration (Table 3). Low-salinity irrigation resulted in the highest NO$_3$ level. Among soil amendments, BC application augmented the soil NO$_3$ level (226.7 ppm), while Si application and no soil amendment had the lowest NO$_3$ concentration (Table 3). Biochar application substantially augmented the soil K concentration as the highest K level (66.3 ppm) was recorded for high-salinity irrigation treatment receiving BC amendment, while soil receiving Si and no amendment (11.5 ppm) had the lowest K irrespective of irrigation salinity level (Figure 2c). The soil Na level was lowest in the low-salinity irrigation treatment, irrespective of soil amendment. However, in high-salinity irrigation, BC application alone or in combination with Si reduced (122 ppm) the soil Na concentration (Figure 2d).

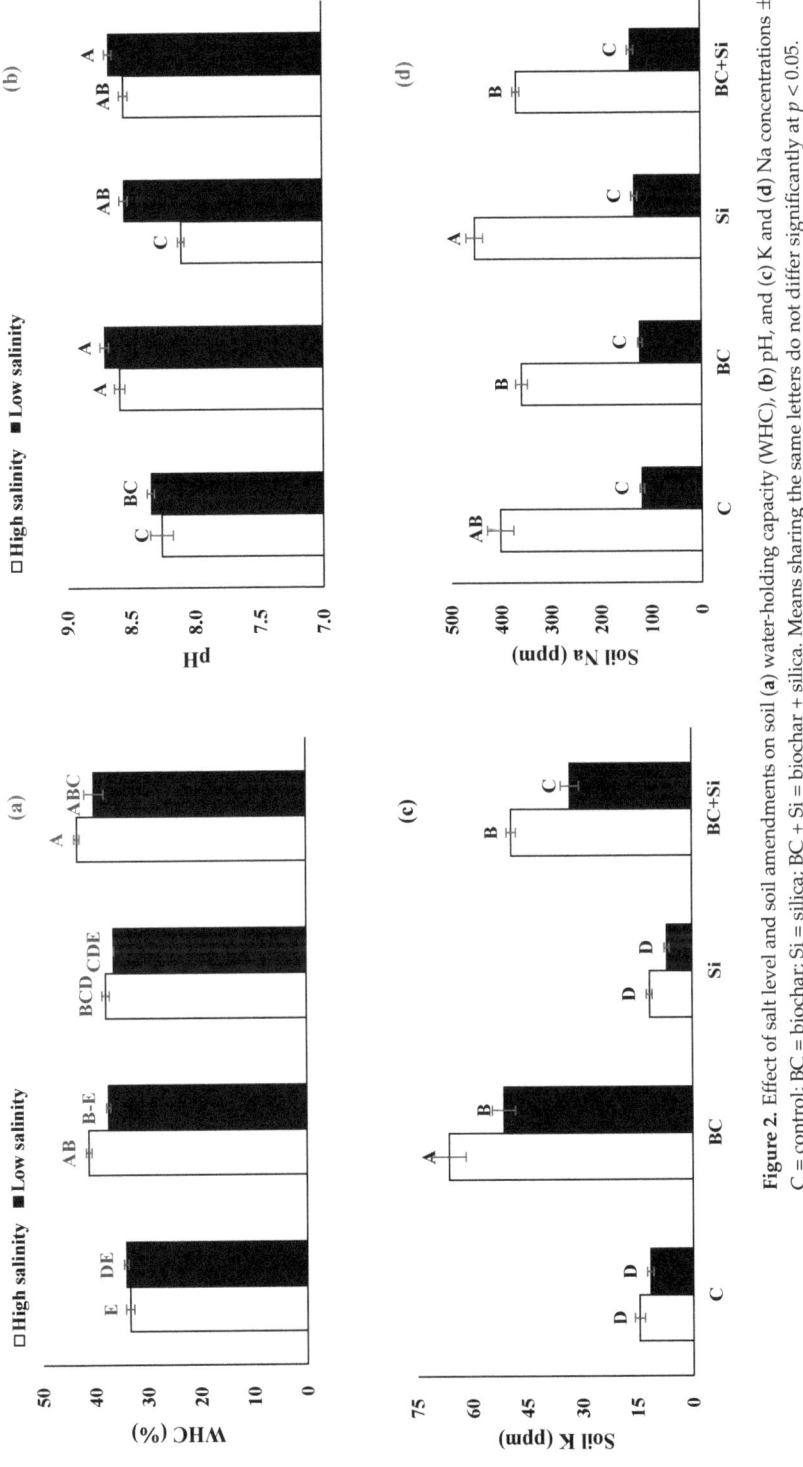

Figure 2. Effect of salt level and soil amendments on soil (**a**) water-holding capacity (WHC), (**b**) pH, and (**c**) K and (**d**) Na concentrations ± S.E. C = control; BC = biochar; Si = silica; BC + Si = biochar + silica. Means sharing the same letters do not differ significantly at *p* < 0.05.

3.2.2. Biomass, Growth, and Sap Nutrient Concentrations

The analysis of variance revealed that cucumber cultivars (C) significantly differed for leaf count, plant fresh weight, sap NO_3, sap Na, and sap K concentration (Table 4). Irrigation salinity level significantly influenced all traits except SPAD value and NO_3 concentration in plant tissues. Soil amendments significantly influenced the measured traits. The interaction of C × S was significant only for leaf number and sap K concentration. The C × T interaction was significant for plant height and leaf K concentration. However, the S × T interaction was significant for all traits except plant height and SPAD value. The three-way interaction C × S × T was only significant for leaf count and sap K concentration (Table 4).

Table 4. Plant height, leaf count, SPAD, biomass, and sap nutrient content and significance levels for Experiment 2.

Treatments	Plant Height	Leaf Count	SPAD	Plant fresh Weight	Plant Dry Weight	Leaf NO₃	Leaf Na	Leaf K
Cucumber Cultivars (C)	(cm)			(g Plant⁻¹)	(g Plant⁻¹)	(ppm)	(ppm)	(ppm)
SV8975 CB	6.87	3.21 B	34.3	3.11 B	0.349	362 B	3063 A	3208 A
Jabbar, F1	6.48	4.05 A	35.5	3.84 A	0.357	511 A	2734 B	2447 B
$p < 0.05$	ns	***	ns	***	ns	**	*	***
Salinity level (S)								
High salt	6.01 B	3.11 B	34.7	2.38 B	0.248 B	406 A	4808 A	2697 B
Low salt	7.33 A	4.16 A	35.1	4.56 A	0.457 A	467 A	988 B	2958 A
$p < 0.05$	***	***	ns	***	***	ns	***	*
Treatments (T)								
Control	7.03 AB	3.15 C	36.5 AB	2.41 C	0.262 B	670 A	3430 A	1954 C
BC	7.33 A	3.88 A	30.1 C	4.24 A	0.410 A	193 B	3235 A	4016 A
Si	6.40 BC	3.56 B	39.9 A	3.24 B	0.334 AB	648 A	2599 B	1971 C
BC + Si	5.96 C	3.95 A	33.1 BC	3.99 A	0.404 A	235 B	2329 B	3550 B
$p < 0.05$	***	***	***	***	***	***	***	***
Interactions ($p < 0.05$)								
C × S	ns	*	ns	ns	ns	ns	ns	***
C × T	*	ns	ns	ns	ns	ns	ns	***
S × T	ns	**	ns	***	**	**	***	***
C × S × T	ns	*	ns	ns	ns	ns	ns	***

Means sharing the same letters in the column do not differ significantly at $p < 0.05$. * = $p < 0.05$; ** = $p < 0.01$; *** = $p < 0.001$. SPAD = soil plant analysis development; BC = biochar; Si = silica; BC + Si = biochar + silica.

The cultivars and irrigation treatment did not influence the SPAD values. However, among the soil amendments, the highest SPAD value was noted with Si application, while the lowest SPAD value was observed with BC application (Table 4). The interaction C × T showed that the tallest cucumber plants (7.63 cm) were noted for cultivar SV8975CB without any soil amendment, while BC + Si combined application reduced (5.87 cm) the plant height in both cultivars (Figure 3a). The highest (five) number of leaves per plant were noted for cultivar Jabbar grown with BC and BC + Si amendment in low-salinity irrigation treatment (Figure 3b). The interactive effect of S × T showed that the highest plant fresh (6.06 g plant⁻¹) and dry (0.58 g plant⁻¹) weights were produced with application of BC alone or in combination with Si in low-salinity irrigation. The lowest values of fresh (1.74 g plant⁻¹) and dry weight (0.17 g plant⁻¹) were noted for control plants grown in high-salinity irrigation treatment (Figure 3c,d). The interaction of S × T revealed that highest leaf NO_3 was noted with Si application with low-salinity irrigation (810 ppm), which was similar to the control in high-salinity, while the lowest NO_3 concentrations was noted for BC and BC + Si in both low (315 ppm) and high (157 ppm) salinity treatments (Figure 3e). High-salinity irrigation substantially increased the Na concentration in plant sap (Table 4). Nevertheless, soil amendments reduced the Na uptake, with Si alone or in combination with BC reducing (742 ppm) the Na accumulation in plant sap (Figure 3f). The interaction C × S × T showed that cultivar SV8975CB had the highest sap K concentration (5367 ppm) under high salinity with BC amendment, while cultivar Jabbar showed the lowest sap K (1100 ppm) under high-saline treatment without BC amendment (Figure 3g).

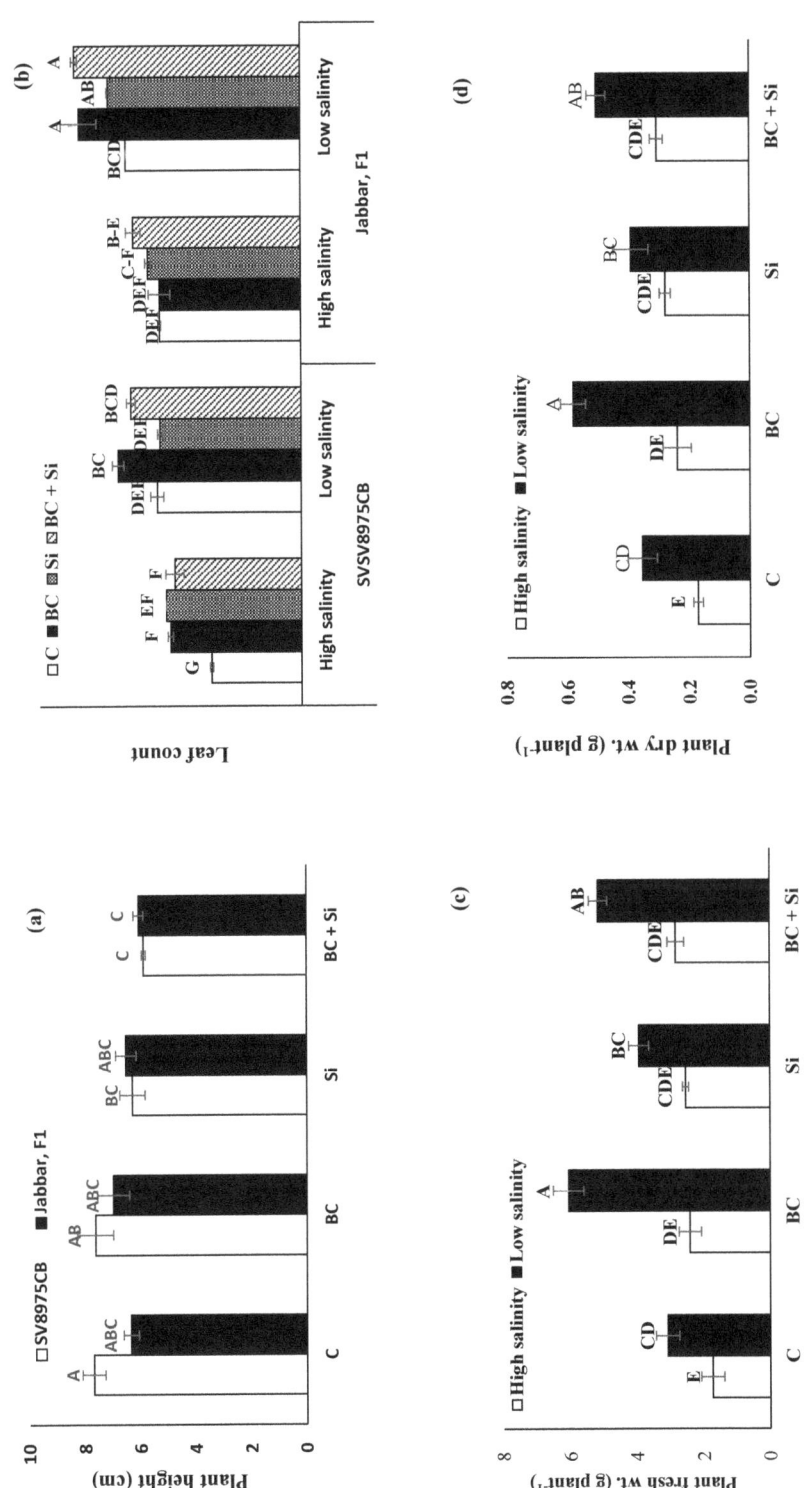

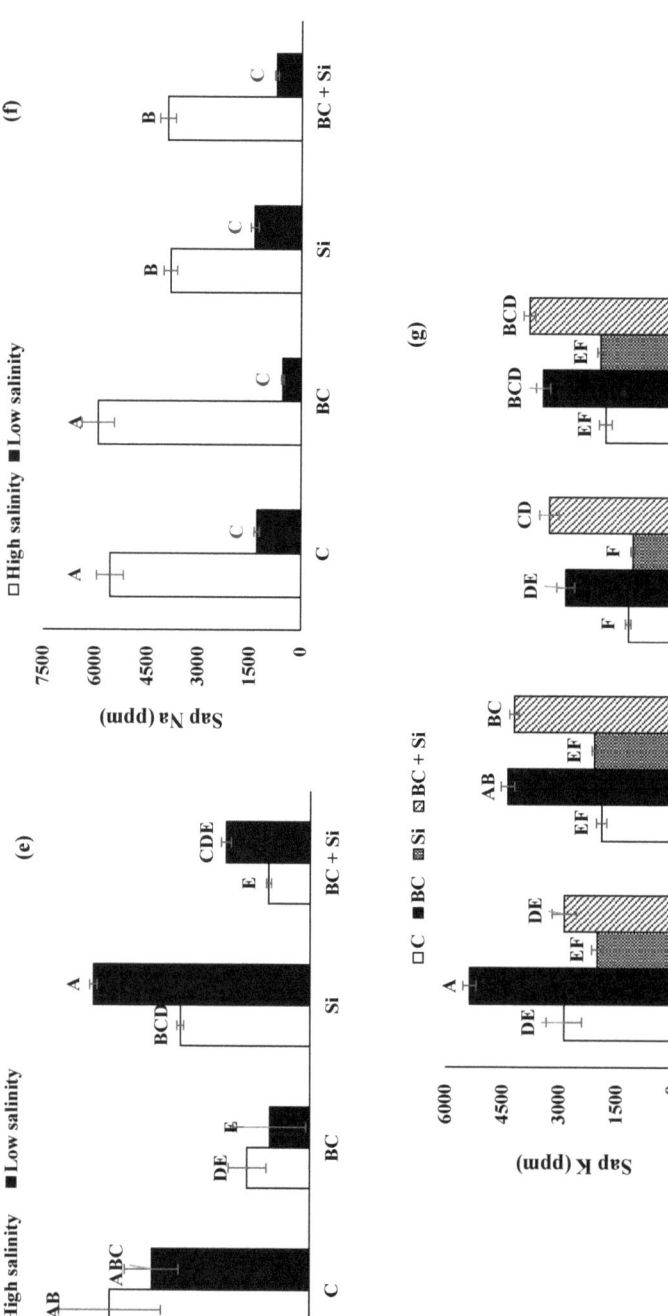

Figure 3. Influence of cultivar (C), soil type (S), and soil amendment (SA) interactions on morphology and sap nutrient concentration of cucumber. Comparison of C × SA interaction for (**a**) plant height; C × S × SA interaction for (**b**) leaf count; S × SA interaction for (**c**) plant fresh weight, (**d**) plant dry weight, (**e**) sap NO₃ concentration, and (**f**) sap Na concentration; C × S × SA interaction for (**g**) sap K concentration ± S.E. C = control; BC = biochar; Si = silica; BC + Si = biochar + silica. Means sharing the same letters do not differ significantly at $p < 0.05$.

4. Discussion

Soil salinization is one of the major threats to crop production globally as it adversely affects soil physicochemical properties and nutrient availability. This study used cucumber as an indicator plant and demonstrated that soil salinity negatively affected the soil properties and plant growth, leading to decreased biomass production as plant growth and yield are greatly contingent on the soil physical properties and nutritional status. However, the addition of soil amendments, such as BC, showed potential in improving plant growth in salt-affected soil.

Soil salinity adversely affected the physicochemical properties in both experiments. In Experiment 1, even though the high EC soil had higher WHC, nutrient levels, and OM due to past fertilization practices, it also had higher EC and soil sodium levels, which negatively affected plant growth. In the first study, the soil OM was enhanced by BC addition, but BC and Si both failed to lower the soluble Na levels of the soil or reduce the uptake of Na measured in the plant sap. Plant growth (germination, vigor, root, or top fresh weight) was not affected by any of the amendments (BC or Si). The OM content was higher in the high-salinity irrigation treatment because it is influenced by two opposing factors, i.e., reduced plant inputs and slower decomposition rate, which could increase soil organic carbon content [27]. However, the addition of BC further enhanced the accumulation of OM in the soil, which can be accredited to the chemical properties of BC, such as the carbon content, as the BC used in this study had higher organic content [21]. Biochar has a porous structure that is more resistant to degradation compared to the original feedstock material [28]. This property allows biochar to significantly improve soil physicochemical and biological properties, which can promote better crop production [29].

In this study, the addition of BC appeared to slightly increase the soluble NO_3^- content of the soil. This is consistent with previous studies that have shown biochar's ability to retain NO_3^- within its pores [30]. However, it should be noted that applying BC to soils may affect the conditions that control nitrification, denitrification [31], and other nitrogen transformation and loss pathways. However, the plant sap levels of NO_3 were lower with BC treatment, indicating reduced uptake by plants. This could be due to a complex interaction between the added biochar and other soil components that influence nutrient availability, uptake, and transport in plants. The lower values of soil EC, NO_3, P, K, Na, and Si in the less saline soil in Experiment 1 compared to high-saline soil were primarily due to the dilution effect and better fertility of high-saline soil. However, the application of BC in high and low-saline soil increased the soil NO_3, P, and K concentrations. This increase in soil nutrient concentration can be attributed to the adsorption of salts and replacement of Na^+ from the exchangeable sites in the soil with other cations, such as the Ca^{2+} and Mg^{2+} [32,33]. The biochar's ability to exchange Na^+ with these other cations can help to reduce the amount of sodium ions in the soil, which can lead to enhanced leaching of sodium ions from salt-affected soils [32]. Furthermore, biochar can also improve soil fertility by increasing the availability of nutrients to plants through its impact on soil pH. Biochar has been shown to have a liming effect, which can increase soil pH, making nutrients such as P and K more available to plants.

Cucumber is highly sensitive to salt stress, and cucumber seeds sown in high-saline soils showed inhibited germination and reduced vigor. The salinity-induced poor emergence and vigor of cucumber seedling was possibly due to lower activities of germination-related enzymes (e.g., as α-amylase) [21] as lower activities of α-amylase is associated with poor seedling emergence [10,21]. In the present study, root fresh weight was found to be lower with salt treatment and highest in the low-salt soil with perlite. The lower root fresh weight in high-saline soil or control treatment was due to root growth inhibition owing to excessive Na intake [34]. These findings indicate that salt stress disturbs the nutrient homeostasis in plants through excessive accumulation of Na in different parts of plants.

High-salinity irrigation in Experiment 2 decreased plant height, leaf count, plant fresh and dry weight, and sap K, while SPAD and sap NO_3 were unaffected by salt level. Sap levels of Na were higher with high-salt treatment but were reduced with the Si soil amendment treatments. The reduction in Na uptake observed in the Si soil amendment treatments is consistent with the well-established role of Si in enhancing salt stress tolerance in plants as Si application has been shown to improve plant growth and alleviate the negative effects of salt stress by reducing Na^+ absorption, improving ion balance, and enhancing osmotic adjustment of the plants [16]. Salt stress disturbs the nutrient homeostasis in plants through excessive accumulation of Na in different parts of plants [1]. However, application of BC enhanced the sap K in both experiments, while Si application substantially reduced the sap Na levels in Experiment 2. Moreover, improvement in plant growth and biomass production with BC amendment can be ascribed to increased availability of nutrients such as N, P, and K from BC itself, which leads to higher levels of K but not NO_3 in sap, or modification in nutrient cycling and retention, which improves plant growth [21,35].

Cucumber cultivars also varied in leaf count, plant fresh weight, and tissue nutrient level. The cultivar Jabbar F1 exhibited higher leaf count, fresh weight, and sap NO_3, along with lower Na and low K. The variation in these traits may be due to genetic differences and different responses to salt stress as Jabbar restricted Na accumulation, while SV8975CB accumulated more Na in plant tissues. In the present study, the higher leaf count, fresh weight, and sap NO_3 levels in Jabbar F1 may be attributed to its ability to maintain better nutrient uptake and assimilation under salt stress conditions. On the other hand, SV8975CB may have accumulated more Na in plant tissues, which may have contributed to its lower growth and nutrient accumulation.

5. Conclusions

Soil salinity inhibited the emergence and growth of cucumber. Soil amendments, particularly biochar, improved soil fertility and physical properties such as soil K, P, organic matter content and, in some cases, soil levels of NO_3 and water-holding capacity but did not improve plant growth in Experiment 1. When salt stress was imposed as irrigation water, the effect of biochar on plant growth resulted in higher levels of sap K but lower levels of sap NO_3 and better plant growth. Silica, but not biochar, reduced plant uptake of Na based on sap Na measurements. The biochar amendment improved plant growth and sap K level under salt stress conditions. Application of biochar and Si can help improve cucumber production on salt-affected soils or under saline irrigation conditions, but there does not appear to be any synergistic effect from applying both at the same time.

Author Contributions: Conceptualization, R.R.J., M.M.K. and M.A.; methodology, M.A.-T., R.R.J. and A.R.; software, R.R.J., A.R. and W.M.A.-B.; formal analysis, M.A.-T. and W.M.A.-B.; investigation, M.A.-T.; resources, R.R.J., M.M.K. and M.A.; data curation, R.R.J. and A.R.; writing—original draft preparation, M.A.-T. and A.R.; writing—review and editing, R.R.J. and A.R.; visualization, M.A.-T. and A.R.; supervision, R.R.J.; project administration, W.M.A.-B.; funding acquisition, R.R.J. All authors have read and agreed to the published version of the manuscript.

Funding: This research received no external funding.

Institutional Review Board Statement: Ethical review and approval were waived for this study because the study did not involve humans or animals.

Data Availability Statement: Data from this study will be shared upon reasonable request to the corresponding author.

Acknowledgments: The authors would like to thank Sultan Qaboos University for in-kind support for use of the laboratory, growth chamber, and other facilities, which made this work possible.

Conflicts of Interest: The authors declare no conflict of interest.

References

1. Munns, R.; Tester, M. Mechanisms of salinity tolerance. *Annu. Rev. Plant Biol.* **2008**, *59*, 651–681. [CrossRef]
2. Walworth, J.L. Soil salinity, fertility, and management: Opportunities for improvement. In Proceedings of the International Conference on Soils and Ground Water Salinization in Arid Countries, Sultan Qaboos University, Muscat, Oman, 11–14 January 2010.

3.   Qadir, M.; Quillérou, E.; Nangia, V.; Murtaza, G.; Singh, M.; Thomas, R.J.; Drechsel, P.; Noble, A.D. Economics of salt-induced land degradation and restoration. *Nat. Resour. Forum* **2014**, *38*, 282–295. [CrossRef]

4.   Shahid, S.A.; Abdelfattah, M.A.; Omar, S.A.; Harahsheh, H.; Othman, Y.; Mahmoudi, H. Mapping and monitoring of soil salinization remote sensing, GIS, modeling, electromagnetic induction and conventional methods—Case studies. In Proceedings of the International Conference on Soils and Groundwater Salinization in Arid Countries, Sultan Qaboos University, Muscat, Oman, 11–14 January 2010.

5.   Hussain, M.I.; Farooq, M.; Muscolo, A.; Rehman, A. Crop diversification and saline water irrigation as potential strategies to save freshwater resources and reclamation of marginal soils—A review. *Environ. Sci. Pollut. Res.* **2020**, *27*, 28695–28729. [CrossRef]

6.   Qin, C.; Ahanger, M.; Zhou, J.; Ahmed, N.; Wei, C.; Yuan, S.; Ashraf, M.; Zhang, L. Beneficial role of acetylcholine in chlorophyll metabolism and photosynthetic gas exchange in *Nicotiana benthamiana* seedlings under salinity stress. *Plant Biol.* **2020**, *22*, 357–365. [CrossRef]

7.   Tokarz, K.M.; Wesołowski, W.; Tokarz, B.; Makowski, W.; Wysocka, A.; Jędrzejczyk, R.J.; Chrabaszcz, K.; Malek, K.; Kostecka-Gugała, A. Stem photosynthesis—A key element of grass pea (*Lathyrus sativus* L.) acclimatization to salinity. *Int. J. Mol. Sci.* **2021**, *22*, 685. [CrossRef]

8.   Tanveer, M.; Shah, A.N. An insight into salt stress tolerance mechanisms of *Chenopodium album*. *Environ. Sci. Pollut. Res.* **2017**, *24*, 16531–16535. [CrossRef]

9.   Hussain, S.; Khaliq, A.; Tanveer, M.; Matloob, A.; Hussain, H.A. Aspirin priming circumvents the salinity-induced effects on wheat emergence and seedling growth by regulating starch metabolism and antioxidant enzyme activities. *Acta Physiol. Plant.* **2018**, *40*, 68–75. [CrossRef]

10.   Shahzad, B.; Rehman, A.; Tanveer, M.; Wang, L.; Park, S.K.; Ali, A. Salt stress in brassica: Effects, tolerance mechanisms, and management. *J. Plant Growth Regul.* **2022**, *41*, 781–795. [CrossRef]

11.   Munns, R.; James, R.A.; Launchli, A. Approaches to increasing the salt tolerance of wheat and other cereals. *J. Exp. Bot.* **2006**, *57*, 1025–1043. [CrossRef]

12.   He, K.; He, G.; Wang, C.; Zhang, H.; Xu, Y.; Wang, S.; Kong, Y.; Zhou, G.; Hu, R. Biochar amendment ameliorates soil properties and promotes Miscanthus growth in a coastal saline-alkali soil. *Appl. Soil Ecol.* **2020**, *155*, 103674. [CrossRef]

13.   Torabian, S.; Farhangi-Abriz, S.; Rathjen, J. Biochar and lignite affect H^+-ATPase and H^+-PPase activities in root tonoplast and nutrient contents of mung bean under salt stress. *Plant Physiol. Biochem.* **2018**, *129*, 141–149. [CrossRef] [PubMed]

14.   Shi, Y.; Wang, Y.C.; Flowers, T.J.; Gong, H.J. Silicon decreases chloride transport in rice (*Oryza sativa* L.) in saline conditions. *J. Plant Physiol.* **2013**, *170*, 847–853. [CrossRef] [PubMed]

15.   Abdel-Haliem, M.E.; Hegazy, H.S.; Hassan, N.S.; Naguib, D.M. Effect of silica ions and nano silica on rice plants under salinity stress. *Ecol. Eng.* **2017**, *99*, 282–289. [CrossRef]

16.   Shen, Z.; Pu, X.; Wang, S.; Dong, X.; Cheng, X.; Cheng, M. Silicon improves ion homeostasis and growth of liquorice under salt stress by reducing plant Na^+ uptake. *Sci. Rep.* **2022**, *12*, 5089. [CrossRef] [PubMed]

17.   Al-Rawahy, S.A.; Ahmed, M.; Hussain, N. Management of salt-affected soils and water for sustainable agriculture: The project. In *Monograph on Management of Salt-Affected Soils and Water for Sustainable Agriculture*; Mushtaque, A., Al-Rawahi, S.A., Hussain, N., Eds.; Sultan Qaboos University: Seeb, Oman, 2010; pp. 1–8.

18.   Al-Dhuhli, H.S.; Al-Rawahy, S.A.; Prathapar, S. Effectiveness of mulches to control soil salinity in sorghum fields irrigated with saline water. In *Monograph on Management of Salt-Affected Soils and Water for Sustainable Agriculture*; Mushtaque, A., Al-Rawahi, S.A., Hussain, N., Eds.; Sultan Qaboos University: Seeb, Oman, 2010; pp. 41–46.

19.   Al-Mulla, Y. Salinity mapping in Oman using remote sensing tools: Status and trends. In *Monograph on Management of Salt-Affected Soils and Water for Sustainable Agriculture*; Mushtaque, A., Al-Rawahi, S.A., Hussain, N., Eds.; Sultan Qaboos University: Seeb, Oman, 2010; pp. 17–24.

20.   Gartley, K.L. Recommended soluble salts tests. In *Recommended Soil Testing Procedures for the Northeastern United States*; Northeastern Regional Publication (493); University of Delaware, College of Agriculture & Natural Resources: Newark, DE, USA, 1995.

21.   Farooq, M.; Rehman, A.; Al-Alawi, A.K.; Al-Busaidi, W.M.; Lee, D.J. Integrated use of seed priming and biochar improves salt tolerance in cowpea. *Sci. Hort.* **2020**, *272*, 109507. [CrossRef]

22.   Rani, B.; Sharma, V. Standarisation of methodology for obtaining the desired salt stress environment for salinity effect observation in rice seedlings. *Int. J. Environ. Sci.* **2015**, *6*, 232.

23.   Estefan, G.; Sommer, R.; Ryan, J. *Methods of Soil, Plant, and Water Analysis: A Manual for the West Asia and North Africa Region*; ICARDA: Beirut, Lebanon, 2013; pp. 170–176.

24.   Altamimi, M.E.; Janke, R.R.; Williams, K.A.; Nelson, N.O.; Murray, L.W. Nitrate-nitrogen sufficiency ranges in leaf petiole sap of *Brassica oleracea* L., pac choi grown with organic and conventional fertilizers. *HortScience* **2013**, *48*, 357–368. [CrossRef]

25.   Farooq, M.; Almamari, S.A.D.; Rehman, A.; Al-Busaidi, W.M.; Wahid, A.; Al-Ghamdi, S.S. Morphological, physiological and biochemical aspects of zinc seed priming-induced drought tolerance in faba bean. *Sci. Hort.* **2021**, *281*, 109894. [CrossRef]

26.   Nelson, D.W.; Sommers, L.E. Total carbon, organic carbon, and organic matter. In *Methods of Soil Analysis*; Bigham, J.M., Ed.; Part 3. Chemical Methods-SSSA Book Series No. 5. Soil Science Society of America and American Society of Agronomy: Madison, WI, USA, 1996; Chapter 34; pp. 1001–1006.

27.   Setia, R.; Gottschalk, P.; Smith, P.; Marschner, P.; Baldock, J.; Setia, D.; Smith, J. Soil salinity decreases global soil organic carbon stocks. *Sci. Total Environ.* **2013**, *465*, 267–272. [CrossRef]

28. Bird, M.I.; Ascough, P.L. Isotopes in pyrogenic carbon: A review. *Org. Geochem.* **2012**, *42*, 1529–1539. [CrossRef]
29. Ahmad, M.; Rajapaksha, A.U.; Lim, J.E.; Zhang, M.; Bolan, N.; Mohan, D.; Vithanage, M.; Lee, S.S.; Ok, Y.S. Biochar as a sorbent for contaminant management in soil and water: A review. *Chemosphere* **2014**, *99*, 19–23. [CrossRef] [PubMed]
30. Padhye, L.P. Influence of surface chemistry of carbon materials on their interactions with inorganic nitrogen contaminants in soil and water. *Chemosphere* **2017**, *184*, 532–547.
31. Liu, Q.; Zhang, Y.; Liu, B.; Amonette, J.E.; Lin, Z.; Liu, G.; Ambus, P.; Xie, Z. How does biochar influence soil N cycle? A meta-analysis. *Plant Soil* **2018**, *426*, 211–225. [CrossRef]
32. Sadegh-Zadeh, F.; Parichehreh, M.; Jalili, B.; Bahmanyar, M.A. Rehabilitation of calcareous saline-sodic soil by means of biochars and acidified biochars. *Land Degrad. Dev.* **2018**, *29*, 3262–3271. [CrossRef]
33. Zhang, J.; Bai, Z.; Huang, J.; Hussain, S.; Zhao, F.; Zhu, C.; Zhu, L.; Cao, X.; Jin, Q. Biochar alleviated the salt stress of induced saline paddy soil and improved the biochemical characteristics of rice seedlings differing in salt tolerance. *Soil Tillage Res.* **2019**, *195*, 104372. [CrossRef]
34. Hu, X.; Wang, D.; Ren, S.; Feng, S.; Zhang, H.; Zhang, J.; Qiao, K.; Zhou, A. Inhibition of root growth by alkaline salts due to disturbed ion transport and accumulation in Leymus chinensis. *Environ. Exp. Bot.* **2022**, *200*, 104907. [CrossRef]
35. Lehmann, J.; Joseph, S. Biochar for environmental management: An introduction. In *Biochar for Environmental Management, Science and Technology*; Lehmann, J., Joseph, S., Eds.; Earthscan: London, UK, 2009; pp. 1–12.

soil systems

MDPI

Article

Robustness of Optimized Decision Tree-Based Machine Learning Models to Map Gully Erosion Vulnerability

Hasna Eloudi [1], Mohammed Hssaisoune [1,2,3,*], Hanane Reddad [4], Mustapha Namous [5], Maryem Ismaili [5], Samira Krimissa [5], Mustapha Ouayah [5] and Lhoussaine Bouchaou [1,3]

[1] Applied Geology and Geo-Environment Laboratory, Faculty of Sciences, Ibn Zohr University, Agadir 80000, Morocco
[2] Faculty of Applied Sciences, Ibn Zohr University, Ait Melloul 86150, Morocco
[3] International Water Research Institute, Mohammed VI Polytechnic University, Ben Guerir 43150, Morocco
[4] Laboratoire d'Ingénierie & de Technologies Appliquées (LITA), École Supérieure de Technologie de Beni Mellal, Sultan Moulay Slimane University, Beni-Mellal 23000, Morocco
[5] Data Science for Sustainable Earth Laboratory (Data4Earth), Sultan Moulay Slimane University, Beni-Mellal 23000, Morocco
*   Correspondence: m.hssaisoune@uiz.ac.ma; Tel.: +212-670929680

Abstract: Gully erosion is a worldwide threat with numerous environmental, social, and economic impacts. The purpose of this research is to evaluate the performance and robustness of six machine learning ensemble models based on the decision tree principle: Random Forest (RF), C5.0, XGBoost, treebag, Gradient Boosting Machines (GBMs) and Adaboost, in order to map and predict gully erosion-prone areas in a semi-arid mountain context. The first step was to prepare the inventory data, which consisted of 217 gully points. This database was then randomly subdivided into five percentages of Train/Test (50/50, 60/40, 70/30, 80/20, and 90/10) to assess the stability and robustness of the models. Furthermore, 17 geo-environmental variables were used as potential controlling factors, and several metrics were examined to evaluate the performance of the six models. The results revealed that all of the models used performed well in terms of predicting vulnerability to gully erosion. The C5.0 and RF models had the best prediction performance (AUC = 90.8 and AUC = 90.1, respectively). However, according to the random subdivisions of the database, these models exhibit small but noticeable instability, with high performance for the 80/20% and 70/30% subdivisions. This demonstrates the significance of database refining and the need to test various splitting data in order to ensure efficient and reliable output results.

Keywords: soil erosion; inventory data; performance; robustness; spatial prediction

Citation: Eloudi, H.; Hssaisoune, M.; Reddad, H.; Namous, M.; Ismaili, M.; Krimissa, S.; Ouayah, M.; Bouchaou, L. Robustness of Optimized Decision Tree-Based Machine Learning Models to Map Gully Erosion Vulnerability. *Soil Syst.* 2023, 7, 50. https://doi.org/10.3390/soilsystems7020050

Academic Editor: Luis Eduardo Akiyoshi Sanches Suzuki

Received: 10 February 2023
Revised: 12 May 2023
Accepted: 14 May 2023
Published: 16 May 2023

1. Introduction

Soil erosion is known as a loosening of sediment from the uplands to the valley floor induced by runoff [1]. This phenomenon is described as a catastrophic global issue with extensive environmental, social, and economic repercussions [2]. Soil erosion endangers water and soil resources, both of which are vital to human existence and the environmental equilibrium. There are several types of soil erosion, the most notable of which is gully erosion (GE) [3]. This type contributes to landscape shaping while also causing significant damage such as the degradation of arable land fertility, damage to water infrastructure and shortening of its life span, and the disruption of countries' economic and societal circumstances [4]. This phenomenon has affected one-third of the world's arable land in the last few decades [5]. According to the literature, soil erosion affects more than 10 million hectares of agricultural land each year, with annual global loss rates of approximately 43 Pg [6]. According to FAO [7], soil losses due to soil erosion are estimated to result in a $1 billion economic loss. Soil erosion affects 40% of Moroccan territory, with annual loss rates ranging from 23 to 55 t/ha/yr on average, and extreme values reaching

524 t/ha/yr in some places [8]. Furthermore, agriculture is the main source of income for the people who live in the mountainous areas of Morocco. However, these areas are heavily affected by soil erosion, which decreases the amount of fertile land, reduces the quality and quantity of water, and has other serious economic and social effects [9]. In this respect, the Lakhdar watershed in the Moroccan High Atlas is one of the regions impacted by significant soil degradation as a result of its very complicated physical features, such as a very high topography and a steep slope occupied by rocks with differing properties. In connection with these factors, the study of gully erosion vulnerability may be a crucial tool for understanding erosive processes in comparable environments. As a result, identifying areas prone to soil erosion is an important step toward good natural resource management and long-term protection and a deeper comprehension of the erosive processes and the factors that influence this phenomenon under current climatic conditions.

Since the 1930s, numerous models have been developed to estimate soil loss rates and qualitatively assess soil erosion sensitivity. Currently, combining remote sensing with geographic information systems makes this task easier and more efficient [10]. According to the literature, there seem to be two distinct methods of soil erosion analysis: Qualitative and quantitative approaches. To assess medium- and long-term soil loss rates, empirical models such as the Universal Soil Loss Equation (USLE), Modified Universal Soil Loss Equation (MUSLE), and Revised Universal Soil Loss Equation (RUSLE) have been employed [11,12]. There are also physical models that can be used to quantify soil loss averages, such as the Water Erosion Prediction Project Model (WEPP) [13], the Chemical Runoff and Erosion for Agricultural Management System (CREAMS) [14], and other models such as The Erosion-Productivity Impact Calculator (EPIC) [15] and the Limburg Soil Erosion Model (LSEM). These models, however, cannot predict gully erosion susceptibility because this type of erosion is controlled by several factors not completely taken into account by their formulas, including topographic, hydraulic, climatic, soil conditions, and morphometric characteristics [16]. Furthermore, quantitative models necessitate calibration and are subject to significant uncertainty in terms of differences between predicted and measured loss rates [17]. In this regard, other bivariate and multivariate statistical models have been developed [18]. Furthermore, hierarchical process analysis (HPA) methods [19], logistic regression models [20], Weight-of-Evidence (WoE) models, and entropy indexes are used to evaluate the sensitivity to gully erosion [21,22]. Moreover, machine learning techniques have proven to be an effective tool for assessing and mapping gully erosion [23,24]. These methods are a subset of the artificial intelligence field that is based on the hypothesis that computer programs can learn from inventory and model input data without the need for human intervention [25]. Presently, the use of machine learning-based approaches has become popular, particularly in the mapping and monitoring of natural hazards, because they produce high-accuracy results in data processing, classification, and prediction [26]. In addition, numerous researchers have tested the high performance of deep learning models in the spatial prediction of vulnerable zones to gullying phenomena [27]. On the basis of this previous investigation, we aim to fill a gap in the analysis of inventory data by investigating various possible subdivisions and proposing to researchers and decision-makers a simple, less expensive, and effective method for predicting soil erosion vulnerability.

The objectives of this investigation are to identify the factors that cause gully erosion and to test several "decision tree" models to develop a gully erosion susceptibility detection and prediction model suitable for mountainous and semi-arid areas. It is therefore essential to evaluate the stability of these models against the variation in training and testing percentages. Because of this, we will test how well these models perform with five different splitting of data on Training and Testing: 90/10%, 80/20%, 70/30%, 60/40%, and 50/50%. The advantage of this kind type of decision tree model is that it determines the relationships between the explanatory variables, the dependent factors, and the occurrence of the phenomenon in a simple tree structure. This makes these models more comprehensive compared to mathematical formulas or correspondence tables. According to the literature, numerous studies have used these models to assess and monitor gully erosion vulnerability [24,28]. Despite this, the use of these combined approaches to predict areas susceptible to gully erosion on the one hand, and their tests under different quantities of input data subdivision on the other hand, remains very limited. Additionally, the combination of different types of decision trees has never been tested in Morocco, lending originality to this research. Finally, the development of these advanced methods to map gully erosion-vulnerable areas is critical because it will support decision-making in terms of planning and implementing sustainable policies and strategies for land management of water and soil resources.

2. Materials and Methods

2.1. Study Area

The Lakhdar watershed is one of the Oum Er-Rbia sub-basins (Figure 1) and is located in the Atlas Mountains axial zone and covers approximately 1600 km². The study area is divided into three geomorphological distinct units: High mountains with altitudes of up to 4000 m, plateaus, and valleys carved deeply by soil erosion. The region has a major hydraulic structure of critical importance in terms of drinking water supply for Marrakech city as well as irrigation of the Haouz plain downstream. Geologically, the Lakhdar watershed is composed of an amalgam of lithologies with a dominance of Jurassic limestones and Permo-Triassic sandstones; its upstream part is primarily characterized by detrital deposits represented essentially by clays, marls, and alluvial deposits of the Quaternary period (Figure 2 and Table 1). The study area is classified as a semi-arid zone with hot summers (from June to August) and cold winters (since December to February). The aridity primarily affects the downstream portion of the watershed area; however, the upstream portion is controlled by high altitudes, resulting in significant spatial differences in rainfall amounts. In general, the average annual rainfall is approximately 450 mm, with maximum values of 600 mm recorded in the upstream portion and minimum values of 300 mm recorded primarily in the downstream areas. The area under investigation has a deteriorated vegetative cover, which is exacerbated by the dynamics and anthropic activities that invade the area. This is supported by a 36% reduction in forest area over the last few decades. As a result, the watershed area serves as a test bed for studying soil erosion processes and comprehending the erosive processes that occur in a semi-arid mountainous area.

Table 1. Description of lithological units in Lakhdar watershed, Morocco.

Class	Description
1	Silurian: Graptolitic shales
2	Stephano-Triassic: Sandstones and red conglomerates
3	Permian-Triassic: Basalt
4	Lower Lias: Limestones, and red clays
5	Lower Lias: Limestones and marls
6	Middle Lias: Limestones
7	Upper Lias: Conglomerates, sandstones, and clays
8	Dogger: Marls and limestones
9	Quaternary: Alluvial and Rockfull

2.2. Methodology

The current study's approach includes several major steps illustrated by the flowchart presented in Figure 3. In addition, the same figure presents an overview of the approach that was developed for probabilistic gully erosion susceptibility using decision tree models (C5.0, XGBoost, treebag, GBM, and Adaboost) to produce accurate Gully Erosion Susceptibility Maps (GESMs).

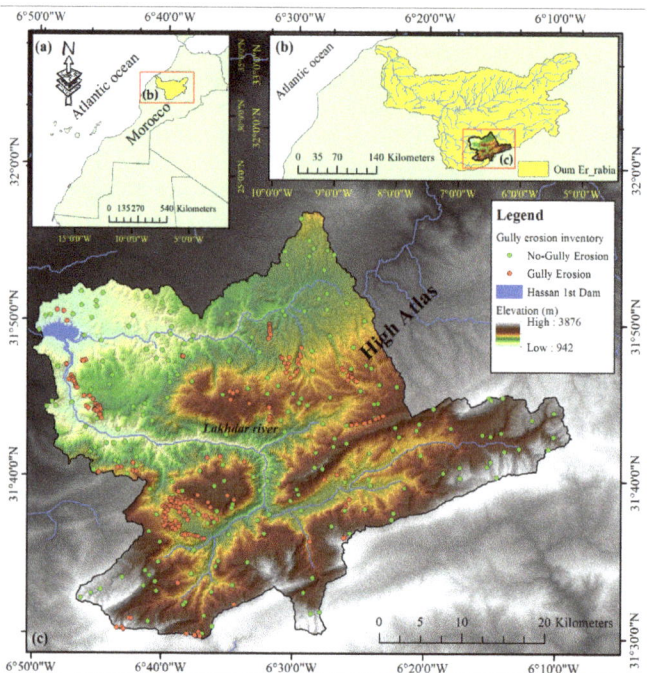

Figure 1. Location maps of the study area (**a**) at Moroccan scale, (**b**) at Oum Er-Rbia watershed scale, and (**c**) DEM of Lakhdar watershed.

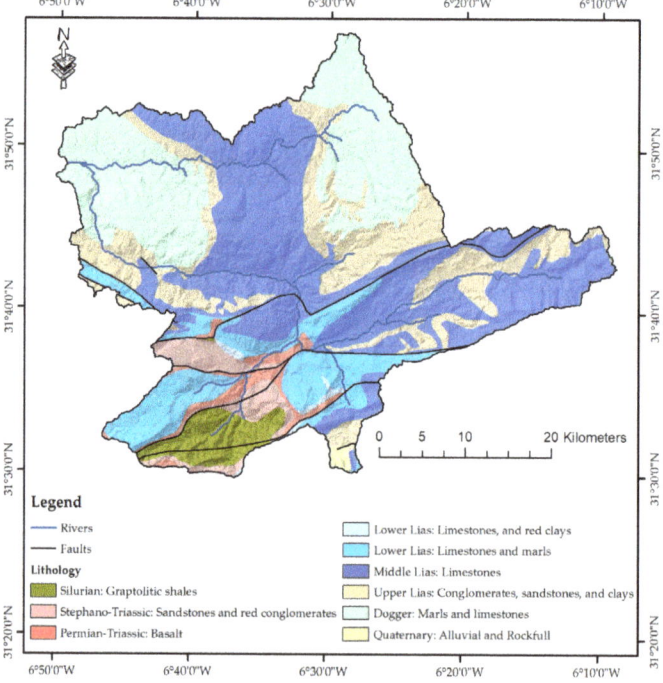

Figure 2. Geological map of the study area.

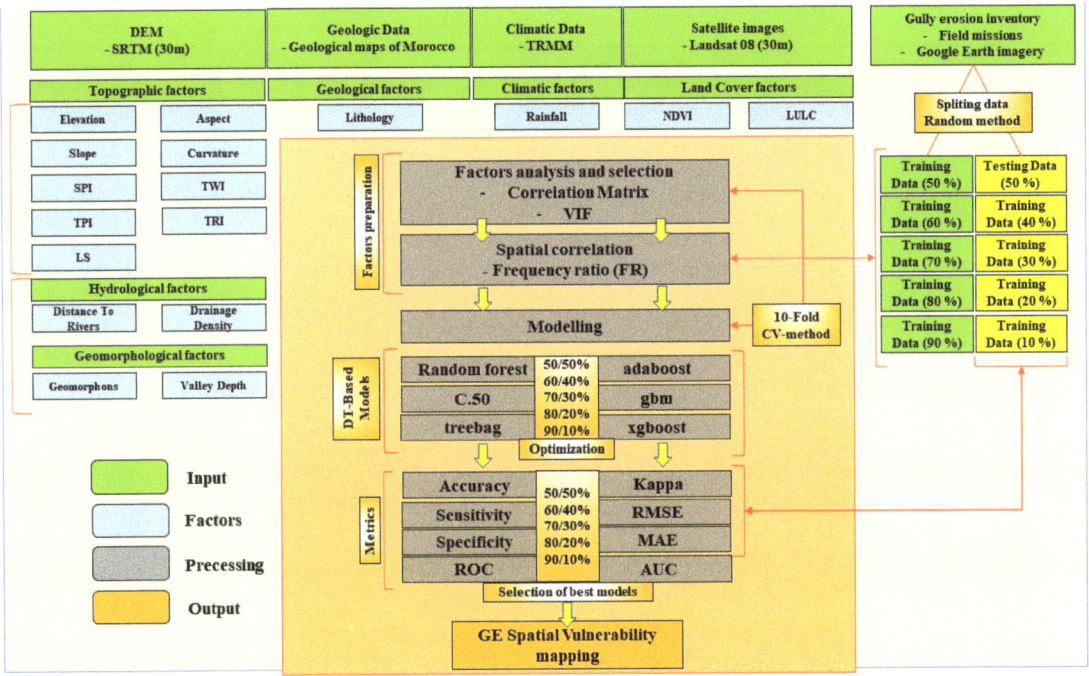

Figure 3. A detailed flowchart shows research methods.

2.2.1. Gully Erosion Inventory Mapping

One of the most important indicators for assessing gully erosion susceptibility is the gully erosion inventory map (GEIM), which presents the spatial locations of the gullies. The distribution of traditional and present gully locations can be used to estimate the potential probability of gully erosion in a region. As a result, it is important to create a gully erosion inventory map in order to estimate the optimal future gully erosion [29]. The GEIM is required for the preparation of GESMs by various predictive models [30] and was used as the dependent variable in this study. For GEIM preparation, gully locations were identified by conducting fieldwork in the study region combined with google earth image analysis. Gully locations were determined using a handheld GPS device. In the study area, 217 gullies were collected (Figure 1). This database was then randomly subdivided into five quantities (50/50%, 60/40%, 70/30%, 80/20%, and 90/10%) to assess the performance, robustness, and stability of the models (Figure 2).

2.2.2. Dataset Preparation for Spatial Modelling

The selection of Gully Erosion Conditioning Factors (GCFs) is a crucial stage in the development of GESMs using several techniques [31]. In this study, 17 geo-environmental variables were used for spatial modelling of gully erosion, including elevation, slope, aspect, rainfall, LandUse-LandCover (LULC), Normalized Difference Vegetation Index (NDVI), distance to rivers, Drainage Density, Valley Depth, Curvature, Lithology, Geomorphons, Topographic Position Index (TPI), Topographic Wetness Index (TWI), Topographic Roughness Index (TRI), Slope Length (LS), and Stream Power Index (SPI) (Table 2), while taking previous literature and multicollinearity into account. Note that for all quantitative factors, the classification is based on the Natural break technique, as suggested by the majority of researchers [32].

Table 2. Data sources used in this study.

Factors	Data Layers	Data Source
Topographic factors	Elevation Slope (°) Stream Power Index (SPI) Topographic Position Index (TPI) Slope Length (LS) Aspect Curvature Topographic Wetness Index (TWI) Topographic Roughness Index (TRI)	SRTM-DEM (Digital Elevation Model) were downloaded from the website of United States Geological Survey (USGS) (http://gdex.cr.usgs.gov/gdex/ (accessed on 2 August 2022)); Pixel size of 30 m × 30 m.
Hydrological factors	Distance To Rivers Drainage Density	
Geomorphological factors	Valley depth Geomorphons	
Geological factors	Lithology	Geologic map of Ouaouizghte-Dades 1/200,000 Bourcart et al., 1942 [32] Geologic map of Demnate-Telouate 1/200,000 Termier, 1941 [32]
Climatic factors	Rainfall (mm)	TRMM data
LAND cover factors	Normalized Difference Vegetation Index (NDVI) LandUse-LandCover (LULC)	LANDSAT-8 OLI TIRS satellite image

The elevation data layer was created using the digital elevation model (DEM) obtained from the USGS (Figure 4a). The study area's altitude was separated into five groups: 942–1513 m, 1504–1947 m, 1937–2379 m, 2381–2866 m, and 2860–3876 m (Figure 4a). The slope has a big effect on how gullies form [33]. The slope map was created in GIS using a DEM and was divided into five groups: 0–9, 10–18, 19–26, 27–36, and 37–71° (Figure 4a). The aspect map, similar to that of the slope map, was created from the DEM and divided into nine classes: Flat, north, northeast, east, southeast, south, southwest, west, and northwest. The curvature is also mapped from DEM using GIS and divided into five classes −24.9 to −2.4, −2.3 to −0.9, −0.8 to −0.4, 0.5 to 2.1, and 2.2 to 30.2 (Figure 4a). The sediment power index (SPI) reveals the discharge, carrying potential, and water erosion energy, which influences the sensitivity to gully erosion [34]. The following Equation (1) was used to obtain the SPI from the DEM:

$$SPI = As \times \tan\beta, \tag{1}$$

where As is the upstream drainage area and β is the slope degree. The SPI was classified into the five sub-categories of 0–443, 444–959, 960–1587, 1588–2547, and 2548–9410 (Figure 4a). The topographic wetness index (TWI) is regarded as a key gully erosion conditioning factor. Using the following Equation (2), the TWI was obtained from DEM data:

$$TWI = \ln(As/\tan\beta), \tag{2}$$

where As is the upstream drainage area and β is the slope degree. The TWI was categorized into five classes: 2–6, 7–8, 9–11, 12–16, and 17–25 (Figure 4a). The slope length (LS) factor was calculated also from the DEM by means of Equation (3).

$$LS = (m + 1) \times [As/22] \times [\sin\beta/0.0896], \tag{3}$$

where As is the upstream drainage area and β is the slope degree. The LS was categorized into five classes: 0–4.16, 4.16–9.02, 9.02–14.58, 14.58–27.76, and 27.76–177 (Figure 4b). The Terrain ruggedness index (TRI) indicates the elevation difference between the surrounding cells of a DEM [35]. The TRI was classified into five classes: 0–4.49, 4.49–8.66, 8.66–13.80, 13.80–22.46, and 22.49–81.83 (Figure 4b). The topographic position index (TPI) is also calculated using DEM; TPI is a terrain classification method in which the altitude of each data point is compared to its neighbors. In a nutshell, we calculate the height difference between each data point, or pixel in a raster DEM, and its immediate surroundings. The TPI was classified into five classes: −13–152, 152–298, 298–459, 459–632, and 632–966 (Figure 4b). Drainage density factors were also used and categorized into five classes: 0.14–0.46, 0.47–0.64, 0.65–0.79, 0.8–0.93, and 0.94–1.3 (Figure 4b). The distance from the river map was prepared by applying the Euclidian distance buffer (EDB) tool in GIS (Figure 4a). It was classed into five sub-classes, namely 0–185 m, 186–419 m, 420–668 m, 669–966 m, and 967–2052 m (Figure 4a).

Despite the fact that gully erosion is highly dependent on the lithology qualities of the exposed material near the earth's surface, lithology indicators play an essential function in assessing gully erosion vulnerability [33]. The lithological map was generated from the available geological data of Morocco and was classified into nine classes numbered one through nine (Figure 4a and Table 1). The NDVI was calculated using the Landsat 8 imagery in a GIS environment following this Equation (4).

$$NDVI = (NIR − R)/(NIR + R), \qquad (4)$$

where NIR is the near-infrared spectrum and R is the red spectrum. The map was categorized into five classes: −0.12 to 0.1, 0.11 to 0.14, 0.15 to 0.2, 0.21 to 0.31, and 0.32 to 0.58 (Figure 4a). The Land Use Land Cover (LULC) map was obtained from Landsat 8 imagery based on the supervised classification process in the GIS environment. Water bodies, soil bare, sparse vegetation, agricultural land, and forest are the LULC classes (Figure 4b).

The geomorphological factors used are Valley depth and Geomorphons. The first was classified into five classes: −13–152, 152–298, 298–459, 459–632, and 632–966. The second was classified into ten classes: Flat, summit, Ridge, shoulder, spur, slope, hollow, footslope, and Valley depression (Figure 4b). Rainfall is a major factor that directly contributes to gully erosion, and annual precipitation data were obtained from the Tropical Rainfall Measuring Mission (TRMM). According to the rainfall map, the annual average rainfall in the study area ranges between 390 and 610 mm.year^{-1}. The most significant values are found in the south, while precipitation decreases sharply in the north (Figure 4a). The rainfall map was subdivided into five classes: 330–395, 395–450, 450–505, 505–552, and 552–610.

2.2.3. Multicollinearity Analysis

The multicollinearity test is an important approach to measure the linear dependency among the specified independent parameters in statistical modelling. This method needs to be applied to machine learning models in order to improve their performance [31]. This study used the correlation matrix and variable inflation factor (VIF) methods to determine the multicollinearity of the Gully erosion factors. Using the correlation between predictor pairs alone has limitations, whether small or large [36].

2.2.4. Decision Tree-Based Approaches
Random Forest (RF)

The random forest (RF) algorithm is a statistical technique for controlling a large number of connected variables [37]. In 2001, Breiman [38] developed the technique as a binary tree decision-making system [39]. RF may also assess dynamic trends and understand nonlinear connections between explanatory and dependent variables. It will also merge multiple data formats due to the lack of a uniform distribution of the data used. RF is ideally suited to geographical studies and is often employed in land movement sensitivity mapping [40]. This method combines many decision trees, with many bootstrap samples obtained from the data and a range of input variables arbitrarily added to each tree. Furthermore, the RF approach categorizes elements according to their relevance. The weights are determined by taking the average decline in forecast accuracy.

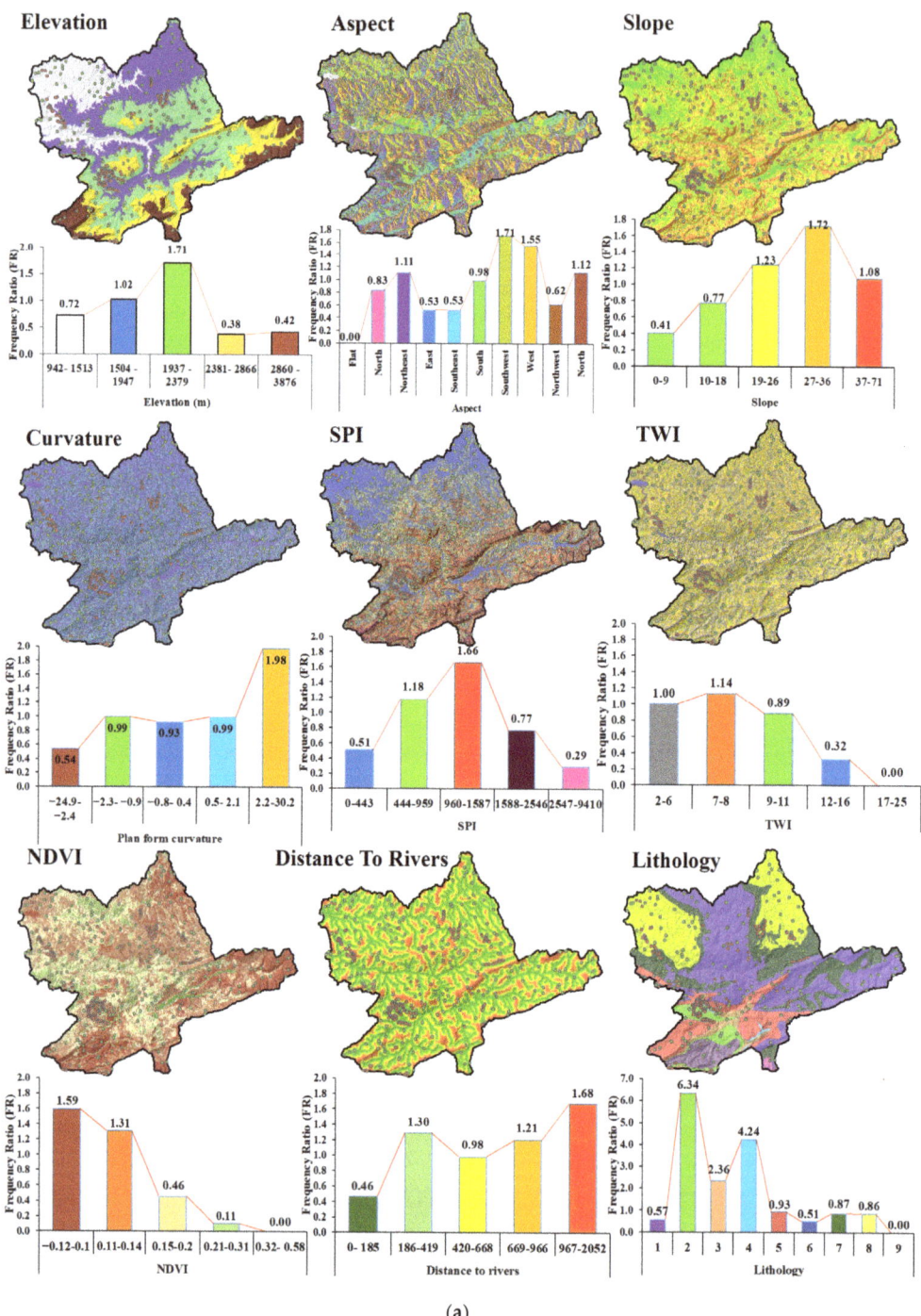

(a)

Figure 4. *Cont.*

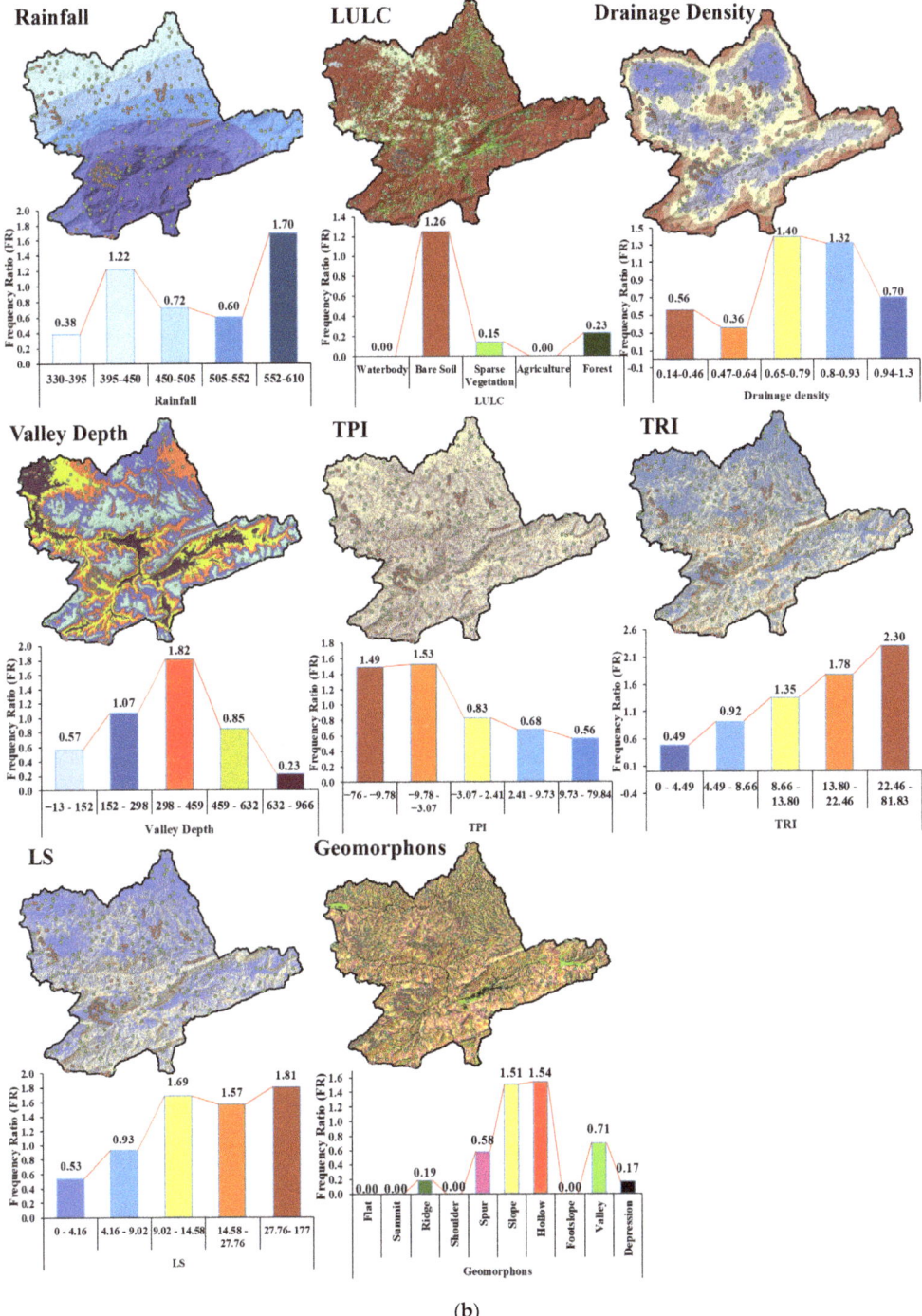

(b)

Figure 4. (**a**,**b**). Conditioning factor maps and spatial correlation between factors and Gully erosion location using FR method.

C5.0

C5.0 is a decision tree technique that works by first testing the classifier to classify unseen data and then using the final decision. Pandya and Pandya [41] demonstrate decisively that C5.0 is an improvement over C4.5 in terms of processing time, memory consumption efficiency, error, and, ultimately, classification accuracy. When compared to more advanced and complicated machine learning models (e.g., neural networks and support vector machines), the C5.0 algorithm decision trees perform almost as well but are considerably easier to understand and use [42].

Adaboost

Adaboost is a method for reducing the error of a weak learning algorithm. In theory, the weak learning algorithm can be any that can generate classifiers that are only marginally better than random guessing [43]. There are two primary issues with boosting: Determining how to modify the training set so that the weak classifier can train using it and how to combine the weak classifiers gained during training to form a strong one. Previous authors [44] developed the Adaboost (adaptive boosting) method, which adjusts the weight without requiring prior information on learner learning. Adaboost has been employed in ensembles to increase prediction performance, most notably in neural networks [45], support vector machines [46], and decision trees [47]. The classifier uses an adaptive resampling strategy to select training samples, which means that a misclassified dataset generated by a prior classifier is chosen more frequently than correctly classified ones, allowing a new classifier to perform well in a fresh dataset. Each iteration gives the dataset a weight so that the following integration concentrates on reweighted datasets that were previously misclassified. In the final classifier, the ensemble predictions are weighted. The Adaboost algorithm can be applied to two-class problems, multi-class single-label issues, multi-class multi-label problems, single-label problem categories, and regression problems [47].

Treebag

Bagging or bootstrap aggregation is an ensemble method developed [48] that involves repeatedly training the same algorithm using different subsets of the training data. After that, the final output forecast is averaged over all sub-model projections. Bagging, in general, increases classification accuracy by lowering the variation of classification incertitude [49]. Freund and Schapire [48] claim that bagging can considerably enhance accuracy if changing the learning set creates a major change in the predictor built. The ensemble's majority vote is used to forecast a test sample [50]. Bagging attempts to reduce the error level owing to the variation of the base classifier by voting on the predictions of each classifier because each ensemble member is trained with a separate set of data [48].

Gradient Boosting Machine (GBM)

The Gradient Boosting Machine (GBM) is a forward-learning ensemble approach developed by [51] that is commonly used in machine learning. It is an effective method for developing predictive models for regression and classification tasks. GBM assists us in obtaining a predictive model in the form of an ensemble of weak prediction models such as decision trees [52]. When a decision tree performs poorly as a learner, the resulting algorithm is known as gradient-boosted trees [30]. Most supervised learning algorithms, in general, rely on a unique predictive model, such as decision trees and regression models. However, some supervised ML algorithms rely on the ensemble, which is a combination of various models. In other words, when multiple base models contribute predictions, boosting algorithms adapt to an average of all predictions. GBM is made up of three components, which are as follows: Weak learners, a loss function, and an additive model.

Extreme Gradient Boosting (Boost)

The gradient boosting theory is the basis for the XGBoost model, which combines a set of weak learners' predictions to create a robust learner through an additive training strategy [53]. The XGBoost model requires a number of parameter selections to predict the model, but the performance is always dependent on the selection of the optimal parameters. Thus, in the modelling process, the user needs to select three key parameters: colsample by tree (the portion of the variables to be used in each tree), subsample (the subsample ratio for the data to be considered in each tree), and nrounds (the maximum number of boosting iterations).

2.2.5. Models' Optimization

Cross-validation is an extremely effective tool in advanced and powerful machine learning models [54]. It allows us to make better use of our data and provides us with much more information about the performance of our algorithms. In this research, we used two approaches: K-fold cross-validation and tuning hyperparameters. For the first approach, the K-fold cross-validation method splits the input dataset into K groups of identical-size samples. The name given to these samples is folds. The prediction process uses k-1 folds for the separate training data and the remaining folds are used for the testing data. This is a popular CV approach because it is simple to understand and produces fewer biased results compared to other techniques. For the second approach, the process of tuning the parameters present as item sets while building ML models is known as hyperparameter tuning. These parameters are defined by us and can be manipulated as desired by the scientist in order for the model to perform well.

2.2.6. Validation and Accuracy Assessment

To assess the robustness of the used ML DT-based models used in the GE modelling process, we employed a number of statistics-based metrics, including sensitivity and specificity. This enables us to assess how the gully modelling predictive skill is employed to classify gully locations; specificity denotes the non-gully areas, while sensitivity denotes the gully area. These methods are relevant to predicting gully and non-gully areas. In addition, the kappa approach is utilized to assess the reliability of a gully erosion model. The values fall within the interval of -1 to 1, with 1 representing the best results. In addition, we used the accuracy, RMSE, and MAE values to assess the performance of the models tested for each data subdivision. A high value of accuracy and lower values of RMSE and MAE indicates better results of gully erosion modelling. Finally, the receiver operating characteristic curve (ROC) is regarded as a standard metric for evaluating the results of using ML models. To evaluate the performance of the modelling process, we use four types of possible metrics: True Positive (TP), False Positive (FP), True Negative (TN), and False Negative (FN). All of the equations used to calculate these parameters are mentioned below:

$$\text{Sensitivity} = \frac{TP}{TP + FN}, \tag{5}$$

$$\text{Specificity} = \frac{TN}{FP + TN}, \tag{6}$$

$$\text{Accuracy} = \frac{TN + TP}{TP + FP + TN + TP}, \tag{7}$$

$$\text{Kappa} = \frac{\text{Accuracy} - B}{1 - B}, \tag{8}$$

$$\text{Where B} = \frac{(TP + FN)(TP + FP) + (FP + TN)(FN + TN)}{\sqrt{TP + TN + FN + FP}}, \tag{9}$$

$$\text{RMSE} = \sqrt{\frac{1}{n} \sum_{i=1}^{n} (X_P - X_A)^2}, \tag{10}$$

$$\mathrm{MAE} = \frac{1}{n}\sum\nolimits_{i=1}^{n}\left|(X_P - X_A)^2\right|, \tag{11}$$

2.2.7. Variable Importance Analysis

We adopted two methods based on the RF model to generate a classification of factors according to their importance. The first is the mean decrease in accuracy and the second is the mean decrease in Gini. The mean decrease in accuracy shows how much the model accuracy loses when a factor is left out. The more the accuracy decreases, the greater the significance of the variable for effective results. The mean decrease in Gini is a measure of variable importance based on the principle that whenever a node is split on variable m, the Gini impurity criterion for the two descendent nodes is lesser compared to the parent node. Adding the Gini reductions for each variable across all trees offers a rapid measure of variable importance [55].

3. Results

3.1. Preliminary Data Analysis

The correlation matrix and the variance inflation factor (VIF) were used to examine the collinearity between the explanatory factors (Figure 5 and Table 3). The correlation matrix shows a high value of 0.8 between the LS factor and the TRI factor, while the collinearity of the remaining factors remains acceptable. The VIF shows a tolerance level with values less than 5: Curvature (1.069), TWI (1.075), and Distance to Rivers have the lowest VIF values (1.088), and the highest value is related to the LS-factor (3.396). As a result of the collinearity test, the LS factor has been omitted from the analysis.

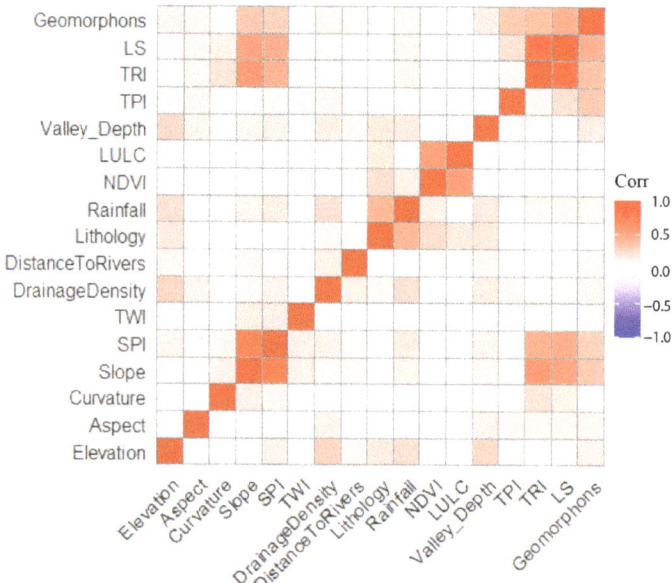

Figure 5. Correlation matrix results between conditioning factors. LandUse-LandCover (LULC), Normalized Difference Vegetation Index (NDVI), Topographic Position Index (TPI), Topographic Wetness Index (TWI), Topographic Roughness Index (TRI), Slope Length (LS), and Stream Power Index (SPI).

Soil Syst. **2023**, *7*, 50

Table 3. Variance inflation factor (VIF) and Tolerance (TOL) results.

Factors	VIF	TOL
Elevation	1.264	0.791
Aspect	1.118	0.895
Curvature	1.069	0.936
Slope	2.805	0.356
SPI	2.461	0.406
TWI	1.075	0.930
Drainage Density	1.214	0.824
Distance To Rivers	1.088	0.919
Lithology	1.366	0.732
Rainfall	1.406	0.711
NDVI	1.537	0.650
LULC	1.464	0.683
Valley Depth	1.207	0.828
TPI	1.217	0.822
TRI	3.301	0.303
LS	3.396	0.295
Geomorphons	1.494	0.669

3.2. Spatial Relationship between Gully Locations and Effective Factors

A bi-variate statistical approach based on the frequency ratio (FR) was used to correlate causative factors with the spatial distribution of gullies (Figure 4a,b). For a given factor, this ratio determines the likelihood of gully occurrence versus non-occurrence [56]. The highest value of FR is 6.34 represented by the lithology class occupied by sandstones and red conglomerates followed by the class of limestones and red clays, which had an FR value of 4.24, and lastly, the Basalts class with an FR of 2.36. The TRI factor and curvature represent a strong spatial correlation with the gullies with an FR value of 2.30 (class 22.46–81.83) and 1.98 (class 2.2–30.2), respectively. The topographic factors also showed a high spatial correlation with an FR value of 1.82 for valleys ranging in depth from 298 m to 459 m followed by the highest class of the LS factor with an FR of 1.81, and then the slope class (27–36°) with an FR of 1.72 and the elevation class, which ranges from 1937–2379 m with an FR value of 1.71. The majority of gullies developed on the southwest-facing slopes, which is represented by the high value of the Aspect factor (FR = 1.71). Rainfall also has a strong concordance with gully development where the highest value of FR (1.70) is given to the maximum rainfall class (552 and 610 mm). Compared with the rest of the factors, the majority of gullies developed in areas where the distance to rivers was more than 552 m (FR = 1.68), areas classified as the moderate SPI class (FR = 1.66), bare soil areas with an FR = 1.26 for the LULC factor, and areas in which the NDVI class ranged between −0.12 and 0.1 (FR = 1.59). Furthermore, the majority of gullies form on slopes and cavity areas, as indicated by the geomorphic factor, in which FR for these classes is 1.54. The TWI naturally correlates with gully formation areas; in the current study, this index has an FR = 1.14 represented by classes 7–8 of the TWI factor.

3.3. Variable Importance Analysis

Two measures were considered to identify the importance of the predictive factors of gully erosion: The average decrease in accuracy and the average decrease in Gini, which is based on the RF model with four subdivisions of the input database (50%, 60%, 70%, 80%, and 90%) (Figure 6). In general, the results of these two measures show that all variables play a role in gully formation. However, some factors were more important in predicting the spatial distribution of GE based on the average decrease in the accuracy index. The results show that the factors lithology, geomorphons, elevation, and LULC are the most important in terms of controlling gully formation. This can be explained by the study area's mountainous and geomorphological characteristics, as well as its continuous active tectonic aspect. These findings also demonstrate the effect of anthropogenic action on gully erosion sensitivity and the role of vegetation cover protection in combating this phenomenon. Furthermore, the average decrease in the Gini index results is in perfect agreement with the previous results, confirming the importance of lithology, geomorphological unit, and vegetation cover protection in the formation of gullies.

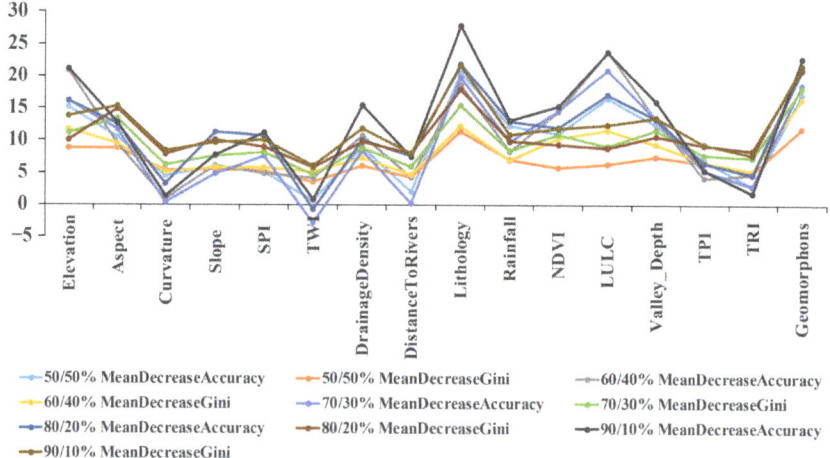

Figure 6. Conditioning factors importance assessment using RF algorithm.

3.4. Gully Erosion Susceptibility Mapping

The gully erosion susceptibility maps (GESMs) allow us to visualize the spatial distribution of gullies and identify the areas vulnerable to gully formation. GESMs were produced using the R interface and reclassified using the natural break method in GIS software (Figure 7). The percentages of the areas occupied by each gully erosion sensitivity class in relation to each model are shown in Figure 8. According to these results, the higher sensitivity classes account for 24% of the total area for the RF and XGBoost models, 23% of the study area for the C5.0 and GMB models, 25% of the area for the treebag model, and 28% of the total area for the Adaboost model. However, the areas with moderate gullying susceptibility range in percentage of the area from 28% for Adaboost to 24% for the RF model, 22% for the C5.0, and 19% for the treebag and GBM models. These findings are consistent with field observations, as the majority of mapped gullies are classified as having high or very high gully erosion sensitivity. Furthermore, all gully erosion sensitivity maps show increasing spatial variation from the very low gully erosion class to the very high gully erosion class, demonstrating the effectiveness of the models used and the reliability of the results.

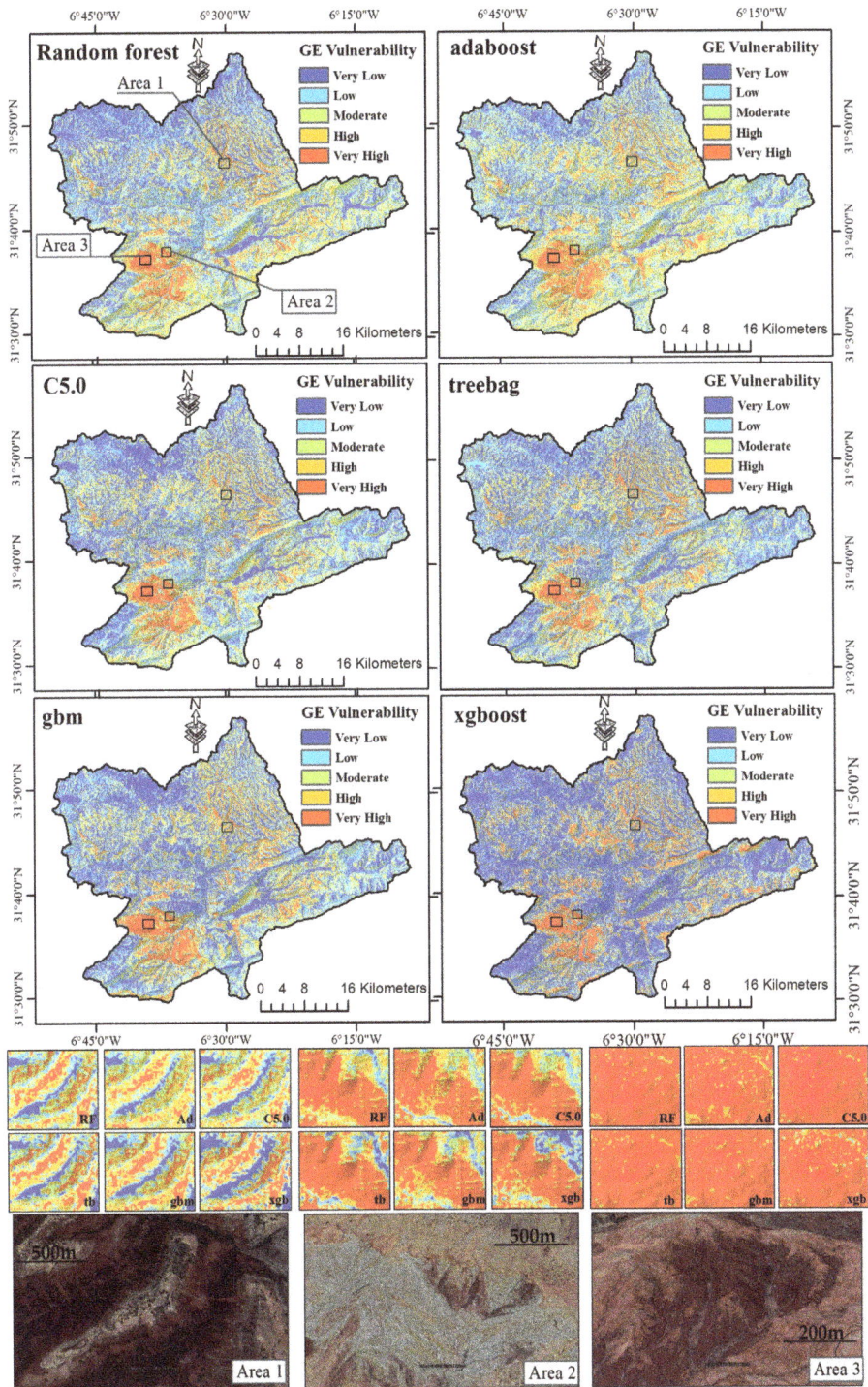

Figure 7. Gully erosion sensitivity maps using Dt-based models.

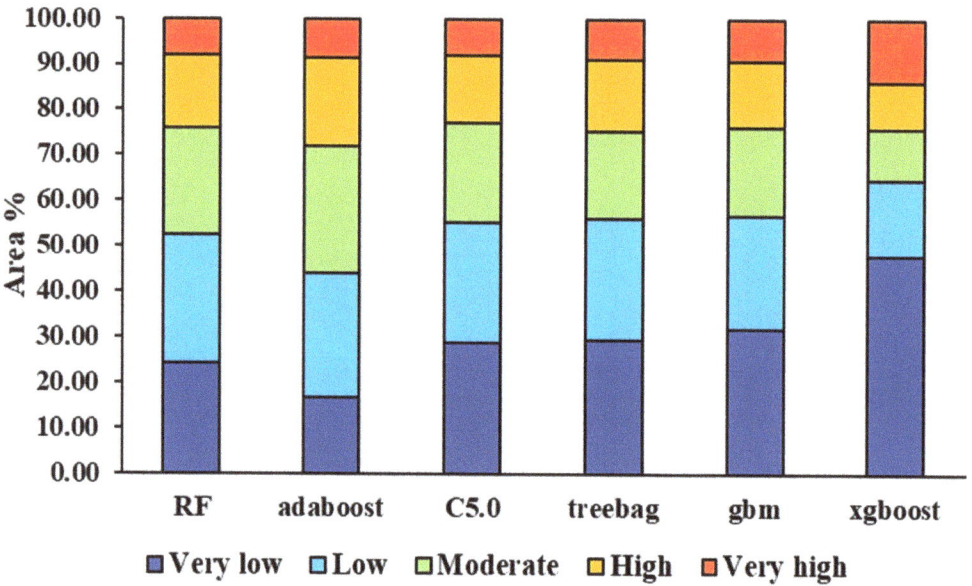

Figure 8. Distribution of gully erosion (GE) susceptibility classes (area %).

3.5. Model Accuracy and Validation Results

In this regard, six models were used, the input data were evaluated by cross-validation ten times for each model, and the accuracy was calculated on the basis of five subdivisions by four parameters, namely Kappa, ROC-AUC, RMSE, and MAE. As a result, we were able to identify the degree of discrimination and reliability, reflecting the performance of the chosen models. Comparing the results obtained (Figures 9–11), the C5.0 model shows a better performance, especially for the 70/30% subdivision, with an AUC value of 90.80% followed by the RF model with an AUC value equal to 90.10% for the 80/20% subdivision, then XGBoost and Adaboost models with an AUC of 90% for the 70/30% subdivision, then the GBM model with an AUC of 88.20% for the 90/10% subdivision, and finally, the treebag model with an AUC of 87.7% for the 70/30% subdivision. This demonstrates that the entire accuracy of the used models is high, particularly at the 70/30% and 80/20% subdivisions for the majority of these models. The average Kappa index values for the RF, C5.0, Adaboost, GBM, treebag, and XGBoost models are 0.58, 0.56, 0.59, 0.55, 0.57, and 0.54, respectively. These results are classified as acceptable to moderate. The average RMSE values range between 0.45 for the RF and Adaboost models, 0.46 for the C5.0 and treebag models, and 0.47 for the GBM and XGBoost models, indicating that the output results are of high quality and reliability. In the 10-fold cross-validation analysis, the prediction models used demonstrated robustness and stability for the calibration and validation datasets. These models also had a high accuracy, which exceeded 80% for the set of random subdivisions used.

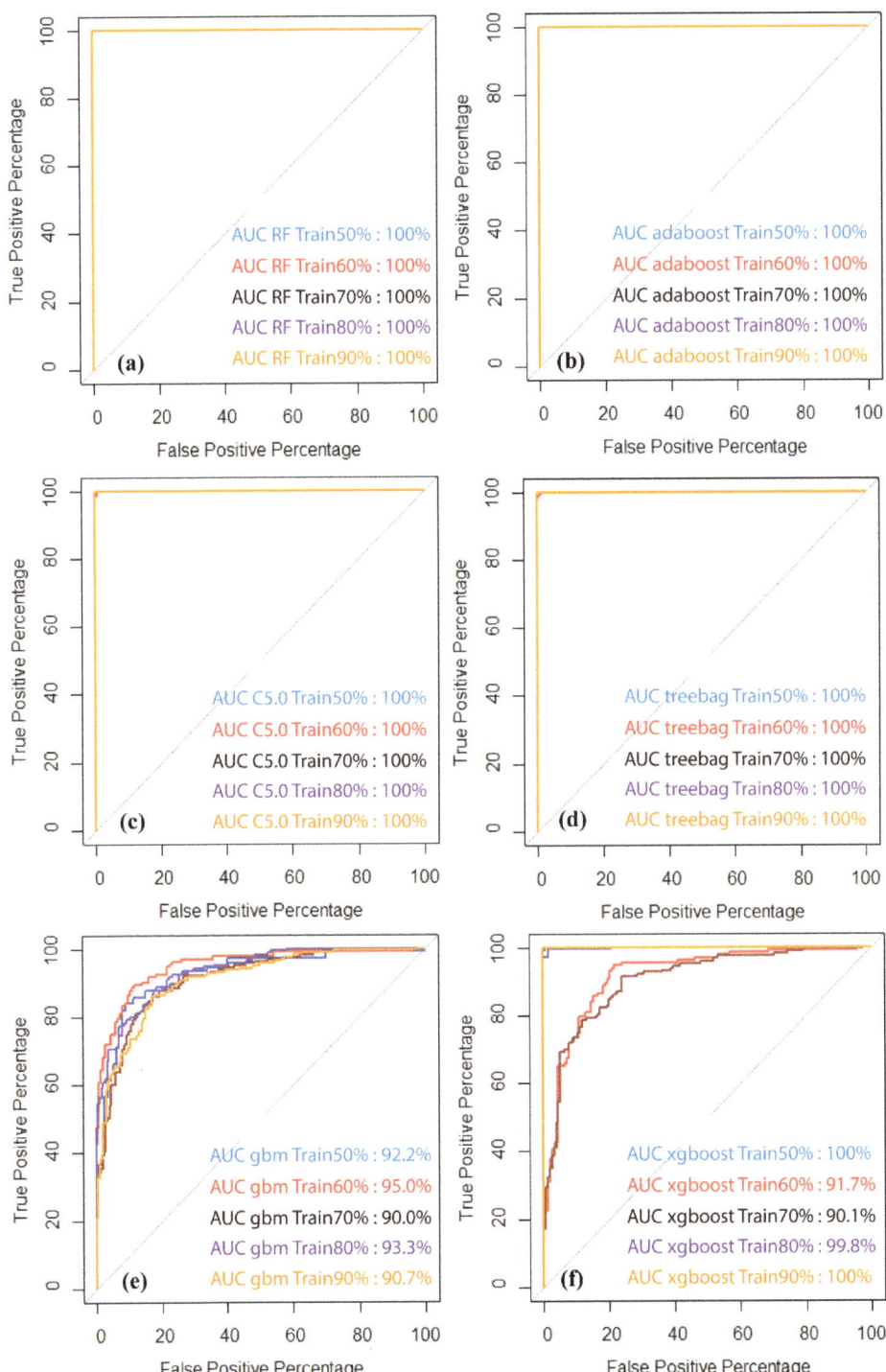

Figure 9. Success rate curve and the area under curve values of the DT-based models using the training dataset: RF (**a**), Adaboost (**b**), C5.0 (**c**), treebag (**d**), (**e**) GBM, and XGBoost (**f**).

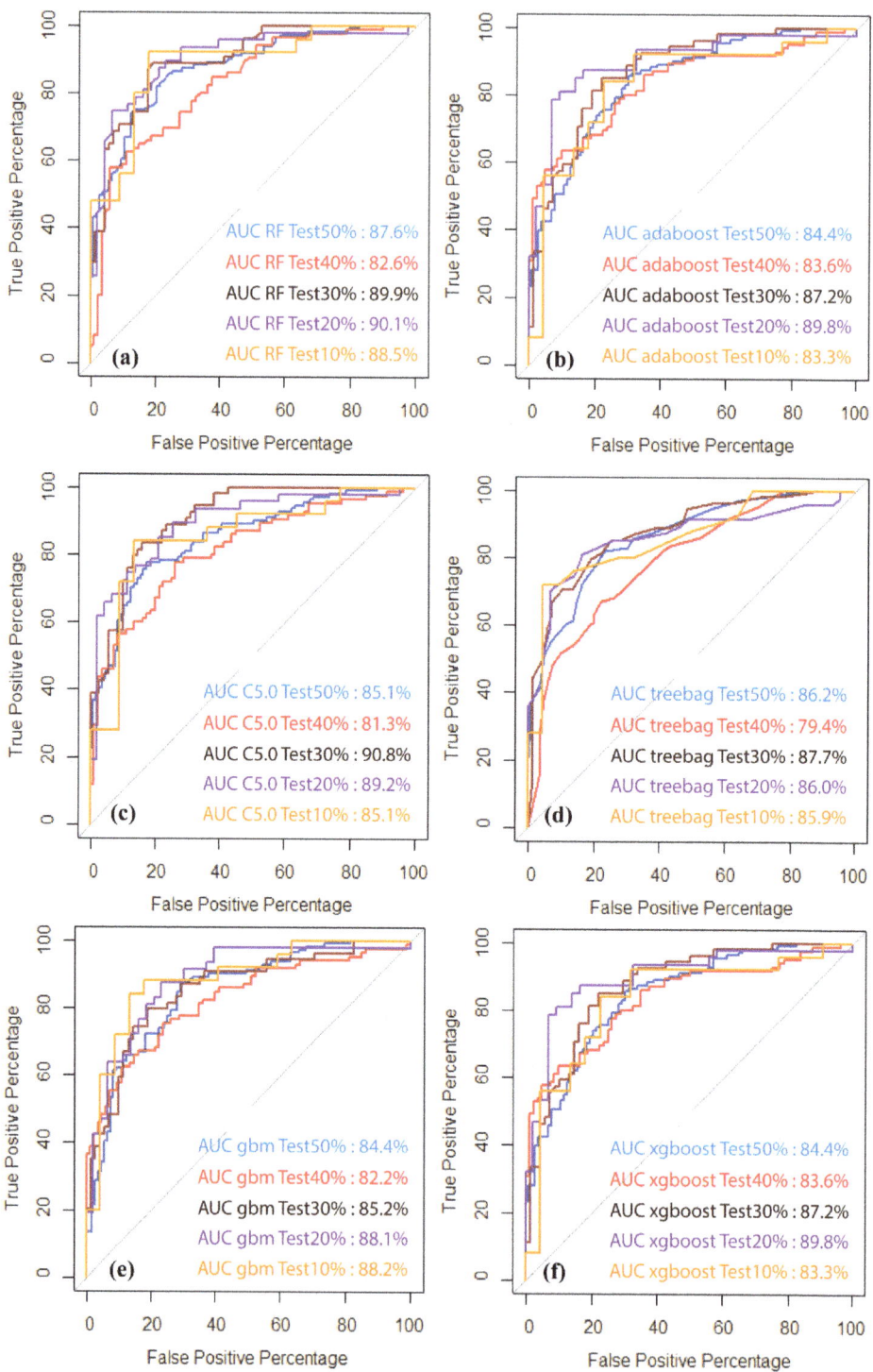

Figure 10. Prediction rate curve and the area under curve of the DT-based models using the testing dataset: RF (**a**), Adaboost (**b**), C5.0 (**c**), treebag (**d**), (**e**) GBM, and XGBoost (**f**).

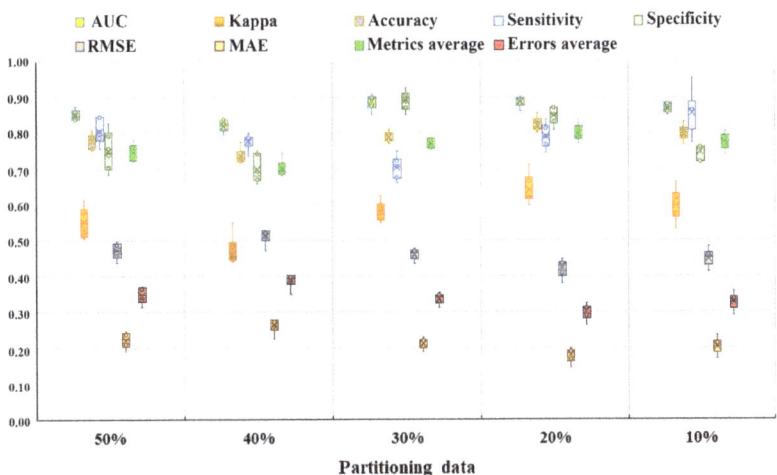

Figure 11. Validation results and metric parameters calculation of all models in each splitting quantity using testing data.

4. Discussion

In this section, the results are discussed in three parts: (i) The analysis of the models' performances; (ii) the investigation of the importance of each geo-environmental factor in the modelling of gully erosion; and (iii) the analysis of gully erosion vulnerability mapping results.

4.1. Accuracy Assessment and Comparison

The performance of the modelling is based on two fundamental aspects: Discrimination and reliability. In this respect, the evaluation of the performance of the GE sensitivity models was carried out according to five random subdivisions (50/50%, 60/40%, 70/30%, 80/20%, and 90/10%) by 10-fold cross-validation through several statistical metrics, namely the Kappa index, AUC, RMSE, and MAE.

In terms of prediction accuracy, the C5.0-70/30% model performed the best (AUC = 90.8), followed by the RF-80/20% model (AUC = 90.1), the Adaboost-70/30% model (AUC = 90), the XGBoost-80/20% model (AUC = 89.8), the GBM-80/20% model (AUC = 88.2), and the treebag-70/30% model (AUC = 87.7). These precision values indicate that all of the models utilized demonstrated a high level of performance and robustness, making them applicable to a variety of study domains and the monitoring and evaluation of natural hazards such as soil erosion, landslides, floods, and others [25,57,58]. To confirm this performance, however, the use of a single accuracy indicator may increase the margin of error, leading to potentially inaccurate results [59]. In this regard, the determination of additional accuracy indices such as RMSE, MAE, and the Kappa index can bolster the validity of the employed models [60]. Although the values of AUC and Kappa in terms of the discrimination index are greater in the present study, the values of RMSE and MAE are lower, indicating that the majority of gully inventory points were recognized on the final gully erosion sensitivity maps, reflecting the accuracy of the used models.

Moreover, despite the fact that these decision tree models are highly intuitive and do not necessitate a great deal of work in the preparation and processing of the database, they do require some effort. The obtained results indicate that the accuracy measures have high sensitivity to random database partitioning. Nonetheless, the majority of models perform better at the 70/30% and 80/20% subdivision levels, indicating that one of the disadvantages of decision tree-based models is that a simple change in the database can result in a change in the general structure of the decision tree and, as a result, model instability. For this reason, it is necessary to test multiple subdivisions in conjunction with 10-fold cross-validation to select a more accurate prediction model.

In the case of managing natural hazards such as gully erosion, the primary goal of the manager is to identify high-risk regions. However, the cost and time required to accomplish this goal are extremely significant. Consequently, the adoption of predictive models can be advantageous in terms of costs and resources mobilized to solve such an issue, since these models enable managers to concentrate on management priorities, thereby enhancing the efficiency of decision making.

4.2. Geoenvironmental Variable Importance Analysis

Several studies have highlighted that a large database is necessary for obtaining accurate results and a more accurate prediction of gully-vulnerable locations [61]. For this purpose, in this work, 17 factors were utilized to build GESMs, including topographical, hydrological, geomorphological, climatological, and soil-property-associated factors. The integration of these parameters with inventory data facilitates the identification of regions with a high risk of gully erosion.

According to five random subdivisions (50, 60, 70, 80, and 90%) of the model training database and using two measures (the average decrease in accuracy and the average decrease in precision), the RF technique determined the importance of the factors. The overall examination of these results revealed that all influencing variables contribute to gully formation. Furthermore, lithology, elevation, geomorphic factors, and LULC are the most significant contributors. This is consistent with the mountainous character of the study area and also demonstrates the visible influence of human interference on natural ecosystems on the acceleration of soil erosion. This is because the combination of highly friable lithologies such as clays and marls, high altitudes, and degraded vegetation cover facilitates gully development, particularly on steep slopes and in places with damaged vegetation cover [62]. Multiple investigations in comparable circumstances have confirmed that these variables effectively regulate the degree of soil particle detachment and gully formation vulnerability [33]. Furthermore, the LULC factor refers to human activities and natural land surface changes. In addition, the lack of a viable alternative economic sector for the local population, other than forest exploitation, significantly exacerbates soil erosion (wood, pasture, etc.). Therefore, people strive to make a living by clearing, overgrazing, and over-exploiting firewood in order to satisfy the significant rise in demand for arable land [63].

In other words, areas covered by friable lithologies such as clays and marls are the most susceptible to soil particle detachment [64]; therefore, vegetation cover protects the soil, and its degradation increases the likelihood of gully formation [62]. In addition, research on the effect of topographic parameters on gully formation in arid and semi-arid contexts has revealed the existence of direct and indirect impacts of topographic circumstances on the evolution of vegetation cover, rainfall, and runoff kinetic energy [65–68]. In reality, topographical features influence the local climate, which is characterized by geographically and temporally localized rainfall events, therefore places with steep slopes, such as hillsides, are characterized by high runoff velocities. This results in soil saturation, a substantial separation of soil particles, and the creation of ravines. The geomorphic element, which is also of major importance, verified this. This feature, which enables the mapping of slope units [69] and demonstrates that the majority of gullies are related to slopes and depressions, validates the effect of topography on the expression of erosive processes in mountainous regions.

Finally, all factors demonstrated significance in predicting and identifying regions with a high vulnerability to gully erosion; however, only the LS component was excluded because it was inconsistent with the other topographic variables.

4.3. Gully Erosion VulnerAbility Maps

Taking into consideration the subdivision where each model performs best, various models were used to develop vulnerability maps. The findings reveal that certain factors influence the spatial variability in vulnerability more strongly than others. Thus, the rise in the proportion of the most susceptible regions from upstream to downstream of a basin is directly attributable to the topographical impact. This is consistent with the substantial geographical link between these sites and classes with slopes above 27 degrees, TRI > 22, as well as slopes and Hollow units in Geomorhons' factor. Moreover, precipitation and LULC seem to be significant elements in regulating gully development, which is why all models anticipate that gully formation will be greatest in regions with high precipitation and degraded vegetation cover. These conclusions are comparable to those of earlier studies conducted in specific localities of the High Atlas, Morocco [70,71].

Comparing the maps generated by the various models, it is evident that the Adaboost model predicts more susceptible regions than the other models, especially in comparison to the XGBoost model, which predicts the fewest vulnerable areas. In general, the differences between the predictions of the models are limited; this is evident in Figure 7, where the areas highly susceptible to gully formation were predicted almost identically by all six models (Figure 7—Areas 2 and 3); however, for the low-vulnerability areas, only minor differences between XGBoost and the other models can be observed (Figure 7—Area 1). In general, the results of this study using RF, C5.0, Adaboost, XGBoost, treebag, and GBM models demonstrates that machine learning methods are capable of producing GESMs with great precision. This can be viewed as a fundamental tool to aid planners and managers in ensuring the sustainable and effective management of soil erosion-affected areas in a semi-arid mountain setting.

5. Conclusions

Gully erosion is a phenomenon of great complexity. To ensure appropriate management of this phenomenon, it is vital to comprehend the geographical distribution of gullies and detect regions with a high possibility of gully formation. Six decision tree models based on machine learning algorithms (Random Forest (RF), C5.0, XGBoost, 18 treebag, Gradient Boosting Machines (GBMs), and Adaboost) were tested to determine the role of 17 parameters in gully formation in a semi-arid environment with a hilly character and to test their stability in response to the changing splitting quantities in input data. The outcome was six erosion vulnerability maps for gullies. The examination of these results demonstrates that all the utilized models are robust and extremely reliable at predicting and identifying the sensitivity to gully erosion and that the most influential factors are Lithology, LandUse-LandCover (LULC), Geomorphons, and Elevation factors. In addition, the analysis of factors and their effects on gully formation and soil degradation revealed that topographical factors, such as geomorphological units and valley depths, play a significant role in the formation of gullies in this mountain environment. The validation of these results is likewise satisfactory, as they demonstrate congruence between the regions predicted by the ML models and the inventory points recovered from the real field data. This substantiates the accuracy of the predicted gullies' future results. The results also confirmed the need to test the performance of the models under many subdivisions of the input data in order to build a more accurate and stable model in terms of prediction. In this semi-arid highland context, the vulnerability maps generated have been shown to be a valuable tool for the sustainable management and planning of gully-erosion-affected areas.

Author Contributions: Conceptualization, H.E., M.H. and H.R.; methodology, H.E., M.H., M.N. and M.I.; software, H.E., M.I., M.O. and S.K.; validation, M.H., H.R. and L.B.; formal analysis, H.E.; investigation, H.E., M.N., S.K. and M.O.; resources, H.E., M.H., H.R. and L.B.; data curation, H.E., M.I., S.K. and M.O.; writing—original draft preparation, H.E.; writing—review and editing, H.E., M.H., H.R. and L.B.; visualization, H.E. and M.N.; supervision, M.H.; project administration, L.B. All authors have read and agreed to the published version of the manuscript.

Funding: This research received no external funding.

Institutional Review Board Statement: Not applicable.

Informed Consent Statement: Not applicable.

Data Availability Statement: Not applicable.

Acknowledgments: This work was carried out within the CHARISMA Project with the assistance of the Hassan II Academy of Science and Technology. This project is supported by GeanTech project funded by OCP Foundation. We acknowledge special support from IRD representation in Morocco.

Conflicts of Interest: The authors declare no conflict of interest.

References

1. Poesen, J.; Nachtergaele, J.; Verstraeten, G.; Valentin, C. Gully erosion and environmental change: Importance and research needs. *Catena* **2003**, *50*, 91–133. [CrossRef]
2. Roy, P.; Chandra Pal, S.; Arabameri, A.; Chakrabortty, R.; Pradhan, B.; Chowdhuri, I.; Tien Bui, D. Novel ensemble of multivariate adaptive regression spline with spatial logistic regression and boosted regression tree for gully erosion susceptibility. *Remote Sens.* **2020**, *12*, 3284. [CrossRef]
3. Li, Z.; Fang, H. Impacts of climate change on water erosion: A review. *Earth-Sci. Rev.* **2016**, *163*, 94–117. [CrossRef]
4. Zabihi, M.; Pourghasemi, H.R.; Motevalli, A.; Zakeri, M.A. Gully erosion modeling using GIS-based data mining techniques in Northern Iran: A comparison between boosted regression tree and multivariate adaptive regression spline. In *Natural Hazards GIS-Based Spatial Modeling Using Data Mining Techniques*; Springer: Cham, Switzerland, 2019; pp. 1–26. [CrossRef]
5. Gupta, G.S. Land degradation and challenges of food security. *Rev. Eur. Stud.* **2019**, *11*, 63. [CrossRef]
6. Borrelli, P.; Robinson, D.A.; Panagos, P.; Lugato, E.; Yang, J.E.; Alewell, C.; Wuepper, D.; Montanarella, L.; Ballabio, C. Land use and climate change impacts on global soil erosion by water (2015–2070). *Proc. Natl. Acad. Sci. USA* **2020**, *117*, 21994–22001. [CrossRef]
7. FAO. *Global Soil Status, Processes and Trends. Status of the World's Soil Resources (SWSR)—Main Report of the Food and Agriculture Organization*; FAO: New York, NY, USA, 2015.
8. Acharki, S.; El Qorchi, F.; Arjdal, Y.; Amharref, M.; Bernoussi, A.S.; Aissa, H.B. Soil erosion assessment in Northwestern Morocco. *Remote Sens. Appl. Soc. Environ.* **2022**, *25*, 100663. [CrossRef]
9. Markhi, A.; Laftouhi, N.; Grusson, Y.; Soulaimani, A. Assessment of potential soil erosion and sediment yield in the semi-arid N′ fis basin (High Atlas, Morocco) using the SWAT model. *Acta Geophys.* **2019**, *67*, 263–272. [CrossRef]
10. Micheletti, N.; Foresti, L.; Robert, S.; Leuenberger, M.; Pedrazzini, A.; Jaboyedoff, M.; Kanevski, M. Machine learning feature selection methods for landslide susceptibility mapping. *Math. Geosci.* **2013**, *46*, 33–57. [CrossRef]
11. Smith, S.J.; Williams, J.R.; Menzel, R.G.; Coleman, G.A. Prediction of sediment yield from southern plains grasslandds with the modified universal soil loss equation. *J. Range Manag.* **1984**, *37*, 295–297. [CrossRef]
12. Renard, K.G.; Foster, G.R.; Weesies, G.A.; Porter, J.P. RUSLE, revised universal soil loss equation. *J. Soil Water Conserv.* **1991**, *46*, 30–33.
13. Flanagan, D.C.; Nearing, M.A. *USDA-Water Erosion Prediction Project: Hill Slope and Watershed Model Documentation. NSERI Report No. 10*; USDA-ARS National Soil Erosion Research Laboratory: West Lafayette, IN, USA, 1995.
14. Wischmeier, W.H.; Smith, D.D. *Predicting Rainfall Erosion Losses: A Guide to Conservation Planning. Agriculture Handbook. 282*; USDA-ARS: Beltsville, MA, USA, 1978.
15. Williams, J.R.; Jones, C.A.; Dyke, P.T. *The EPIC Model. United States Department of Agriculture (USDA) Teachnical Bulletin No. 1768*; United States Department of Agriculture: Washington, DC, USA, 1990.
16. Gayen, A.; Saha, S. Application of weights-of-evidence (WoE) and evidential belief function (EBF) models for the delineation of soil erosion vulnerable zones: A study on Pathro river basin, Jharkhand, India. *Model. Earth Syst. Environ.* **2017**, *3*, 1123–1139. [CrossRef]
17. Alewell, C.; Borrelli, P.; Meusburger, K.; Panagos, P. Using the USLE: Chances, challenges and limitations of soil erosion modelling. *Int. Soil Water Conserv. Res.* **2019**, *7*, 203–225. [CrossRef]
18. Luca, F.; Conforti, M.; Robustelli, G. Comparison of GIS-based gullying susceptibility mapping using bivariate and multivariate statistics: Northern Calabria, South Italy. *Geomorphology* **2011**, *134*, 297–308. [CrossRef]
19. Svoray, T.; Michailov, E.; Cohen, A.; Rokah, L.; Sturm, A. Predicting gully initiation: Comparing data mining techniques, analytical hierarchy processes and the topographic threshold. *Earth Surf. Process. Landf.* **1991**, *37*, 607–619. [CrossRef]
20. Conoscenti, C.; Angileri, S.; Cappadonia, C.; Rotigliano, E.; Agnesi, V.; Marker, M. Gully erosion susceptibility assessment by means of GIS-based logistic regression: A case of Sicily (Italy). *Geomorphology* **2014**, *204*, 399–411. [CrossRef]
21. Dube, F.; Nhapi, I.; Murwira, A.; Gumindoga, W.; Goldin, J.; Mashauri, D.A. Potential of weight of evidence modelling for gully erosion hazard assessment in Mbire District—Zimbabwe. *Phys. Chem. Earth* **2014**, *67*, 145–152. [CrossRef]
22. Zakerinejad, R.; Maerker, M. An integrated assessment of soil erosion dynamics with special emphasis on gully erosion in the Mazayjan basin, southwestern Iran. *Nat. Hazards* **2015**, *79*, 25–50. [CrossRef]
23. Du Plessis, C.; Van Zijl, G.; Van Tol, J.; Manyevere, A. Machine learning digital soil mapping to inform gully erosion mitigation measures in the Eastern Cape, South Africa. *Geoderma* **2020**, *368*, 114287. [CrossRef]
24. Zhao, X.; Chen, W. Gis-based evaluation of landslide susceptibility models using certainty factors and functional trees-based ensemble techniques. *Appl. Sci.* **2020**, *10*, 16. [CrossRef]

25. Sahour, H.; Gholami, V.; Vazifedan, M. A comparative analysis of statistical and machine learning techniques for mapping the spatial distribution of groundwater salinity in a coastal aquifer. *J. Hydrol.* **2020**, *591*, 125321. [CrossRef]
26. Marjanović, M.; Kovačević, M.; Bajat, B.; Voženílek, V. Landslide susceptibility assessment using SVM machine learning algorithm. *Eng. Geol.* **2011**, *123*, 225–234. [CrossRef]
27. Chen, W.; Lei, X.; Chakrabortty, R.; Pal, S.C.; Sahana, M.; Janizadeh, S. Evaluation of different boosting ensemble machine learning models and novel deep learning and boosting framework for head-cut gully erosion susceptibility. *J. Environ. Manag.* **2021**, *284*, 112015. [CrossRef]
28. Alaboz, P.; Dengiz, O.; Demir, S.; Şenol, H. Digital mapping of soil erodibility factors based on decision tree using geostatistical approaches in terrestrial ecosystem. *Catena* **2021**, *207*, 105634. [CrossRef]
29. Pal, S.C.; Chakrabortty, R.; Arabameri, A.; Santosh, M.; Saha, A.; Chowdhuri, I.; Roy, P.; Shit, M. Chemical weathering and gully erosion causing land degradation in a complex river basin of Eastern India: An integrated field, analytical and artificial intelligence approach. *Nat. Hazards* **2022**, *110*, 847–879. [CrossRef]
30. Saha, S.; Roy, J.; Arabameri, A.; Blaschke, T.; Tien Bui, D. Machine Learning-Based Gully Erosion Susceptibility Mapping: A Case Study of Eastern India. *Sensors* **2020**, *20*, 1313. [CrossRef]
31. Pourghasemi, H.R.; Sadhasivam, N.; Kariminejad, N.; Collins, A.L. Gully erosion spatial modelling: Role of machine learning algorithms in selection of the best controlling factors and modelling process. *Geosci. Front.* **2020**, *11*, 2207–2219. [CrossRef]
32. Tiwari, A.; Arun, G.; Vishwakarma, B.D. Parameter importance assessment improves efficacy of machine learning methods for predicting snow avalanche sites in Leh-Manali Highway, India. *Sci. Total Environ.* **2021**, *794*, 148738. [CrossRef]
33. Rahmati, O.; Tahmasebipour, N.; Haghizadeh, A.; Pourghasemi, H.R.; Feizizadeh, B. Evaluation of different machine learning models for predicting and mapping the susceptibility of gully erosion. *Geomorphology* **2017**, *298*, 118–137. [CrossRef]
34. Conforti, M.; Aucelli, P.; Robustelli, G.; Scarciglia, F. Geomorphology and GIS analysis for mapping gully erosion susceptibility in the Turbolo Stream catchment (Northern Calabria, Italy). *Nat. Hazards* **2011**, *56*, 881–898. [CrossRef]
35. Sharma, M.; Garg, R.D.; Badenko, V.; Fedotov, A.; Min, L.; Yao, A. Potential of airborne LiDAR data for terrain parameters extraction. *Quat. Int.* **2021**, *575*, 317–327. [CrossRef]
36. Holloway, J.; Rudy, A.; Lamoureux, S.; Treitz, P. Determining the terrain characteristics related to the surface expression of subsurface water pressurization in permafrost landscapes using susceptibility modelling. *Cryosphere* **2017**, *11*, 1403–1415. [CrossRef]
37. Gutiérrez, Á.G.; Schnabel, S.; Contador, F.L. Gully erosion, land use and topographical thresholds during the last 60 years in a small rangeland catchment in SW Spain. *Land Degrad. Dev.* **2009**, *20*, 535–550. [CrossRef]
38. Breiman, L. Random forests. *Mach. Learn.* **2001**, *45*, 5–32. [CrossRef]
39. Ravì, D.; Bober, M.; Farinella, G.M.; Guarnera, M.; Battiato, S. Semantic segmentation of images exploiting DCT based features and random forest. *Pattern Recognit.* **2016**, *52*, 260–273. [CrossRef]
40. Zhang, G.; Cai, Y.; Zheng, Z.; Zhen, J.; Liu, Y.; Huang, K. Integration of the Statistical Index Method and the Analytic Hierarchy Process technique for the assessment of landslide susceptibility in Huizhou, China. *Catena* **2016**, *142*, 233–244. [CrossRef]
41. Pandya, R.; Pandya, J. C5. 0 algorithm to improved decision tree with feature selection and reduced error pruning. *Int. J. Comput. Appl.* **2015**, *117*, 18–21. [CrossRef]
42. Putra, F.; Sitanggang, I. Classification model of air quality in Jakarta using decision tree algorithm based on air pollutant standard index. *IOP Conf. Ser. Earth Environ. Sci.* **2020**, *528*, 012053. [CrossRef]
43. Pham, B.T.; Nguyen, M.D.; Nguyen-Thoi, T.; Ho, L.S.; Koopialipoor, M.; Kim Quoc, N.; Armaghani, D.J.; Le, H.V. A novel approach for classification of soils based on laboratory tests using Adaboost, Tree and ANN modeling. *Transp. Geotech.* **2021**, *27*, 100508. [CrossRef]
44. Freund, Y.; Schapire, R.E. Experiments with a new boosting algorithm. *ICML* **1996**, *96*, 148–156.
45. West, D.; Dellana, S.; Qian, J. Neural network ensemble strategies for financial decision applications. *Comput. Oper. Res. Appl. Neural Netw.* **2005**, *32*, 2543–2559. [CrossRef]
46. Wang, S.; Mathew, A.; Chen, Y.; Xi, L.; Ma, L.; Lee, J. Empirical analysis of support vector machine ensemble classifiers. *Expert Syst. Appl.* **2009**, *36*, 6466–6476. [CrossRef]
47. Hong, H.; Liu, J.; Bui, D.T.; Pradhan, B.; Acharya, T.D.; Pham, B.T.; Zhu, A.-X.; Chen, W.; Ahmad, B.B. Landslide susceptibility mapping using J48 Decision Tree with AdaBoost, Bagging and Rotation Forest ensembles in the Guangchang area (China). *Catena* **2018**, *163*, 399–413. [CrossRef]
48. Breiman, L. Bagging predictors. *Mach. Learn.* **1996**, *24*, 123–140. [CrossRef]
49. Chan, J.C.-W.; Paelinckx, D. Evaluation of Random Forest and Adaboost tree-based ensemble classification and spectral band selection for ecotope mapping using airborne hyperspectral imagery. *Remote Sens. Environ.* **2008**, *112*, 2999–3011. [CrossRef]
50. Banfield, R.E. *Learning on Complex Simulations*; University of South Florida: Tampa, FL, USA, 2007.
51. Friedman, J.H. Greedy Function Approximation: A Gradient Boosting Machine. *Ann. Stat.* **2001**, *29*, 1189–1232. Available online: https://www.jstor.org/stable/2699986 (accessed on 10 May 2023). [CrossRef]
52. Sahin, E.K. Assessing the predictive capability of ensemble tree methods for landslide susceptibility mapping using XGBoost, gradient boosting machine, and random forest. *SN Appl. Sci.* **2020**, *2*, 1308. [CrossRef]

53. Chen, T.; Guestrin, C. XGBoost: A Scalable Tree Boosting System. In Proceedings of the 22nd ACM SIGKDD International Conference on Knowledge Discovery and Data Mining, KDD '16, San Francisco, CA, USA, 13–17 August 2016; Association for Computing Machinery: New York, NY, USA; pp. 785–794. [CrossRef]
54. Ramezan, C.A.; Warner, T.A.; Maxwell, A.E. Evaluation of sampling and cross-validation tuning strategies for regional-scale machine learning classification. *Remote Sens.* **2019**, *11*, 185. [CrossRef]
55. Breiman, L.; Cutler, A. A deterministic algorithm for global optimization. *Math. Program.* **1993**, *58*, 179–199. [CrossRef]
56. Lee, S.; Pradhan, B. Landslide hazard mapping at Selangor, Malaysia using frequency ratio and logistic regression models. *Landslides* **2006**, *4*, 33–41. [CrossRef]
57. Guo, Z.; Shi, Y.; Huang, F.; Fan, X.; Huang, J. Landslide susceptibility zonation method based on C5. 0 decision tree and K-means cluster algorithms to improve the efficiency of risk management. *Geosci. Front.* **2021**, *12*, 101249. [CrossRef]
58. Masselink, R.H.; Temme, A.J.A.M.; Giménez Díaz, R.; Casalí Sarasíbar, J.; Keesstra, S.D. Assessing hillslope-channel connectivity in an agricultural catchment using rare-earth oxide tracers and random forests models. *Cuad. Investig. Geográfica* **2017**, *43*, 19–39. [CrossRef]
59. Tehrany, M.S.; Pradhan, B.; Mansor, S.; Ahmad, N. Flood susceptibility assessment using GIS-based support vector machine model with different kernel types. *Catena* **2015**, *125*, 91–101. [CrossRef]
60. Pham, B.T.; Prakash, I.; Singh, S.K.; Shirzadi, A.; Shahabi, H.; Tran, T.-T.T.; Tien Bui, D. Landslide susceptibility modeling using reduced error pruning trees and different ensemble techniques: Hybrid machine learning approaches. *Catena* **2019**, *175*, 203–218. [CrossRef]
61. Romer, C.; Ferentinou, M. Shallow landslide susceptibility assessment in a semiarid environment—A quaternary catchment of KwaZulu-Natal, South Africa. *Eng. Geol.* **2016**, *201*, 29–44. [CrossRef]
62. Arabameri, A.; Tiefenbacher, J.P.; Blaschke, T.; Pradhan, B.; Tien Bui, D. Morphometric analysis for soil erosion susceptibility mapping using novel gis-based ensemble model. *Remote Sens.* **2020**, *12*, 874. [CrossRef]
63. Bouzekraoui, H.; El Khalki, Y.; Mouaddine, A.; Lhissou, R.; El Youssi, M.; Barakat, A. Characterization and dynamics of agroforestry landscape using geospatial techniques and field survey: A case study in central High-Atlas (Morocco). *Agrofor. Syst.* **2016**, *90*, 965–978. [CrossRef]
64. Azareh, A.; Rahmati, O.; Rafiei-Sardooi, E.; Sankey, J.B.; Lee, S.; Shahabi, H.; Ahmad, B.B. Modelling gully-erosion susceptibility in a semi-arid region, Iran: Investigation of applicability of certainty factor and maximum entropy models. *Sci. Total Environ.* **2019**, *655*, 684–696. [CrossRef]
65. Nazari Samani, A.; Ahmadi, H.; Jafari, M.; Ghoddousi, J. Geomorphic threshold conditions for gully erosion in Southwestern Iran (Boushehr-Samal watershed). *J. Asian Earth Sci.* **2009**, *35*, 180–189. [CrossRef]
66. Bochet, E.; García-Fayos, P. Factors controlling vegetation establishment and water erosion on motorway slopes in Valencia, Spain. *Restor. Ecol.* **2004**, *12*, 166–174. [CrossRef]
67. Wang, L.; Wei, S.; Horton, R.; Shao, M.A. Effects of vegetation and slope aspect on water budget in the hill and gully region of the Loess Plateau of China. *Catena* **2011**, *87*, 90–100. [CrossRef]
68. Beullens, J.; Van de Velde, D.; Nyssen, J. Impact of slope aspect on hydrological rainfall and on the magnitude of rill erosion in Belgium and northern France. *Catena* **2014**, *114*, 129–139. [CrossRef]
69. Luo, W.; Liu, C.C. Innovative landslide susceptibility mapping supported by geomorphon and geographical detector methods. *Landslides* **2018**, *15*, 465–474. [CrossRef]
70. Barakat, A.; Rafai, M.; Mosaid, H.; Islam, M.S.; Saeed, S. Mapping of Water-Induced Soil Erosion Using Machine Learning Models: A Case Study of Oum Er Rbia Basin (Morocco). *Earth Syst. Environ.* **2022**, *7*, 151–170. [CrossRef]
71. Meliho, M.; Khattabi, A.; Mhammdi, N. A GIS-based approach for gully erosion susceptibility modelling using bivariate statistics methods in the Ourika watershed, Morocco. *Environ. Earth Sci.* **2018**, *77*, 655. [CrossRef]

soil systems

MDPI

Article

Changes in Soil Water Retention and Micromorphological Properties Induced by Wetting and Drying Cycles

Luiz F. Pires

Laboratory of Physics Applied to Soils and Environmental Sciences, Department of Physics, State University of Ponta Grossa, Ponta Grossa 84030-900, Brazil; lfpires@uepg.br; Tel.: +55-42-32203044

Abstract: Wetting and drying (W-D) cycles are responsible for significant changes in soil structure. Soil often undergoes irreversible changes affecting infiltration and solute retention through W-D cycles. Thus, it becomes essential to evaluate how soils under natural conditions are altered by W-D cycles. This study analyzed two non-cultivated (from grassland and secondary forest) Oxisols (Typic Hapludox and Rhodic Hapludox) of different textures under 0 and 6 W-D cycles. The main results obtained showed that soil water retention was mainly affected in the driest regions (smaller pore sizes). The contribution of residual pores to total porosity increased with 6 W-D and transmission pores decreased in both soils. The Rhodic Hapludox presented differences in water content at field capacity (increase), while the Typic Hapludox showed alterations at the permanent wilting point (increase), affecting the amount of free water (Rhodic Hapludox) and water available to plants (Typic Hapludox). Both soils showed increases in imaged porosity with 6 W-D. Variations in the contribution of small and medium rounded pores, mainly large and irregular (with an increase in both soils not significant in the Rhodic Hapludox), could explain the results observed. The micromorphological properties were mainly influenced by changes in the number of pores, in which smaller pores joined, forming larger ones, increasing the areas occupied by larger pores. Overall, this study showed that the investigated soils presented pore systems with adequate water infiltration and retention capacities before and after continuous W-D cycles.

Keywords: Oxisols; image analysis; pore shape; pore size distribution; soil pore system; soil water retention curve

Citation: Pires, L.F. Changes in Soil Water Retention and Micromorphological Properties Induced by Wetting and Drying Cycles. *Soil Syst.* **2023**, *7*, 51. https://doi.org/10.3390/soilsystems7020051

Academic Editor: Luis Eduardo Akiyoshi Sanches Suzuki

Received: 27 March 2023
Revised: 10 May 2023
Accepted: 11 May 2023
Published: 17 May 2023

1. Introduction

A good soil structure is fundamental for ideal agricultural development and for the prevention of environmental damage [1]. When the soil has a good structure, water infiltrates adequately, and this soil has a proper capacity to hold water for the plants [2]. Suitable water drainage is vital to avoid processes that lead to sediment transport, such as erosion [3]. Such a good structure is also fundamental for the appropriate development of the root system of crops [4]. Naturally, soils are subjected to numerous processes involving rainfall, temperature variations, wind action, and the decomposition of organic material, among many others [5]. All these processes modify the soil over time and provoke changes mainly in its structure. In the broadest sense, soil structure relates to how the soil components (primary particles, organic material, iron and aluminum oxides, carbonates, etc.) are arranged [5]. This results in an arrangement containing particles (matrix) and pores, usually filled by the soil solution and gases [6].

Observing the pore system is vital for understanding the processes that occur in the soil. This pore system results from the arrangement of particles and aggregates within the soil; therefore, different pore size distributions are related to particular soils [7]. Pore sizes are fundamental to the retention and movement of solutes in the soil profile [8]. In addition to different pore sizes, their shapes also influence the dynamics of solutes in the soil [9,10]. In micromorphology studies, pores can usually be classified as rounded, elongated, and complex (irregular), with each shape having a distinct origin and exerting different influences on the soil's processes [11–13]. Wetting and drying (W-D) cycles are among the processes that cause changes in soil structure. These cycles occur naturally through rainfall but can also be artificially induced when irrigating the soil. Many studies have reported that soil when subjected to numerous W-D cycles, can undergo alterations that are often non-reversible due to the modifications that occur in the soil's pore system [14–17]. These changes often cause pore size distribution and shape alterations, impacting infiltration and water retention [18–20].

Hussein and Adey [14] demonstrated the influence of W-D cycles in soil pore systems with the shape of pores changing from planar to compound due to the wetting effect on water dynamics. An et al. [16] showed that the proportion of micropores and mesopores decreased under W-D cycles with an increasing number of macropores in granite soils. Those authors claimed that changes in the clay microstructure explained the observed results. Pardini et al. [17] found that W-D cycles increased porosity resulting from the formation of large cracks and fissures. Those authors also observed increases in the number of pores after 3 W-D cycles. Xia et al. [18] noticed decreases in the saturated and residual water contents with increased alternate W-D cycles and increases in the saturated hydraulic conductivity. The soil structure degradation was pointed out by them as the cause of changes under W-D cycles. Thus, it has become crucial to analyze how the pore system of natural soils behaves when subjected to W-D cycles. It is known that when managed, the soil structure suffers substantial changes. Therefore, analyzing how these cycles affect natural soils can provide insights into their susceptibility to modifications under W-D.

One tool successfully used to characterize the soil pore system is the analysis of resin-impregnated blocks [9,21–24]. In this technique, two-dimensional (2D) images of sections in the soil blocks allow for the analysis of numerous morphological properties on the micrometer scale [25]. By using image analysis, it is possible to quantify changes in pore shape and size distribution. Numerous scientific papers have shown that alterations in soil morphological properties can affect the proper development of crops and modify solute and gas dynamics [4,26–31]. Thus, micromorphological analysis has become a useful method to check how soils behave when subjected to W-D cycles. Therefore, the objectives of this paper are two-fold: namely, to analyze how the water-holding capacity of two Oxisols under natural conditions is affected after repeated W-D cycles and to evaluate how the micromorphological properties of the soils are influenced by W-D cycles. Concerning the first objective, water retention data were employed to generate some indices related to soil quality and to assess the contribution of pores based on their roles in total porosity.

2. Materials and Methods

2.1. Soil Sampling

This research was conducted using soil samples collected in 2013 at experimental areas of the University of São Paulo (USP) research farm (22°72′ S, 47°62′ W) and Agronomic Institute of Campinas (IAC) research station (22°70′ S, 47°64′ W), both located in Piracicaba, Brazil. The soils collected were classified as Typic Hapludox (USP) and Rhodic Hapludox (IAC) [32]. The former has a sandy clay loam texture (69% sand, 10% silt, and 21% clay), while the latter has a clay loam texture (22% sand, 28% silt, and 50% clay) (USDA soil texture triangle). The Typic Hapludox samples were collected in an area covered with natural grass, while the Rhodic Hapludox samples were collected in a secondary forest area. The organic carbon content in the experimental sites was c. 16.2 g dm^{-3} (Typic Hapludox) and c. 26.9 g dm^{-3} (Rhodic Hapludox), respectively.

Undisturbed soil cores were collected in the topsoil layer (0–10 cm) using stainless steel cylinders (c. 5 cm in diameter and c. 3 cm in height) for the soil–water retention curve (SWRC) and micromorphological analysis. The Kopeck ring method was used to collect the samples in which a woody castle was employed, with the cylinder inserted into the soil surface using a rubber hammer [33]. The cylinder was slowly introduced into the soil to avoid damage to its structure. After cylinder insertion, the surrounding soil was carefully excavated with trowels to remove the cylinder. The excess soil outside the cylinder was removed with a palette knife to leave the soil volume equal to the internal volume of the cylinder. Next, the samples were wrapped in plastic film and taken to the laboratory.

The samples were collected with soil moisture near field capacity to avoid damage to the soil structure due to sampling. In the laboratory, the samples used for the micromorphological analysis were left to dry in the air for a few weeks, and close to the impregnation process, they were placed in an oven (forced air circulation) and dried at 40 °C for 48 h. A total of 44 undisturbed soil samples were collected for this study.

2.2. Wetting and Drying (W-D) Cycles

The capillary rise process was used to wet the samples [34]. This procedure was performed by placing a 1 cm layer of water around the cylinders, and then every hour, 0.5 cm of water was poured out up to approximately half the height of the cylinder. After a period of 24 h, 0.5 cm of water was again placed until approximately 9/10 of the height of the cylinder had been filled. Samples were left in water for 24 h to ensure they were saturated. Drying was achieved by placing the samples in Richards's chamber and subjecting them to a pressure head (h) of 40 kPa. After the thermodynamic hydraulic equilibrium was achieved, the samples were submitted to a new wetting and drying process. Thus, samples not subjected to W-D cycles (0 W-D) and subjected to six W-D cycles (6 W-D) were analyzed. Two sample sets were prepared for this study, one for SWRC analysis (24 samples—6 samples × 2 soils × 2 W-D cycles) and another for micromorphological analysis (20 samples − 5 samples × 2 soils × 2 W-D cycles).

2.3. Water Retention Measurement

The SWRC determination was carried out using a suction table (Eijkelkamp Sandbox for pF determination) and low- to high-pressure Richards's chambers (Soil Moisture Equipment Co., Goleta, CA, USA). Suctions of 3, 6, 9, and 10 kPa (suction table) and pressures of 30, 50, 150, 500, and 1500 kPa (Richards's chambers) were selected for this study. The thermodynamic hydraulic equilibrium was defined when the water stopped leaving the undisturbed soil samples [35]. Thus, the samples previously subjected to 0 and 6 W-D were also submitted to the pressure heads (suctions and pressures) described above. After applying all the pressure heads, the samples were oven dried at 105 °C for 24–48 h. The gravimetric water content was obtained by the ratio between the wet soil mass (for each suction and pressure) and the dry soil mass. The volumetric water content (θ) was determined by considering the relation between the gravimetric water content (G), soil bulk density (ρ_s), and water density (ρ_w): $\theta = G(\rho_s/\rho_w)$ [5,36].

The measured θ data related to each h were fitted to the van Genuchten–Mualem (VGM) model [37] according to:

$$\theta(h) = \theta_r + \frac{\theta_s - \theta_r}{\left[1 + (\alpha h)^n\right]^{1-(1/n)}} \tag{1}$$

where θ_r and θ_s are the residual and saturated water contents, and α and n are the VGM model fitting parameters. The fitting of Equation (1) was performed in the SWRC Fit program [38]. The coefficient of determination (r^2) and root mean square error (RMSE) was used to assess the experimental data fit quality.

The equivalent pore diameter (Equation (2)) was determined based on the Young–Laplace equation [39]:

$$d = \frac{4\sigma\cos(\varphi)}{\rho_w g h} \cong \frac{298}{h} \tag{2}$$

where d is the equivalent pore diameter (μm), h is the pressure head (kPa), σ is the water surface tension, ρ_w is the water density, g is the acceleration of gravity, and φ is the water-pore contact angle.

The water content at saturation was measured after the samples had undergone the capillary rise saturation procedure. Water contents at the field capacity (θ_{fc}) and permanent wilting point (θ_{pwp}) were obtained with the samples submitted to pressure heads of 10 kPa and 1500 kPa, respectively [1]. The gravitational or free-soil water content (θ_{fw}), plant-available soil water (θ_{aw}), and critical plant-available water (θ_{cwc}) were calculated based on the following set of equations [40,41]:

$$\theta_{fw} = \theta_s - \theta_{fc} \tag{3}$$

$$\theta_{aw} = \theta_{fc} - \theta_{pwp} \tag{4}$$

$$\theta_{cwc} = 0.75(\theta_{fc}) \tag{5}$$

The free-soil water content and plant-available soil water are also sometimes known as bulk soil air capacity (AC) and plant-available water capacity (PAWC) [36,40].

2.4. Micromorphological Analysis

Before the impregnation procedure, the undisturbed samples were very carefully taken out of the volumetric rings by exerting pressure on one of their surfaces. Next, the samples were impregnated under vacuum using a non-saturated polyester resin (Crystic SR 17449) mixed with a styrene monomer. To facilitate image segmentation, fluorescent pigments (Unitex OB) were added to the resin. After the impregnated blocks had hardened (about two months), they were cut with special tools (diamond cut-off saw), and one of the sides was polished [25]. Finally, a block of each sample with a thickness of approximately 1 cm and an area of approximately 4.5×3.0 cm^2 was extracted from the impregnated blocks.

Digital images were acquired using a CCD camera with 1024×768-pixel resolution coupled to a petrographic microscope with the optical lens at ten ($\times 10$) times magnification. Blacklight lamps illuminated the blocks during image acquisition [42]. The two-dimensional (2D) images were processed and analyzed using Noesis-Visilog® 5.4 software. Each 2D image was related to an area of approximately 1.4×1.1 cm^2. Three images were obtained for each impregnated block. The imaged porosity (P) was determined by dividing the total area occupied by the voids (pores) by the total area of the image (ROI—region of interest) [43].

The areas occupied by the pores in the 2D images were also characterized based on their shapes and sizes. The pores were classified into the following shapes: rounded, elongated, and complex [22]. Two indices (Equations (6) and (7)) were employed to classify the pores in terms of shape [25]. Concerning size, the pores were grouped into the following classes: 20–50, 50–100, 100–200, 200–300, 300–400, 400–500, 500–1000, and >1000 µm:

$$\Gamma_1 = \frac{Pe^2}{4\pi A} \tag{6}$$

$$\Gamma_2 = \frac{(1/z)\sum_i (N_I)_i}{(1/v)\sum_j (D_F)_j} \tag{7}$$

where Pe and A represent the perimeter and the area of the pore, N_I is the number of intercepts of the object in the direction i ($i = 0°, 45°, 90°$, and $135°$), D_F is the diameter of Feret of an object in the direction j ($j = 0°$ and $90°$), and z and v are the number of i and j directions, respectively.

Table 1 shows the values of indices Γ_1 and Γ_2, which were used to discriminate the pores according to their shape.

Table 1. Classification of pores according to shapes based on indices Γ_1 (Equation (6)) and Γ_2 (Equation (7)).

Pore Shapes	$\Gamma1$	$\Gamma2$
Rounded (Round)	≤ 5	-
Elongated (Elon)	$5 < \Gamma1 \leq 25$	≤ 2.2
Irregular (Irr)	$5 < \Gamma1 \leq 25$ or >25	>2.2

The pores that were classified according to shape were also classified as small (0.00016 to c. 0.016 mm^2), medium (>0.016 to c. 0.16 mm^2), and large (>0.16 mm^2).

2.5. Statistical Analysis

The variance analysis statistical model was applied to compare the treatments (samples submitted to W-D cycles). Assumptions of residual normality and homoscedasticity were verified by the Shapiro–Wilk and Bartlett tests and the F-test was employed. Mean values were compared using Student's *t*-test ($p < 0.05$). All the statistical data processing was performed using the PAST software (version 3.20) [44].

3. Results

3.1. Water Retention Measurements

The soil water retention and air-filled porosity curves of both soils are illustrated in Figure 1. The VGM model was a good fit for the SWRC data [37]. The coefficient of determination and root mean square error was used to verify the fitting data quality. The lowest r^2 was 0.94 (Rhodic Hapludox), while the highest was 0.98 (Typic Hapludox). The root mean square error presented minimum and maximum values of 0.005 and 0.014, respectively (Typic Hapludox).

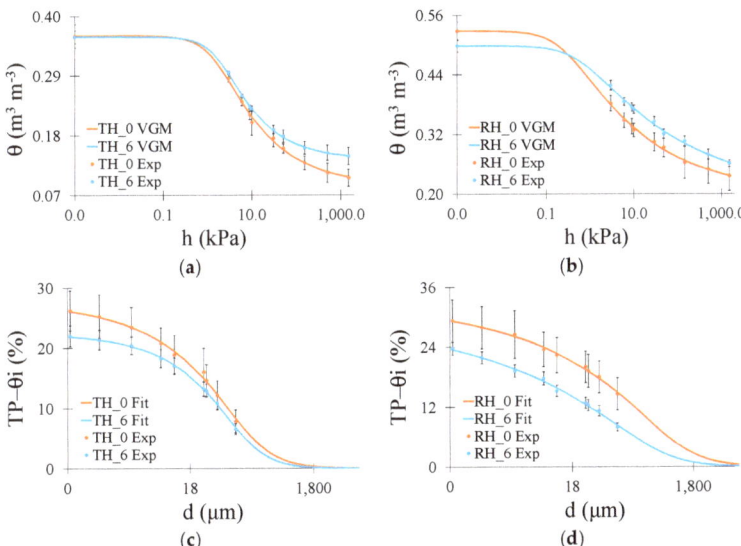

Figure 1. (**a**) Soil water retention curve (SWRC) obtained for Typic Hapludox (TH); (**b**) SWRC for Rhodic Hapludox (RH); (**c**) Air-filled porosity (TP–θi) for Typic Hapludox; (**d**) TP–θi for Rhodic Hapludox. Numbers 0 and 6 indicate that the soil samples were submitted to 0 and 6 wetting and drying (W-D) cycles. Bars are the standard deviations. VGM: van Genuchten–Mualem model. Exp: experimental data. Fit: fitting.

In the Typic Hapludox, the most remarkable differences occurred in the driest region (micropores) of the SWRC (Figure 1a). The W-D cycles increased θ at the permanent wilting point by c. 40%, which was also confirmed by a higher residual θ following the cycles (Table 2). The parameters n and α (VGM model) varied by only a minor extent ($p > 0.05$), with cycles indicating similarities in the SWRC shape and air-entry region (Table 2). For Rhodic Hapludox (Figure 1b), a higher θ was observed for the largest pore sizes for 0 W-D cycles. The saturation θ adjusted by the VGM model was reduced by 0.030 m^3 m^{-3} after the W-D cycles, which resulted in significant differences (Table 2), whereas $θ_{pwp}$ increased by c. 11% (no significant differences compared to 0 W-D cycles). The slight differences in n indicate similarities in the shape of SWRCs similar to the Typic Hapludox (Table 2). Concerning the α parameter (Table 2), related to the point at the largest pores where air could enter the soil [1,5], it decreased by c. 54% with 6 W-D cycles.

Table 2. van Genuchten–Mualem (VGM) mathematical model parameters [37] used to fit the soil water retention data for the two soils (Typic Hapludox and Rhodic Hapludox) subjected to 0 and 6 wetting and drying (W-D) cycles. Different lowercase letters indicate significant differences between W-D cycles (same soil) at $p < 0.05$.

VGM Parameters	Typic Hapludox		Rhodic Hapludox	
	0 W-D	6 W-D	0 W-D	6 W-D
$θ_s$ (m^3 m^{-3})	0.364	0.362	0.528 a	0.498 b
$θ_r$ (m^3 m^{-3})	0.087 a	0.136 b	0.195	0.190
α (kPa^{-1})	0.613	0.493	3.677 a	1.695 b
n	1.416	1.533	1.244	1.183
m [1]	0.294	0.348	0.196	0.155

[1] Parameter m was calculated as: m = 1 − (1/n) [37].

The air-filled porosity curves (Figure 1c,d) showed only slight differences in the region of the largest pores in both soils. However, as the soil dried, samples under 0 W-D began to exhibit greater volumes of air in the region from mesopores to micropores in the Rhodic Hapludox. In the Typic Hapludox, TP–θ varied by only a minor extent ($p > 0.05$) within the W-D cycles. For example, the application of W-D cycles decreased the air-filled porosity by c. 16% (Typic Hapludox) and c. 19% (Rhodic Hapludox) for the smallest pore size (0.2 μm) analyzed. Aiming to complement the results of SWRC and air-filled porosities, an analysis of the water retention for different pore sizes based on their functions is presented in Figure 2.

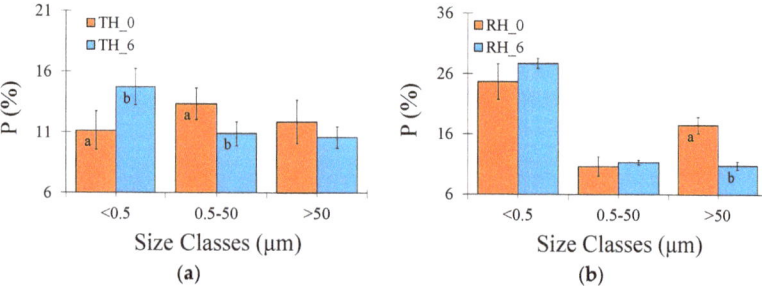

Figure 2. *Cont.*

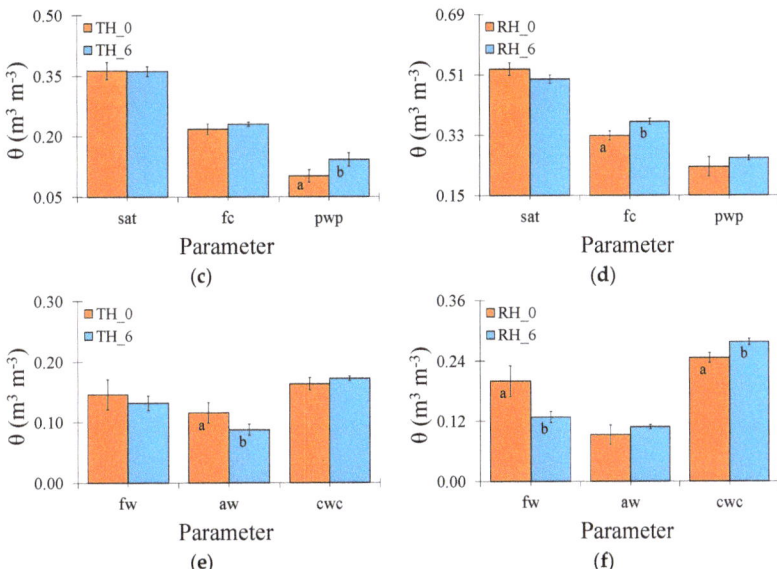

Figure 2. (**a**) Porosity (P) variation as a function of pore sizes in the Typic Hapludox (TH); (**b**) P variation as a function of pore sizes in the Rhodic Hapludox (RH); (**c**) Water content at saturation (sat), field capacity (fc), and permanent wilting point (pwp) in the Typic Hapludox; (**d**) Water content at saturation, field capacity, and permanent wilting point in the Rhodic Hapludox; (**e**) Free-soil water content (fw), plant-available soil water (aw), and critical plant-available water (cwc) in the Typic Hapludox; (**f**) Free-soil water content, plant-available soil water, and critical plant-available water in the Rhodic Hapludox. Numbers 0 and 6 indicate that the soil samples were submitted to 0 and 6 wetting and drying (W-D) cycles. Bars are the standard deviations. Different lowercase letters on the bars indicate a significant difference between W-D cycles (same soil) at $p < 0.05$.

The contribution of different pore sizes to the total porosity was obtained based on the classification suggested by Greenland [45]. According to that author, pores <0.5 μm are classified as residual or bonding pores, between 0.5 and 50 μm are storage pores, and >50 μm are transmission and macropores. In the Typic Hapludox, residual pores increased by c. 32% while the storage ones decreased by c. 19% after 6 W-D cycles (Figure 2a). Regarding θ at saturation, field capacity, and permanent wilting point, the latter increased by c. 40% after 6 W-D cycles with only minor differences for θ_{sat} and θ_{fc} ($p > 0.05$) (Figure 2c). The W-D cycle application decreased the plant-available water content by c. 24% (Figure 2e). In the Rhodic Hapludox, transmission pores decreased by c. 38% after 6 W-D cycles, while the other pore types varied by only a minor extent ($p > 0.05$) (Figure 2b). The analysis of θ for specific pressure heads showed differences for θ_{fc}, which increased by c. 13% (Figure 2d); θ_{fw} decreased by c. 36%, and θ_{cwc} increased by c. 13% after 6 W-D cycles (Figure 2f).

3.2. Micromorphological Soil Properties

The soil pore system was also studied using 2D image data with a micrometer resolution (Figure 3). This analysis was carried out to complement the measurements based on water retention.

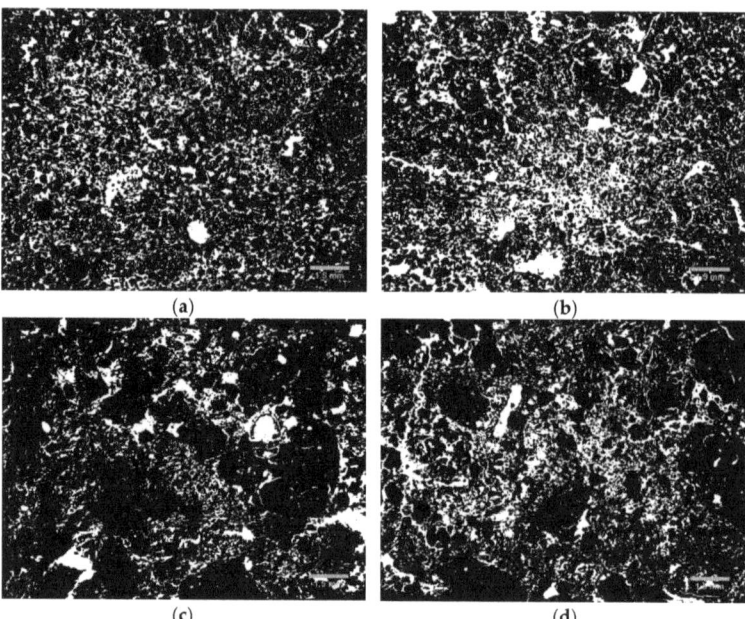

Figure 3. 2D binary images (pores appear in white and solid matrix in black) for: (**a**) Typic Hapludox and 0 wetting and drying cycles; (**b**) Typic Hapludox and 6 W-D cycles; (**c**) Rhodic Hapludox and 0 W-D cycles; (**d**) Rhodic Hapludox and 6 W-D cycles.

The 2D binary images show that both soils suffered changes in their soil pore system with 6 W-D cycles. In the Typic Hapludox (Figure 3a,b), sequences of W-D cycles seemed to increase the soil porosity, which is probably associated with the connection of small pores. Large and complex pores could be seen in this soil after 6 W-D cycles. The Rhodic Hapludox (Figure 3c,d) also indicated an increase in soil porosity with the cycles. The increment in the number of small pores, the appearance of medium-sized pores, and the connection of pores were evident after 6 W-D cycles in this soil. Thus, aiming to complement the qualitative image analysis, the result of the micromorphological analysis (pore size and shape distribution) of the Typic Hapludox is presented in Figure 4. It is worth pointing out that for these samples, porosity was obtained by image analysis, i.e., named here as imaged porosity [9,10].

Concerning the pore shape and size in the Typic Hapludox (Figure 4a,b), small and medium rounded pores decreased by c. 18% and c. 23%, medium elongated pores decreased by c. 27%, and irregular-shaped pores increased by c. 59% after 6 W-D cycles. Considering all these pore types and sizes, the imaged porosity had an increment of c. 14% following the W-D cycles. Regarding pore size distribution and shape (Figure 4c,d), rounded pores decreased by c. 20% (size intervals from 20 to 200 µm) and c. 28% (200–300 µm), elongated pores decreased by c. 21% (50–100 µm), and irregular-shaped pores increased by c. 96% (>1000 µm) after 6 W-D cycles. When the total number of pores (TNP) was analyzed (Figure 4e,f), only the rounded-shaped pores comprising between 20 and 400 µm showed significant differences ($p < 0.05$). This pore type decreased by c. 17% (size intervals from 20 to 300 µm) and c. 40% (300–400 µm) after 6 W-D cycles, respectively. When all the pore types were analyzed, TNP was seen to decrease by c. 18% after the application of W-D cycles.

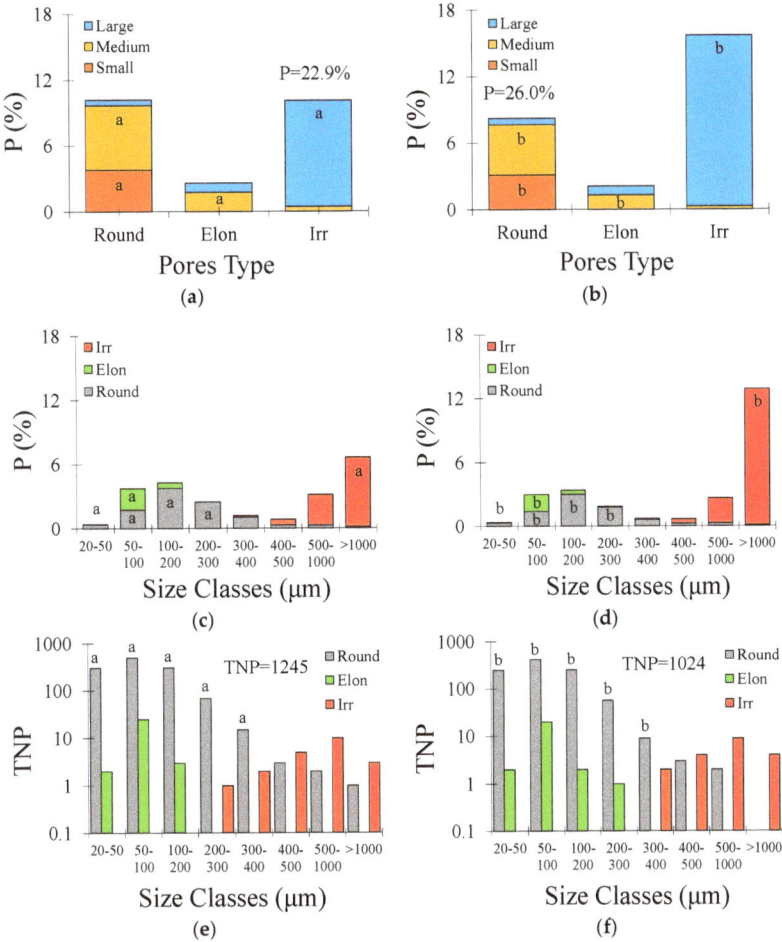

Figure 4. (**a**) Contribution of different pore shapes and sizes to porosity (P) in the Typic Hapludox submitted to 0 wetting and drying (W-D) cycles; (**b**) Contribution of different pore shapes and sizes to P after 6 W-D; (**c**) Pore size distribution as a function of shape and size after 0 W-D; (**d**) Pore size distribution as a function of shape and size after 6 W-D; (**e**) Total number of pores (TNP) as a function of shape and size after 0 W-D; (**f**) TNP as a function of shape and size after 6 W-D. Round: rounded-shaped pores; Elon: elongated-shaped pores; Irr: irregular-shaped pores. Different lowercase letters indicate significant differences between W-D cycles at *p* < 0.05.

The effect of W-D cycles applied on the pore size and shape distribution of the Rhodic Hapludox samples is shown in Figure 5.

The pore shape and size analyses (Figure 5a,b) showed that only the small rounded and medium elongated pores exhibited significant differences (*p* < 0.05) between 0 W-D and 6 W-D in the Rhodic Hapludox. The former increased by c. 54% while the latter increased by c. 90% after 6 W-D cycles, respectively. The other pore types varied by only a minor extent (*p* > 0.05) following the W-D cycles. Similar to the findings of the Typic Hapludox, imaged porosity had an increment of c. 23% after 6 W-D cycles. The distribution of pores based on their size and shape (Figure 5c,d) exhibited differences for some rounded pores (increases of c. 52%—50–100 μm and c. 46%—100–200 μm), elongated pores (increase of c. 71%—50–100 μm), and irregular-shaped pores (increase of c. 2.3 times—500–1000 μm). The total number of pores showed differences (*p* < 0.05) only for rounded pores up to 300 μm and elongated pores between 50 and 100 μm (Figure 5e,f). Unlike the Typic Hapludox, the Rhodic Hapludox had an increment of c. 53% in TNP.

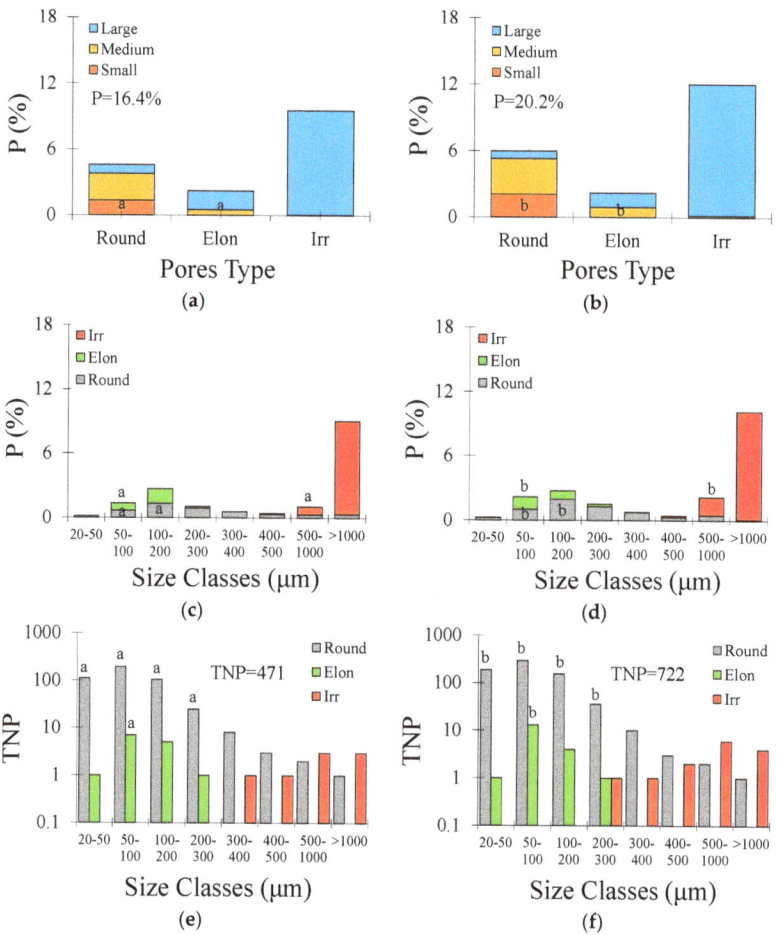

Figure 5. (**a**) Contribution of different pore shapes and sizes to porosity (P) in the Rhodic Hapludox submitted to 0 wetting and drying (W-D) cycles; (**b**) Contribution of different pore shapes and sizes to P after 6 W-D; (**c**) Pore size distribution as a function of shape and size after 0 W-D; (**d**) Pore size distribution as a function of shape and size after 6 W-D; (**e**) Total number of pores (TNP) as a function of shape and size after 0 W-D; (**f**) TNP as a function of shape and size after 6 W-D. Round: rounded-shaped pores; Elon: elongated-shaped pores; Irr: irregular-shaped pores. Different lowercase letters indicate significant differences between W-D cycles at $p < 0.05$.

4. Discussion

This study aimed to evaluate how soils under natural conditions change when submitted to W-D cycles based on the idea that a few cycles provoke modifications in their micromorphological and water retention properties. The water retention and air-filled porosity curves (Figure 1) presented a similar behavior between the soils for the driest region (micropores). In the Typic Hapludox (Figure 1a,c), an increase in the contribution of smaller pores to soil porosity explained the results obtained. The detachment of sand particles (more weakly bound to other particles) from small aggregates under continuous wetting might cause the appearance of smaller pores resulting in a large amount of water retained in high-pressure heads [46]. The migration of soil sediments under W-D cycles had been reported by Pires et al. [47], indicating particle detachment from aggregates. On the other hand, the presence of organic materials at the topsoil, which enhances soil aggregation, also helped to keep a more stable structure under W-D cycles, as verified here by SWRC shape similarities [48]. Pires et al. [22], working with both tilled and untilled soils, demonstrated that even sandy soils were subject to changes in the textural and structural pore regions under continuous W-D cycles [49,50]. In another study [51], the same authors reported increases in the water retained at high-pressure heads, corroborating with the results presented here. The Rhodic Hapludox (Figure 1b,d) showed a similar behavior to that of Typic Hapludox for high-pressure heads, but with differences in θ in the region of mesopores (30–100 µm) and micropores (<30 µm) [6]. Despite differences in textural soil compositions and, consequently, in mineralogy (not evaluated in this study) [52], the water retention and air-filled porosity curves showed similar behavior between the soils. Lu et al. [53] reported that variations in the amount of clay affect the distribution of pores, mainly the smallest ones, explaining the differences between the soils (Table 2). In addition, the presence of organic materials (barks, leaves, etc.) in the secondary forest area was expected to affect water retention, as demonstrated by Kodešová et al. [54] and Cooper et al. [55] in the Rhodic Hapludox. Due to the high amount of clay, a more stable structure was expected in this soil under W-D cycles. However, as the samples were confined in cylinders, the creation of larger pores (mainly in the Rhodic Hapludox) after W-D cycles might press the soil against the cylinder walls contributing to the appearance of smaller pores, as observed here [56]. Peng et al. [57] demonstrated that W-D cycles affect larger pores increasing the total porosity, whereas Leij et al. [58] found that the cycles reduce structural porosity due to the coalescence of soil aggregates. In the Rhodic Hapludox, the soil structure changes observed caused a reduction in the aeration capacity after W-D cycles, which was influenced by variations in the regions of small to medium pore sizes.

The changes verified in the SWRC following W-D cycles increased the contribution of residual and bonding pores to the total porosity and decreased the importance of storage pores in the Typic Hapludox [1,5] (Figure 2a). This finding meant that the sandy soil maintained a good water transmission capacity even after W-D cycles. According to Greenland [45], fine pores (<0.5 µm) are responsible for the retention and diffusion of ions and for supporting major forces between the soil particles. Pores with sizes between 0.5 and 50 µm play a vital role in the retention of water against gravity and release, while pores >50 µm act in the air movement and drainage of excess water. The increase in the frequency of residual and bonding pores could be directly related to a higher θ_{pwp} after 6 W-D cycles verified in the Typic Hapludox [41] (Figure 2c). As a consequence, θ_{aw} decreased after 6 W-D cycles (Figure 2e); this could be explained by the differences observed in θ_{pwp} and θ_{fc}. Nonetheless, the results of θ_{fw} and θ_{aw} indicated adequate soil aeration capacity before and after W-D cycles, as pointed out by Reynolds [2,36]. However, θ_{aw} was close to the limits defined as droughty for root development, though the reference value of <0.10 m^3 m^{-3} (poor soil capacity to store and provide water to plants) might not be adequate for sandy soils [59]. The application of W-D cycles decreased the contribution of transmission pores to the total porosity in the Rhodic Hapludox (Figure 2b). This reduction might affect air movement and soil water infiltration, as demonstrated by Cooper et al. [55]. The water content at field capacity increased with the application of 6 W-D cycles (Figure 2d), which was mainly associated with the contribution of residual and storage pores, as indicated by Tarawally et al. [41]. The free water content decreased with 6 W-D cycles (Figure 2f), which was mainly influenced by the reduction in transmission pores, indicating a decrease in the soil aeration capacity after W-D cycles. However, θ_{fw} measured is still considered adequate for root development [2,36]. The critical plant-available water followed the same trend of θ_{fc}, which could be explained by variations in storage and residual pores after 6 W-D cycles.

The resin impregnated in the soil blocks was employed to complement the results of the SWRC data. However, it is important to mention that the 2D micromorphological analysis (area-based analysis) was not performed on the same samples of SWRCs (volume-based analysis). In the Typic Hapludox, the most important changes in pore distribution occurred for the rounded (small and medium), elongated (medium), and irregular (large) shaped pores after 6 W-D cycles (Figure 4a,b). Rounded pores, associated with soil air trapping following drying and the influence of biological activity (grass roots and soil fauna), created chambers and channels [60] and decreased their contribution to the imaged porosity with W-D cycles (Figure 3a,b). In addition, a decrease in the number of these pore types was noticed after 6 W-D cycles, which were probably converted into large irregular-shaped pores [61,62] (Figure 4c–f). The increase in the contribution of complex pores (irregular) to the imaged porosity might have resulted from the connection of small pores, as the number of rounded pores (from 20 to 400 μm) was reduced after 6 W-D cycles [61] (Figure 4e,f). Pardini et al. [17] reported that W-D cycles produce fissures, increasing the contribution of large pores to porosity, similar to the findings observed here (Figure 3b). These interconnected macropores (drainage pores), usually found in tropical soils, play an important role in water dynamics [8,21,63]. The Rhodic Hapludox exhibited differences only for small rounded and medium elongated-shaped pores with a negligent influence of the latter to imaged porosity after 6 W-D cycles (Figures 3c,d and 5a,b). This result is mainly associated with the increased contribution of pore sizes between 50 and 200 μm to imaged porosity after W-D cycles (Figure 5c,d). In addition, an increase in the number of rounded-shaped pores from 20 to 300 μm and elongated-shaped pores between 50 and 100 μm also helped to explain the increment of the contribution of these pore types to imaged porosity [64,65] (Figure 5e,f). The increased contribution of elongated-shaped pores to imaged porosity following W-D cycles could probably be associated with the appearance of fine fissures, while that of large (500–1000 μm) irregular-shaped pores was due to a slight increase in the number of these pore types [16,19,60].

Finally, the imaged porosity allowed to classify both soils (Figures 4a,b and 5a,b) as moderately porous (10–25%) to highly porous (25–40%), indicating suitable soil structures to water infiltration [66]. This kind of finding has been reported by many other authors [67,68]. Pires et al. [22] found increases in porosity for tropical tilled and untilled soils, especially after 9 W-D cycles. Hussein and Adey [14] showed similar results after 4 W-D cycles, employing capillarity as one of the wetting procedures. However, imaged porosity results contradicted those of SWRC (Figure 1a,b). One possible explanation for these contradictory results is that micromorphological analysis allowed to access only pores with sizes >20 μm. In both soils, residual (<0.5 μm) and storage (0.5–50 μm) pores showed a considerable contribution to the total porosity (Figure 2a,b). The great contribution of larger pores (>500 μm) to imaged porosity can also be associated with an adequate soil structure to plant root development and water drainage (Figures 4c,d and 5c,d). However, when the influence of these large pores exceeded 70–80% of the soil porosity, it indicates poor soil structure [60,69]. In the Typic Hapludox, an increase in the contribution of larger pores to the imaged porosity after 6 W-D cycles were observed, whereas in the Rhodic Hapludox, the contribution remained stable, but these numbers were <60%.

5. Conclusions

The results obtained in this study have implications for understanding the way that soils under natural conditions (secondary forest and grassland) can change under continuous wetting and drying cycles. Water retention was influenced by W-D cycles, mainly in the region of micropores (Typic Hapludox) and mesopores to micropores (Rhodic Hapludox), with an increase in the soil water holding capacity following the cycles. As a consequence of these changes, the contribution of residual (increase) and storage (decrease) pores to total porosity was affected by W-D cycles in the Typic Hapludox, while in the Rhodic Hapludox, only transmission pores (decrease) were influenced. Water content at the permanent wilting point increased after 6 W-D cycles in the Typic Hapludox, while the water available to plants decreased. In the Rhodic Hapludox, the water retained at field capacity increased after 6 W-D, affecting the free-soil water content (decrease) and critical plant-available water (increase). These results highlight that even under natural conditions, the soil water retention capacity was modified by sequences of W-D cycles. The main implications of the variations observed in the water retained at different pore sizes concerned changes in the amount of water available to the plants (Typic Hapludox) and soil aeration (Rhodic Hapludox). However, even after 6 W-D cycles, both soils kept soil structures that were adequate for plant root development and were capable of retaining reasonable amounts of water available to plants based on the water retention parameters measured.

The pore size and shape distributions based on 2D images presented significant alterations in the rounded (small and medium), elongated (medium), and irregular (large) shaped pores in the Typic Hapludox after 6 W-D cycles. In the Rhodic Hapludox, only the small rounded and medium elongated pores were modified significantly by the cycles. However, the large (>500 μm) irregular-shaped pores increased their contribution to porosity after 6 W-D while also contributing to an increase in the imaged porosity in the Rhodic Hapludox and Typic Hapludox. This finding indicates that both soils improved their aeration capacity and water flux after the cycles. In the Typic Hapludox, rounded-shaped pores (20–300 μm) decreased their contribution to imaged porosity, followed by a decrease in the number of these pore types. On the other hand, the opposite was found in the Rhodic Hapludox, with an increase in the contribution of rounded pores (50–200 μm) to imaged porosity followed by an increase in the number of these pore types (50–300 μm). Overall, the two Oxisols were characterized by significant contributions of rounded pores to imaged porosity (before and after the cycles), indicating intense biological activity. Finally, it seems relevant to mention that the micromorphological properties of the soils under natural conditions were changed by W-D cycles with impacts on the distribution of pore sizes and shapes responsible for water retention and movement.

Funding: This research was partially funded by the Brazilian National Council for Scientific and Technological Development (CNPq) (Grant 304925/2019-5).

Data Availability Statement: All data are available upon reasonable request to lfpires@uepg.br.

Acknowledgments: The author is thankful to "Laboratório de Análise Morfológica do Solo da Universidade de São Paulo (USP/ESALQ)" (Soil Morphological Analysis Laboratory of the University of São Paulo) for the infrastructure related to the micromorphological analysis.

Conflicts of Interest: The authors declare no conflict of interest.

References

1. Lal, R.; Shukla, M.K. *Principles of Soil Physics*; Marcel Dekker, Inc.: New York, NY, USA, 2004.
2. Reynolds, W.D.; Bowman, B.T.; Drury, C.F.; Tan, C.S.; Lu, X. Indicators of Good Soil Physical Quality: Density and Storage Parameters. *Geoderma* **2002**, *110*, 131–146. [CrossRef]
3. Correchel, V.; Bacchi, O.O.S.; De Maria, I.C.; Dechen, S.C.F.; Reichardt, K. Erosion Rates Evaluated by the [137]Cs Technique and Direct Measurements on Long-Term Runoff Plots under Tropical Conditions. *Soil Tillage Res.* **2006**, *86*, 199–208. [CrossRef]
4. Tracy, S.R.; Black, C.R.; Roberts, J.A.; Sturrock, C.; Mairhofer, S.; Craigon, J.; Mooney, S.J. Quantifying the Impact of Soil Compaction on Root System Architecture in Tomato (*Solanum lycopersicum*) by X-Ray Micro-Computed Tomography. *Ann. Bot.* **2012**, *110*, 511–519. [CrossRef] [PubMed]
5. Hillel, D. *Environmental Soil Physics*; Academic Press: San Diego, CA, USA, 1998.
6. Reichardt, K.; Timm, L.C. *Soil, Plant and Atmosphere: Concepts, Processes and Applications*; Springer Nature: Cham, Switzerland, 2020.
7. Nimmo, J.R. *Porosity and Pore Size Distribution*; Reference Module in Earth Systems and Environmental Sciences; Elsevier: Amsterdam, The Netherlands, 2013.
8. Juhász, C.E.P.; Cooper, M.; Cursi, P.R.; Ketzer, A.O.; Toma, R.S. Savanna Woodland and Soil Micromorphology Related to Water Retention. *Sci. Agric.* **2007**, *64*, 344–354. [CrossRef]
9. Bouma, J.; Jongerius, A.; Boersma, O.; Jager, A.; Schoonderbeek, D. The Function of Different Types of Macropores During Saturated Flow through Four Swelling Soil Horizons. *Soil Sci. Soc. Am. J.* **1977**, *41*, 945–950. [CrossRef]
10. Fox, D.M.; Bryan, R.B.; Fox, C.A. Changes in Pore Characteristics with Depth for Structural Crusts. *Geoderma* **2004**, *120*, 109–120. [CrossRef]
11. De Pierri Castilho, S.C.; Cooper, M.; Simões da Silva, L.F. Micromorphometric Analysis of Porosity Changes in the Surface Crusts of Three Soil in the Piracicaba Region, São Paulo State, Brazil. *Acta Sci. Agron.* **2015**, *37*, 385–395. [CrossRef]

12. Cooper, M.; Vidal-Torrado, P. Caracterização Morfológica, Micromorfológica e Físico-Hídrica de Solos com Horizonte B Nítico. *Rev. Bras. Ciênc. Solo* **2005**, *29*, 581–595. [CrossRef]
13. Pagliai, M.; Marsili, A.; Servadio, P.; Vignozzi, N.; Pellegrini, S. Changes in Some Physical Properties of a Clay Soil in Central Italy Following the Passage of Rubber Tracked and Wheeled Tractors of Medium Power. *Soil Tillage Res.* **2003**, *73*, 119–129. [CrossRef]
14. Hussein, J.; Adey, M.A. Changes in Microstructure, Voids and b-Fabric of Surface Samples of a Vertisol Caused by Wet/Dry Cycles. *Geoderma* **1998**, *85*, 63–82. [CrossRef]
15. Tang, C.S.; Cui, Y.J.; Shi, B.; Tang, A.M.; Liu, C. Desiccation and Cracking Behaviour of Clay Layer from Slurry State under Wetting-Drying Cycles. *Geoderma* **2011**, *166*, 111–118. [CrossRef]
16. An, R.; Zhang, X.; Kong, L.; Liu, X.; Chen, C. Drying-Wetting Impacts on Granite Residual Soil: A Multi-Scale Study from Macroscopic to Microscopy Investigations. *Bull. Eng. Geol.* **2022**, *81*, 447. [CrossRef]
17. Pardini, G.; Vigna Guidi, G.; Pini, R.; Regüés, D.; Gallart, F. Structure and Porosity of Smectitic Mudrocks as Affected by Experimental Wetting—Drying Cycles and Freezing—Thawing Cycles. *Catena* **1996**, *27*, 149–165. [CrossRef]
18. Xia, J.; Zhang, L.; Ge, P.; Lu, X.; Wei, Y.; Cai, C.; Wang, J. Structure Degradation Induced by Wetting and Drying Cycles for the Hilly Granitic Soils in Collapsing Gully Erosion Areas. *Forests* **2022**, *13*, 1426. [CrossRef]
19. Louati, F.; Trabelsi, H.; Jamei, M.; Taibi, S. Impact of Wetting-Drying Cycles and Cracks on the Permeability of Compacted Clayey Soil. *Eur. J. Environ. Civ.* **2018**, *25*, 696–721. [CrossRef]
20. Ng, C.W.W.; Peprah-Manu, D. Pore Structure Effects on the Water Retention Behaviour of a Compacted Silty Sand Soil Subjected to Drying-Wetting Cycles. *Eng. Geol.* **2023**, *313*, 106963. [CrossRef]
21. Momoli, R.S.; Cooper, M.; de Pierri Castilho, S.C. Sediment Morphology and Distribution in a Restored Riparian Forest. *Sci. Agric.* **2007**, *64*, 486–494. [CrossRef]
22. Pires, L.F.; Cooper, M.; Cássaro, F.A.M.; Reichardt, K.; Bacchi, O.O.S.; Dias, N.M.P. Micromorphological Analysis to Characterize Structure Modifications of Soil Samples Submitted to Wetting and Drying Cycles. *Catena* **2008**, *72*, 297–304. [CrossRef]
23. Lima, H.V.; Silva, A.P.; Santos, M.C.; Cooper, M.; Romero, R.E. Micromorphology and Image Analysis of a Hardsetting Ultisol (Argissolo) in the State of Ceará (Brazil). *Geoderma* **2006**, *132*, 416–426. [CrossRef]
24. Lipiec, J.; Walczak, R.; Witkowska-Walczak, B.; Nosalewicz, A.; Słowińska-Jurkiewicz, A.; Sławiński, C. The Effect of Aggregate Size on Water Retention and Pore Structure of Two Silt Loam Soils of Different Genesis. *Soil Tillage Res.* **2007**, *97*, 239–246. [CrossRef]
25. Cooper, M.; Vidal-Torrado, P.; Chaplot, V. Origin of Microaggregates in Soils with Ferralic Horizons. *Sci. Agric.* **2005**, *62*, 256–263. [CrossRef]
26. Hobson, D.; Harty, M.; Tracy, S.R.; McDonnell, K. The Effect of Tillage Depth and Traffic Management on Soil Properties and Root Development during Two Growth Stages of Winter Wheat (*Triticum aestivum* L.). *Soil* **2022**, *8*, 391–408. [CrossRef]
27. Dal Ferro, N.; Sartori, L.; Simonetti, G.; Berti, A.; Morari, F. Soil Macro- and Microstructure as Affected by Different Tillage Systems and their Effects on Maize Root Growth. *Soil Tillage Res.* **2014**, *140*, 55–65. [CrossRef]
28. Budhathoki, S.; Lamba, J.; Srivastava, P.; Williams, C.; Arriaga, F.; Karthikeyan, K.G. Impact of Land Use and Tillage Practice on Soil Macropore Characteristics Inferred from X-ray Computed Tomography. *Catena* **2022**, *210*, 105886. [CrossRef]
29. Helliwell, J.R.; Sturrock, C.J.; Grayling, K.M.; Tracy, S.R.; Flavel, R.J.; Young, I.M.; Whalley, W.R.; Mooney, S.J. Applications of X-ray Computed Tomography for Examining Biophysical Interactions and Structural Development in Soil Systems: A Review. *Eur. J. Soil Sci.* **2013**, *64*, 279–297. [CrossRef]
30. Kravchenko, A.N.; Negassa, W.C.; Guber, A.K.; Rivers, M.L. Protection of Soil Carbon within Macro-Aggregates depends on Intra-Aggregate Pore Characteristics. *Sci. Rep.* **2015**, *5*, 16261. [CrossRef]
31. Koestel, J.; Schlüter, S. Quantification of the Structure Evolution in a Garden Soil over the Course of Two Years. *Geoderma* **2019**, *338*, 597–609. [CrossRef]
32. Soil Survey Staff. *Simplified Guide to Soil Taxonomy*; USDA Natural Resources Conservation Service, National Soil Survey Center: Lincoln, NE, USA, 2013.
33. Booman, G.; Leiker, S. *Soil Sampling Guide*; Document ID: RND_SSG_001; Regen Network Development, Inc.: Northfield, MA, USA, 2021.
34. Klute, A. Water Retention: Laboratory Methods. In *Methods of Soil Analysis. Part 1: Physical and Mineralogical Methods*; Black, C.A., Ed.; Soil Science Society of America: Madison, WA, USA, 1986; pp. 635–662.
35. Dane, J.H.; Hopmans, J.W. Pressure Plate Extractor. In *Methods of Soil Analysis. Part 4: Physical Methods*; Dane, J.H., Topp, G.C., Eds.; Soil Science Society of America: Madison, WA, USA, 2002; pp. 688–690.
36. Reynolds, W.D.; Drury, C.F.; Yang, X.M.; Tan, C.S. Optimal Soil Physical Quality Inferred Through Structural Regression and Parameter Interactions. *Geoderma* **2008**, *146*, 466–474. [CrossRef]
37. Van Genuchten, M.T. A Closed-Form Equation for Predicting the Hydraulic Conductivity of Unsaturated Soils. *Soil Sci. Soc. Am. J.* **1980**, *44*, 892–898. [CrossRef]
38. Seki, K. SWRC Fit—A Nonlinear Fitting Program with a Water Retention Curve for Soils Having Unimodal and Bimodal Pore Structure. *Hydrol. Earth Syst. Sci. Discuss.* **2007**, *4*, 407–437.
39. Jury, W.A.; Horton, R. *Soil Physics*; Willey: New Jersey, NJ, USA, 2004.
40. Reynolds, W.D.; Drury, C.F.; Tan, C.S.; Fox, C.A.; Yang, X.M. Use of Indicators and Pore Volume-Function Characteristics to Quantify Soil Physical Quality. *Geoderma* **2009**, *152*, 252–263. [CrossRef]

41. Tarawally, M.A.; Medina, H.; Frómeta, M.E.; Alberto Itza, C. Field Compaction at Different Soil-Water Status: Effects on Pore Size Distribution and Soil Water Characteristics of a Rhodic Ferralsol in Western Cuba. *Soil Tillage Res.* **2004**, *76*, 95–103. [CrossRef]
42. Ringrose-Voase, A.J.; Bullock, P. The Automatic Recognition and Measurement of Soil Pore Types by Image Analysis and Computer Programs. *J. Soil Sci.* **1984**, *35*, 673–684. [CrossRef]
43. Huf dos Reis, A.M.; Armindo, R.A.; Pires, L.F. Physical Assessment of a Haplohumox Soil Under Integrated Crop-Livestock System. *Soil Tillage Res.* **2019**, *194*, 104294. [CrossRef]
44. Hammer, Ø.; Harper, D.A.T.; Ryan, P.D. PAST: Paleontological Statistics Software Package for Education and Data Analysis. *Palaeont. Elect.* **2001**, *4*, 1–9.
45. Greenland, D.J. Soil Damage by Intensive Arable Cultivation: Temporary or Permanent? *Philos. Trans. R. Soc. B* **1977**, *281*, 193–208.
46. Cássaro, F.A.M.; Pires, L.F.; dos Santos, R.A.; Gimenez, D.; Reichardt, K. Funil de Haines Modificado: Curvas de Retenção de Solos Próximos à Saturação. *Rev. Bras. Ciênc. Solo* **2008**, *32*, 2555–2562. [CrossRef]
47. Pires, L.F.; Villanueva, F.C.A.; Dias, N.M.P.; Bacchi, O.O.S.; Reichardt, K. Chemical Migration During Soil Water Retention Curve Evaluation. *Annu. Acad. Bras. Cienc.* **2011**, *83*, 1097–1107. [CrossRef]
48. Dapla, P.; Hriník, D.; Hrabovský, A.; Simkovic, I.; Zarnovican, H.; Sekucia, F.; Kollár, J. The Impact of Land-Use on the Hierarchical Pore Size Distribution and Water Retention Properties in Loamy Soils. *Water* **2020**, *12*, 339.
49. Kutílek, M.; Jendele, L.; Panayiotopoulos, K.P. The Influence of Uniaxial Compression upon Pore Size Distribution in Bi-Modal Soils. *Soil Tillage Res.* **2006**, *86*, 27–37. [CrossRef]
50. Bodner, G.; Scholl, P.; Kaul, H.-P. Field Quantification of Wetting–Drying Cycles to Predict Temporal Changes of Soil Pore Size Distribution. *Soil Tillage Res.* **2013**, *133*, 1–9. [CrossRef] [PubMed]
51. Pires, L.F.; Bacchi, O.O.S.; Reichardt, K. Assessment of Soil Structure Repair due to Wetting and Drying Cycles through 2D Tomographic Image Analysis. *Soil Tillage Res.* **2007**, *94*, 537–545. [CrossRef]
52. Testoni, S.A.; de Almeida, J.A.; da Silva, L.; Pugliese Andrade, G.R. Clay Mineralogy of Brazilian Oxisols with Shrinkage Properties. *Rev. Bras. Ciênc. Solo* **2017**, *41*, e0160487. [CrossRef]
53. Lu, S.-G.; Malik, Z.; Chen, D.-P.; Wu, C.-F. Porosity and Pore Size Distribution of Ultisols and Correlations to Soil Iron Oxides. *Catena* **2014**, *123*, 79–87. [CrossRef]
54. Kodešová, R.; Pavlů, L.; Kodeš, V.; Žigová, A.; Nikodem, A. Impact of Spruce Forest and Grass Vegetation Cover on Soil Micromorphology and Hydraulic Properties of Organic Matter Horizon. *Biologia* **2007**, *62*, 565–568. [CrossRef]
55. Cooper, M.; Dalla Rosa, J.; Medeiros, J.C.; de Oliveira, T.C.; Toma, R.S.; Juhász, C.E.P. Hydro-Physical Characterization of Soils under Tropical Semi-Deciduous Forest. *Sci. Agric.* **2012**, *69*, 152–159. [CrossRef]
56. Tang, C.-S.; Cheng, Q.; Gong, X.; Shi, B.; Inyang, H.I. Investigation on Microstructure Evolution of Clayey Soils: A Review Focusing on Wetting/Drying Process. *J. Rock Mech. Geotech. Eng.* **2023**, *15*, 269–284. [CrossRef]
57. Peng, X.; Horn, R.; Smucker, A. Pore Shrinkage Dependency of Inorganic and Organic Soils on Wetting and Drying Cycles. *Soil Sci. Soc. Am. J.* **2007**, *71*, 1095–1104. [CrossRef]
58. Leij, F.J.; Ghezzehei, T.A.; Or, D. Modeling the Dynamics of the Soil Pore-Size Distribution. *Soil Tillage Res.* **2002**, *64*, 61–78. [CrossRef]
59. Hall, D.G.M.; Reeve, M.J.; Thomasson, A.J.; Wright, V.F. *Water Retention, Porosity and Density of Field Soils*; Soil Survey Technical Monograph; Rothamsted: Harpenden, UK, 1977; Volume 9.
60. Pagliai, M.; La Marca, M.; Lucamante, G. Micromorphometric and Micromorphological Investigations of a Clay Loam Soil in Viticulture under Zero and Conventional Tillage. *J. Soil Sci.* **1983**, *34*, 391–403. [CrossRef]
61. Ringrose-Voase, A.J. Measurements of Soil Macropore Geometry by Image Analysis of Sections Through Impregnated Soil. *Plant Soil* **1996**, *183*, 27–47. [CrossRef]
62. Cooper, M.; Medeiros, J.C.; Dalla Rosa, J.; Soria, J.E.; Toma, R.S. Soil Functioning in a Toposequence under Rainforest in São Paulo, Brazil. *Rev. Bras. Ciênc. Solo* **2013**, *37*, 392–399. [CrossRef]
63. Huf dos Reis, A.; Auler, A.C.; Armindo, R.A.; Cooper, M.; Pires, L.F. Micromorphological Analysis of Soil Porosity under Integrated Crop-Livestock Management Systems. *Soil Tillage Res.* **2021**, *205*, 104783. [CrossRef]
64. Wen, T.; Chen, X.; Shao, L. Effect of Multiple Wetting and Drying Cycles on the Macropore Structure of Granite Residual Soil. *J. Hydrol.* **2022**, *614*, 128583. [CrossRef]
65. Diel, J.; Vogel, H.J.; Schlüter, S. Impact of Wetting and Drying Cycles on Soil Structure Dynamics. *Geoderma* **2019**, *345*, 63–71. [CrossRef]
66. Pagliai, M. Soil porosity aspects. *Int. Agrophys.* **1988**, *4*, 215–232.
67. Sartori, G.; Ferrari, G.A.; Pagliai, M. Changes in Soil Porosity and Surface Shrinkage in a Remolded, Saline Clay Soil Treated with Compost. *Soil Sci.* **1985**, *139*, 523–530. [CrossRef]
68. Pagliai, M.; La Marca, M.; Lucamante, G. Changes in Soil Porosity in Remoulded Soils Treated with Poultry Manure. *Soil Sci.* **1987**, *144*, 128–140. [CrossRef]
69. Pagliai, M.; Guidi, G.; La Marca, M.; Giachetti, M.; Lucamante, G. Effect of sewage sludges and composts on soil porosity and aggregation. *J. Environ. Qual.* **1981**, *10*, 556–561. [CrossRef]

soil systems

Article

Long-Term Integrated Systems of Green Manure and Pasture Significantly Recover the Macrofauna of Degraded Soil in the Brazilian Savannah

Carolina dos Santos Batista Bonini [1,*], Thais Monique de Souza Maciel [2], Bruno Rafael de Almeida Moreira [3], José Guilherme Marques Chitero [2], Rodney Lúcio Pinheiro Henrique [2] and Marlene Cristina Alves [2,*]

[1] Department of Plant Production, School of Agronomic and Technological Sciences, São Paulo State University (Unesp), Dracena 17900-000, São Paulo, Brazil
[2] Department of Plant Protection, Rural Engineering and Soils, Faculty of Engineering, São Paulo State University (Unesp), Ilha Solteira 15385-000, São Paulo, Brazil
[3] Department of Engineering and Mathematical Sciences, School of Veterinarian and Agricultural Sciences, São Paulo State University (Unesp), Jaboticabal 14884-900, São Paulo, Brazil
* Correspondence: carolina.bonini@unesp.br (C.d.S.B.B.); marlene.alves@unesp.br (M.C.A.); Tel.: +55-(18)-3821-7488 (C.d.S.B.B.); +55-(18)-99694-0221 (M.C.A.)

Abstract: Healthy soil biota is the key to meeting the world population's growing demand for food, energy, fiber and raw materials. Our aim is to investigate the effect of green manure as a strategy to recover the macrofauna and the chemical properties of soils which have been anthropogenically degraded. The experiment was a completely randomized block design with four replicates. Green manure, *Urochloa decumbens*, with or without application of limestone and gypsum, composed the integrated systems. The macroorganisms as well as the soil fertility were analyzed after 17 years of a process of soil restoration with the aforementioned systems. The succession of *Stizolobium* sp. with *Urochloa decumbens*, with limestone and gypsum, was teeming with termites, beetles and ants. This integrated system presented the most technically adequate indexes of diversity and uniformity. Multivariate models showed a substantial increase in the total number of individuals due to the neutralization of harmful elements and the gradual release of nutrients by limestone and plaster. These conditioners have undergone multiple chemical reactions with the substrate in order to balance it chemically, thus allowing the macroinvertebrates to grow, develop, reproduce and compose their food web in milder microclimates. It was concluded that the integration of green manure together with grass is an economical and environmentally correct strategy to restore the macrofauna properties of degraded soil in the Brazilian savannah.

Keywords: *Cajanus* sp.; *Canavalia* sp.; *Urochloa decumbens*; macroorganisms

Citation: Bonini, C.d.S.B.; Maciel, T.M.d.S.; Moreira, B.R.d.A.; Chitero, J.G.M.; Henrique, R.L.P.; Alves, M.C. Long-Term Integrated Systems of Green Manure and Pasture Significantly Recover the Macrofauna of Degraded Soil in the Brazilian Savannah. *Soil Syst.* **2023**, *7*, 56. https://doi.org/10.3390/soilsystems7020056

Academic Editor: Adriano Sofo

Received: 1 March 2023
Revised: 20 May 2023
Accepted: 23 May 2023
Published: 30 May 2023

1. Introduction

Experts rely on a healthy soil biota as the key to meeting the increasing world population's demand for food, energy, fiber and raw materials [1]. Soil fauna consists of micro-, meso- and macroorganisms. Macroorganisms are living things larger than 2 mm in body diameter. The groups of macroinvertebrates proposed by specialists because of their ability to modify soil aggregates are bioturbators, reorganizers and weathering agents of clay minerals. Bioturbators refer to mobilizing and structuring agents of organic and mineral compounds. Ants, beetles, earthworms and termites are highly effective bioturbators in the production and molding of biogenic aggregates through their habits of foraging, tunneling, digging and nesting. Ecologists recognize them as the greatest soil engineers. Reorganizers are likely to alter soil structure by redefining internal organizational patterns of primary particles. Annelids and termites, in particular, process clay minerals via mandibular crushing and intestinal transition, respectively, consequently transforming them into pellet-shaped feces strengthened by chemical and physical bonds. Saliva and mucus membranes reorient precast micro-, meso- and macroaggregates, thus collaborating with the rehabilitation of soil properties; aging, drying and coating make newly formed biogenic aggregates naturally stronger. Weathering bioagents refer to macroinvertebrates able to modify clay mineralogy, making them either expansive, adsorptive or absorptive due to shifts in the ratio of tetrahedral to octahedral blades [2–7].

Irrespective of functional group, soil-dwelling macroinvertebrates offer a wide range of valuable ecological services that continuously improve agriculture, for instance, cycling of nutrients, decomposition and preservation of long-lasting pools of soil organic matter, mitigation of greenhouse gases threatening the environment and public health, promotion of microbial symbiosis between crop plants and microorganism and improvement of absorption and retention of water and gaseous exchanges through pores. These benefits are extremely relevant to optimize the functioning of the soil as global support for the growth and development of floral and faunal components [8].

Abundance, diversity, evenness and richness are paramount ecological aspects of macrofauna. They vary drastically with soil biological, chemical and physical properties. Hence, any disturbance by atypical pedoclimatic or anthropogenic forces on the environment causes several negative impacts on trophic relationships among ants, beetles, earthworms, termites and other aboveground and underground macroinvertebrates. Macroorganisms are sensitive to land use and management practice, so they are useful bioindicators of degradation events by erosion, salinization, pollution, desertification, etc. [9–13].

The presence of cover vegetation on the soil provides greater food availability and, consequently, the establishment of taxonomic groups aiming to colonize the soil, which improve the physical structure and act in the initial decomposition processes of plant residues. With the increase in the colonization of edaphic fauna in the areas under rehabilitation, the organisms reach a standard that is very close or even higher than that of being under native vegetation [14]. These authors report that in the Brazilian savannah research, the use of grass might have favored the emergence of organisms due to the greater layer of dry matter in different degrees of decomposition, sheltering organisms with different survival strategies.

Food and energy top the list of mankind's earliest and most essential goods and services. Hydroelectric power, especially, offers key benefits to civilizations from around the world, mainly including primary production of renewable energy and improvement of quality of life. Yet, large-scale plants for the generation of electricity in both low-income and high-income countries can easily devastate natural resources, such as soil and water. When the soil is under degradation by either pedoclimatic or anthropogenic forces, a study of physical, chemical and biological properties is necessary to select suitable management practices to restore it. The rehabilitation of degraded soil, traditionally performed using mechanical, physical or chemical techniques, may be expensive and detrimental to the environment. The development and implementation of alternative strategies are, therefore, necessary for both economic and environmental reasons [15].

Our multidisciplinary team from the School of Engineering at São Paulo State University (Unesp) has, for a long time, been studying and developing environmentally friendly and cost-effective technologies to monitor and mitigate events of degradation in the Brazilian savannah. This mixed woodland–grassland landscape is the country's second largest biome in terms of area, behind the Amazon jungle. In the past years, there have been studies in order to develop the promising agricultural frontiers of crop–livestock–forest frameworks in tropical zones [16,17]. There have been, nonetheless, very few studies reported in the literature focusing on the recovery of degraded sites [18–25].

When planning the recovery of degraded soil, the main objective to be achieved is the establishment of an A horizon so that, from there, the process is catalyzed by the biosphere, giving the possibility of the emergence of other horizons, according to natural conditioning. In soil restoration work, the initial activity is to identify and characterize the active degradation processes and analyze their environmental consequences. Thus, it is necessary to use indicators that allow us to qualify and quantify the degree of existing degradation in the area to be restored as well as to monitor the evolution of soil rehabilitation through these indicators afterwards.

With the aim of studying the influence of bioindicators on soil recovery and proposing an appropriate use for them, this work intended, with macroorganisms as indicators, to evaluate the biological properties of an Oxisol that has been in the process of recovery for 17 years with the use of liming, gypsum and plant species. The hypotheses of this work were that the combination of the use of liming, gypsum and plant species in a long run would restore the soil with a high degree of degradation and that the soil macroorganisms would be good indicators of the progress in the rehabilitation of this soil.

2. Materials and Methods

The field experiment was carried out at the Teaching, Research and Extension Farm of the School of Engineering, at São Paulo State University (Unesp), located in the municipality of Selvíria, state of Mato Grosso do Sul, Midwest Brazil, coordinates 20°22′ S and 51°22′ W and elevation of 317 m. According to the Brazilian Soil Classification System, the soil of the area is a sesquioxides-rich, eutrophic red Latosol, with sandy–loamy texture, corresponding to an Oxisol [26,27].

The earthworks and dam foundation required to install the Ilha Solteira Hydroelectric Power Plant made the site under investigation drastically degraded. The soil substantially excavated at an 8.6 m depth remained exposed since the 1970s. In the early 1990s, after a long period of natural rehabilitation, native species of grasses, shrubs and trees spontaneously grew at low density on the mixed woodland–grassland biome.

The tests planning to restore the degraded soil consisted of combinations of genera of green manures with a tropical species of pasture, with or without administration of limestone and/or agricultural gypsum, as detailed in Table 1.

Table 1. Characterization of an integrated system planned to recover physical, chemical and macrofaunal properties of an anthropogenically degraded site in the Brazilian savannah.

Code	Integrated System *
NC	Exposed soil; negative control
SMB	Soil under native vegetation with cultivation of *Urochloa decumbens*
MPB	Succession of *Stizolobium* sp. and *U. decumbens*
GFPB	Succession of *Cajanus* sp., *Canavalia* sp. and *U. decumbens*
CMPB	Succession of *Stizolobium* sp. and *U. decumbens* with limestone
CGFPB	Succession of *Cajanus* sp., *Canavalia* sp. and pasture of *U. decumbens* with limestone
CGeMPB	Succession of *Stizolobium* sp. and *U. decumbens* with limestone and gypsum
CGeGFPB	Succession of *Cajanus* sp., *Canavalia* sp. and *U. decumbens* with limestone and gypsum
PC	Forest; positive control

* Each plot was 10 m in length by 10 m in width, totaling 400 m^2 per test.

A border of 2 m on all sides of each 10 m × 10 m plot was disregarded (assessed area 64 m^2), and replications were 2 m apart from each other. Each test comprised four replicates and the experiment was a completely randomized block design. Exposed soil and native forest were references for potential contrasting.

Before the installation of the experimental field, the conventional preparation consisted of subsoiling, plowing and harrowing. From 1992 to 1999, for the sowing of green manure, the soil was prepared with plowing by using plow and leveling harrows. The chemical and physical characterization consisted of carrying out sampling at the depths of 0.00–0.20 and 0.20–0.40 m, according to Quaggio et al. [28] and Teixeira et al. [29].

The amendment of natural soil fertility of the experimental field consisted in carrying out the application of dolomitic limestone of 70% neutralizing power and agricultural gypsum at 1850 and 580 kg per hectare, respectively, aiming to improve the saturation of exchangeable cations to approximately 70%. The limestone was incorporated into the soil with the use of a harrow. In mid-1992, the sowing and the cultivation of the genera of green fertilizers, *Cajanus* sp., *Canavalia* sp. and *Stizolobium* sp., at a density of 10 seeds per meter was performed. In order to produce biomass faster for the recovery of the biological and chemical properties of the soil, the cutting-off of the N-fixing plants at the beginning of flowering was performed while maintaining as much agricultural residues as possible on the soil surface. In 1996, the liming of the experimental plots with a saturation of exchangeable cations out of the critical range of 50–70%, reported by Quaggio et al. [28], for crop plants growing in tropical zones was carried out. In 1999, the sowing of *U. decumbens* on no-tillage systems consisting of agricultural residues gradually released from the green manures throughout the harvest seasons running from 1992 to 1999 was finally performed. The management of pasture consisted of cutting it off to avoid overgrazing.

Annually, the soil physical and chemical properties and the dry mass plant productivity were analyzed. With 17 years of the influence of the treatments, in addition to these attributes of the soil and plants, the evaluation of soil macrofauna was added to the investigation. This biological attribute characterizes a dynamic property of the soil; therefore, the objective was the study in space. Because it is dynamic and greatly influenced by temperature and humidity conditions, the proposal was the relative analysis of the results and the use of two controls (native vegetation and soil with a high degree of degradation under natural recovery). The evaluation was carried out in winter and summer. With 14 to 15 years of implementation of treatments for soil recovery, with the influence of treatments, native tree plant species began to appear spontaneously in the study area. This aspect was an indicator of the improvement of soil conditions, and we decided to add the analysis of a soil biological attribute, in this case its macrofauna. At 17 years old, native tree vegetation grew at a high density, indicating a change in the edaphic conditions of the soil, which could favor and/or influence the appearance of species of organisms. In the periods of winter and summer, with 17 years of soil under the recovery system, the sampling of soil-dwelling macroinvertebrates was formally performed, according to Velasquez and Lavelle [30]. In order to collect macrofauna, square-shaped traps (0.3 m length by 0.3 m width by 0.3 m height) were placed at 0.00–0.05, 0.05–0.10 and 0.10–0.20 m depths. Afterwards, the collected monoliths were carefully introduced into polyethylene flasks containing alcoholic solution at 200 mL L^{-1}, and then they were transferred to the Laboratory of Soil Science, Unesp, for further analytical procedures of macrofauna technical assessment. The sampling of experimental plots, exposed soil and native forest for characterization of chemical properties was performed simultaneously, according to Quaggio et al. [28].

The technical assessment of macroinvertebrates comprised the visual count and the taxonomic identification of orders or classes as well as the calculation of total abundance and indexes of Shannon, Simpson and Pielou [31,32].

Firstly, the Shapiro–Wilk and Bartlett procedures were run to check if the data set was normal and the distribution and homogeneous were in random variance, respectively. The effects of integrated systems of green manure species with tropical pasture with or without application of limestone and gypsum on the soil macrofauna and chemical properties were tested using a one-way analysis of variance. Then, the treatments were compared by mean values by post-hoc Tukey's HSD test. Other methods of applying non-traditional statistics to track and understand multivariate patterns included the Pearson product-moment correlation test, PCA and MRA. The Pearson product-moment correlation test measured the strength and direction of linear associations between macrofaunal and chemical properties. Prior to running the PCA, the Bartlett's test of sphericity tested if the original data set was reliable. Then, the Kaiser–Meyer–Olkin procedure was used to test the significance of eigenvalues of principal components needed to reduce the dimensionality of the original data set while preserving as much statistically understandable variability as possible into orthogonally rotated subsets with absence of multicollinear variables and ambiguities. A factorial map was customized to contrast macrofaunal and chemical properties of the soil before and after recovery by long-term integrated systems of species of green manures with tropical pasture. MRA was performed to figure out the relationship between independent variables and predictors; the criteria for selection of significant multiple regression models were AIC (Akaike information criterion), BIC (Bayesian information criterion) and Radj2 (Adjusted R^2). The software used for statistical computing and graphics was R software [33].

3. Results

The chemical composition of the exposed soil mainly consisted of high contents of H^+ and Al^{+3} and low contents of Presin, K, Ca, Mg and SOM. These properties caused the saturation of exchangeable cations to be predictably low in the environment before starting the process of restoration (Table 2). Additionally, the physical composition was replete with micropores, thus presenting a high degree of compaction by mechanical forces of earthworks and dam foundations visible on the deforested landscape. The highest apparent density was at the deepest depth of 0.20–0.40 m.

Table 2. Chemical and physical properties of soil degraded by earthworks and dam foundations for the establishment of a hydroelectric power plant in the Brazilian savannah, before the installation of the experiment, in 1992.

Property	Depth (m)	
	0.00–0.20	0.20–0.40
Presin (mg dm^{-3})	1.00	0.00
Organic matter (g dm^{-3})	7.00	4.00
pH	4.0	4.20
K (cmolc dm^{-3})	0.20	0.20
Ca (cmolc dm^{-3})	2.00	2.00
Mg (cmolc dm^{-3})	1.00	1.00
Potential acidity (cmolc dm^{-3})	20.00	20.00
Sum of exchangeable cations (cmolc dm^{-3})	3.20	3.20
Cation-exchange capacity (cmolc dm^{-3})	23.20	23.10
Saturation of exchangeable cations (%)	14.00	14.00
Total porosity (m^3 m^{-3})	0.34	0.33
Macroporosity (m^3 m^{-3})	0.09	0.07
Microporosity (m^3 m^{-3})	0.25	0.26
Soil bulk density (kg m^{-3})	1.60	1.74

The integrated systems, exposed soil and native forest, collectively, yielded about 2210 macroinvertebrates making up the orders or classes Aranae, Coleoptera, Dermaptera, Diplopoda, Isoptera, Haplotaxida, Hemiptera, Hymenoptera and Orthoptera; eggs and larval stages of arthropods taxonomically unidentified were other elementary components of heterogeneous macrofauna of the site under investigation.

Figure 1A–D show the numbers of individuals and orders, organized by treatments and seasons of the year studied. Regardless of the treatment, a greater number of termites (Winter: 359, 133, 1333; Summer: 419, 438, 438) were found in the soil depths of 0.00–0.05, 0.05–0.10 and 0.10–0.20 m, respectively.

In the winter, after 17 years of influence of the treatments, the soil with native vegetation cover with subsequent cultivation of *U. decumbens* had an inexpressive number of spiders, adult millipedes and ants but an impressive number of termites at a depth of 0.00–0.05 m depth. These macroinvertebrates collaboratively caused the SMB to yield a total abundance greater than that of the exposed soil, where macroorganisms under harsh microclimate were notably scarce, regardless of the depth. Similarly, succession of *Stizolobium* sp. with *U. decumbens*, without conditioning by regular administration of limestone and gypsum, also yielded a great number of termites but an insignificant number of beetles. These macroinvertebrates provided the MPB with acceptable indexes of diversity, dominance and evenness, in comparison to other integrated systems and native forest, where spiders, adult millipedes and crickets were absent beneath litterfall. The integrated system of *Cajanus* sp., *Canavalia* sp. and *U. decumbens*, without limestone and gypsum, was seemingly microclimatically selective for termites, ants, beetles and stinkbugs. Moderate indexes of diversity and evenness reflected the GFPB's ecological aspects, thus distinguishing it from the others. Technically similar to SMB, MPB and GFPB, the succession of *Stizolobium* sp. with *U. decumbens*, without limestone and gypsum, also had a remarkable abundance of termites but inexpressive counts of ants, spiders, beetles, white grubs and adult millipedes. Yet, CMPB had a satisfactory diversity of bioturbators, reorganizers and weathering agents of clay minerals.

Winter

(A)

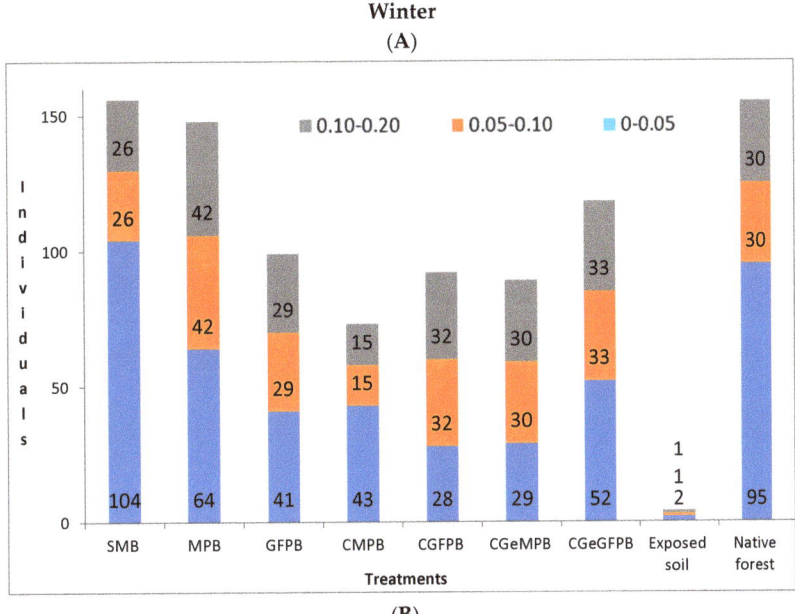

(B)

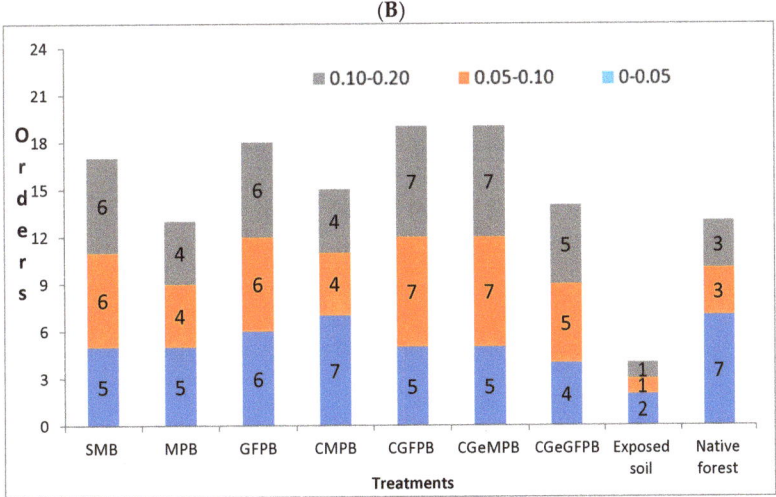

Figure 1. *Cont.*

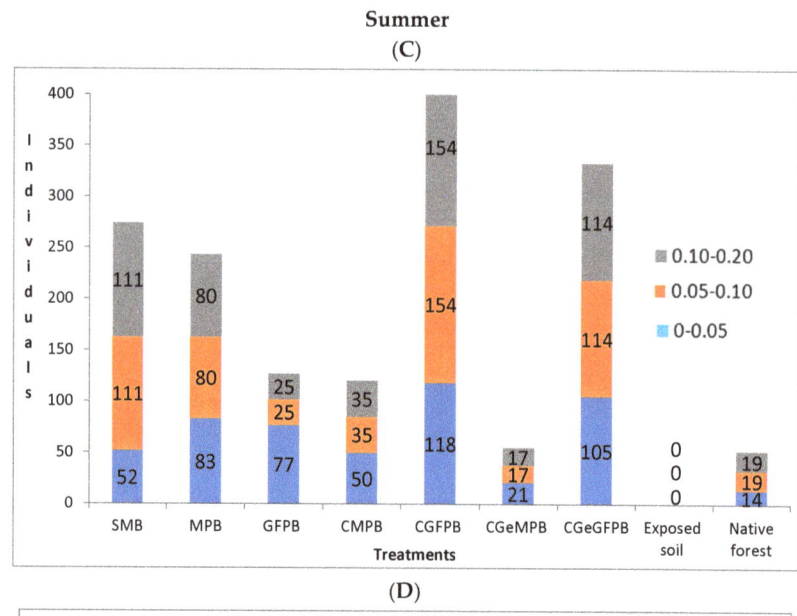

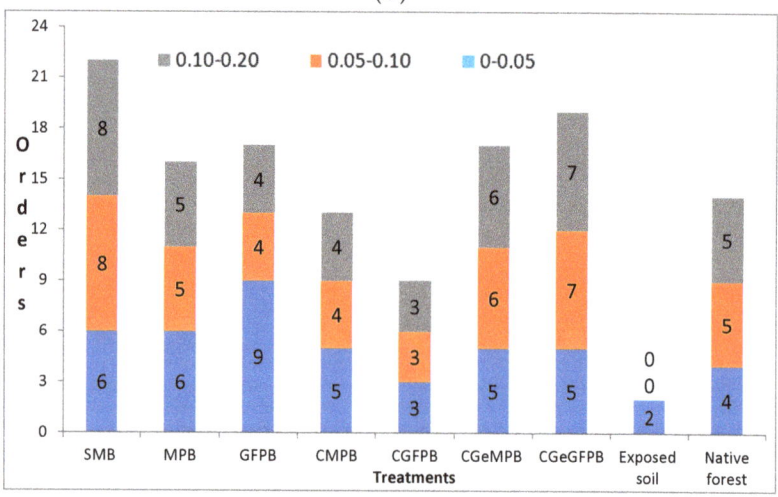

Figure 1. Time–space profiles of macroinvertebrates at different depths for (**A**) individuals—winter; (**B**) orders—winte, (**C**) individuals—summer; and (**D**) orders—summer at a site of Brazilian savannah recovered for 17 years by integrated systems of green fertilizers with pasture, with or without application of limestone and gypsum.

The integrated system of *Cajanus* sp., *Canavalia* sp. and *U. decumbens* with limestone was also microenvironmentally suitable for the growth, development and reproduction of termites and ants. This succession had moderate indexes of diversity and evenness. The integrated system of *Stizolobium* sp. with *U. decumbens*, with limestone and gypsum, interestingly was richer in termites, eggs and larval stages of arthropods taxonomically unidentified than other integrated systems but poorer in beetles, white grubs and adult centipedes. Evidently, the composition of the net of macroinvertebrates was dependent on the microclimate. The integrated system of *Cajanus* sp. *Canavalia* sp. and *U. decumbens* with limestone and gypsum also predominantly consisted of termites and had a relatively low count of adult centipedes and spiders, which are typical natural predators of mesoinvertebrates and macroinvertebrates in tropical, subtropical and temperate zones. The native forest was the genuine habitat of termites, earthworms, beetles, ants and stinkbugs. These macroinvertebrates grew at a relatively similar quantity. Generally, in the winter after 17 years in recovery, the integrated system of *Stizolobium* sp. with *U. decumbens* with limestone and gypsum provided the macrofauna from the top depth with the most technically pleasing indexes of diversity, dominance and evenness (Table 3).

Table 3. Ecological features of macrofauna of the Brazilian savannah site restored by integrated systems of green manures with pasture with or without application of limestone and gypsum after 17 years in recovery.

Integrated System	Period of Sampling							
	Winter				Summer			
Index	Diversity	Dominance	Evenness	Total Abundance (Ind m^{-2})	Diversity	Dominance	Evenness	Total Abundance (Ind m^{-2})
Exposed soil	0.10 e	1.00 a	1.00 a	0.15 c	0.00 h	0.00 h	0.00 f	0.00 i
SMB	1.61 b	0.11 b	0.23 b	3.58 a	0.97 c	0.40 c	0.54 b	5.70 c
MPB	1.61 b	0.15 b	0.29 b	2.90 b	0.85 d	0.35 d	0.47 c	5.35 d
GFPB	0.79 d	0.67 a	0.82 a	2.05 b	1.20 b	0.48 b	0.55 b	3.65 e
CMPB	1.95 a	0.23 b	0.37 b	1.93 b	0.65 e	0.23 e	0.40 c	2.60 f
CGFPB	1.61 b	0.54 a	0.70 a	3.18 a	0.20 g	0.30 d	0.18 e	7.93 a
CGeMPB	1.61 b	0.68 a	0.85 a	1.55 b	1.41 a	0.74 a	0.88 a	1.68 g
CGeGFPB	1.39 c	0.15 b	0.34 b	2.45 b	0.39 f	0.12 f	0.24 d	5.95 b
Native forest	1.95 a	0.33 b	0.44 b	3.58 a	0.60 e	0.35 d	0.43 c	1.08 h
F (5%)	4000 *	1190 *	1403 *	0910 *	6447 *	4828 *	3430 *	4591 *
CV (%)	2.14	7.69	5.85	14.37	6.60	12.20	7.46	0.62

Mean values followed by the same letters in the column do not differ according to post-hoc Tukey's HSD test; * significative by the post-hoc Tukey's HSD test at $p < 0.05$. CV (%): coefficient of variation.

In the summer, with 17 years of influence of the treatments, soil with cover of native vegetation with subsequent cultivation of *U. decumbens* retained a great number of termites and insignificant numbers of ants, white grubs, adult centipedes and beetles at a depth of 0.00–0.05 m. Unlike in the winter, with 17 years of influence of treatments, SMB had no counts of spiders, earwigs, adult millipedes, earthworms, stinkbugs and crickets, probably due to the morphophysiological sensitiveness of these macroinvertebrates to the constantly increasing temperature in the substrate. The succession of *Stizolobium* sp. with *U. decumbens* continued to have a predominance of termites over beetles, white grubs, ants and earwigs, making up the smaller insect orders, Coleoptera, Hymenoptera and Dermaptera, respectively. These macroorganisms enabled the MPB to have a higher total abundance compared to native forest and exposed soil, where macrofauna did not exist. Indeed, soil degraded by mechanical forces of earthworks and dam foundations was the harshest microclimate for the growth, development, reproduction and residence of functional groups of bioturbators, reorganizers and weathering agents of clay minerals. Technically similar to MPB, the integrated system of *Cajanus* sp., *Canavalia* sp. and *U. decumbens* also appeared to be microclimatically suitable for termites, geophages definitely predominant over beetles, white grubs, adult centipedes, earwigs, earthworms, stinkbugs, ants and crickets.

For all indices (diversity, dominance, uniformity, abundance) and two seasons of the year studied, there was a statistical difference. The exposed soil had less diversity and abundance of individuals, unlike the native forest and treatments with combinations of green manure + limestone + gypsum.

In the summer period, there was a greater number of termites in the superficial soil layer, followed by a smaller number of ants, lacrals and beetles. The results obtained in the exposed soil treatment characterize this as the least favorable environment for the development of soil fauna; consequently, the others were favorable and resilient in the recovery process.

Regarding the seasons, summer, due to the higher incidence of solar radiation and abundant precipitation (tropical climate), favored a greater number of individuals and the abundance of some treatments studied, such as combinations of green manure + limestone + gypsum.

The integrated system of *Cajanus* sp., *Canavalia* sp., *Stizolobium* sp. and *U. decumbens* with limestone and gypsum had significant indexes of Shannon, Simpson and Pielou, which were dependent on the specific abundance of termites. The native forest provided the macrofauna with the highest indexes of diversity and evenness in the winter. However, it was not technically and ecologically efficient in protecting spiders, white grubs, adult millipedes, earthworms and crickets from potential adversities of climatic changes in the summer. Apparently, similarly to the integrated systems, in the native forest there were also termites at the highest population density, possibly due to the acidity of the soil beneath litterfall. Globally, in the summer, *Stizolobium* sp. with *U. decumbens* with limestone and gypsum remained the most effective succession to provide the macroinvertebrates dwelling through the depth of 0.00–0.05 m with the most balanced indexes of diversity, dominance and evenness. Irrespective of succession, termites have been also persistently predominant at the depths of 0.05–0.10 and 0.10–0.20 m. The presence of termites in the treatments studied, including in the forest, is due to the genetic characteristic of the savannah soil, which is acidic, and termites are an indicator organism of soil acidity.

Other relevant groups dwelling abundantly through deep depths were beetles, white grubs and ants. By contrasting the periods of winter and summer, macroinvertebrates significantly moved from the top towards the deep depths of the soil. The migratory behavior of macroorganisms to deeper layers of the soil in the summer probably occurred because the more superficial layers reached higher temperatures.

The integrated systems significantly improved the chemical properties of the soil, in contrast to the reference (Table 4) 0–0.05 m depth with the largest availability of Presin, while the exposed soil had the lowest content of this macronutrient after several years of experimentation. In addition, CGeGFPB, CGFPB and MPB were apparently the most efficient integrated systems in concentrating Presin in the deepest depths of 0.05–0.10 and 0.10–0.20 m. The integrated systems had a relatively similar content of organic matter, regardless of the depth. Soil under the process of recovery had higher absolute values of pH in comparison to exposed soil and native forest, certainly due to the neutralization of H^+ by limestone undergoing multiple reactions with the soil to balance it chemically. The integrated systems of either *Canavalia* sp. or *Stizolobium* sp. with *U. decumbens* caused the top depth of 0.00–0.05 m to have the highest K content. Meanwhile, soil under *Stizolobium* sp. with *U. decumbens* with limestone and gypsum had lower K content when compared to the soil beneath litterfall from native forest.

Irrespective of soil depth, the native forest was the microclimate with the highest absolute values of Al^{+3}, potential acidity and CEC, lowest absolute values of SEC and saturation of exchangeable cations. Soils in the Brazilian savannah are generally high in potential acidity.

The Pearson product-moment correlation test accurately tracked significant correlations between macrofaunal and chemical properties of the site in the Brazilian savannah that were successfully recovered by successions of green manures and pasture with and without conditioning by limestone and gypsum (Figure 2). Amongst the main linear relationships, stinkbugs had positive correlations with termites ($r = 0.50$), earthworms ($r = 0.75$), beetles ($r = 0.60$), earwigs ($r = 0.80$), eggs ($r = 0.80$) and total of individuals ($r = 0.45$) but negative correlation with spiders ($r = -0.45$). Termites had a positive correlation with total of individuals ($r = 0.95$). The total of orders negatively correlated with potential acidity ($r = -0.95$). The saturation of exchangeable cations had a negative correlation with Al^{3+} ($r = -0.75$). SOM positively correlated with Presin ($r = 0.85$), K ($r = 0.95$), Mg ($r = 0.85$) and SEC ($r = 0.65$). The total of individuals had positive linear relationships with pH ($r = 0.50$), Presin ($r = 0.50$), K ($r = 0.50$), Mg ($r = 0.70$) and Ca ($r = 0.50$) but a negative linear relationship with Al^{3+} ($r = -0.50$). In line with the multivariate patterns recognized through the Pearson product-moment correlation test, principal component analysis robustly divided the high complexity net of linear associations between macrofaunal and chemical properties of restored soil into the subsets PCI, PCII, PCIII, PCIV and PCV. The explanations for these components are chemical neutralization of toxic elements, diversity of food sources on the soil, biological decomposition of organic matter, unavailability of aboveground biomass and predatory activity, respectively. These components collaboratively retained about seventy-five percent variance within orthogonally rotated subsets without the presence of collinear variables (Table 5).

Table 4. Chemical properties of Brazilian savannah soil restored by integrated systems of green manures with pasture with or without application of limestone and gypsum after 17 years in recovery.

Integrated System	P_{resin}	SOM	pH	K	Ca	Mg	$H + Al^{3+}$	Al^{3+}	SEC	CEC	SEC/CEC
	(mg dm^{-3})	(g dm^{-3})		(mmolc dm^{-3})							%
Soil Depth (m)	**0.00–0.10**										
Exposed soil	3.00 D	4.75 C	5.00 A	0.25 C	2.00 C	2.75 A	19.75 A	0.75 A	5.50 A	25.25 C	22.75 AB
SMB	6.75 AB	10.75 AB	5.00 A	1.00 AB	5.00 AB	4.25 B	23.75 A	1.75 A	10.25 AB	34.00 BC	29.50 A
MPB	7.25 AB	11.00 AB	5.00 A	1.25 B	5.5 AB	5.25 C	25.25 A	1.75 A	12.00 AB	37.25 BC	32.00 A
GFPB	7.75 AB	10.00 AB	5.00 A	0.75 AB	5.25 AB	4.25 B	24.75 A	1.75 A	10.25 AB	35.00 BC	28.75 A
CMPB	8.00 AB	11.75 B	5.00 A	1.00 AB	6.25 AB	5.00 BC	25.25 A	1.75 A	12.25 B	37.50 BC	32.50 A
CGFPB	7.00 AB	11.00 B	5.00 A	1.00 AB	5.50 A	4.50 B	23.5 A	1.25 A	12.00 AB	35.50 BC	33.50 A
CGeMPB	8.75 A	10.75 AB	4.75 A	1.00 AB	5.50 A	4.50 B	25.00 A	2.00 A	11.00 AB	36.00 BC	30.25 A
CGeGFPB	7.25 AB	9.00 AB	5.00 A	1.00 AB	4.00 ABC	3.25 A	24.50 A	2.25 A	8.25 AB	32.70 AB	24.50 AB
Native forest	4.25 C	12.75 B	4.00 B	1.00 AB	2.50 AB	2.75 A	36.50 B	6.75 B	5.75 AB	42.25 A	13.00 C
F-value	2.83 *	3.78 *	16.00 *	2.70 *	7.33 *	2.19 *	14.99 *	11.70 *	3.94 *	6.95 *	5.49 *
CV (%)	15.17	12.00	3.43	9.90	10.58	14.28	9.19	13.97	13.27	9.87	11.89
MSD	1.01	0.95	0.40	0.33	0.59	0.77	5.60	0.58	1.04	8.31	1.37
	0.10–0.20										
Exposed soil	3.00 A	3.25 B	4.25 AB	0.25	1.00 B	1.00 A	22.50 BC	2.25 A	2.25 B	19.70 A	9.25 BC
SMB	3.00 A	5.00 B	4.50 AB	0.20	4.25 A	2.00 C	20.75 AB	1.50 A	6.25 A	23.50 A	24.00 A
MPB	3.75 A	6.25 AB	4.25 AB	0.25	3.75 A	2.00 C	24.25 A	2.75 A	6.00 AB	21.7 A	20.75 AB
GFPB	3.00 A	6.75 AB	5.00 B	0.25	3.50 A	1.50 B	22.00 BC	1.75 A	5.25 AB	22.70 A	19.50 ABC
CMPB	4.75 B	5.75 AB	4.75 AB	0.25	3.75 A	2.00 C	23.00 BC	2.25 A	6.00 AB	22.70 A	21.25 AB
CGFPB	3.75 A	5.75 AB	4.75 AB	0.25	3.75 A	2.50 D	21.50 BC	1.50 A	6.50 A	24.5 A	23.75 A
CGeMPB	3.00 A	4.75 AB	5.00 B	0.25	3.50 A	1.75 BC	21.25 B	1.25 A	5.50 AB	22.20 A	20.75 AB
CGeGFPB	5.00 B	4.50 AB	4.75 AB	0.25	2.25 AB	1.25 A	18.75 A	2.25 A	3.75 AB	21.70 A	12.50 ABC
Native forest	3.00 A	8.00 A	4.00 A	0.25	1.00 B	1.00 A	32.75 D	7.75 B	2.25 B	30.20 B	6.75 C
F-value	0.43	4.11 *	2.84 *	1.12	9.15 *	2.64 *	68.16 *	7.83 *	6.36 *	7.79 *	10.83 *
CV (%)	19.05	10.76	9.15	18.88	11.45	12.00	4.07	17.31	12.60	9.04	11.89
MSD	0.98	0.67	1.00	0.50	0.55	0.47	2.32	0.77	0.73	5.00	1.23
	0.20–0.40										
Exposed soil	2.00 A	3.00 A	5.00 AB	0.20	1.00 B	1.00 A	21.50 A	2.00 A	2.00 A	23.50 A	9.25 AB
SMB	3.00 AB	3.75 ABC	4.50 AB	0.17	2.50 AB	1.25 A	21.00 A	1.50 A	3.75 AB	24.75 A	16.00 ABC
MPB	3.00 AB	4.50 BC	4.75 AB	0.12	2.50 AB	1.50 A	21.00 A	1.50 A	4.00 AB	25.00 A	16.75 ABC
GFPB	2.00 A	3.25 AB	5.00 AB	0.17	1.75 AB	1.25 A	21.75 A	2.25 A	3.00 AB	24.75 A	12.75 ABC
CMPB	2.00 A	4.00 ABC	5.20 AB	0.20	3.25 A	2.00 B	21.25 A	1.75 A	5.25 AB	26.50 A	20.00 BC
CGFPB	3.00 AB	5.00 C	5.00 AB	0.20	3.50 A	2.00 B	20.75 A	1.50 A	5.75 AB	26.50 A	22.00 C
CGeMPB	2.00 A	3.50 AB	5.00 AB	0.12	2.25 AB	1.25 A	21.00 A	1.75 A	3.50 AB	24.50 A	14.75 ABC
CGeGFPB	2.00 A	4.00 ABC	5.00 AB	0.17	2.00 AB	1.25 A	21.75 A	2.25 A	3.25 AB	25.00 A	13.75 ABC
Native forest	4.00 B	7.00 D	4.00 A	0.22	1.00 B	1.00 A	31.00 B	7.75 B	2.00 A	33.00 B	7.00 BC
F-value	5.40 *	22.80 *	2.67 *	1.99	4.08 *	1.26 *	21.38 *	28.22 *	3.58 *	13.45 *	5.41 *
CV (%)	14.34	10.29	9.50	27.20	13.72	12.92	6.33	9.16	14.51	5.90	11.38
MSD	1.03	1.23	1.10	0.12	0.58	0.48	3.40	0.40	0.74	3.68	1.05

Mean values superscripted by the same capital letters are not significantly different according to post-hoc Tukey's HSD test; * $p < 0.05$. CV (%): coefficient of variation; MSD: minimal significant difference.

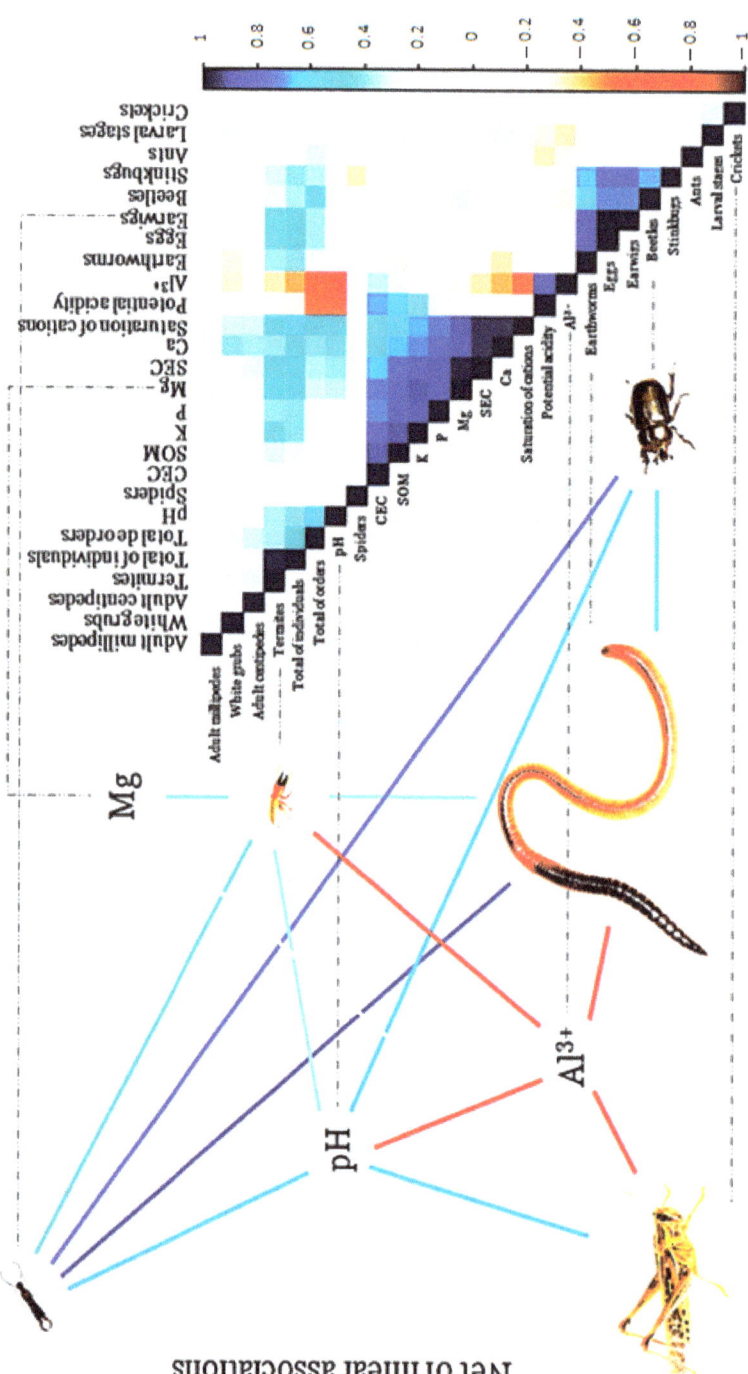

Figure 2. Correlogram for the linear associations between macrofaunal and chemical properties of the Brazilian savannah restored soil by long-term integrated systems of green manures with pasture with or without application of limestone and gypsum. Blue cells show positive linear associations between analytical variables placed along with the lines and columns in the data-to-viz card, while the red color represents negative linear associations. A few of the linkages of the net of linear associations are hard to read, but solid lines exist for macrofaunal and chemical properties at statistically understandable correlations.

Table 5. Principal component (PC) analysis for the macrofaunal and chemical properties of the Brazilian savannah soil restored by integrated systems of green manures with pasture with or without application of limestone and gypsum.

	Bartlett's Test of Sphericity				
Chi-square	8500				
Degree of freedom	325				
p-value	<0.05 *				
	Kaiser–Mayer–Olkin Test				
Index/variable	Principal Component				
	PCI	PCII	PCIII	PCIV	PCV
Eigenvalue	7.83	5.53	3.62	1.78	1.43
Percentage of variance	30.12	21.29	13.93	6.84	5.50
Cumulative percentage of variance	30.12	51.41	65.33	72.17	77.67
	Loading				
Spiders	0.06	−0.25	−0.16	0.13	0.04
Beetles	0.35	0.63 *	0.32	0.40 *	0.20
Adult millipedes	0.24	−0.08	−0.22	0.32	0.41 *
Earthworms	−0.10	0.78 *	0.50 *	0.01	0.06
Adult centipedes	0.32	−0.06	−0.17	−0.06	−0.76 *
Larval stages	0.14	0.09	−0.46 *	−0.14	0.68 *
Eggs	0.16	0.80 *	0.53 *	0.08	0.03
Termites	0.65 *	0.37	0.18	−0.48 *	−0.05
Earwigs	0.16	0.80 *	0.53 *	0.08	0.03
Ants	0.17	0.20	−0.28	−0.21	−0.12
White grubs	0.15	−0.01	−0.52 *	0.55 *	0.06
Stinkbugs	0.27	0.66 *	0.43 *	0.19	−0.02
Crickets	0.01	0.10	−0.16	−0.77 *	0.29
Total of individuals	0.69 *	0.45 *	0.09	−0.42 *	0.03
Total of orders	0.49 *	0.58 *	−0.34	0.24	−0.14
Presin	0.84 *	−0.35	0.12	0.06	0.03
SOM	0.72 *	−0.55 *	0.37	0.01	−0.04
pH	0.47 *	0.43 *	−0.48 *	−0.07	−0.17
K	0.82 *	−0.42 *	0.27	−0.01	0.03
Ca	0.90 *	−0.22	−0.31	0.06	0.01
Mg	0.96 *	−0.18	0.11	0.03	0.04
Potential acidity	−0.02	−0.69 *	0.67 *	0.05	0.05
Al^{3+}	−0.42 *	−0.57 *	0.63 *	−0.01	0.02
SEC	0.96 *	−0.21	−0.07	0.03	0.02
CEC	0.60 *	−0.55 *	0.48 *	−0.02	0.09
Saturation of exchangeable cations	0.95 *	0.01	−0.23	0.04	0.03
	Percentage of Contribution				
Spiders	0.04	1.13	0.71	0.95	0.09
Beetles	1.57	7.07	2.78	9.04	2.87
Adult millipedes	0.75	0.11	1.33	5.90	11.53
Earthworms	0.00	10.94	6.92	0.00	0.23
Adult centipedes	1.33	0.06	0.82	0.21	40.64
Larval stages	0.24	0.13	5.91	1.07	32.36
Eggs	0.32	11.53	7.65	0.39	0.07
Termites	5.35	2.43	2.11	13.13	0.18
Earwigs	0.32	11.53	7.65	0.39	0.07
Ants	0.38	0.71	2.11	2.48	0.93
White grubs	0.28	0.00	7.44	16.92	0.24
Stinkbugs	0.93	7.98	5.08	2.04	0.03
Crickets	0.00	0.18	0.72	33.28	5.91
Total of individuals	6.06	3.72	0.23	9.87	0.05

Table 5. *Cont.*

	Bartlett's Test of Sphericity				
Total of orders	0.49 *	0.58 *	−0.34	0.24	−0.14
Presin	0.84 *	−0.35	0.12	0.06	0.03
SOM	0.72 *	−0.55 *	0.37	0.01	−0.04
pH	0.47 *	0.43 *	−0.48 *	−0.07	−0.17
K	0.82 *	−0.42 *	0.27	−0.01	0.03
Ca	0.90 *	−0.22	−0.31	0.06	0.01
Mg	0.96 *	−0.18	0.11	0.03	0.04
Potential acidity	−0.02	−0.69 *	0.67 *	0.05	0.05
Al^{3+}	−0.42 *	−0.57 *	0.63 *	−0.01	0.02
SEC	0.96 *	−0.21	−0.07	0.03	0.02
CEC	0.60 *	−0.55 *	0.48 *	−0.02	0.09
Saturation of exchangeable cations	0.95 *	0.01	−0.23	0.04	0.03
	Percentage of Contribution				
Spiders	0.04	1.13	0.71	0.95	0.09
Beetles	1.57	7.07	2.78	9.04	2.87
Adult millipedes	0.75	0.11	1.33	5.90	11.53
Earthworms	0.00	10.94	6.92	0.00	0.23
Adult centipedes	1.33	0.06	0.82	0.21	40.64
Larval stages	0.24	0.13	5.91	1.07	32.36
Eggs	0.32	11.53	7.65	0.39	0.07
Termites	5.35	2.43	2.11	13.13	0.18
Earwigs	0.32	11.53	7.65	0.39	0.07
Ants	0.38	0.71	2.11	2.48	0.93
White grubs	0.28	0.00	7.44	16.92	0.24
Stinkbugs	0.93	7.98	5.08	2.04	0.03
Crickets	0.00	0.18	0.72	33.28	5.91
Total of individuals	6.06	3.72	0.23	9.87	0.05
Total of orders	3.11	6.04	3.17	3.25	1.42
Presin	9.01	2.26	0.39	0.24	0.05
SOM	6.69	5.55	3.75	0.01	0.12
pH	2.79	3.33	6.31	0.24	2.12
K	8.59	3.17	1.97	0.00	0.08
Ca	10.28	0.84	2.67	0.20	0.00
Mg	11.84	0.60	0.32	0.06	0.09
Potential acidity	0.01	8.65	12.33	0.14	0.19
Al^{3+}	2.27	5.83	10.88	0.01	0.03
SEC	11.87	0.78	0.13	0.05	0.02
CEC	4.57	5.45	6.29	0.02	0.60
Saturation of exchangeable cations	11.41	0.00	1.51	0.09	0.08
	Chemical neutralization of toxic elements	Diversity of food sources on the soil surface	Biological decomposition of organic matter	Unavailability of aboveground biomass	Predatory activity

* significative at 5% (p-value < 0.05).

The first component shows the integrated systems with limestone and gypsum to be positively correlated with the total abundance of macroinvertebrates, termites, Presin, SOM, K, Mg, Ca, SEC, CEC and, obviously, saturation of exchangeable cations. The chemical properties SEC, Mg, SEC/CEC, Ca and Presin had the strongest contributions to interpretable variance of chemical neutralization of toxic elements, followed by the macrofaunal properties, the total of individuals and specific abundance of termites, in that order. The second component was the proof of the native forest microclimate positively correlated with beetles, earthworms, eggs, stinkbugs, total of orders and pH, but negatively correlated with SOM, K, potential acidity, Al^{3+} and CEC. The macrofaunal properties earthworms, earwigs and eggs of arthropods taxonomically unidentified contributed nearly equally to the variance of diversity of food sources on the soil surface, followed by stinkbugs and beetles. The third component had positive loadings with earthworms, earwigs, eggs, Al^{3+} and potential acidity but a negative linear relationship with pH. The potential acidity had the largest percentage of contribution to variance of decomposition of organic matter, followed by Al^{3+} and the macrofaunal properties white grubs, earwigs and earthworms. Crickets and white grubs had positive and negative correlations with the fourth component, respectively; these groups of macroinvertebrates jointly explained the largest portion of variance of unavailability of aboveground biomass. Finally, larval stages of arthropods taxonomically unidentified, adult centipedes and adult millipedes had positive and negative correlations with the fifth component, respectively. Adult millipedes and larval stages jointly explained the largest portion of variance of predatory activity. Globally, potential acidity was the most suppressive chemical factor over the macrofauna in the exposed soil.

The multivariate regression analysis concretely enabled the outcomes of the Pearson product-moment correlation test and the principal component analysis for the taxon–microclimate relationship to fit. Despite the fluctuation in the values of AIC, BIC and Radj2, the regression models described the commendable improvements in soil macrofaunal properties, mainly depending on the neutralization of potential acidity, enhancement of availability of food sources and suitability of microhabitats under integrated systems with regular administration of limestone and gypsum as conditioners (Table 6). The lower the absolute values of AIC and BIC and the higher the value of Radj2, the more reliable and accurate the regression model is to predict how much the integrated systems of green manures and pasture changed the soil to make it chemically and biologically healthier.

Table 6. Parameters and goodness-of-fit of multivariate regression models for the taxon–microclimate relationship.

Variable	Fitted Regression Model †	AIC	BIC	Radj2
White grubs	White grubs (ind m^{-2}) = 0.25 + 0.45 Ca *	108.40	112.30	0.12
Adult centipedes	Adult centipedes (ind m^{-1}) = −0.12 + 0.09 Ca *	27.50	31.38	0.10
Termites	Termites (ind m^{-2}) = 24.09 − 11.48 Ca * − 5.61 K + 20.67 Mg * − 4.75 Al	245.20	252.90	0.32
Total of individuals	Total of individuals = 21.68 * + 8.98 Mg ** − 4.65 Al *	246.90	252.10	0.38
Total of orders	Total of orders = 2.18 − 34.21 SEC + 19.01 Potential acidity *	94.69	99.87	0.46
Soil organic matter	SOM (g dm^{-3}) = 2.78 ** + 0.19 P + 0.01 Ca + 6.55 K **	97.43	103.90	0.82
Spiders	Spiders (ind m^{-2}) = 0.69 ** − 0.30 Stinkbugs *	53.46	57.35	0.07
Termites	Termites (ind m^{-2}) = 18.81 ** + 10.73 Earthworms *	250.10	254.00	0.11
Eggs	Eggs (ind m^{-2}) = −0.01 + 0.19 Earthworms **	−45.63	41.74	0.75
Earwigs	Earwigs (ind m^{-2}) = −0.01 + 0.19 Earthworms **	−45.63	41.74	0.75
Beetles	Beetles (ind m^{-2}) = 0.90 ** + 1.07 Earthworms **	95.64	99.53	0.31
Stinkbugs	Stinkbugs (ind m^{-2}) = 0.08 + 0.54 Earthworms **	−46.10	49.99	0.43
Earwigs	Earwigs (ind m^{-2}) = −0.05 + 0.04 Beetles * + 0.15 Stinkbugs **	37.49	32.31	0.67
Beetles	Beetles (ind m^{-2}) = 0.84 ** + 1.57 Stinkbugs **	89.51	93.40	0.45

AIC: Akaike information criterion; BIC: Bayesian information criterion; Radj2: Adjusted R^2. * significative at 5% (*p*-value < 0.05); ** significative at 1% (*p*-value < 0.01).

The regression models assisted us in drafting the following insights on the taxon–microclimate relationship:

If the Ca content increases marginally, the visual count of white grubs would increase by 0.25 individuals per area unit. Therefore, the addition of Ca to the soil by applying limestone and gypsum combined with the decomposition of agricultural residues would be beneficial for the growth and development of young Coleoptera.

If the Ca content increases marginally, the visual count of adult centipedes would increase by 0.09 units. Therefore, microclimates richer in Ca would be more suitable for adult Dermaptera than those poorer in this macronutrient, whose major sources are certainly limestone and gypsum.

If the Al^{+3} content increases and the Mg content increases, both marginally, the visual count of termites would increase by about 25.4 individuals per area unit. The administration of gypsum or chemical calcium sulphate ($CaSO_4.2H_2O$) would boost the population density of termites by complexing aluminum, thus protecting them from eventual adversities of this element, which was the most toxic physical property for both the floral and faunal components of the soil,

If the contents of P, K and Ca increase marginally, the soil organic carbon content would increase by 6.75 g dm^{-3}. Thus, the improvement in the availability of macronutrients from the conditioners and the degradation of mulching by genera of N-fixing plants throughout the years would concentrate organic matter. It is important to have technologies that protect the storage of organic carbon in the soil, considering its quick decomposition in tropical climates due the humidity and high temperatures.

If the number of earthworms increases marginally, the quantities of termites and beetles would both increase by about 10.75 and 1.1 units, respectively. Any change in soil structure caused by Haplotaxida would be beneficial for Isoptera and Coleoptera and even for earthworms at very low specific abundance.

4. Discussion

The soil bulk density and its total porosity, as well as the contents of Presin, SOM, K, Ca and Mg of exposed soil, in the field experiment were predictably out of the critical ranges described in the literature [34]. The integrated systems with and without the administration of limestone and gypsum greatly improved the physical and chemical properties of the site under investigation. After 17 years of hard experimentation, Presin content was significantly higher in successions of *Cajanus* sp., *Stizolobium* sp. and *U. decumbens* than in exposed soil. Symbiotic relations between the genera of N-fixing plants and plant growth-promoting microorganisms, such as endophytic and free-living bacteria and arbuscular mycorrhizal fungi, as well as the biological decomposition of agricultural residues by macro- and mesoinvertebrates were probably factors that positively influenced the availability of phosphorus in the depths of 0.00–0.10, 0.10–0.20 and 0.20–0.40 m. Additionally, increased contents of K, Ca, Mg and SOM, as well as increased SEC, CEC and saturation of exchangeable cations, could be the result of gradual reactions of the soil conditioners with H^+ and Al^{3+}, thus neutralizing their potential adverse impacts on the dynamics of organic and inorganic elements for the growth and development of faunal and floral components.

The soil pH for the treatments used for restoration had similar behavior, differing in the Brazilian savannah's superficial layer (the soils are naturally acidic). The similar pH in the superficial layer was due to the contribution of *U. decumbens* in the addition of organic matter in all experimental plots. Similar results were found by Franchini et al. [35]. They report that the presence of certain plant materials is capable of enhancing the effect of liming, mobilizing the so-called front alkaline. These organic compounds have the ability to complex and mobilize Ca and Mg, raise the pH and neutralize Al. Similar results were verified by Fonseca et al. [36]. The pH increase in their study with the phytomass decomposition of green manure may explain the increase in pH, mainly due to the contribution of organic matter. Raij et al. [37] reported that the contents of Presin and SOM in tropical soils in Brazil typically range from 13 to 30 mg dm^{-3} and from 16 to 30 mg dm^{-3}, respectively. The results of this work for the contents of Presin and SOM were lower than the ones reported in the literature. Despite the substantial improvement of chemical properties in the Brazilian savannah site, the integrated systems were not efficient enough when it came to Presin and SOM. Considering the macrofaunal patterns reported in the literature, the relatively low total count of macroinvertebrates was probably due to an unsatisfactory availability of Presin and SOM. Chemical properties certainly had a crucial influence on the time–space distribution and ecological features of macroorganisms dwelling through the depths of 0.00–0.05, 0.00–0.10 and 0.10–0.20 m. Therefore, one could expect the macrofauna to reach a great total count in microclimates where suitable food sources are available. Organic matter and nutrients are vital to meet the global energy demand for the food web of macroinvertebrates, regardless of climate [38–41].

The soil-dwelling macrofauna of the Brazilian savannah site consisted mostly of termites; beetles, white grubs and ants made up the other relevant smaller insect orders. These macroinvertebrates were more sensitive to edaphoclimatic changes during the process of restoring the degraded soil than termites, beetles, white grubs and ants. The macrofauna taxonomy was indeed dependent on the availability and quality of food sources and weather conditions, varying drastically with the microclimate, period of sampling and soil depth.

Marchão et al. [34] carried out a scientific study on the ecological aspects of macroinvertebrates under integrated crop–livestock systems in the Brazilian savannah. The authors found 194 morphospecies of macroinvertebrates belonging to 30 groups, orders or families, including mostly Coleoptera, Diplopoda, Formicidae, Isoptera and Oligochaeta. They asserted that conventional and integrated fields of soybean yielded exactly 105 and 102 morphospecies, respectively, while the permanent pastures had only 37 morphospecies, probably due to uncontrollable and unpredictable conditions for growth, development and reproduction. The authors still reported a total of about 4790 ind m^{-2} for native vegetation and 980 ind m^{-2} for integrated systems, which had lower species diversity and abundance. Complementarily, the most predominant groups of macroinvertebrates were Isoptera and Formicidae, with significant values of specific abundance of roughly 4340 and 245 ind m^{-2}, respectively; Formicidae was the most substantial family with 60 morphospecies of Myrmicinae, Formicinae and Panerinae, which are subfamilies naturally occurring in forests and grasslands. Lammel et al. [42] conducted a scientific study on the microbiological and macrofaunal properties of conventional and organic systems of coffee crop. The authors reported that integration of coffee crop with *U. decumbens* and Arachis pintoi had macrofauna which heterogeneously consisted of Coleoptera, Hymenoptera, Isoptera, Mollusca and Oligochaeta. Contextually, a scientific study was performed on the spatial profile of macroinvertebrates and on soil properties [43]. The authors reported that the macrofauna of riparian forest mostly consisted of Aranae, Coleoptera, Dermaptera, Diptera, Haplotaxida, Hymenoptera and Mollusca, all of which jointly yielded a total abundance of 43.1 ind m^{-2}. They still stated that natural forests were richer in macroinvertebrates when compared to agroecosystems and artificial woodlands, which is in line with the results of this work, as the native forest was the most receptive microhabitat for the coexistence of Aranae, Coleoptera, Dermaptera, Diplopoda, Isoptera, Haplotaxida, Hemiptera, Hymenoptera and Orthoptera.

According to Sithole et al. [7], who performed a scientific study on the impacts of conservationist agriculture on soil properties and productive yield of corn crop (*Zea mays* L.), Aranae, Chilopoda and Gastropoda made up the macrofauna under a conventional system, whereas Coleoptera, Diplopoda and Isoptera composed the macrofauna under a no-tillage system, with a total abundance ranging from as low as 30 to upwards of 40 ind m^{-2}. The authors pointed out that plant material remaining on the soil surface improved the patterns of macrofauna by not only controlling the temperature and relative humidity but also by releasing food sources and offering worthy microhabitats for growth, development and reproduction. From the perspective of Suárez et al. [44], who performed a scientific study on the soil macrofaunal properties under land uses in the Colombian Amazon, Isoptera and Hymenoptera were predominant over other groups of macroinvertebrates, including mostly Aranae, Chilopoda, Diplura and Pseudoscorpionida. Out of all 7854 individuals, 2937 and 2241 were termites and ants, respectively, which is consistent with the predominance of termites over other macroorganisms in this study. The authors identified 21 taxonomic groups, which is inconsistent with the results of this study. A study on the effects of the preparation and management of agricultural systems for the production of corn crop on the chemical and physical properties of the soil macrofauna has been conducted [31]. The authors asserted that Aranae, Carabidae, Chilopoda, Enicocephalidae, Latridiidae, Lumbricidae, Scarabaeidae and Staphylinidae composed the highly heterogenous macrofauna of no-tillage and conventional systems, with or without the removal of plant material.

Wang et al. [45] performed a scientific study on the responses of macroinvertebrates to water stress in agroforest systems. The authors found macrofauna of 218 individuals belonging to 13 genera, 5 orders and 11 families; *Drawida* sp. and *Eisenia* sp. were the genera of earthworms with the highest percentages of 48.2 and 21.1% of the total samples from co-cropping systems of *Glycine max*, *Capsicum annuum* and *Zanthoxylum bungeanum*. They emphasized the physicochemical composition of plant material and relative humidity of the soil as relevant factors determining the composition of the food web of macroinvertebrates. Plant tissues of lower carbon to nitrogen ratio benefited the macroinvertebrates; in contrast, plant alkaloids offered potential risks to macroorganisms, especially earthworms, eggs and larval stages of arthropods. Velasquez and Lavelle [30] carried out a scientific study on macroinvertebrates as bioindicators of ecosystem services in agricultural landscapes. The authors reported that soil-dwelling macrofauna mostly consisted of Isoptera and Oligochaeta, with significant values of specific abundance of 1461 and 408 ind m^{-2}, respectively; other relevant groups were Coleoptera and Myriapoda, with lower values of specific abundance of 263 and 256 ind m^{-2}, respectively. They pointed out that the pastures improved the macrofaunal properties by cooling down the microclimate by shading, thus biologically preventing the macroinvertebrates from potential body dehydration due to high temperatures. It is worth citing this reference in this work in order to better understand how the morphological architecture and physiology of the pasture species *U. decumbens* could have assisted the integrated systems to restore macrofaunal properties of the Brazilian savannah site, mostly with regard to beetles and white grubs.

Webster et al. [32] performed a scientific study on macrofauna as a bioindicator of soil functioning in agroecosystems. The authors recorded an impressive 20 orders and 60 families of macroinvertebrates in pastures. Coleoptera, Hymenoptera, Isoptera and Oligochaeta were the most predominant groups; ants, earthworms, beetles and termites accounted for 37.8, 16.8, 15.4 and 12.4% of the total macrofauna, respectively. They still reported that pastures under degradation caused the earthworms to have a lower specific abundance of 27.3 ind m^{-2}, as compared to managed pastures with 56.9 ind m^{-2}, which is inconsistent with the results of this work. In the particular case of this study, Haplotaxida practically did not exist in integrated systems of green manures with *U. decumbens*, regardless of the application of limestone and agricultural gypsum. Globally, despite dissimilarities in diversity, richness, evenness and total abundance, orders and classes of macroinvertebrates reported through this work were relatively in agreement with the taxonomic aspects of the macrofauna reported in the literature. The taxonomic aspects of macrofauna vary drastically with land use, management practices and weather conditions as well as with other biotic and abiotic factors influencing the growth, development and reproduction of macroinvertebrates [46–48].

The long-term integrated systems of green manures with tropical pasture with or without the application of limestone and agricultural gypsum proved to be technically viable to recover soil macrofaunal properties of formerly degraded sites by earthworks and dam foundations. Irrespective of period of sampling and soil depth, the succession of *Stizolobium* sp. with *U. decumbens* with the application of soil conditioners had an insignificant count of macroinvertebrates and, consequently, a lower total abundance. Yet, this integrated system provided the macrofauna of the recovered site of Brazilian savannah with the most balanced indexes of diversity, dominance and evenness. Hence, *Stizolobium* sp. was simultaneously the most suitable N-fixing plant for the growth, development and reproduction of Aranae, Coleoptera, Dermaptera, Diplopoda, Hymenoptera, Isoptera, and eggs and larval stages of arthropods taxonomically unidentified under subsequent cultivation of *U. decumbens*. The next most ecologically suitable integrated system was the succession of *Cajanus* sp., *Canavalia* sp. and *U. decumbens* with limestone. In fact, the greater the availability of sources of organic and mineral elements released from the crop plants and soil conditioners during the process of recovery, the more appropriate the indexes of Shannon, Simpson and Pielou. The higher the diversity and evenness and the lower the dominance, the more versatility and advantages the macrofauna can provide for plant growth and development, such as improvement in the mechanical resistance of soil to penetration, formation and stabilization of biogenic aggregates, infiltration and retention of water, gaseous exchanges, cycling of nutrients, decomposition of organic matter and the controlling of temperature. In addition to availability, the quality of plant material was another relevant factor determining the patterns of soil-dwelling macrofauna. The C:N ratio may have influenced the decomposition of organic matter, making it readily assimilable for spiders, beetles, white grubs, adult millipedes, earwigs, adult centipedes, ants and termites, thus satisfying their respective energy demands. Further scientific studies on how physicochemical composition of plant material of the genera of green fertilizers *Cajanus* sp., *Canavalia* sp. and *Stizolobium* sp. influences the time–space distribution, diversity, dominance, evenness and total abundance of macrofauna of the Brazilian savannah are, therefore, necessary to validate the hypothesis.

In the winter, after 17 years of implementation of the research and monitoring of the properties of the soil under study, the exposed soil, interestingly, had the highest indexes of dominance and evenness, exclusively due to the inexpressive presence of termites making up the greatest insect order, Isoptera. Such a finding is of high importance to documenting how devastating the installation of a hydroelectric power plant is when it comes to the diversity and richness of the Brazilian savannah macrofauna and how termites are impressively resilient to the biologically, chemically and physically degraded soil by mechanical forces. In the summer, after 17 years of implementation of the research and monitoring of the properties of the soil under study, the total absence of macrofauna, including termites, in exposed soil meant that the anthropogenically degraded microclimate became much more sensitive to internal and external biotic and abiotic factors, suppressing the macrofauna. The noticeable scarcity of macroinvertebrates in the exposed soil was probably due to the continuously rising temperature, the low relative humidity of the soil and the low availability of food sources because of the low density of native vegetation in the process of natural recovery. Vegetation is evidently highly important in restoring and protecting the soil biota against stressing weather agents. Technically and ecologically, the more adequate indexes of diversity, dominance and evenness of macroinvertebrates dwelling through the depths of 0.00–0.05, 0.05–0.10 and 0.10–0.20 m in integrated systems and native forest proved how relevant the flora, either natural or artificial, was in the recovery of the macrofaunal properties of the heavily degraded Brazilian savannah site in the beginning of the field experiment that was conducted for 17 years. Effectively, the soil under mechanical degradation by earthworks and dam foundations was the harshest microclimate for the food web of macroinvertebrates. Increased contents of H^+ and Al^{3+} and decreased contents of SOM, Presin, K, Ca, Mg, in combination with heightened apparent density and microporosity, predictably declined the growth, development and reproduction of macroinvertebrates. Therefore, the macrofauna was dependent on the physical and chemical properties of the soil.

In the summer, after 17 years of implementation of the research and monitoring of the properties of the soil under study, the macrofauna increased to a higher density at the depths of 0.05–0.10 and 0.10–0.20 m. The high temperature and low relative humidity of the soil probably forced the macroinvertebrates to migrate from the top to deeper depths, where microhabitats were theoretically milder for the growth, development and reproduction of spiders, beetles, white grubs, adult millipedes, earwigs, adult centipedes, stinkbugs, earthworms, ants, termites, crickets, eggs and larval stages of arthropods taxonomically unidentified. The spatial distribution of the macrofauna was evidently dependent on the floral composition of the integrated system's physical and chemical properties of the substrate and on the weather agents. The rainfall, temperature, solar irradiance, relative humidity of the air, habits of dispersion and colonization top the list of the most relevant biotic and abiotic factors ubiquitously influencing the time–space profile of macroinvertebrates, as pointed by Marchão et al. [34]. The authors stated that savannah forest, pastures, conventional fields of soybean and integrated crop–livestock systems had a greater number of macroinvertebrates in the first 0.30 m; Coleoptera and Oligochaeta, with percentages of

85% and 75% of the total macrofauna, respectively, appeared to be more predominant at a depth of 0.00–0.10 m, beneath on forestry litterfall, whereas Diptera and Formicidae were present at higher abundance at a depth of 0.10–0.30 m, mostly in co-cultivation of legumes with pasture. Soil depth determines the spatial distribution of macrofauna because of the availability of food sources and gaseous exchanges through the pores. In a compacted soil, low oxygen flow often makes the macrofauna decline [44,49,50]. In sites heavily degraded by pedoclimatic or anthropogenic forces, macroinvertebrates commonly concentrate in deeper depths as a function of their adaptive mechanisms to stressing agents, such as high temperature and low moisture [45,51–53]. The escaping behavior of the macroinvertebrates was, therefore, a reliable bioindicator of macrofauna stressed by mechanical compaction and depletion of chemical properties, such as Presin and SOM.

According to Rampelotto et al. [54], who conducted a scientific study on the changes in abundance, diversity and structure of soil biota in the Brazilian savannah, dissimilarities in patterns of macrofauna under natural and artificial environments are normal. The authors asserted that the land uses (sugarcane field, pasture and forest) determined different trophic relations and behavior habits of bioturbators, reorganizers and weathering agents of clay minerals due to the particularities in the availability and quality of food sources as well as the physical and chemical properties of the substrate. Franco et al. [55] performed a scientific study on the association between soil structure and the abundance of macroinvertebrates under economic exploitations. The authors pointed out that 58% of the total macrofauna in pastures were termites, which is in line with the predominance of termites over other groups of macroinvertebrates reported in this work. They still stated that termites were the greatest soil engineering bioagents at performing foraging, tunneling and nesting; these habits provide the soil functioning with a wide range of ecological services, such as cycling of nutrients, decomposition of organic matter, formation and stabilization of biogenic aggregates, improvement of hydraulic flow and gaseous exchanges. These references are worth citing in this work to evidence how particular ecological aspects of integrated systems, exposed soil, and forest could have influenced in such exclusive ways the macrofauna of the Brazilian savannah and how the predominance of termites could have assisted the biosystems in rehabilitating the soil health. The genera of N-fixing plants, *Cajanus* sp., *Canavalia* sp. and *Stizolobium* sp., the species of pasture, *U. decumbens*, and termites were apparently complementary—the tropical crop plants offered worthy conditions for the food and reproduction habits of termites, while termites positively changed the microclimates, making them biologically, physically and chemically suitable for plant growth and development. The habits of termites may have assisted with the process of soil recovery in which green fertilizers and tropical pasture growing in harsh environments are used. Winter and summer pastures are biosystems technically effective at mitigating the degradation of soil biological, chemical and physical properties [56–61].

Kamau et al. [1] performed a scientific study on the effects of dominant tree species and gradient soil degradation on macrofauna. The authors reported a total abundance of 14 to 389 ind m^{-2} for earthworms, 82 ind m^{-2} for termites and 8 ind m-2 for spiders in artificial forestry systems of *Z. gilletii*, *Eucalyptus grandis* and *Croton megalocarpus*. They incisively asserted that the time of cultivation, availability and quality of plant materials, and soil chemical properties, mostly including total carbon and total N, were the most important factors influencing the trophic relationships, intensity, quality and regularity of ecological functions of macroinvertebrates. In the particular case of this work, the native forest was replete with termites, with a moderate number of beetles, earthworms and ants, which is inconsistent with the ecological features of the macrofauna reported in the literature. Effectively, the macrofauna is dependent on the forest ecosystem. In the Brazilian savannah, soils are naturally more acidic, richer in Al^{3+} and poorer in SOM and mineral elements. Such chemical properties could explain the inferiority of earthworms in relation to termites and beetles. Annelids are often more sensitive to the quality of substrate due to their more fragile body structure. Mechanization, pesticides and fertilizers pose risks to earthworms, termites, beetles, spiders, crickets and millipedes [1,62,63].

According to Melman et al. [31], the production of corn crop under a no-tillage system with preservation of plant material on the soil surface caused the macrofauna to increase its total abundance of 637 ind m^{-2} and body biomass of earthworms of 50.4 g m^{-2}, compared to conventional systems, with and without the removal of straw, which were 218 ind m^{-2} and 7.7 g m^{-2}, respectively. The authors attributed the results to improved structure, infiltration and retention of water, pH, electric conductivity, availability of organic carbon, total N and Presin. They complementarily argued that crop plants growing in symbiosis with plant-growth-promoting microbes, such as N-fixing bacteria and arbuscular mycorrhizal fungi, were often less responsive to earthworms. Considering the reference, a relatively lower count of earthworms under integrated systems consisting of genera of green fertilizers with tropical pasture with or without the application of limestone and gypsum was probably due to effective symbiotic relationships between *Cajanus* sp., *Canavalia* sp. and *Stizolobium* sp., and diazotrophic bacteria naturally existing in the soil.

The positive linear associations between stinkbugs, termites, earthworms, beetles and earwigs probably meant that these macroinvertebrates did not spend a substantial portion of energy on competing interspecifically for ecological niches and food sources. In particular, stinkbugs, beetles and earwigs were typical components of aboveground macrofauna, while the earthworms and termites were underground macroinvertebrates predominantly existing in the drilosphere and termitosphere, respectively. Foraging, tunneling and nesting habits by termites and earthworms may have positively influenced the stinkbugs and earwigs by indirectly increasing the availability of food sources, as well as by conditioning of the microclimate. Therefore, the greater the count of Coleoptera, Haplotaxida and Isoptera, the greater the count of Dermaptera and Hemiptera. The negative correlation between spiders and stinkbugs meant that Aranae was the natural enemy of Hemiptera. Therefore, the higher the predatory pressure of spiders on stinkbugs, the smaller the insect group of Hemiptera. This finding is of high importance to the development of biological control. The positive correlation between termites and the total of individuals was solely due to the predominance of Isoptera over other orders or classes of macroinvertebrates.

The negative correlation between the total of orders and potential acidity meant that the higher the cumulative content of H$^+$ and Al^{3+}, the lesser the spatial presence of Aranae, Coleoptera, Dermaptera, Diplopoda, Haplotaxida, Hemiptera, Hymenoptera, Isoptera and Orthoptera. This linear association proved that the exposed soil was indeed the harshest microclimate for the growth, development and reproduction of spiders, beetles, white grubs, adult millipedes, earwigs, adult centipedes, earthworms, ants, termites and crickets because of its high potential acidity. The higher SOM content and greater count of macroinvertebrates in integrated systems with nutritional supplementation of limestone and gypsum was in line with the correlations between macronutrients, organic matter and macrofauna. We confidently advocate the use of limestone and gypsum as the most affordable management practice to help the genera of N-fixing plants *Cajanus* sp., *Canavalia* sp. and *Stizolobium* sp. and the species of pasture *U. decumbens* to mitigate situations of soil degradation in the Brazilian savannah.

In line with the patterns recognized through the Pearson correlation test, the primary component referring to the chemical neutralization of toxic elements proved that the limestone and gypsum are technically effective in assisting integrated systems of green fertilizers with pasture to restore chemical and macrofaunal properties of sites formerly degraded by earthworks and dam foundations. The soil conditioners offered key benefits to the successions, including, but not limited to, the amendment of potential acidity by neutralizing H$^+$ and Al^{3+}, improvement of availability of minerals readily assimilable by crop plants, such as Ca, Mg and S, as well as hypothetical acceleration of decomposition of native organic matter and agricultural residues remaining on the soil surface after harvesting at the flowering stage. A high density of macroinvertebrates, specially termites, great availability of Presin, K, Ca, Mg and SOM and high SEC/CEC ratio were the most sensitive and reliable indicators of chemically balanced soil. The secondary component correlated positively with the total of the taxonomic orders but negatively with the potential acidity and SOM content, which meant the more diverse the sources of organic and mineral elements on the soil surface, the more heterogeneous the food web of macroinvertebrates dwelling in the soil beneath the litterfall. The ternary component described the earthworms and larval stages of arthropods taxonomically unidentified, as well as pH and potential acidity, as the most important biological and chemical indicators as to the degree of decomposition of SOM in the aluminum-rich, naturally acidic soil of the Brazilian savannah. With significant loadings between eigenvectors and PCIII, it is possible to point out that the more acidic the soil beneath the litterfall, the more intensive the natural decomposition of organic matter by earthworms, as well as the smaller the count of white grubs and larval arthropods. Macroinvertebrates in the beginning of the cycle of life are much more sensitive to acidic substrates than those at older stages. The quaternary component figured out that crickets and termites were the most sensitive and most reliable bioindicators of the unavailability of aboveground biomass. Thus, the lower the productive yield of biomass of integrated systems of green fertilizers and *U. decumbens*, the smaller the count of crickets and termites; in the particular case of termites, the unavailability of aboveground biomass may affect their habits of foraging and nesting. The quintenary component was in line with our expectation of adult millipedes to be natural predators of larval stages of arthropods taxonomically unidentified.

Franco et al. [55] carried out a scientific study on the relation between the visual structure of the soil status and the abundance of soil engineering invertebrates depending on different land uses. The authors reported a positive correlation between the total abundance and visual evaluation scoring as well as a negative correlation between Oligochaeta and Isoptera. Such linear features proved that the less structurally consistent the soil is, the higher the count of termites and the total abundance of macroinvertebrates, as well as the smaller the number of samples of earthworms in disturbed substrate. In this study, Oligochaeta and Isoptera contrasted each other in terms of spatial representativeness, proving that earthworms and termites were bioindicators of ecologically constant microclimates as well as their reestablishment, respectively.

According to Kamau et al. [1], earthworms correlated positively with P and K and negatively with lignin content in mulching; adult centipedes correlated positively with the ratios of C:N, C:P:lignin:N and polyphenols:N and negatively with N, P, Ca and Mg; termites correlated positively with Mg and negatively with N and P; earthworms and adult centipedes positively correlated with the physicochemical quality of root tissues, while the beetles, adult millipedes, ants and spiders did not correlate with physical and chemical properties of the aboveground biomass and root system.

When analyzing the reference systematically, the predominance of termites over other groups of macroinvertebrates dwelling through the depths of 0.00–0.05, 0.05–0.10 and 0.10–0.20 m in integrated systems of green fertilizers with tropical pasture, with application of limestone and agricultural gypsum, was probably due to the release of Mg and Ca from the soil conditioners rather than due to the mineralization of N-rich agricultural residues of *Cajanus* sp., *Canavalia* sp. and *Stizolobium* sp. Fitting multiple regression models such as the ones relating termites to chemical properties of the soil could support this analytical inference.

From the perspective of [7], adult centipedes prefer C-rich plant materials, while earthworms prefer lower ratios of C:P, C:K and lignin:N. The physicochemical quality of the root system of *U. decumbens*, which hypothetically is rich in C, may, therefore, be more suitable for the food habits of adult centipedes, thus supporting the predominance of Diplopoda over Haplotaxida in integrated systems. Additionally, specific abundance of the species of earthworms, Drawida sp. and Eisenia sp., positively correlates with $N-NH_4^+$; the authors also emphasize that ammonium makes the soil acidic and increases the count of earthworms. The relatively higher density of Haplotaxida in the native forest reported in this work, compared to exposed soil and integrated systems, was probably due to the higher availability of $N-NH_4^+$ from litterfall decomposition, as the earthworms and pH did not correlate with each other.

According to Webster et al. [32], patterns of macrofauna had significantly positive correlations with clay content and yield of mulching and negative correlations with percentage of exposed soil. The authors apprised the higher degree of compaction and the lower organic matter content as physical and chemical factors suppressing the macroinvertebrates in exposed soil. Therefore, the substantially low total abundance reported through these works for macroinvertebrates in exposed soil was not only due to high Al^{3+} content but also due to high apparent density, low macroporosity and low SOM content in the anthropogenically degraded site of the Brazilian savannah at the beginning of the experiment. These stress factors commonly reduce infiltration and retention of water as well as gaseous exchanges through the pores. Furthermore, the negative correlation between the macrofauna and the percentage of exposed soil reported in the literature is paramount to realizing how beneficial plant cover is in the process of recovering soil macrofauna and its chemical properties. The availability and quality of food sources, temperature of the soil and oxygenation are factors which influence the ecological characteristics of macrofauna, as reported in the scientific study by Abail and Whalen [41] on the dynamics of earthworm populations.

In this study, Aranae did not distinguish adequately restored soil from degraded soil, since spiders were present in integrated systems of green fertilizers with tropical pasture, exposed soil and native forest, despite their lower density, as compared to termites, beetles and ants. The most desirable characteristic for a bioindicator is being a generalist; other relevant characteristics include prediction accuracy and cost-effectiveness. Velasquez and Lavelle [30] argued that the Oligochaeta and Isoptera associated with old-growth forest, also termed as primary forest, and Coleoptera associated with fallow and pasture, thus corresponding very well with the results of this study for beetles and white grubs as well as for earthworms at higher densities in the presence of *U. decumbens* and native forest, respectively. The authors still described the Myriapoda as an order of detritivore macroinvertebrates in tight association with perennial crops and that Formicidae are associated with preserved, explored or burnt forestry systems. In line with the reference, ants did not perform as a bioindicator of neither the exposed soil nor the biologically restored soil in this work. Ants were also not reliable to study and assess the process of restoring degraded soil in the Brazilian savannah. For this purpose, the use of termites is confidently recommended. The next most suitable options would be beetles and white grubs, as the Coleoptera and integrated systems of green fertilizers with *U. decumbens* were strongly associated with each other.

5. Conclusions

The long-term integration of the genera of green fertilizers *Cajanus* sp., *Canavalia* sp. and *Stizolobium* sp. with the species of pasture *Urochloa decumbens* is an inexpensive, environmentally friendly strategy for the restoration of the macrofaunal properties of a Brazilian savannah site following anthropogenic degradation due to mechanical forces of earthworks and dam foundations after the installation of a hydroelectric power plant.

The integrated systems, with and without the administration of limestone and gypsum, improved the soil chemical properties of the site under investigation.

The macrofauna was a good indicator of soil recovery.

After 17 years of the effects of these treatments for soil restoration, the soil macrofauna is either similar to or even better than the natural soil conditions.

Author Contributions: M.C.A. conceived and designed the experiments; T.M.d.S.M. performed the experiments; R.L.P.H., B.R.d.A.M. and J.G.M.C. analyzed the data; M.C.A., T.M.d.S.M. and C.d.S.B.B. contributed reagents/materials/analysis tools; T.M.d.S.M. and M.C.A., methodology; C.d.S.B.B., writing—review and editing. All authors have read and agreed to the published version of the manuscript.

Funding: This research received no external funding.

Institutional Review Board Statement: Not applicable.

Informed Consent Statement: Not applicable.

Data Availability Statement: The data presented in this study are available on request from the corresponding authors.

Acknowledgments: To FAPESP for the master's and doctoral scholarship (Proc. 2009/54804-8 and 2008/50853-1) and financial support for research project (2009/50066-2).

Conflicts of Interest: The authors declare no conflict of interest.

References

1. Reyes-Sánchez, L.B.; Horn, R.; Costantini, E.A.C. (Eds.) *Sustainable Soil Management as a Key to Preserving Soil Biodiversity and Stopping Its Degradation;* International Union of Soil Sciences (IUSS): Vienna, Austria, 2022.
2. Kamau, S.; Barrios, E.; Karanja, N.K.; Ayuke, F.O.; Lehmann, J. Soil macrofauna abundance under dominant tree species increases along a soil degradation gradient. *Soil Biol. Biochem* **2017**, *112*, 35–46. [CrossRef]
3. Lavelle, P.; Decaëns, T.; Aubert, M.; Barot, S.; Blouin, M.; Bureau, F.; Margerie, P.; Mora, P.; Rossi, J.-P. Soil invertebrates and ecosystem services. *Eur. J. Soil Biol.* **2006**, *42*, S3–S15. [CrossRef]
4. Obrycki, J.F.; Karlen, D.L. Is Corn Stover Harvest Predictable Using Farm Operation, Technology, and Management Variables? *J. Agron.* **2018**, *110*, 749–757. [CrossRef]
5. Ranaivoson, L.; Naudin, K.; Ripoche, A.; Affholder, F.; Rabeharisoa, L.; Corbeels, M. Agro-ecological functions of crop residues under conservation agriculture. A review. *Agron. Sustain. Dev.* **2017**, *37*, 26. [CrossRef]
6. Sawyer, J.E.; Woli, K.P.; Barker, D.W.; Pantoja, J.L. Stover Removal Impact on Corn Plant Biomass, Nitrogen, and Use Efficiency. *J. Agron.* **2017**, *109*, 802–810. [CrossRef]
7. Sithole, N.J.; Magwaza, L.S.; Mafongoya, P.L.; Thibaud, G.R. Long-term impact of no-till conservation agriculture on abundance and order diversity of soil macrofauna in continuous maize monocropping system. *Acta Agric. Scand. B Soil Plant Sci.* **2018**, *68*, 220–229. [CrossRef]
8. Elbasiouny, H.; El-Ramady, H.; Elbehiry, F.; Rajput, V.D.; Minkina, T.; Mandzhieva, S. Plant Nutrition under Climate Change and Soil Carbon Sequestration. *Sustainability* **2022**, *14*, 914. [CrossRef]
9. Chenu, C.; Angers, D.A.; Barré, P.; Derrien, D.; Arrouays, D.; Balesdent, J. Increasing organic stocks in agricultural soils: Knowledge gaps and potential innovations. *Soil Tillage Res.* **2019**, *188*, 41–52. [CrossRef]
10. Jouquet, P.; Chintakunta, S.; Bottinelli, N.; Subramanian, S.; Caner, L. The influence of fungus-growing termites on soil macro and micro-aggregates stability varies with soil type. *Appl. Soil Ecol.* **2016**, *101*, 117–123. [CrossRef]
11. Kaiser, D.; Lepage, M.; Konaté, S.; Linsenmair, K.E. Ecosystem services of termites (*Blattoidea: Termitoidae*) in the traditional soil restoration and cropping system Zaï in northern Burkina Faso (West Africa). *Agric. Ecosyst. Environ.* **2017**, *236*, 198–211. [CrossRef]
12. Lima, S.S.; Pereira, M.G.; Pereira, R.N.; de Pontes, R.M.; Rossi, C.Q.; de Lima, S.S.; Pereira, M.G.; Pereira, R.N.; de Pontes, R.M.; Rossi, C.Q. Termite Mounds Effects on Soil Properties in the Atlantic Forest Biome. *Rev. Bras. Cienc. Solo* **2018**, *42*. [CrossRef]
13. Sarker, J.R.; Singh, B.P.; Dougherty, W.J.; Fang, Y.; Badgery, W.; Hoyle, F.C.; Dalal, R.C.; Cowie, A.L. Impact of agricultural management practices on the nutrient supply potential of soil organic matter under long-term farming systems. *Soil Tillage Res.* **2018**, *175*, 71–81. [CrossRef]
14. Kitamura, A.E.; Tavares, R.L.M.; Alves, M.C.; Souza, Z.M.; Siqueira, D.S. Soil macrofauna as bioindicator of the restoration of degraded Cerrado soil. *Cienc. Rural* **2020**, *50*, 8. [CrossRef]
15. Harit, A.; Moger, H.; Duprey, J.-L.; Gajalakshmi, S.; Abbasi, S.A.; Subramanian, S.; Jouquet, P. Termites can have greater influence on soil properties through the construction of soil sheetings than the production of above-ground mounds. *Insect. Soc.* **2017**, *64*, 247–253. [CrossRef]
16. Oliveira, C.C.; Alves, F.V.; Almeida, R.G.; Gamarra, É.L.; Villela, S.D.J.; Almeida Martins, P.G.M. Thermal comfort indexes assessed in integrated production systems in the Brazilian savannah. *Agrofor. Syst.* **2018**, *92*, 1659–1672. [CrossRef]
17. Sone, J.S.; Oliveira, P.T.S.; Zamboni, P.A.P.; Vieira, N.O.M.; Carvalho, G.A.; Macedo, M.C.M.; Araujo, A.R.; Montagner, D.B.; Alves Sobrinho, T. Effects of long-term crop-livestock-forestry systems on soil erosion and water infiltration in a Brazilian Cerrado site. *Sustainability* **2019**, *11*, 5339. [CrossRef]
18. Alves, M.C.; Suzuki, L.G.A.S.; Suzuki, L.E.A.S. Densidade do solo e infiltração de água como indicadores da qualidade física de um Latossolo Vermelho distrófico em recuperação. *Rev. Bras. Cienc. Solo* **2007**, *31*, 617–625. [CrossRef]
19. Campos, F.S.; Alves, M.C. Uso de lodo de esgoto na reestruturação de solo degradado. *Rev. Bras. Cienc. Solo* **2008**, *32*, 1389–1397. [CrossRef]
20. Bonini, C.B.S.; Alves, M.C. Aggregate stability of a degraded Oxisol in restoration with green manure, lime and gypsum. *Rev. Bras. Cienc. Solo* **2011**, *35*, 1263–1270. [CrossRef]
21. Bonini, C.D.S.B.; Alves, M.C.; Montanari, R. Recuperação da estrutura de um Latossolo vermelho degradado utilizando lodo de esgoto. *Rev. Ciênc. Agron* **2015**, *10*, 34–42. [CrossRef]
22. Monreal, C.M.; Alves, M.C.; Schnitzer, M.; Filho, S.N.S.; Batista Bonini, C.D.S. Mass spectrometry of organic matter influenced by long-term pedogenesis and a short-term reclamation practice in an Oxisol of Brazil. *Can. J. Soil Sci.* **2016**, *96*, 64–85. [CrossRef]

23. Giácomo, R.G.; Souza, R.C.; Alves, M.C.; Pereira, M.G.; Garcia de Arruda, O.; Paz González, A. Soil fauna: Bioindicator of soil restoration in Brazilian savannah. *Rev. Ciênc. Agron.* **2017**, *12*, 236–243.

24. Neto, A.B.; Bonini, C.D.S.B.; Bisi, B.S.; dos Reis, A.R.; Coletta, L.F.S. Artificial neural network for classification and analysis of degraded soils. *IEEE Lat. Am. Trans.* **2017**, *15*, 503–509. [CrossRef]

25. Tseng, C.L.; Alves, M.C.; Crestana, S. Quantifying physical and structural soil properties using X-ray microtomography. *Geoderma* **2018**, *318*, 78–87. [CrossRef]

26. Santos, H.G.; Jocomine, P.K.T.; Anjos, L.H.C.; Oliveira, V.A.; Lumbrearas, J.F.; Coelho, M.R.; Almeida, J.Á.; Filho, J.C.A.; Oliveira, J.B.; Cunha, T.J.F. *Sistema Brasileiro de Classificação de Solos*, 5th ed.; Embrapa: Rio de Janeiro, Brasil, 2018; p. 286.

27. Soil Survey Staff. *Keys to Soil Taxonomy | NRCS*, 12th ed.; EUA: Washington, DC, USA, 2014.

28. Quaggio, J.A.; Raij, B.; Malavolta, E. Alternative use of the SMP-buffer solution to determine lime requirement of soils. *Commun. Soil Sci. Plant Anal.* **1985**, *16*, 245–260. [CrossRef]

29. Teixeira, P.C.; Donagemma, G.K.; Fontana, A.; Teixeira, W.G. *Manual de Métodos de Análise de Solo*, 3rd ed.; Embrapa: Brasilia, Brazil, 2017; p. 557.

30. Velasquez, E.; Lavelle, P. Soil macrofauna as an indicator for evaluating soil-based ecosystem services in agricultural landscapes. *Acta Oecol.* **2019**, *100*, 103446. [CrossRef]

31. Melman, D.A.; Kelly, C.; Schneekloth, J.; Calderón, F.; Fonte, S.J. Tillage and residue management drive rapid changes in soil macrofauna communities and soil properties in a semiarid cropping system of Eastern Colorado. *Appl. Soil Ecol.* **2019**, *143*, 98–106. [CrossRef]

32. Webster, E.; Gaudin, A.C.M.; Pulleman, M.; Siles, P.; Fonte, S.J. Improved Pastures Support Early Indicators of Soil Restoration in Low-input Agroecosystems of Nicaragua. *Environ. Manag.* **2019**, *64*, 201–212. [CrossRef]

33. R Core Team. *R: A Language and Environment for Statistical Computing*; R Foundation for Statistical Computing: Vienna, Austria, 2018; Available online: https://www.R-project.org/ (accessed on 8 February 2023).

34. Marchão, R.L.; Lavelle, P.; Celini, L.; Balbino, L.C.; Vilela, L.; Becquer, T. Soil macrofauna under integrated crop-livestock systems in a Brazilian Cerrado Ferralsol. *Pesqui Agropecu Bras.* **2009**, *44*, 1011–1020. [CrossRef]

35. Franchini, J.C.; Hoffmann-Campo, C.B.; Torres, E.; Miyazawa, M.; Pavan, M.A. Organic composition of green manure during growth and its effect on cation mobilization in an acid oxisol. *Comm. Soil Sci. Plant Anal.* **2003**, *34*, 2045–2058. [CrossRef]

36. Fonseca, W.S.; Martins, S.V.; Villa, P.M. Green Manure as an Alternative for Soil Restoration in a Bauxite Mining Environment in Southeast Brazil. *Floresta E Ambiente* **2023**, *30*, e20220041. [CrossRef]

37. Raij, B.; Cantarella, H.; Quaggio, J.A.; Hiroce, R.; Furlani, M.C. *Recomendações de Adubação e Calagem para o Estado de São Paulo*; IAC: São Paulo, Brasil, 1986.

38. Bowles, T.M.; Jackson, L.E.; Loeher, M.; Cavagnaro, T.R. Ecological intensification and arbuscular mycorrhizas: A meta-analysis of tillage and cover crop effects. *J. Appl. Ecol.* **2017**, *54*, 1785–1793. [CrossRef]

39. Fox, J.T.; Zook, A.N.; Freiss, J.; Appel, B.; Appel, J.; Ozsuer, C.; Sarac, M. Thermal conversion of blended food production waste and municipal sewage sludge to restoreable products. *J. Clean. Prod.* **2019**, *220*, 57–64. [CrossRef]

40. Murphy, B.W. Impact of soil organic matter on soil properties—A review with emphasis on Australian soils. *Soil Res.* **2015**, *53*, 605–635. [CrossRef]

41. Abail, Z.; Whalen, J.K. Corn residue inputs influence earthworm population dynamics in a no-till corn-soybean rotation. *Appl. Soil Ecol.* **2018**, *127*, 120–128. [CrossRef]

42. Lammel, D.R.; Azevedo, L.C.B.; Paula, A.M.; Armas, R.D.; Baretta, D.; Cardoso, E.J.B.N.; Lammel, D.R.; Azevedo, L.C.B.; Paula, A.M.; Armas, R.D.; et al. Microbiological and faunal soil attributes of coffee cultivation under different management systems in Brazil. *Braz. J. Biol.* **2015**, *75*, 894–905. [CrossRef]

43. Gholami, S.; Sayad, E.; Gebbers, R.; Schirrmann, M.; Joschko, M.; Timmer, J. Spatial analysis of riparian forest soil macrofauna and its relation to abiotic soil properties. *Pedobiologia* **2016**, *59*, 27–36. [CrossRef]

44. Suárez, L.R.; Josa, Y.T.P.; Samboni, E.J.A.; Cifuentes, K.D.L.; Bautista, E.H.D.; Salazar, J.C.S.; Suárez, L.R.; Josa, Y.T.P.; Samboni, E.J.A.; Cifuentes, K.D.L.; et al. Soil macrofauna under different land uses in the Colombian Amazon. *Pesqui Agropecu Bras.* **2018**, *53*, 1383–1391. [CrossRef]

45. Wang, S.; Pan, K.; Tariq, A.; Zhang, L.; Sun, X.; Li, Z.; Sun, F.; Xiong, Q.; Song, D.; Olatunji, O.A. Combined effects of cropping types and simulated extreme precipitation on the community composition and diversity of soil macrofauna in the eastern Qinghai-Tibet Plateau. *J. Soils Sediments* **2018**, *18*, 3215–3227. [CrossRef]

46. Baretta, D.; Brescovit, A.D.; Knysak, I.; Cardoso, E.J.B.N. Trap and soil monolith sampled edaphic spiders (*Arachnida: Araneae*) in *Araucaria angustifolia* forest. *Sci. Agric.* **2007**, *64*, 375–383. [CrossRef]

47. Bartz, M.L.C.; Pasini, A.; Brown, G.G. Earthworms as soil quality indicators in Brazilian no-tillage systems. *Appl. Soil Ecol.* **2013**, *69*, 39–48. [CrossRef]

48. Bottinelli, N.; Jouquet, P.; Capowiez, Y.; Podwojewski, P.; Grimaldi, M.; Peng, X. Why is the influence of soil macrofauna on soil structure only considered by soil ecologists? *Soil Tillage Res* **2015**, *146*, 118–124. [CrossRef]

49. Pauli, N.; Barrios, E.; Conacher, A.J.; Oberthür, T. Soil macrofauna in agricultural landscapes dominated by the Quesungual Slash-and-Mulch Agroforestry System, western Honduras. *Appl. Soil Ecol.* **2011**, *47*, 119–132. [CrossRef]

50. Santos, D.C.; Guimarães Júnior, R.; Vilela, L.; Pulrolnik, K.; Bufon, V.B.; de S. França, A.F. Forage dry mass accumulation and structural characteristics of Piatā grass in silvopastoral systems in the Brazilian savannah. *Agric. Ecosyst. Environ.* **2016**, *233*, 16–24. [CrossRef]

51. Gongalsky, K.B.; Persson, T. Restoration of soil macrofauna after wildfires in boreal forests. *Soil Biol. Biochem.* **2013**, *57*, 182–191. [CrossRef]

52. Mariotte, P.; Le Bayon, R.-C.; Eisenhauer, N.; Guenat, C.; Buttler, A. Subordinate plant species moderate drought effects on earthworm communities in grasslands. *Soil Biol. Biochem.* **2016**, *96*, 119–127. [CrossRef]

53. Potapov, A.M.; Goncharov, A.A.; Semenina, E.E.; Korotkevich, A.Y.; Tsurikov, S.M.; Rozanova, O.L.; Anichkin, A.E.; Zuev, A.G.; Samoylova, E.S.; Semenyuk, I.I.; et al. Arthropods in the subsoil: Abundance and vertical distribution as related to soil organic matter, microbial biomass and plant roots. *Eur. J. Soil Biol.* **2017**, *82*, 88–97. [CrossRef]

54. Rampelotto, P.H.; de Siqueira Ferreira, A.; Barboza, A.D.M.; Roesch, L.F.W. Changes in Diversity, Abundance, and Structure of Soil Bacterial Communities in Brazilian Savanna Under Different Land Use Systems. *Microb. Ecol.* **2013**, *66*, 593–607. [CrossRef]

55. Franco, A.L.C.; Cherubin, M.R.; Cerri, C.E.P.; Guimarães, R.M.L.; Cerri, C.C. Relating the visual soil structure status and the abundance of soil engineering invertebrates across land use change. *Soil Tillage Res.* **2017**, *173*, 49–52. [CrossRef]

56. Geraei, D.S.; Hojati, S.; Landi, A.; Cano, A.F. Total and labile forms of soil organic carbon as affected by land use change in southwestern Iran. *Geoderma Reg.* **2016**, *7*, 29–37. [CrossRef]

57. Hurisso, T.T.; Culman, S.W.; Horwath, W.R.; Wade, J.; Cass, D.; Beniston, J.W.; Bowles, T.M.; Grandy, A.S.; Franzluebbers, A.J.; Schipanski, M.E.; et al. Comparison of Permanganate-Oxidizable Carbon and Mineralizable Carbon for Assessment of Organic Matter Stabilization and Mineralization. *J. Soil Sci. Soc. Am. J.* **2016**, *80*, 1352–1364. [CrossRef]

58. Korboulewsky, N.; Perez, G.; Chauvat, M. How tree diversity affects soil fauna diversity: A review. *Soil Biol. Biochem.* **2016**, *94*, 94–106. [CrossRef]

59. Moura, E.G.; Aguiar, A.C.F.; Piedade, A.R.; Rousseau, G.X. Contribution of legume tree residues and macrofauna to the improvement of abiotic soil properties in the eastern Amazon. *Appl. Soil Ecol.* **2015**, *86*, 91–99. [CrossRef]

60. Sun, F.; Pan, K.; Tariq, A.; Zhang, L.; Sun, X.; Li, Z.; Wang, S.; Xiong, Q.; Song, D.; Olatunji, O.A. The response of the soil microbial food web to extreme rainfall under different plant systems. *Sci. Rep.* **2016**, *6*, 37662. [CrossRef] [PubMed]

61. Zhu, X.; Zhu, B. Diversity and abundance of soil fauna as influenced by long-term fertilization in cropland of purple soil, China. *Soil Tillage Res.* **2015**, *146*, 39–46. [CrossRef]

62. Tsiafouli, M.A.; Thébault, E.; Sgardelis, S.P.; de Ruiter, P.C.; van der Putten, W.H.; Birkhofer, K.; Hemerik, L.; de Vries, F.T.; Bardgett, R.D.; Brady, M.V.; et al. Intensive agriculture reduces soil biodiversity across Europe. *Glob. Chang. Biol.* **2015**, *21*, 973–985. [CrossRef]

63. de Vries, F.T.; Thébault, E.; Liiri, M.; Birkhofer, K.; Tsiafouli, M.A.; Bjørnlund, L.; Jørgensen, H.B.; Brady, M.V.; Christensen, S.; Ruiter, P.C.; et al. Soil food web properties explain ecosystem services across European land use systems. *Proc. Natl. Acad. Sci. USA* **2013**, *110*, 14296–14301. [CrossRef]

soil systems

MDPI

Article

Long Term of Soil Carbon Stock in No-Till System Affected by a Rolling Landscape in Southern Brazil

Edivaldo L. Thomaz [1,*] and Julliane P. Kurasz [2]

[1] Soil Erosion Laboratory, Department of Geography, Universidade Estadual do Centro-Oeste, UNICENTRO, Élio Antonio Dalla Vecchia, 838-Bairro-Vila Carli, Guarapuava 85040-080, Brazil
[2] Department of Geography, Universidade Estadual do Centro-Oeste, UNICENTRO, Élio Antonio Dalla Vecchia, 838-Bairro-Vila Carli, Guarapuava 85040-080, Brazil
* Correspondence: thomaz@unicentro.br

Abstract: In the 1960s, a conservationist agricultural practice known as a "no-tillage system" was adopted. Several benefits such as soil erosion reduction and soil carbon sequestration, among others, could be ascribed to no-till systems. Therefore, it is important to evaluate the long-term sustainability of this agricultural system in different environments. This study has the objective to evaluate the soil organic carbon dynamic in a no-till system (40-year) and on a rolling landscape in Southern Brazil. A systematic grid with four transversal–longitudinal transects was used for soil sampling. Soil samples from 0–20, 20–40, and 40–60 cm depths were collected (16 trenches × 3 depths × 1 sample per soil layer = 48), and a forest nearby was used as control (4 trenches × 3 depths × 1 sample = 12). The soil at the forest site showed 20% more carbon stock than no-till at the 0–20 cm soil depth. However, the entire no-till soil profile (0–60 cm) showed similar soil carbon as forest soil. The soil carbon stock (0–20 cm) in no-till was depleted at a rate of 0.06 kg C m^{-2} year^{-1}, summing up to a carbon loss of 2.43 kg C m^{-2}. In addition, the non-uniform hillslope affected the soil carbon redistribution through the landscape, since the convex hillslope was more depleted in carbon by 37% (15.87 kg C m^{-2}) when compared to the concave sector (25.27 kg C m^{-2}). On average, the soil carbon loss in the subtropical agroecosystem was much lower than those reported in literature, as well as our initial expectations. In addition, the no-till system was capable of preserving soil carbon in the deepest soil layers. However, presently, the no-till system is losing more carbon in the topsoil at a rate greater than the soil carbon input.

Keywords: soil-geomorphology; oxisols; soil conservation; sustainability; food system; food security

Citation: Thomaz, E.L.; Kurasz, J.P. Long Term of Soil Carbon Stock in No-Till System Affected by a Rolling Landscape in Southern Brazil. *Soil Syst.* **2023**, *7*, 60. https://doi.org/10.3390/soilsystems7020060

Academic Editor: Luis Eduardo Akiyoshi Sanches Suzuki

Received: 24 February 2023
Revised: 23 May 2023
Accepted: 1 June 2023
Published: 7 June 2023

1. Introduction

Contemporary society is challenged with the crucial paradox of producing more food, fiber, and biofuel to supply for an increasing global population, especially on increasingly degraded soils and agroecosystems worldwide [1–3]. The conversion of forest and natural grassland to a cropping system is detrimental to soil carbon stock [4–6]. Opting for an intensive agricultural system can increase crop production. However, this may decrease the ecosystem services including water quality, biodiversity, and carbon sequestration. A cropland with restored ecosystem services may mitigate the effects of agriculture intensification, and guarantee sustainable development [7,8].

In the 1960s, the no-tillage system was adopted as a conservationist agricultural practice. The basic background principles of this system are: (1) avoid soil disturbance; (2) keep the residue (mulching) over the topsoil; (3) plan crop rotations (i.e., not only double crops such as wheat–soybean); and (4) contour tillage practices and soil conservation according to the terrain [9,10]. Many benefits were ascribed to the no-till system, especially when compared to conventional tillage (CT), such as enhancement of soil aggregate stability [11,12], soil organic matter improvement and carbon sequestration [13,14], and reduction of runoff and soil loss [8,15].

Soil organic matter and carbon stock are of the utmost importance to soil functions, ecosystem productivity, and soil carbon sequestration. Despite the disagreement about the capacity of the no-till system to sequester carbon [16,17], the no-till system is recognized as an agricultural conservationist practice that ensures conservation of carbon, soil, water, and sustainability of the agroecosystem [10,18].

Soil carbon is very sensitive to land conversion and can therefore be lost in a few decades in temperate regions, and in a few years in tropical regions [19–21]. Even with advances in the past few years regarding soil carbon dynamics in agricultural systems, some research priorities need to focus on long-term soil resilience on soil organic carbon (depletion or conservation), assessment of actual carbon stock in space-time, assessment of soil carbon stock beyond the topsoil limit (0–10 or 0–20 cm), and effect of soil management and soil erosion on carbon stock [6,22,23]. In addition, it is necessary to understand the carbon dynamics across the landscape, i.e., source, transport, deposition, and export [24,25]. Therefore, study of the hillslope system under long-term land conversion and the no-till system is crucial to understand the soil carbon depletion–maintenance interaction in the soil system.

We conducted this research to address some of these priorities or scarcity of studies pointed out above, particularly the long-term effect of land conversion on soil carbon stock. The objectives of this study are (1) to estimate the soil organic carbon stock on the long-term (40-year) no-till system; (2) to explore the effect of landforms on soil organic carbon redistribution; and (3) to put the local no-till system in a long-term (69-year) land conversion context through a literature review.

Herein, explanation of soil carbon dynamics on the long-term no-till system is important to evaluate the sustainability of the agroecosystem in subtropical regions, as well as to support soil conservation practices. Moreover, the study area is one of the oldest under the no-till system in Southern Brazil.

2. Materials and Methods

The study area is located at the center-south of the state of Paraná, Brazil at 1120 m (above sea level). The zero-order catchment (hollow) is around 5 ha and is located at the Agricultural Foundation for Agricultural Research–FAPA (Figure 1). The hillslopes are gently ranging from 3% to 5% inclination, while the slope is 5% along the thalweg with 270 m. The soil consists of brown aluminum Oxisols (Ferralsols are high-weathering, resulting in oxic horizon [26]) developed over basalt rock. Kaolinite is the main type of clay, followed by subsidiary gibbsite and iron oxides including hematite and goethite (clay 657 g kg^{-1}, silt 266 g kg^{-1}, and sand 77 g kg^{-1}). Additionally, the pristine horizon is rich in soil organic carbon >40 g kg^{-1} [27].

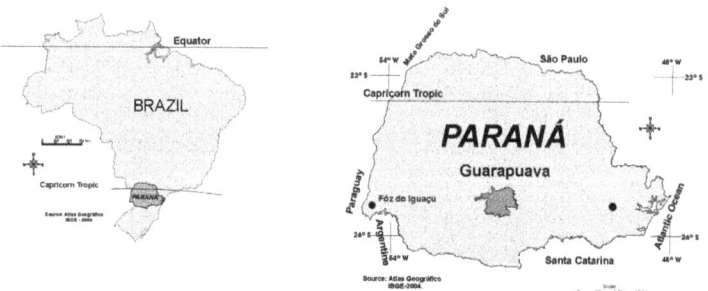

Figure 1. Study area in the context of Southern Brazil and the southern center of the state of Paraná, municipality of Guarapuava.

The annual rainfall ranges from 1800 to 2000 mm. The rain is distributed along the year (i.e., there is no seasonality), with the lowest rainfall during winter (August, 80–100 mm) and the highest rainfall during spring (October, 200–220 mm). The annual temperature is around 17 °C–18 °C. During winter (June to August), the average temperature is 13.5 °C, and during summer (December to February), the average temperature is 21.5 °C [28].

2.1. Soil Sampling

Soil samples were collected from an area with an agricultural land use history of 69 years. In addition, several experiments and case studies were revisited to put the long-term soil carbon history into context (see Table 1). The sampling area has been transited from different levels of land conversion and intensification (Figure 2). Herein, the focus is on the last phase of conversion from conventional tillage to a no-till system in a 40-year term. From 1968 to 1978 (10 years), the area was cultivated with conventional soil preparation such as plowing and harrowing. The crops were grown in simple succession (i.e., double crops), with wheat cultivated during winter and soybeans cultivated during summer [29]. Since 1978, the area has been cultivated with seeding a mulch-based cropping system (DMC), and thereafter, the no-till system was adopted. Within 40 years of cultivation, a total of 82 summer–winter harvests occurred in the area. During summer, soybean (*Glycine max*) (80%) is the most cultivated crop followed by corn (*Zea mays*) (20%). During winter, the most cultivated crop or cover crops are wheat (*Triticum aestivum*) (30%), barley (*Hordeum vulgare*) (26%), oat (*Avena sativa*) (18%), vetch (*Vicia vilosa*) (13%), and turnip (*Raphanus raphanistrum*) (13%) [30].

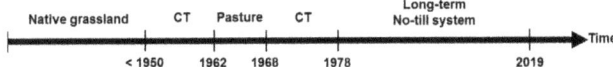

Figure 2. Long-term land use history of the study area modified from Jaster et al. (1993) [29] and the present study. Note: CT (Conventional Tillage).

A systematic grid with four transversal–longitudinal transects was used for soil sampling (Figure 3). In total, 16 trenches of 50 cm length and 60 m depth were dug. Soil samples were collected using a metal ring with $100 \, \text{cm}^3$ (~5.03 cm of diameter and 5.03 cm of height) at the depths of 0–20 cm, 20–40 cm, and 40–60 cm for analysis of soil bulk density and carbon. A soil sample was collected from each depth (16 trenches × 3 depths × 1 sample per soil layer = 48). Through the systematic grid along the area, nine trenches were dug over convex hillslopes, while seven trenches were dug over concave hillslopes (Figure 3). Four trenches from a secondary forest in the same pedogeomorphic unit ~850 m of distance were used as a reference site. Native grass was not found near the study area to serve as a reference site. However, elsewhere, both areas of forest and native pasture might have an equivalent soil carbon stock at 0–25 cm soil depth $12.8 \, \text{kg m}^{-2}$ and $12.2 \, \text{kg m}^{-2}$, respectively [31]. Furthermore, they were used for a general context and not for a direct comparison with the study area.

Total soil organic carbon in g kg^{-1} was determined according to the Walkley and Black method [32], and the soil carbon stock was estimated (Equation (1)) [33].

$$SCS = (TOC \times D \times D)/100 \tag{1}$$

where

SCS = Soil carbon stock (kg m^{-2});
TOC = Total soil organic carbon at the sampled soil layer (g kg^{-1});
BD = Soil bulk density of the soil layer (kg m^{-3});
D = Soil layer sampled thickness (cm);
Soil carbon loss was estimated through (Equation (2)).

$$CL = \frac{CA - NT}{Time\ of\ conversion} \tag{2}$$

where

CL = carbon loss in kg C m^{-2} year^{-1}, CA = carbon stock in control area in kg C m^{-2} year^{-1}, and NT = carbon stock in no-till according to time of implantation in kg C m^{-2} year^{-1}.

Moreover, the soil carbon stock in the soil was used as a proxy to infer and estimate the long-term soil loss in the no-till system, since when erosion occurs, it transports sediment and carbon as well.

Figure 3. Systematic grid distribution of the soil collection points (16 trenches).

2.2. Data Analysis

In a previous study (i.e., literature survey), soil carbon content was collected up to 10 cm soil depth in the study area. Through an empirical model developed for the study area, it was estimated that the soil carbon decreases by 0.326 kg for every cm of soil depth increment (Figure 4). Overall, 75–85% of the total carbon from 0–20 cm depth is concentrated at 0–10 cm soil depth, and approximately 25% of the carbon is reduced at the deepest soil layer [30]. Here, this model was tested in two soil carbon profiles of the Southern Paraná region [12,34], and the model showed a similar soil carbon distribution pattern (data not reported). We used this model to transform (estimate) the total soil carbon content in 0–10 cm to 0–20 cm soil layer equivalent (see Table 1). This strategy was used to apply the no-till system in the local and regional contexts when the soil carbon content or stock was displayed only at the 0–10 cm soil layer.

Therefore, the no-till system was evaluated in two ways: (a) whole soil profile 0–60 cm depth, through the experiment of this study; (b) topsoil 0–20 cm depth, from the data of the present study and the data from the literature survey, as in most cases, soil carbon data were only obtained at the soil depth of 0–10 cm or 0–20 cm.

The Mann–Whitney U test was used to compare independent samples. In addition, the critical *p*-value established in the comparisons was unrestricted, and the maximum of $p \leq 0.05$ was adopted as significant.

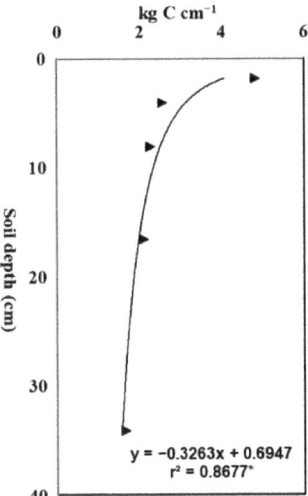

Figure 4. Soil carbon distribution in the study area according to soil depth modified from Silva (2013) [30]. (n = 3 soil profiles); * $p \leq 0.05$.

3. Results and Discussion

In this study, the soil organic carbon stock on the long-term (40-year) no-till system decreased, especially at 0–20 cm, but the carbon depletion rate was below our initial expectations, as well as that reported in the literature. Despite a long-term (69-year) conversion from grassland to a cropping system, the study area showed a soil carbon average above a critical limit.

However, we observed that landforms (such as rolling landscape) affect soil erosion and soil carbon redistribution. Hillslopes (convex sectors) are more depleted in carbon (37%) than valley bottom (concave sectors). Presently, in the study area, the soil loss was estimated around 1.13 ± 0.17 Mg ha^{-1} year^{-1}, and the long-term soil carbon loss rate at topsoil was 0.06 kg C m^{-2} year^{-1}. The local conservationist agriculture is facing a critical phase related to soil carbon conservation, particularly at the topsoil.

3.1. Total Soil Carbon and Soil Carbon Stock on A 40-Year No-till System

The forest showed a higher concentration of total soil carbon at all depths compared to no-till (Figure 5). Overall, the soil carbon decreased with respect to the soil depth (Figure 5). In the no-till system, the soil surface displayed 17% and 28% more carbon when compared to the deepest soil layers at 20–40 and 40–60 cm, respectively ($p \leq 0.05$). However, the soil carbon at 20–40 and 40–60 cm was similar. Soil carbon content exhibits great variation along the area depending on the soil depth. At topsoil (0–20 cm), the soil carbon content ranged from 15.66 g kg^{-1} to 28.51 g kg^{-1}; at 20–40 cm depth, the soil carbon content ranged from 13.66 g kg^{-1} to 24.84 g kg^{-1}; and at 40–60 cm, the soil carbon content ranged from 12.87 g kg^{-1} to 23.80 g kg^{-1} (Table S1). For the three depths evaluated, the minimum and maximum soil carbon content maintained a constant amplitude ratio of 1.8 times.

Ribas (2010) evaluated a large sample of soil carbon content (n = 6534, soil depth 0–20 cm) from the southern state of Paraná, and about 85% of the soil samples came from areas with a no-till system. Overall, the soil in this region showed a total soil carbon ranging from 22.7 to 24.7 g kg^{-1}, with an average of 23.6 g kg^{-1}. In addition, 92.6% of the samples displayed a total soil carbon lower than 29.0 g kg^{-1}. Similarly, in the Guarapuava municipality, a total of 1212 samples was evaluated. The total soil carbon ranged from 23.5 to 24.9 g kg^{-1}, with an average of 23.7 g kg^{-1} [35]. Surprisingly, in this study, the 40-year no-till system showed a total soil carbon content (23.53 ± 3.63 g kg^{-1}) such as those of regional and local soils, considering the lag time of a decade than that in the study by Ribas.

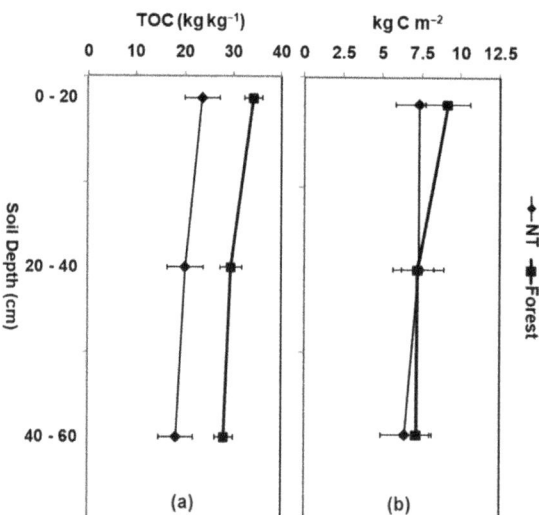

Figure 5. Total soil organic carbon (**a**) and soil carbon stock (**b**) in no-till and secondary forest according to soil depth. Note: no-till, $n = 16$ and forest, $n = 4$.

Despite a lower total carbon content (25% to 35%) compared to that of the soil forest ($p \leq 0.05$), the differences in soil carbon stock in the no-till system and forest along the profile were not significant (Figure 5). The soil bulk density in the no-till system was around 25% denser when compared to the forest soil, particularly in the deepest soil layers. The superior bulk density in the no-till system was compensated by the soil carbon stock. However, the forest showed 20% more carbon stock than the no-till system at the 0–20 cm soil layer. In contrast, at 20–40 cm and 40–60 cm, there was an equivalent soil carbon stock. Finally, considering the entire profile (0–60 cm), the no-till system showed similar soil carbon stock as the forest (Figure 5).

Here, the forest soil carbon stocks in the study area are consistent with those found in other regional studies by Sá et al. (2014) (0–40 cm, 15.9 kg C m^{-2}) and Pereira (2017) (0–20 cm, 8.72 kg C m^{-2}). Similarly, soil carbon stocks in no-till are comparable with other studies by Costa et al. (2004b) (0–20 cm, 7.34 kg C m^{-2}) and Sá et al. (2014) (0–40 cm, 13.7 kg C m^{-2}).

3.2. Long-Term Soil Carbon Stocks in No-till System: Local Context

A proper comparison of soil carbon content on a different agricultural system is fraught with difficulties, particularly due to methods of soil sampling and analysis, as well as depth assessment of soil layers [6]. Many studies have evaluated several distinct soil layers at 0–10 cm [36,37], 0–20 cm [14,35], and 0–30 cm [21,38]. Moreover, studies with soil collected at certain increment depths are typical [12,36]. During the use of the no-till system, the crop residue management causes the topsoil (0–10 cm) to be enriched with soil carbon [10,22] or even smaller depths at 0–5 cm [39]. Sometimes, in a layer of 30 cm, more than 70% of the total soil carbon could be concentrated on the first soil centimeters (e.g., <10 cm) (see Figure 3).

In Table 1, 6- to 40-year no-till systems were grouped to evaluate the carbon dynamics through periods of time. The data were obtained on the same pedogeomorphic climate landscape, as well as nearby the study area. Moreover, the local context no-till system was implemented in different phases, and several measurements were performed over time. At the yearly stage (6–15 years), the soil carbon was lower; however, at >21 years, the soil carbon in the no-till system increased by ~15% ($p \leq 0.05$). Data of soil carbon for the no-till system from 7 to 24 years were based on the topsoil (0–10 cm) [36,40] (Table 1).

The average total soil carbon (0–20 cm) in the no-till system was 23% lower than that of the forest ($p \leq 0.01$) (Table 1). In contrast, soil carbon content reached its peak at 21–25 years of the no-till system being implemented, which was similar to that of the forest. However, an absolute reduction by 12% in soil carbon stock was observed in the older no-till system (40 years) than the no-till system in the previous period ($p = 0.10$). On average, each soil cm of the no-till system was 0.412 kg C cm^{-1} lower than the forest soil (16%).

Over time, soil from the study area, i.e., the 40-year no-till system, lost 25% pristine soil carbon stock. The carbon loss was greater at the early phase (6–15 years), with the loss of 0.194 kg C cm^{-2} year^{-1}. The soil carbon loss ratio decreased drastically to 0.045 kg C cm^{-2} year^{-1} (>16 years). The soil carbon stock in the no-till system (40-year) was depleted at a rate of 0.06 kg C m^{-2} year^{-1}, summing up to a carbon loss of 2.43 kg C m^{-2}. The total organic carbon in the no-till system (40-year) was 16% lower when compared to the CT system applied before the conversion in year 1978 (Table 1).

Table 1. Context of the local long-term soil carbon dynamic in no-till system at topsoil (0–20 cm) based on the literature survey.

No-Till (Year)	Average (Year)	TOC (g kg^{-1})	N	[1] TOC (g kg^{-1}) (0–20 cm)	[2] Soil Carbon Stock kg C m^{-2}	kg C cm^{-1}	Source
6–10	8	§ 34.75 ± 5.40	5	26.07	8.18	0.409	[36,41]
11–15	13	§ 31.48 ± 6.43	6	23.61	7.41	0.370	[36]
16–20	18	§ 36.83 ± 6.28	12	27.62	8.67	0.433	[36]
21–25	23	§ 40.88 ± 3.46	5	30.66	9.62	0.481	[27,34,40]
[3] 31	-	§§ 27.26	1	27.26	8.55	0.428	[30]
38	-	§§ 25.00 ± 1.26	4	25.00	7.84	0.392	[41]
40	-	§§ 23.53 ± 3.63	16	23.53	7.33	0.366	this study
Forest	-	§§ 34.72 ± 4.29	11	34.72	9.82	0.491	[29,39] and this study
[4] Conventional Tillage	-	§§ 28.00	1	28.00	-	-	[29]

Note: § (soil depth mostly 0–10 cm); §§ (soil depth 0–20 cm); [1] total soil carbon estimated for a 0–20 cm soil layer through local empirical model (Figure 4); [2] soil bulk density used to estimate soil carbon stock (no-till 0.91 g cm^{-3} and forest 0.82 g cm^{-3}); [3] composite sample; [4] soil carbon content in the conventional tillage previous to the conversion to no-till system in 1978 (composite sample).

3.3. Long-Term Land Conversion and Its Effect on Soil Carbon

Land use changes, especially the intensification of agriculture, affect the ecosystem functions and services. Soil carbon is one of the most sensible properties that responds to land use conversion and intensification [20,22,37]. The conversion of the natural ecosystem to permanent agriculture in temperate zones can cause 50% of the original organic matter loss in the first 25 years of cultivation. In a tropical ecosystem, the loss of soil organic matter can occur within 5 years of cultivation [20]. In this study, the depletion of soil carbon did not follow this pattern (i.e., time and amount).

Here, before year 1950, the area was covered by native grassland. Afterwards, it was converted to conventional tillage (year 1950–1962) and pasture (year 1962–1968). Again, the area was converted from pasture to conventional tillage (year 1968–1978). Finally, the area was converted from conventional tillage to the no-till system, which remained the soil management system since 1978. In the present study, despite being in a subtropical region, the several conversions and intensification phases of the system seem to preserve a great amount of the soil organic carbon stock. Moreover, the soil carbon loss ratio did not follow the pattern suggested in literature [20,21]. If that had happened in any land conversion phase, most of the study area should have <17 g kg^{-1} total organic carbon.

It is difficult to explain the soil carbon dynamics in the study area prior to year 1978, and this is because there is no available soil data. However, some insights are discussed. The conversion from native grassland (i.e., extensive pasture) to conventional tillage with cultivation of rice (summer) and wheat (winter) may have caused the soil carbon loss from the year 1950–1962. Guo and Gifford (2002) estimated a 59% carbon loss on this type of land use change, and an improvement of 20% when crop is converted to pasture. The conversion from conventional tillage to pasture (year 1962–1968) may yield a gain in soil carbon [19].

Moreover, the pasture in the study area experienced an improvement with the cultivation of white clover (*Trifolium repens*) and winter grasses (the type was not defined) [29]. Again, from the year 1968 to 1978, the pasture was converted to a conventional crop system, with a succession of soybean (summer) and wheat (winter). Since year 1978, the area has been cultivated using the no-till system, with cultivation of wheat and cover crops such as oat (*Avena sativa*), barley (*Hordeum vulgare*), vetch (*Vicia vilosa*), and turnip forage (*Raphanus sativus*) during winter; during summer, corn and soybean are mainly cultivated [30,42].

Since the conventional tillage conversion to no-till, several local studies report an improvement in soil quality including the increase in soil carbon and aggregate stability [27,34,40], soil porosity, soil water retention and infiltration [30,43], soil temperature decrease and crop productivity enhancement [29,36], and soil erosion reduction [44]. The benefits of this system are recognized worldwide as a conservationist system that ensures soil erosion control, as well as water and soil carbon conservation [10,45,46].

In the study area, following a long-term cultivation, the soil carbon content was slightly above the critical limit. A concentration of 20 g kg^{-1} (~2% soil organic carbon) is recognized as the critical point to soil productivity and functions (e.g., microbial diversity) [17]. Therefore, the carbon input into the soil should be superior to its loss due to soil erosion and oxidation of organic matter.

Globally, the no-till system has the potential to sequester 0.030 kg C m^{-2} year^{-1} [47] in Southern Brazil at the rate of 0.068 kg C m^{-2} year^{-1} [48], as well as at the rate in a site with a similar pedogeomorphic climate landscape as the study area which was 0.043 kg C m^{-2} year^{-1} [13]. Lastly, in the study area, a lower rate was registered as 0.015 kg C m^{-2} year^{-1} [34].

Ferreira et al. (2018) argued that areas with a predominance of soybean in the cropping system characterized by a poor soil fertility management have a smaller soil carbon recovery rate [49]. In the study area, soybean (*Glycine max*) is the most cultivated in the cropping system. Moreover, when soybean is practiced as monoculture, without cover crops, it can cause a decline in soil organic matter, especially on the labile fractions [39]. This condition could be the most detrimental scenario to soil carbon depletion, but this is not the study area case. However, if our estimation was reasonable, the soil carbon in a 40-year no-till system (0–20 cm) would have been depleted (i.e., 0.06 kg C m^{-2} year^{-1}) above the rate of carbon input [49].

3.4. Effects of Landforms on Soil Erosion and Soil Carbon Redistribution

Here, and in most parts of Southern Brazil, the basalt rocks form assemblages of landforms known as plateaus. Regionally, a plateau is a flat terrain; however, at a local scale (i.e., farm-land level), the basalt rocks produce a rolling landscape with a non-uniform slope (e.g., convex, straight, concave, hollow) [50] (see Figures S1 and S2). In addition, the geomorphological surface was stable for a long geological time, and the weathering operated mostly on transport-limited conditions, developing deep soils such as Oxisols and Nitisols, and soils with moderate depth such as Cambisols [51].

Consequently, no-till is usually practiced on non-uniform hillslopes. Moreover, the slope shape is prone to different shear stress due to interrill and rill initiation, as well as due to soil erosion rate. For instance, uniform and convex-linear slopes produce 3.4 to 4 times more sediment, respectively, than concave-linear slopes [52]. Therefore, in a rolling landscape, the sediment carbon redistribution occurs through a net with different transport and deposition rates.

In the study area, the hollows (zero-order catchment) are frequently activated as ephemeral streams within the year, especially during prolonged duration and a large volume of rainfall (personal observation). If the concave slope and hollows are sensible to produce convergent runoff in a basaltic plateau landscape [53], then this may affect the local soil carbon redistribution and exportation from the system as well.

In Brazil, a no-tillage system reduced the runoff and soil loss by 70% and 90%, respectively, when compared to that of conventional tillage [48]. Between the years 2001 and 2012, the conservation agriculture in Brazil reduced soil erosion by 20%. However, the national soil erosion average (>20 Mg ha^{-1} year^{-1}) was above the soil loss tolerance threshold (T value 10 Mg ha^{-1} year^{-1}) [54]. Despite the benefits of conservation agriculture, FAO global assessment indicated that the main threats to Brazilian soils are soil erosion and soil organic carbon change [55].

Soil erosion involving aggregates and particle detachment of soil mass, aggregate breakdown during transport, as well as transport and deposition is an important mechanism for lateral flux of carbon redistribution through the landscape [24,25]. In the present study, the convex slope was more depleted in carbon (15.34 to 19.80 g kg^{-1}) compared to the concave slope (20.79 to 20.43 g kg^{-1}) ($p \leq 0.05$). The soil carbon redistribution followed the rolling landscape, that is, non-uniform slopes. Similarly, the convex hillslope (15.87 to <20.57 kg C m^{-2}) showed lower carbon stock in the entire 60 cm profile than the concave hillslope (>20.57 kg C m^{-2}) (Figure 6).

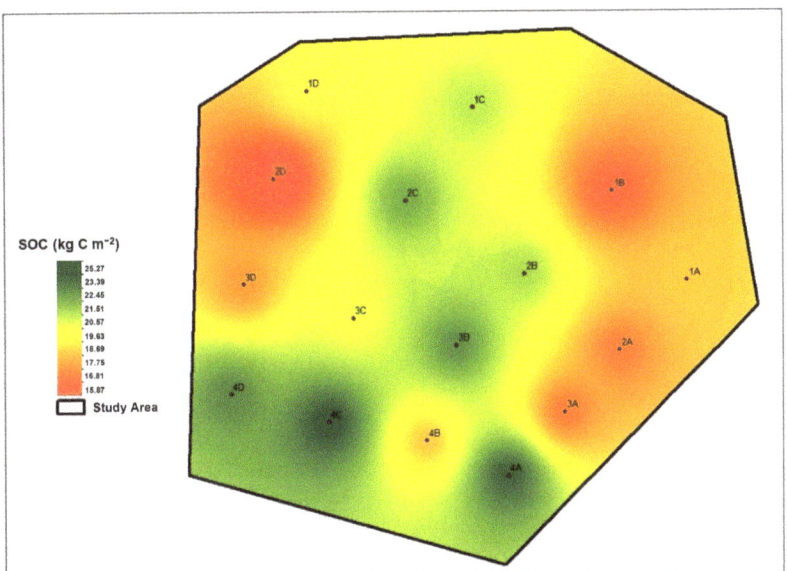

Figure 6. Effects of landforms on soil carbon redistribution in no-till system (0−60 cm soil profile). Note: convex slopes predominate in red to orange colors, while concave slopes (i.e., valley bottom) predominate in yellow to green colors.

The study area is suffering from soil erosion and lateral flux of soil carbon. The long-term soil loss during the land conversion which depletes the soil carbon is estimated at 0.06 kg C m^{-2} year^{-1}. In the southern state of Paraná, the no-tillage system showed a soil loss ranging from 0.4−1.0 Mg ha^{-1} year^{-1}, with an average of 0.73 Mg ha^{-1} year^{-1} [44]. Generally, soil loss rates under conservation tillage on moderate slopes are very low (<1 t ha^{-1} year^{-1}) [8]. We estimated that the long-term soil erosion in the study area was 1.13 ± 0.17 Mg ha^{-1} year^{-1}, considering the soil carbon content average in the forest ($n = 11$) as a baseline (see Table 1).

In this study, the soil loss was slightly greater than its counterpart's average [8,44]. Globally, cropland shows the highest soil carbon loss annually (0.023 kg C m^{-2} year^{-1}). Considering the upper limit for crop and grassland, the soil carbon loss may range from 0.03 kg C m^{-2} year^{-1} to 0.05 kg C m^{-2} year^{-1} [5]. The rate of soil carbon loss in the study area was above that reported in meta-analysis. However, our data were estimated within a 40-year term, while the data from Abdalla et al. (2020) were registered in the field by means of experimental plots operating no longer than 14 years.

Probably, the conservationist agriculture (i.e., no-till) in Southern Brazil is facing a critical phase, since the agriculture in the region is vulnerable to climate change, and there is a tendency of rising temperature and rainfall in the region [56]. Therefore, climate change could affect the soil erosion rates, soil carbon dynamics, and stock [6,10]. In a climate change scenario, conservationist agriculture, soil management practices, and food system adaptations (e.g., climate-smart agriculture) are needed more than ever [8,57].

Farms, stakeholders, and scientists should highlight the basic principles of the conservationist agriculture in order to avoid disturbing the topsoil (no-till); keep mulching on the ground; plan a crop rotation system and not only a crop succession (e.g., wheat–soybean); replace fallow by cover crops; and use contour tillage and terraces wherever necessary. These practices are of the utmost importance, particularly on the rolling landscape such as that of the study area and elsewhere, especially in Southern Brazil.

4. Conclusions

In this study, the objectives were to estimate the soil organic carbon stock on a long-term no-till system, and to apply it to the land conversion historical context over a rolling landscape. We observed that land conversion reduced the soil organic carbon stock at the topsoil (0–20 cm) by 20%. Overall, soil carbon has been depleted by 0.06 kg C m^{-2} $year^{-1}$ above the rate of carbon input. Soil carbon loss in the subtropical agroecosystem was much lower than our initial expectations, as well as that reported in literature. In addition, the no-till system was able to conserve soil carbon in the deepest soil layers.

Land conversion from conventional tillage to the no-till system seems to increase the soil organic carbon up to 25 years or <30 years. Possibly, the no-till system is currently losing more carbon due to soil erosion and organic matter mineralization at a rate greater than the soil carbon input. However, over the long term, the no-till system acted as a buffer, reducing the accelerated soil carbon loss, supposing the conventional tillage was kept in use.

The rolling landscape (i.e., hillslopes) affected the total soil organic carbon content and, consequently, the soil carbon stock. The convex hillslope was more depleted in carbon than the concave sectors. Therefore, at the farm or catchment level, carbon redistribution could show sites with critical limits of soil carbon content. The hollow sites are hydrogeomorphologically dynamic, and the soil conservation should be cautious of this sort of terrain. Moreover, studies about validation of the cropping system and soil carbon dynamic evaluation are mostly conducted on homogeneous or uniform slopes that are not coinciding with a rolling landscape. It is therefore difficult to extrapolate data from experimental areas in space-time to a rolling landscape.

Supplementary Materials: The following supporting information can be downloaded at: https://www.mdpi.com/article/10.3390/soilsystems7020060/s1, Figure S1. The study area MDT greyscale displaying the hollow (zero-order catchment), notice the ephemeral rill in the center; Figure S2. Aspect of the rolling landscape; Table S1. Soil Carbon Forest.

Author Contributions: E.L.T., Thomaz—Conceptualization, methodology, writing—original draft preparation, writing—review and editing; J.P.K., Kurasz—investigation, data collection and preliminary analysis, project administration. All authors have read and agreed to the published version of the manuscript.

Funding: The Brazilian Research and Development Council (CNPq) for the productivity fellowship to the first author (Grant 301665/2017-6).

Institutional Review Board Statement: Not applicable.

Informed Consent Statement: Not applicable.

Data Availability Statement: All data availability in Supplementary Materials.

Acknowledgments: This study was supported by the Coordenação de Aperfeiçoamento de Pessoal de Nível Superior—Brasil (CAPES)—Finance Code 001 (second author) and Fapa—Fundação Agrária de Pesquisa Agropecuária which permitted this research on its experimental area to be carried out.

Conflicts of Interest: The authors declare no conflict of interest.

References

1. Prăvălie, R. Exploring the multiple land degradation pathways across the planet. *Earth-Sci. Rev.* **2021**, *220*, 103689. [CrossRef]
2. Pimentel, D.; Burgess, M. Soil erosion threatens food production. *Agriculture* **2013**, *3*, 443–463. [CrossRef]
3. Thomaz, E.L.; Marcatto, F.S.; Antoneli, V. Soil erosion on the Brazilian sugarcane cropping system: An overview. *Geogr. Sustain.* **2022**, *3*, 129–138. [CrossRef]
4. Aghapour Sabbaghi, M.; Nazari, M.; Araghinejad, S.; Soufizadeh, S. Economic impacts of climate change on water resources and agriculture in Zayandehroud river basin in Iran. *Agric. Water Manag.* **2020**, *241*, 106323. [CrossRef]
5. Abdalla, K.; Mutema, M.; Hill, T. Soil and organic carbon losses from varying land uses: A global meta-analysis. *Geogr. Res.* **2020**, *58*, 167–185. [CrossRef]
6. Gross, C.D.; Harrison, R.B. The case for digging deeper: Soil organic carbon storage, dynamics, and controls in our changing world. *Soil Syst.* **2019**, *3*, 28. [CrossRef]
7. Foley, J.A.; DeFries, R.; Asner, G.P.; Barford, C.; Bonan, G.; Carpenter, S.R.; Chapin, F.S.; Coe, M.T.; Daily, G.C.; Gibbs, H.K. Global consequences of land use. *Science* **2005**, *309*, 570–574. [CrossRef]
8. Govers, G.; Merckx, R.; Van Wesemael, B.; Van Oost, K. Soil conservation in the 21st century: Why we need smart agricultural intensification. *Soil* **2017**, *3*, 45–59. [CrossRef]
9. Kassam, A.; Friedrich, T.; Derpsch, R. Overview of the global spread of conservation agriculture. *Sustain. Dev. Org. Agric. Hist. Perspect.* **2015**, *8*, 53–68.
10. Blanco-Canqui, H.; Lal, R. *Principles of Soil Conservation and Management*; Springer Science & Business Media: Berlin/Heidelberg, Germany, 2010; ISBN 1402087098.
11. Castro Filho, C.; Lourenço, A.; Guimarães, M.D.F.; Fonseca, I.C.B. Aggregate stability under different soil management systems in a red latosol in the state of Parana, Brazil. *Soil Tillage Res.* **2002**, *65*, 45–51. [CrossRef]
12. De Moraes Sá, J.C.; Tivet, F.; Lal, R.; Briedis, C.; Hartman, D.C.; dos Santos, J.Z.; dos Santos, J.B. Long-term tillage systems impacts on soil C dynamics, soil resilience and agronomic productivity of a Brazilian Oxisol. *Soil Tillage Res.* **2014**, *136*, 38–50. [CrossRef]
13. Amado, T.J.C.; Bayer, C.; Conceição, P.C.; Spagnollo, E.; de Campos, B.-H.C.; da Veiga, M. Potential of Carbon Accumulation in No-Till Soils with Intensive Use and Cover Crops in Southern Brazil. *J. Environ. Qual.* **2006**, *35*, 1599–1607. [CrossRef] [PubMed]
14. Bayer, C.; Dieckow, J.; Amado, T.J.C.; Eltz, F.L.F.; Vieira, F.C.B. Cover crop effects increasing carbon storage in a subtropical no-till sandy Acrisol. *Commun. Soil Sci. Plant Anal.* **2009**, *40*, 1499–1511. [CrossRef]
15. De Freitas, P.L.; Landers, J.N. The Transformation of Agriculture in Brazil Through Development and Adoption of Zero Tillage Conservation Agriculture. *Int. Soil Water Conserv. Res.* **2014**, *2*, 35–46. [CrossRef]
16. Du, Z.; Angers, D.A.; Ren, T.; Zhang, Q.; Li, G. The effect of no-till on organic C storage in Chinese soils should not be overemphasized: A meta-analysis. *Agric. Ecosyst. Environ.* **2017**, *236*, 1–11. [CrossRef]
17. McBratney, A.B.; Stockmann, U.; Angers, D.A.; Minasny, B.; Field, D.J. Challenges for soil organic carbon research. In *Soil Carbon*; Springer: Berlin/Heidelberg, Germany, 2014; pp. 3–16.
18. Bai, Z.; Caspari, T.; Gonzalez, M.R.; Batjes, N.H.; Mäder, P.; Bünemann, E.K.; de Goede, R.; Brussaard, L.; Xu, M.; Ferreira, C.S.S.; et al. Effects of agricultural management practices on soil quality: A review of long-term experiments for Europe and China. *Agric. Ecosyst. Environ.* **2018**, *265*, 1–7. [CrossRef]
19. Guo, L.B.; Gifford, R.M. Soil carbon stocks and land use change: A meta analysis. *Glob. Chang. Biol.* **2002**, *8*, 345–360. [CrossRef]
20. Matson, P.A.; Parton, W.J.; Power, A.G.; Swift, M.J. Agricultural intensification and ecosystem properties. *Science* **1997**, *277*, 504–509. [CrossRef]
21. Robert, M. Global change and carbon cycle: The position of soils and agriculture. In *Soil Erosion and Carbon Dynamics*; Roose, E.J., Lal, R., Feller, C., Barthes, B., Stewart, B.A., Eds.; CRC Press: Boca Raton, FL, USA, 2006; pp. 3–12.
22. Hartemink, A.E.; McSweeney, K. *Soil Carbon*; Springer Science & Business Media: Berlin/Heidelberg, Germany, 2014; ISBN 3319040847.
23. Powers, J.S.; Corre, M.D.; Twine, T.E.; Veldkamp, E. Geographic bias of field observations of soil carbon stocks with tropical land-Use changes precludes spatial extrapolation. *Proc. Natl. Acad. Sci. USA* **2011**, *108*, 6318–6322. [CrossRef]
24. Van Oost, K.; Quine, T.A.; Govers, G.; De Gryze, S.; Six, J.; Harden, J.W.; Ritchie, J.C.; McCarty, G.W.; Heckrath, G.; Kosmas, C.; et al. The impact of agricultural soil erosion on the global carbon cycle. *Science* **2007**, *318*, 626–629. [CrossRef]
25. Lal, R. Soil erosion and the global carbon budget. *Environ. Int.* **2003**, *29*, 437–450. [CrossRef] [PubMed]
26. IUSS Working Group WRB. *World Reference Base For Soil Resources*; FAO: Rome, Italy, 2006; p. 103.
27. Costa, F.S.; Albuquerque, J.A.; Bayer, C.; Fontoura, S.M.V.; Wobeto, C. Physical properties of an Oxisol affected by no-tillage and conventional tillage systems (in Portuguese). *Rev. Bras. Ciência Solo* **2003**, *27*, 527–535. [CrossRef]
28. Nitsche, P.R.; Caramori, P.H.; Ricce, W.S.; Pinto, L.F.D. *Climate Atlas of the State of Paraná*; IAPAR: Londrina, Brazil, 2019; 210p. (In Portuguese)

29. Jaster, F.; Eltz, F.L.F.; Fernandes, F.F.; Merten, G.H.; Gaudêncio, C.d.A.; de Oliveira, M.C.N. *Grain Yield in Different Tillage and Soil Management Systems*; Report 61; CNPSo-Centro Nacional de Pesquisa de Soja: Londrina, Brazil, 1993; 37p. (In Portuguese)
30. Da Silva, F.R. Thirty-One Years of Management Systems in Oxisol: Physical and Chemical Attributes and Crop Yield. Ph.D Thesis, UDESC—Universidade do Estado de Santa Catarina, Florianópolis, Brazil, 2013. (In Portuguese).
31. Bayer, C.; Dick, D.P.; Ribeiro, G.M.; Scheuermann, K.K. Carbon stocks in organic matter fractions as affected by land use and soil management, with emphasis on no-tillage effect. *Ciência Rural* **2002**, *32*, 401–406. (In Portuguese) [CrossRef]
32. Walkley, A.; Black, I.A. An Examination of The Degtjareff Method for Determining Soil Organic Matter, and a Proposed Modification of the Chromic Acid Titration Method. *Soil Sci.* **1934**, *37*, 29–38. [CrossRef]
33. Veldkamp, E. Organic Carbon Turnover in Three Tropical Soils under Pasture after Deforestation. *Soil Sci. Soc. Am. J.* **1994**, *58*, 175–180. [CrossRef]
34. Costa, F.d.S.; Bayer, C.; Albuquerque, J.A.; Fontoura, S.M.V. Increase of organic matter in a brown oxisol under no-tillage. *Ciência Rural* **2004**, *34*, 587–589. (In Portuguese) [CrossRef]
35. Ribas, C. Characterization of the Current Fertility of Soils in the Region of Guarapuava-Pr. Master's Thesis, UNICENTRO—Universidade Estadual do Centro-Oeste, Guarapuava, Brazil, 2010. (In Portuguese).
36. Fontoura, S.M.V.; Bayer, C. Nitrogen fertilization for high corn yield in no-tillage in the south-central region of Paraná. *Rev. Bras. Ciência Solo* **2009**, *33*, 1721–1732. (In Portuguese) [CrossRef]
37. Roose, E.; Barthès, B. Soil carbon erosion and its selectivity at the plot scale in tropical and Mediterranean regions. In *Soil Erosion and Carbon Dynamics*; CRC Press: Boca Raton, FL, USA, 2006; pp. 55–72.
38. Ogeh, J.S. Soil Organic Carbon Stocks Under Plantation Crops and Forest in the Rainforest Zone of Nigeria. In *Soil Carbon*; Springer International Publishing: Berlin/Heidelberg, Germany, 2014; pp. 467–473. [CrossRef]
39. Beltrán, M.J.; Sainz-Rozas, H.; Galantini, J.A.; Romaniuk, R.I.; Barbieri, P. Cover crops in the Southeastern region of Buenos Aires, Argentina: Effects on organic matter physical fractions and nutrient availability. *Environ. Earth Sci.* **2018**, *77*, 428. [CrossRef]
40. Albuquerque, J.A.; Mafra, Á.L.; Fontoura, S.M.V.; Bayer, C.; dos Passos, J.F.M. Evaluation of tillage and liming systems in an aluminum Latosol. *Rev. Bras. Ciência Solo* **2005**, *29*, 963–975. (In Portuguese) [CrossRef]
41. Pereira, A.A. Soil Erodibility in No-Tillage System Increases with Management Time. Ph.D Thesis, UEPG—Universidade Estadual de Ponta Grossa, Ponta Grossa, Brazil, 2017. (In Portuguese).
42. Ciotta, M.N.; Bayer, C.; Ernani, P.R.; Fontoura, S.M.V.; Albuquerque, J.A.; Wobeto, C. Acidification of an Oxisol under no-tillage (in Potuguese). *Rev. Bras. Ciência Solo* **2002**, *26*, 1055–1064. [CrossRef]
43. Albuquerque, J.A.; Argenton, J.; Bayer, C.; Do Prado Wildner, L.; Kuntze, M.A.G. Relationship of soil attributes with aggregate stability of a hapludox under distinct tillage systemsand summer cover crops. *Rev. Bras. Cienc. Solo* **2005**, *29*, 415–424. (In Potuguese) [CrossRef]
44. Merten, G.H.; de Araújo, A.G.; de Cesare Barbosa, G.M. *Erosion in the State of Paraná: Fundamentals, Experimental Studies and Challenges*; Instituto Agronômico do Paraná: Londrina, Brazil, 2016. (In Portuguese)
45. Tornquist, C.G.; Giasson, E.; Mielniczuk, J.; Cerri, C.E.P.; Bernoux, M. Soil Organic Carbon Stocks of Rio Grande do Sul, Brazil. *Soil Sci. Soc. Am. J.* **2009**, *73*, 975–982. [CrossRef]
46. Huggins, D.R.; Reganold, J.P. No-till: The quiet revolution. *Sci. Am.* **2008**, *299*, 70–77. [CrossRef] [PubMed]
47. Powlson, D.S.; Stirling, C.M.; Jat, M.L.; Gerard, B.G.; Palm, C.A.; Sanchez, P.A.; Cassman, K.G. Limited potential of no-till agriculture for climate change mitigation. *Nat. Clim. Chang.* **2014**, *4*, 678–683. [CrossRef]
48. Bernoux, M.; Cerri, C.C.; Cerri, C.E.P.; Neto, M.S.; Metay, A.; Perrin, A.-S.; Scopel, E.; Tantely, R.; Blavet, D.; de Piccolo, M.C. Cropping systems, carbon sequestration and erosion in Brazil: A review. *Sustain. Agric.* **2009**, *26*, 75–85.
49. De Oliveira Ferreira, A.; Amado, T.J.C.; Rice, C.W.; Ruiz Diaz, D.A.; Briedis, C.; Inagaki, T.M.; Gonçalves, D.R.P. Driving factors of soil carbon accumulation in Oxisols in long-term no-till systems of South Brazil. *Sci. Total Environ.* **2018**, *622–623*, 735–742. [CrossRef]
50. Gerrard, J. *Rocks and Landforms*; Unwin Hyman Ltd.: London, UK, 1988; ISBN 9401159831.
51. Birkeland, P.W. *Soils and Geomorphology*; Oxford University Press: Oxford, UK, 1999; ISBN 0195033981.
52. Rieke-Zapp, D.H.; Nearing, M.A. Slope shape effects on erosion: A laboratory study. *Soil Sci. Soc. Am. J.* **2005**, *69*, 1463–1471. [CrossRef]
53. Dos Reis Castro, N.M.; Auzet, A.; Chevallier, P.; Leprun, J. Land use change effects on runoff and erosion from plot to catchment scale on the basaltic plateau of Southern Brazil. *Hydrol. Process.* **1999**, *13*, 1621–1628. [CrossRef]
54. Borrelli, P.; Robinson, D.A.; Fleischer, L.R.; Lugato, E.; Ballabio, C.; Alewell, C.; Meusburger, K.; Modugno, S.; Schütt, B.; Ferro, V.; et al. An assessment of the global impact of 21st century land use change on soil erosion. *Nat. Commun.* **2017**, *8*, 2013. [CrossRef]
55. Keesstra, S.D.; Bouma, J.; Wallinga, J.; Tittonell, P.; Smith, P.; Cerdà, A.; Montanarella, L.; Quinton, J.N.; Pachepsky, Y.; Van Der Putten, W.H.; et al. The significance of soils and soil science towards realization of the United Nations sustainable development goals. *Soil* **2016**, *2*, 111–128. [CrossRef]

56. PBMC. *Scientific Basis of Climate Change. Contribution of Working Group 1 of the Brazilian Panel on Climate Change to the First Report of the National Assessment on Climate Change*; COPPE. Universidade Federal do Rio de Janeiro: Rio de Janeiro, Brazil, 2014; ISBN 9788528502077. (In Portuguese)

57. Paustian, K.; Lehmann, J.; Ogle, S.; Reay, D.; Robertson, G.P.; Smith, P. Climate-smart soils. *Nature* **2016**, *532*, 49–57. [CrossRef] [PubMed]

soil systems

Article

Revealing the Combined Effects of Microplastics, Zn, and Cd on Soil Properties and Metal Accumulation by Leafy Vegetables: A Preliminary Investigation by a Laboratory Experiment

John Bethanis [1,2] and Evangelia E. Golia [1,*]

[1] Soil Science Laboratory, School of Agriculture, Faculty of Agriculture, Forestry and Natural Environment, Aristotle University of Thessaloniki, University Campus, 54124 Thessaloniki, Greece
[2] Department of Planning and Regional Development, University of Thessaly, Pedion Areos, 38334 Volos, Greece
* Correspondence: egolia@auth.gr; Tel.: +30-23-1099-8809

Abstract: A pot experiment was carried out to investigate the effects of polyethylene (PE), a broadly utilized polymer type, on soil properties and lettuce growth. Two Zn- and Cd-contaminated soil samples were obtained from urban and rural areas of Greece, respectively. PE fragments (<5 mm) were added at different concentrations (2.5%, 5% w/w). Lettuce seeds were then planted in the pots in a completely randomized experiment. Plant growth patterns and tissue metal accumulation were investigated. The presence of PE in soils resulted in a reduction in pH, significantly enhanced the organic matter content, and increased the cation-exchange capacity. The availability of both metals was also increased. Metal migration from soil to plant was determined using appropriate tools and indexes. A higher metal concentration was detected in lettuce roots compared with that in the edible leaves. The presence of PE MPs (2.5% w/w) increased the amount of available Zn more than that of Cd in highly contaminated soils. When PE MPs were added to agricultural soil, Zn concentrations increased in the plant leaves by 9.1% (2.5% w/w) and 21.1% (5% w/w). Considering that both metals and microplastics cannot be easily and quickly degraded, the fact that the less toxic metal is more available to plants is encouraging. Taking into account the physicochemical soil features, decision makers may be able to limit the risks to human health from the coexistence of heavy metals and microplastics in soils.

Keywords: contamination factor; lettuce; uptake; agricultural and urban soils

Citation: Bethanis, J.; Golia, E.E. Revealing the Combined Effects of Microplastics, Zn, and Cd on Soil Properties and Metal Accumulation by Leafy Vegetables: A Preliminary Investigation by a Laboratory Experiment. *Soil Syst.* **2023**, 7, 65. https://doi.org/10.3390/soilsystems7030065

Academic Editor: Luis Eduardo Akiyoshi Sanches Suzuki

Received: 19 June 2023
Revised: 14 July 2023
Accepted: 16 July 2023
Published: 17 July 2023

1. Introduction

In recent years, plastics have been used in many applications in our daily lives, as they are an easy and economical solution for everyday issues in our home and workplace. Given their massive global production and indiscriminate use, the amount of plastic entering the human environment, particularly the soil, is exceptionally high [1]. Plastic waste accumulates in most parts of the world, taking many years to decompose, as it is corrosion-resistant, chemically stable, and difficult to degrade [2]. When plastics reach the soil, they are frequently broken down into smaller (<5 mm) particles known as microplastics (MPs) due to physical and chemical erosion as well as the impact of UV light [3]. The term microplastics (MPs) covers a large group of plastic materials that include a wide range of polymers with varying chemical compositions, sizes, and dimensions [4]. Microplastics refer to plastic particles, fragments, films, or fibers with a diameter less than 5 mm [5]. MPs are divided into two types: primary MPs, which are intentionally made in sizes less than 5 mm, and secondary MPs, which are formed through the fragmentation of larger plastics or primary MPs [6]. MPs have been proven to influence both physical and chemical soil attributes, as well as the soil's microbial composition and health. The chemical composition of MPs, together with their size and shape, all have a significant impact on the variability of soils' physicochemical properties [7]. Multiple studies have been carried out regarding the effects of MPs on soil environmental variables, yielding often opposing results [8]. Previous studies have shown that the addition of polyethylene, polypropylene, and polystyrene reduces the bulk density of soil [9]. Furthermore, alterations of soil acidity owing to the presence of MPs or even smaller nanoplastics (NPs) have been repeatedly reported. The acidity of soils is a crucial parameter that defines the distinct properties and functionality of soil systems. Numerous studies have shown that the application of MPs or NPs has the potential to increase soil pH; however, a decrease in soil acidity or even a moderate effect has been observed in other instances. The researchers Wang et al. [10] showed that the presence of polylactic acid (PLA) and high-density polyethylene could increase soil pH, while Boots et al. [11] found that the long-term persistence of high-density polyethylene (HDPE) decreased soil pH values. It is widely recognized that in acidic soils, there is reduced adsorption and increased metal mobility. This phenomenon can be attributed to the increased competition between hydrogen cations and dissolved metals for negatively charged surfaces, and this could lead to an increase in the availability of both toxic and trace elements in soils and plants. On the other hand, in alkaline soils, the formation of strong organometallic complexes is favored, resulting in a reduction in the mobility and therefore the availability of these elements, which could lead to severe nutrient deficiencies for plants [11]. The chemical reactions occurring between metals and plastic particles appear to exhibit similarities to those observed between metals and soil organic matter [3]. By altering soil acidity, MPs and NPs can affect the mobility and adsorption of metals and metalloids, the formation and stability of organometallic complexes, bond strength, and the chemical selectivity of soil components [7]. MPs and NPs also have variable effects on soil organic matter, which is essential for both soil fertility and plant nutrition [8]. Studies have shown that the addition of polyethylene decreased soil organic carbon (OC) content [12], while the addition of 2% (w/w) PLA significantly increased soil OC [13]. These contradictory results suggest that the effects of MPs on the soil environment may be regulated by several factors, such as the MP properties (e.g., polymer type, concentration, particle size, and shape), as well as different soil and local climatic conditions [14].

1.1. Identity and Distinguishing Features of the Most Common Microplastics in Soil Environments

Shi and his colleagues in their study [15] explored the impact of three types of microplastics—polyethylene (PE), polystyrene (PS), and polypropylene (PP)—on the early growth of tomato seeds using a combined approach of oxidative stress and nutritional quality analysis. The outcomes indicated that all the different MPs exhibited negative impacts on the tomato seedlings' physiological and metabolic functions, root development, seed germination percentage, and germination index. PE was proven to be the most toxic, while PP was the least toxic. Considering the two most common plastics in soils, PE and PP, it has been documented that polyethylene is carcinogenic at the class 3 level, according to the World Health Organization [16]. Both PE and PP are frequently and widely dispersed in the agricultural environment and in diverse types of microplastics, as claimed by Hu et al. [17]. In their study, Qi et al. [17] investigated the effects of low-density polyethylene (LDPE) and biodegradable starch-based microplastic films of different sizes on potted wheat. It was found that biodegradable-plastic soil-cover films had a particularly significant effect on wheat growth compared with PE. In another study conducted by Machado et al. [18], it was found that polyester fibers and polyamide beads had a positive influence on the growth of spring onions compared with other polymer types (polyethylene, polyester terephthalate, polypropylene, and polystyrene). The environmental pathway of microplastics in soils was also revealed by Gan et al. [19] in a study about their classification, sources, and fate, as well as their effects on cultivated plants.

1.2. Effects of Microplastics on Metal Uptake by Plants

To prevent potential risks, the first thought is to remove microplastic residues from green vegetables using appropriate washing techniques so that they cannot be passed on to the human body [20]. Further efforts are underway to understand the mechanisms of the coexistence of microplastics, toxic elements, and trace elements in the soil [9]. It has been revealed that the presence of PEs together with Cd in soil is able to cause synergistic toxicity with regard to plant growth, such as the suppression of photosynthesis and increased oxidative damage [21]. Various MP polymer types can significantly increase Cd accumulation in plant shoots and roots, while PE appears to have a higher promotive effect. That is mainly explained by the fact that PEs cause a decrease in pH followed by an increase in Cd bioavailability [22]. Huang et al. [23] noticed a significant increase in Cd availability by adding various quantities of PEs to soils. Furthermore, it was observed that the addition of MPs modified the physicochemical properties of the soils, causing an increase in Cd bioavailability and uptake by the plants [7]. In their experiments, Wang et al. [24] examined the effect of HDPE (high-density PE) and PS on maize plants and found that PS caused greater phytotoxicity to the plants. Both HDPE and PS increased the DTPA-extractable Cd content in the soil. Compared with HDPE, PS appears to have more pronounced effects on Cd bioavailability and plant growth inhibition, indicating a higher risk in soil–plant systems [19]. The effects of various MPs with different chemical compositions on soil properties and the availability of Cu, Zn, Cd, and Pb were investigated by Wen et al. [25]. Linear low-density polyethylene (LLDPE), polyamide (PA), polyurethane (PU), polystyrene (PS), and low-density polyethylene (LDPE) MPs caused a decrease in soil pH, while they increased soil organic matter and cation-exchange capacity. These alterations in both soil parameters and the metal adsorption and desorption on the solid surface of MPs determine their availability [26]. It is well known that the addition of certain materials, organic or inorganic, can modulate the mobility and metal availability to plants [27–29]. At the same time, the concentration of MPs in the soil also plays a decisive role in metal mobility. The absorption of harmful metals or beneficial trace elements by plants cultivated in MP-contaminated soils in varying quantities and chemical compositions has been at the epicenter of recent research [30–33].

In the present study, laboratory experiments were conducted to investigate the effects of polyethylene on the metal intake and accumulation by lettuce plants. The primary objectives and aims of the present study were to detect the impact of PE MPs on the physicochemical soil properties and to reveal their effects on metal availability and concentration in moderately and heavily contaminated soils in Mediterranean urban and rural areas.

2. Materials and Methods

2.1. Sampling Areas and Soil Sample Preparation

Two different alkaline soil types from rural and urban areas of central and northern Greece, respectively, were used for our study. Urban soil samples were collected from Thessaloniki city's metropolitan area, whereas rural samples were taken from Almyros town [34] in the region of Thessaly [35].

Specific handling techniques were used during the soil sampling to measure the content of heavy metals. More specifically, a wooden shovel and a two-meter radius were used to obtain six sub-samples from each main sample. The samples were then transferred to the Soil Science Laboratory at the Aristotle University of Thessaloniki, where they were air-dried and prepared for further physical and chemical analysis.

2.2. Preparation of Microplastics

Polyethylene (PE), a common polymer type present in the environment, was selected as a plastic contaminant for the purpose of this research. Plastic particles were obtained by manually cutting commercially available transparent polyethylene bags into smaller pieces [8]. These pieces were then ground and separated by sieving using a 0.5 mm sieve to obtain microplastics of appropriate dimensions (<5 mm). The MPs were then washed with

NaClO 0.01 M solution in order to remove any impurities and to inhibit potential microbial activity. The MPs were subsequently incorporated into the soil samples and left to incubate for several days.

2.3. Pot Experiments

Equal quantities of the two metal-contaminated soil samples were placed in plastic pots with a base area of 26 cm^2 and a height of 16 cm. To achieve the best possible homogeneity, microplastics were added at two different concentrations (2.5% and 5% w/w) and mechanically mixed into the soil. Irrigation (up to 70% saturation) was carried out to maintain an appropriate level of moisture content, and the pots were left to incubate. Ten days later, lettuce seedlings that had been cultivated in the same soil were transferred and planted in the pots. A completely randomized experiment was set up, consisting of 2 soils (urban and rural) × 3 MP treatments (0%, 2.5%, and 5% w/w) × 4 replicates for a total of 24 pots. Experiments were carried out in February 2023 at the Aristotle University of Thessaloniki's Soil Science Laboratory and the University of Thessaly's Analytical Chemistry Laboratory. After 45 days, the lettuce plants were harvested, dried in an oven, and subjected to chemical analysis.

2.4. Analyses of Soil

After air drying, soil samples were ground and passed through a 2 mm sieve. They were subjected to the following soil analyses [36,37]: The soil reaction (pH) and electrical conductivity (EC) values were determined using a mixture fixed by soil and distilled water at a ratio of 1:1. The soil mechanical composition was evaluated by the Bougioukos method, and the percentage (%) of CaCO$_3$ by the Bernard method. The modified Walkley–Black method was used to calculate the percentage of organic matter in the soils. After extraction with 1 N ammonium acetate solution (pH 7.0), a Sherwood's flame photometer was used to determine the percentage of exchangeable cations, while the sum of them was used for calculating the soil Cation-Exchange Capacity value. For the evaluation of Zn and Cd, water-soluble concentrations of a dilute CaCl$_2$ solution were used [29]. The available and total metal concentrations were determined using soil-extraction methods with DTPA and HCl:HNO$_3$ in a 3:1 ratio (Aqua Regia) solution [38]. Metals were quantified using a Perkin Elmer Analyst 700 atomic absorption spectrometer (AAS). The CRM 141R soil standard was used for the method validation. Metal analysis accuracy ranged from 9.1% to 11%, with detection limits of 0.01 and 0.85 mg/kg for Cd and Zn, respectively.

2.5. Chemical Analyses of Lettuce Tissues

Lettuce samples were weighed to determine the fresh weight and then put in paper bags to dry. Following an extraction using the Aqua Regia method in a closed digestion system for 4.5 h, Zn and Cd were determined for each sample's root and aboveground portion [27,29]. Using an atomic absorption spectrophotometer, the metals were quantified after being diluted in 25 mL volumetric flasks. The analytical procedure was validated using a NIST-certified standard tomato sample.

2.6. Statistical Analysis, Soil Pollution Indices and Metal Mobility Indicators

To assess and characterize the contamination level of both soils used in the present study, three typical soil pollution indices were used to classify the soil samples, namely, Contamination Factor (CF), Geo-accumulation Index (Igeo), and Bioavailability Factor (BF) [29]. Furthermore, to reveal the metal behavior and distribution between the soil–plant systems under exposure to MPs, three appropriate indicators were calculated: Transfer Co-efficient (TC), Bioaccumulation Factor (BAF), and Translocation Factor (TF) [30]. Microsoft Office Excel statistical packages (Microsoft, Redmond, WA, USA) and SPSS statistical software (IBM, Armonk, NY, United States) were used for data management and processing. For each data group, the mean, median, minimum, and maximum values, as well as the standard deviation, were calculated. Identifying statistically significant differences at the

0.01 and 0.05 levels was accomplished using the ANOVA method. Additionally, the data were analyzed using the *t*-test by repeatedly comparing value pairs.

3. Results and Discussion

3.1. Influence of Microplastics on Soil Chemical Properties

The physical and chemical properties of the study's soil samples are shown in Table 1. Urban and agricultural soils both have an alkaline pH and comparable levels of electrical conductivity. The application of fertilizers during agricultural activities and crop cultivation is likely responsible for the agricultural soil's comparatively high electrical conductivity rating.

Table 1. Physicochemical properties of the soil samples.

	pH	EC (Electric Conductivity) (µS/cm)	OM (Organic Matter) (%)	CEC (Cation-Exchange Capacity) (cmol$_c$/kg)	Clay (%)	Texture	Cd (mg/kg)	Zn (mg/kg)
Soil 1 (Agricultural)	7.4 ± 0.3	2234 ± 54	2.8 ± 0.2	30.4	47 ± 1.1	CL (Clay Loam)	0.9 ± 0.1	74 ± 1.2
Soil 2 (Urban)	8.1 ± 0.5	2093 ± 49	2.1 ± 0.4	26.5	46.5 ± 2.1	CL	1.1 ± 0.3	79 ± 0.6

However, urban soil is usually affected by a variety of anthropogenic activities, including horticultural decorative landscaping in city flowerbeds [34,36]. The presence of microplastics in the soil samples was the decisive factor in altering the values of their chemical properties. Zhao et al. [37] investigated the effect of different polymers with different shapes on soil pH by conducting a 21-day incubation experiment. A decrease in the pH value was initially observed when polymers were incorporated into the soil samples; however, the pH value increased over time. In the current investigation, the addition of both amounts of PE MPs to soil samples resulted in a decrease in soil reactivity. The pH value of agricultural soil 1 was reduced by 5.4% and 2.9%, respectively, after 2.5% and 5% (w/w) PE MPs were applied, as determined immediately before planting the lettuce seedlings. The corresponding percentages for the urban soil sample were 6.2% and 2.6%, respectively.

Figure 1 depicts the changes in the soils' properties after the addition of PE MPs. The pH of both soils was checked again at the completion of the experiment, and an increase was detected. This is consistent with reports from other studies [37]. Gharahi and Zamani-Ahmadmahmoodi [8] observed that after a 30-day exposure of two soil samples to PET, the pH value decreased. In another study, it was found that soil pH was not significantly affected by MPs when these were added at a low dose; however, the acidity decreased with the addition of high doses of PE and PS MPs and increased with high doses of PLA and PHB MPs [13]. When PE MPs were applied at a lower dose, the pH value decreased more in both soils. The soil pH value is a key factor for achieving metal mobility and for forecasting heavy metal pollution [34].

During the soil organic matter evaluation, a comparable outcome was obtained. When lower concentrations of PE MPs were added to the soils, a considerable increase of 17.9% and 23.8% was detected. However, one would anticipate that as the amount of microplastics in the soil increased, so would its organic content. The soil organic matter content is increased by adding sludge or wheat straw residues [27,35], which is desirable as it enhances soil fertility. The Cation-Exchange Capacity (CEC) value increased significantly in the first and second soil samples: by 8.6% and 9.8%, respectively.

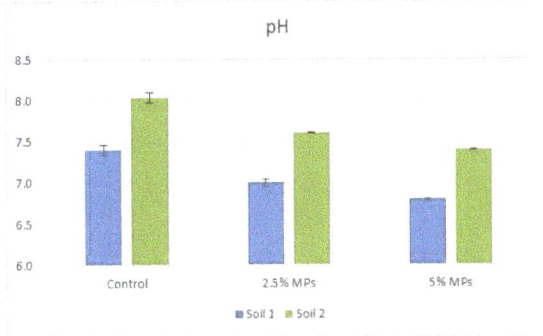

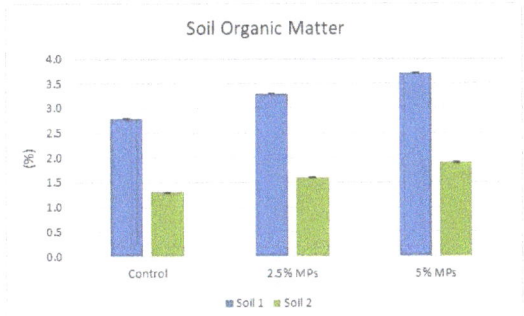

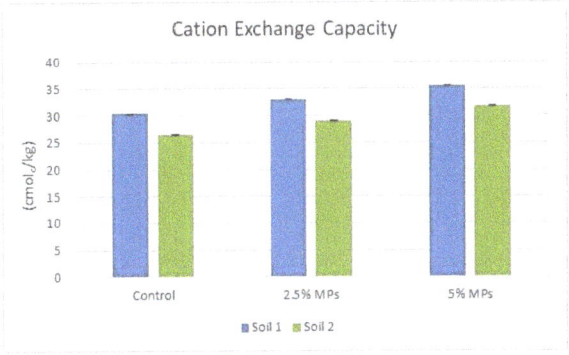

Figure 1. Effect of polyethylene microplastic (PE MP) levels on the physicochemical properties of the study's soil samples.

3.2. Influence of Microplastics on Metal Availability

Figure 2 depicts the variations in water-soluble, available, and pseudo-total Zn and Cd content in the soil samples studied. In both soils, the water-soluble content of Zn increased in excess compared with Cd. When 2.5% *w/w* PE MPs were added to the first soil, the increase in water-soluble Zn content reached 57.3%, while the Cd concentration increased by more than 33.3%. When 2.5% *w/w* PE MPs were introduced to soil 2, the water-soluble concentrations of Zn and Cd increased by 38% and 16.6%, respectively. In a relevant study, it was found that the methods using pure water were highly correlated with each other and showed the strongest correlation with agronomic effectiveness [38–40].

As a result, given the current study's slightly alkaline soils, the water-soluble concentration of metals may indicate the concentration that plants can absorb. The metals' water-soluble concentrations increased; however, it is promising that Zn increased at a significantly higher rate compared to the more hazardous Cd. According to other studies, the addition of microplastics, particularly PE, to the soil at high mixing ratios increased the availability of bivalent lead because it rendered the big aggregates unstable and reduced the rate at which Pb adsorbs onto them [26,41].

Figure 2 also demonstrates the impact of PE MP addition on the extractable quantities of Zn and Cd using the DTPA solution and Aqua Regia. As the total amount of microplastics in both soils rises, so do the amounts of both metals. This is consistent with prior research. In a relevant study, the Cd availability in clay- and sand-based soils was examined following the addition of PU and PP microplastics [42]. In clay soils, accessible Cd was considerably negatively correlated with dissolved organic carbon and pH, whereas in sandy soils, available Cd was strongly negatively correlated with Fe (II). The synergistic toxicity generated by the presence of PE MPs and metals in the soil samples was out of proportion to the amount of MPs supplied [13]. In both rural and urban soil, the addition of microplastics does not seem to have an impact on the pseudo-total concentration.

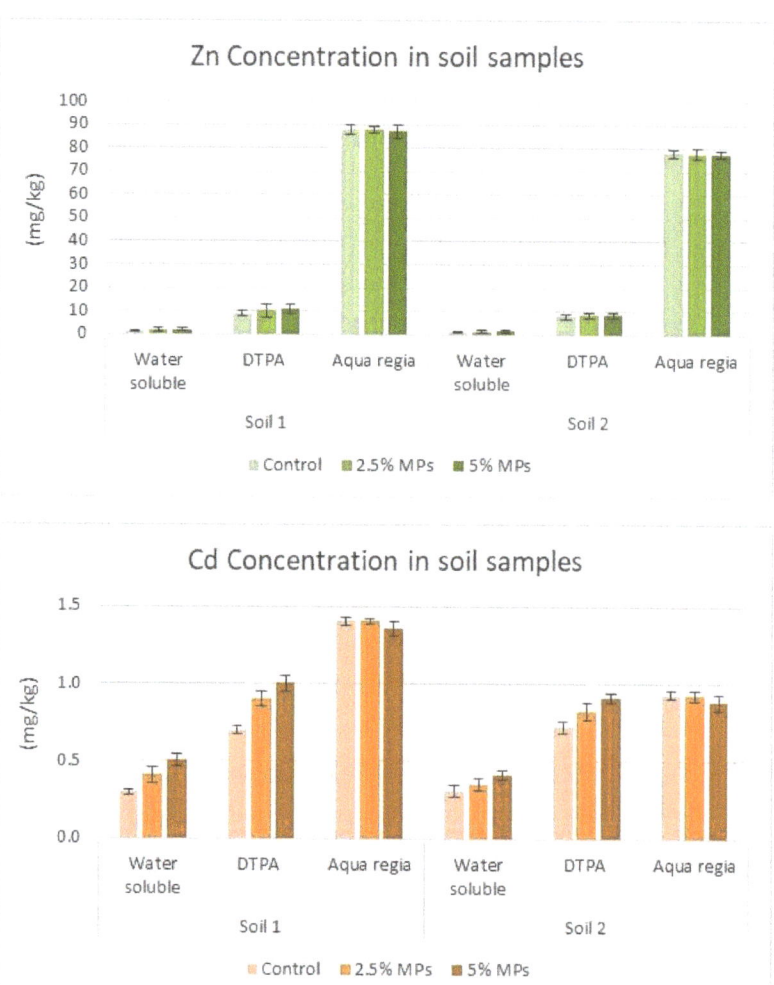

Figure 2. Effect of polyethylene microplastic (PE MP) levels on Zn and Cd water-soluble, DTPA-extractable (available) and Aqua Regia-extractable (total) concentrations in the studied soil samples.

3.3. Effects of Microplastics on Zn and Cd Levels in Lettuce Plants

Figure 3 depicts the impact of various polyethylene microplastic concentrations on the levels of Zn and Cd in lettuce plants grown in the studied soils. The level of both hazardous and nutritious (trace elements) metals absorbed by the cultivated plants is critical [31,43].

In agricultural soil 1, the Zn concentration in lettuce roots was higher than that in the leaves. When PE MPs were added to the soil, Zn concentrations increased in the plant's roots and leaves by 11.5% and 9.1% (2.5% w/w) and 26.6% and 21.1% (5% w/w), respectively. Furthermore, in the trials with the second (urban) soil sample, which was less Zn- and Cd-contaminated, metals accumulated more in the lettuce roots than in the leaves. Following the addition of PE MPs, the Zn concentration in the roots and leaves rose by 4.1% and 3.4% (2.5% w/w) and by 0.2% and 6.8% (5% w/w), respectively. It is well known that the amount, chemical composition, shape, and size of microplastics in soils determine their impacts and, consequently, influence the absorption of hazardous or nutritious compounds by cultivated plants [8,44]. It is widely known that the addition of several materials, as well as the modification of certain soil parameters, can affect the plants' uptake of metals [29,45,46].

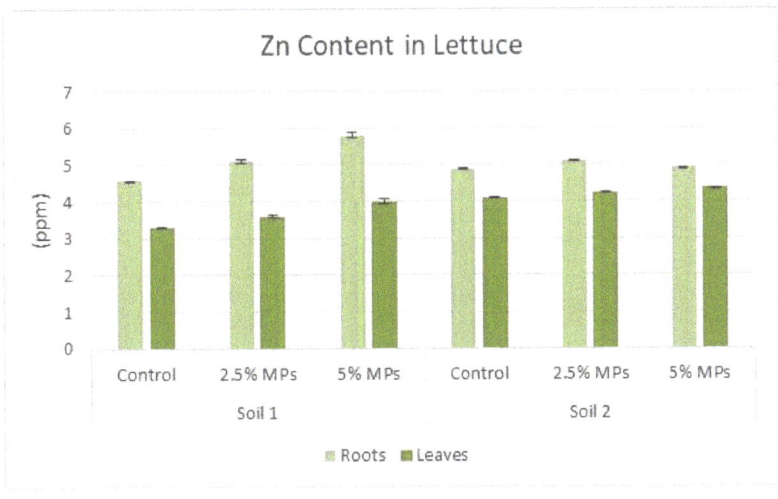

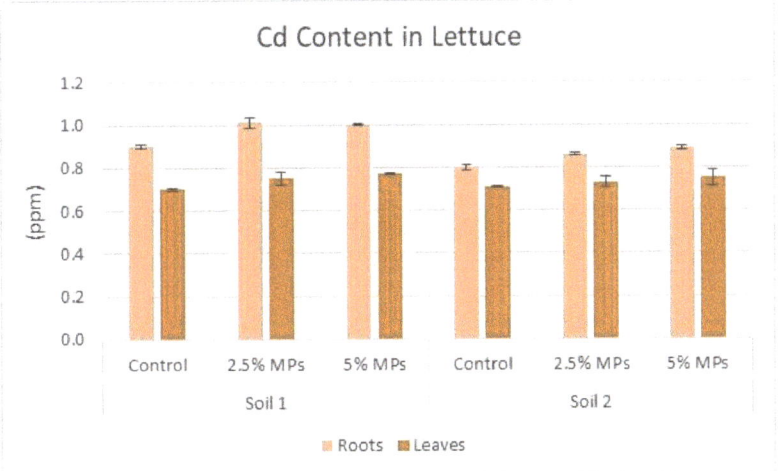

Figure 3. Effect of different polyethylene microplastic (PE MP) amounts on the levels of Zn and Cd in lettuce plants grown in the two soil samples.

The effects of microplastics on heavy metal or trace element uptake by plants are of great concern in the scientific community. MPs may increase the accumulation of heavy metals by altering the rhizosphere microorganisms in lettuce plants [47]. Additionally, in research conducted on strawberry plants, it was found that the increased bioavailability of Cd caused by the presence of microplastics was responsible for the observed negative effects on soil properties and plant performance [48].

In the current investigation, MP addition to the first soil sample at a ratio of 2.5% (w/w) resulted in an increase in the Cd content by 11.1% and 7.3% in the roots and leaves, respectively. At a higher ratio of 5% (w/w), the corresponding increase was 11.2% and 10%.

The variation of Cd content in the lettuce roots and leaves in the second soil sample was 7.5% and 2.8% (2.5% w/w) and 11.3% and 5.6% (5% w/w), respectively.

Numerous research investigations have demonstrated that metals accumulate at varying levels in different parts of the plant [28,30,49,50].

It is important to note that, in both the rural and urban soils examined, the increase in the total amount of metal accumulation imposed by the presence of microplastics was greater in the roots. However, the leaves, which are consumable plant parts, are also impacted but to a lesser degree.

Although the precise mechanisms that define the behavior of microplastics in soils or plants are not well understood, preliminary data suggest that they might enhance metal mobility and translocation [11]. Although several metals are hazardous, there are essential minerals for the growth of plants, i.e., trace elements. Therefore, the presence of microplastics could enhance nutrient absorption, resulting in increased plant growth [13]. However, high microplastic concentrations may induce toxicity to plants, which could be further aggravated by synergistic effects caused by the coexistence of microplastics and heavy metals [18,26].

3.4. The Impact of Polyethylene Microplastics on Soil Pollution Indices

Figure 4 depicts the variations in the soil pollution indices as well as the indices related to the metal content between the soil and the plants.

The CF index values do not appear to change statistically significantly after microplastic addition, since the values used in the estimation do not change. It is well known that the value of the total metal concentration is taken into account when calculating the value of the contamination factor. The first level of polyethylene microplastics, or 2.5% w/w concentration, had no effect on the CF value of zinc in either of the study's examined soil samples. The values of the index place both rural and urban samples in group II, or moderately contaminated soils, based on the index categorization. It was found by Yu et al. [12] that a possible decrease in the bioavailability of soil heavy metals caused by the presence of microplastics varies across aggregate levels, leading to the conclusion that conflicting factors define the outcome of nutrient and pollutant intake.

In a comparable manner, as no alteration in the total concentration was observed alongside the addition of 2.5% w/w PE MPs, no change in the corresponding soil indicators was observed during the resulting calculation of CF indicators for Cd in both soil samples. In terms of Cd pollution, both soils were classified as class I, or virgin, soils.

A slight, but not statistically significant, decrease in the total concentration and, consequently, in the CF values for Zn and Cd in both soil samples was observed by the addition of increased amounts of microplastics, i.e., at a rate of 5% w/w. The samples are nonetheless still categorized in the relevant pollution categories to which they formerly adhered.

The Igeo indices of Zn and Cd revealed the same accomplishments, as the addition of both doses of polyethylene microplastics had no effect on their values.

The availability index (BF) of both metals increased substantially with the addition of microplastics at their highest ratio (5% w/w). A greater increase is observed in agricultural soils (soil 1), whereas urban soils exhibit a modest increase in metal availability. When the higher quantity of polyethylene microplastics is added, the value of the availability index, or the concentration of Zn extracted with the DTPA solution relative to the total, shows the least increase and equals 9.25% in urban soils.

In general, the incorporation of microplastics in soils increased the availability of both the hazardous Cd and the less toxic Zn, in accordance with other studies [6]. It is crucial, however, that the DTPA-extractable concentration be greater rather than the water-soluble concentration. This might be due to a chemical interaction between polyethylene and the DTPA solution, i.e., diethylenetriaminepentaacetic acid. It is well known that metals are able to bind to soil organic compounds as they form chemical complexes along them [27,35]. This should be further investigated, as the increase in the values of the DTPA-extractable amounts of Zn and Cd does not correspond to the increase in the levels of the metals in the plants. On the contrary, the water-soluble concentrations of the metals in the soil samples appear to have a better correlation. It has been indicated by Dioses-Salinas et al. [30] that the heterogeneity presented in the described methods in the literature sometimes makes the results uncomparable. Furthermore, novel methods need to overcome important frontiers and challenges.

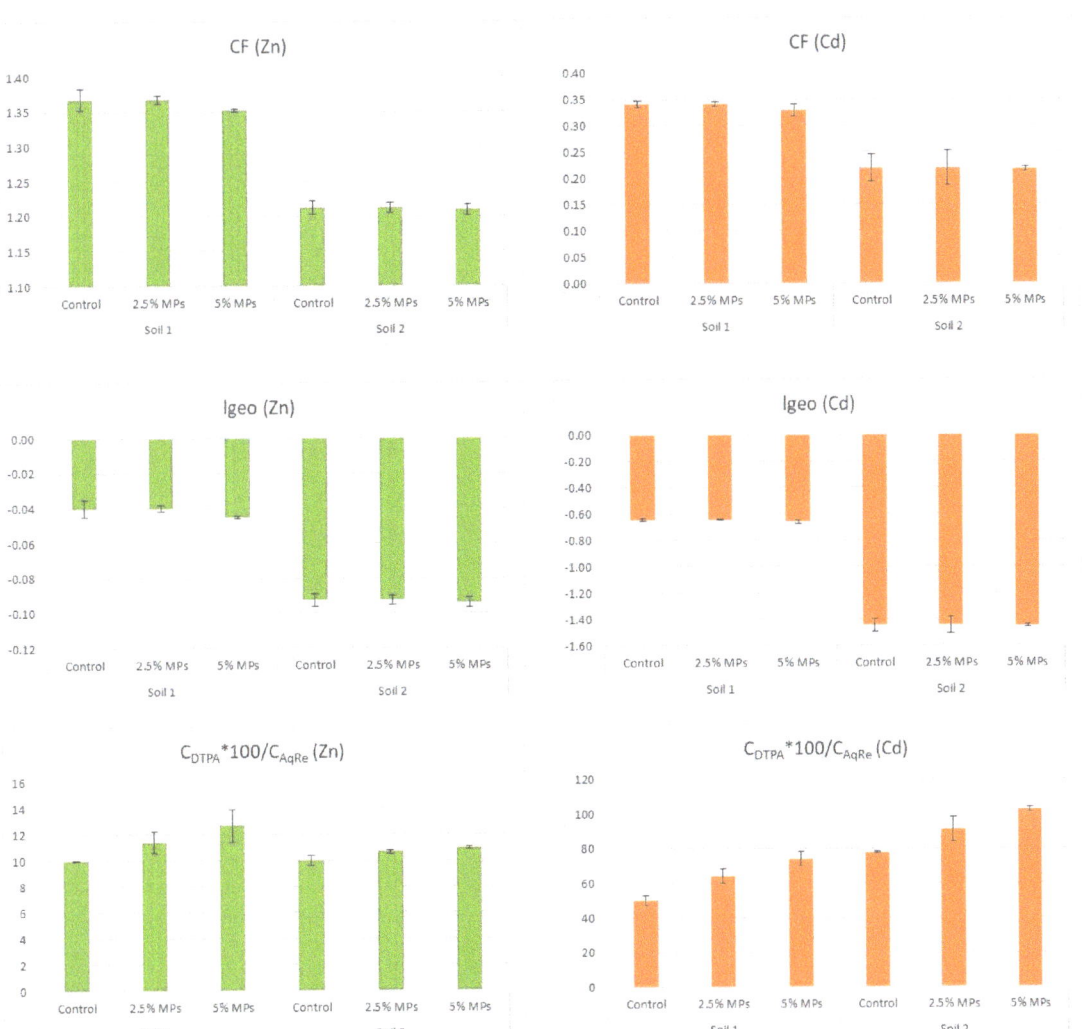

Figure 4. Effect of different polyethylene microplastic (PE MP) amounts on Zn and Cd soil contamination indices.

3.5. The Impact of Polyethylene Microplastics on Soil-to-Plant System Indices

Important inferences about the possible risk of metals to the environment and human health may be derived from the study of indices that represent the metal mobility within the soil–plant system.

Figure 5 depicts three indicators that were examined with respect to their responses to change when the two amounts of polyethylene microplastics were added. Sun and his colleagues [51] studied the foliar uptake and leaf-to-root translocation of plastics with different coating charges in maize plants.

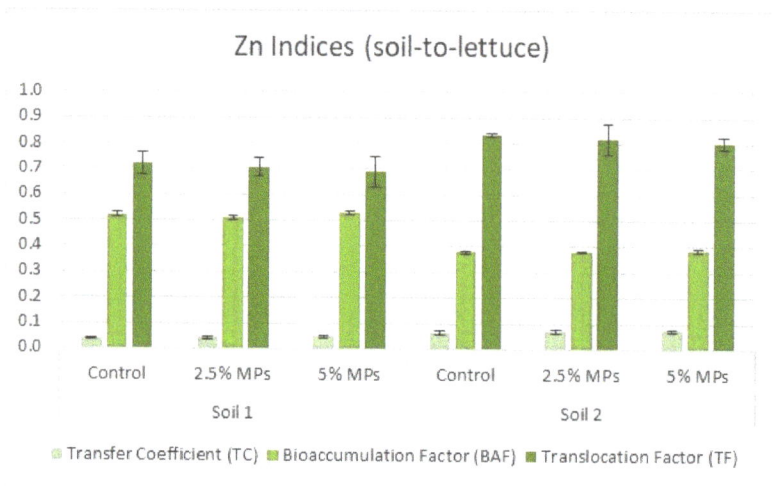

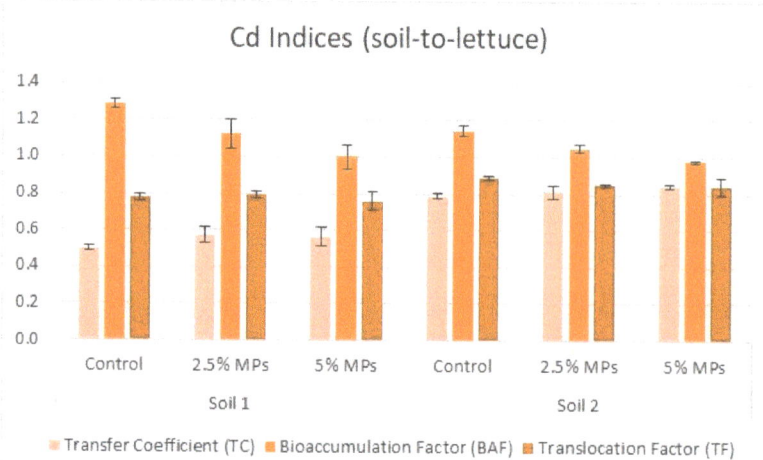

Figure 5. Effect of different polyethylene microplastic amounts on Zn & Cd soil-to-lettuce indices.

In the present study, microplastics appear to dramatically enhance the Transfer Coefficient, which measures the concentration of metals in the aerial portion of the soil sample in relation to the overall concentration in the matching soil sample. Hu et al. [52] also investigated the distribution of micro- and mesoplastics in agricultural soils across China, resulting in alterations to their environmental impacts via soils. When PE MPs are added to both soil types, the BAF index, which reflects the percentage of concentration in the roots compared to the amount of DTPA extractable with the solution, decreases. The TF index, which measures the proportion of concentration in the plant's aboveground vs. subterranean portions, appears to be rising. To explore and assess the phenomenon of microplastic interactions with soil-based metals, it is helpful and essential to apply all three indices. However, it is essential to concentrate our scientific attention on the necessary criteria to meet. Gharahi and Zamani-Ahmadmahmoodi [8] and Zhou et al. [33] found that microplastics appear to have conflicting effects on three distinct crops in the field. In other words, MPs under specific circumstances may improve some growth characteristics of plants.

According to Figure 4, the greatest decrease in the BAF index is observed in the case of Cd in rural soil, followed by Cd in urban soil. A decrease is also observed in the BAF value of Zn in urban soil. The value of BAF in agricultural soil increases with a greater input of microplastics. According to Huang et al. [44] microplastics may influence soil nutrient cycling by affecting the dominant bacteria phyla or genes and enzymes involved in the carbon, nitrogen, and phosphorus cycles. Considering that for the hazardous Cd, the value of BAF decreases overall, for the less toxic Zn, the presence of microplastics seems to contribute to its absorption by plants [22]. Future research in this area must be focused and take into account how time affects the changes in the chemical behavior of plastics in soils. Given that knowledge in this field is still in its infancy, the incubation duration of microplastics in soil and the presence of more metals or other ions should improve the efficacy of this study and provide useful findings.

4. Conclusions

The current study was an initial attempt to assess the impact of microplastics on plants cultivated in soils containing Zn and Cd. Polyethylene, a plastic that is commonly found in high quantities in both agricultural and urban soils, was chosen to obtain microplastics of dimensions less than 5 mm. Lettuce, the most well-known and highly consumed vegetable globally, was chosen as an experimental plant to assess the potential risks. Metal pollution was severe in agricultural soils, while moderate contamination with Zn and Cd was detected in urban soils. Microplastics were added to soil samples in two distinct quantities of 2.5% and 5% w/w. The effects of polyethylene microplastics on the soil's physicochemical characteristics and metal concentrations were investigated. Furthermore, the impact of MPs on the Zn and Cd availability and distribution in the soil–plant system was assessed.

The addition of PE MPs resulted in a decrease in soil pH. On the contrary, an increase in both organic matter and soil cation-exchange capacity was detected. Furthermore, MPs enhanced the available concentrations of both metals. In the heavily polluted agricultural soil, the addition of microplastics at 2.5% w/w increased the readily available concentration of Zn more than that of the hazardous Cd. The presence of PE MPs in both the rural and urban soil samples resulted in a higher metal accumulation in lettuce roots than in the edible above-ground parts of the plants. The coexistence of metals and microplastics in soils may pose risks to soil, plants, and even human health. However, with careful study and understanding of the mechanisms that catalyze synergistic toxicity and appropriate management, it is possible to reduce such risks.

Author Contributions: Conceptualization, J.B. and E.E.G.; methodology, J.B. and E.E.G.; software, J.B. and E.E.G.; validation, J.B. and E.E.G.; formal analysis, J.B.; investigation, J.B. and E.E.G.; resources, E.E.G.; data curation, J.B. and E.E.G.; writing—original draft preparation, J.B. and E.E.G.; writing—review and editing, J.B. and E.E.G.; visualization, J.B. and E.E.G.; supervision, E.E.G.; project administration, E.E.G.; funding acquisition, E.E.G. All authors have read and agreed to the published version of the manuscript.

Funding: This research received no external funding.

Institutional Review Board Statement: Not applicable.

Informed Consent Statement: Not applicable.

Data Availability Statement: Data supporting the findings of this study are available from the corresponding author upon reasonable request.

Conflicts of Interest: The authors declare no conflict of interest.

References

1. Wu, X.; Lu, J.; Du, M.; Xu, X.; Beiyuan, J.; Sarkar, B.; Bolan, N.; Xu, W.; Xu, S.; Chen, X.; et al. Particulate Plastics-Plant Interaction in Soil and Its Implications: A Review. *Sci. Total Environ.* **2021**, *792*, 148337. [CrossRef]
2. Allouzi, M.M.A.; Tang, D.Y.Y.; Chew, K.W.; Rinklebe, J.; Bolan, N.; Allouzi, S.M.A.; Show, P.L. Micro (Nano) Plastic Pollution: The Ecological Influence on Soil-Plant System and Human Health. *Sci. Total Environ.* **2021**, *788*, 147815. [CrossRef] [PubMed]
3. Yu, H.; Zhang, Y.; Tan, W.; Zhang, Z. Microplastics as an Emerging Environmental Pollutant in Agricultural Soils: Effects on Ecosystems and Human Health. *Front. Environ. Sci.* **2022**, *10*, 217. [CrossRef]
4. Horton, A.A.; Walton, A.; Spurgeon, D.J.; Lahive, E.; Svendsen, C. Microplastics in Freshwater and Terrestrial Environments: Evaluating the Current Understanding to Identify the Knowledge Gaps and Future Research Priorities. *Sci. Total Environ.* **2017**, *586*, 127–141. [CrossRef] [PubMed]
5. Thompson, R.C.; Olson, Y.; Mitchell, R.P.; Davis, A.; Rowland, S.J.; John, A.W.G.; McGonigle, D.; Russell, A.E. Lost at Sea: Where Is All the Plastic? *Science* **2004**, *304*, 838. [CrossRef]
6. Akdogan, Z.; Guven, B. Microplastics in the Environment: A Critical Review of Current Understanding and Identification of Future Research Needs. *Environ. Pollut.* **2019**, *254*, 113011. [CrossRef]
7. Wang, F.; Wang, X.; Song, N. Polyethylene Microplastics Increase Cadmium Uptake in Lettuce (*Lactuca sativa* L.) by Altering the Soil Microenvironment. *Sci. Total Environ.* **2021**, *784*, 147133. [CrossRef]
8. Gharahi, N.; Zamani-Ahmadmahmoodi, R. Effect of Plastic Pollution in Soil Properties and Growth of Grass Species in Semi-Arid Regions: A Laboratory Experiment. *Environ. Sci. Pollut. Res.* **2022**, *29*, 59118–59126. [CrossRef]
9. Lozano, Y.M.; Rillig, M.C. Effects of Microplastic Fibers and Drought on Plant Communities. *Environ. Sci. Technol.* **2020**, *54*, 6166–6173. [CrossRef]
10. Wang, F.; Wang, Q.; Adams, C.A.; Sun, Y.; Zhang, S. Effects of Microplastics on Soil Properties: Current Knowledge and Future Perspectives. *J. Hazard. Mater.* **2022**, *424*, 127531. [CrossRef]
11. Boots, B.; Russell, C.W.; Green, D.S. Effects of Microplastics in Soil Ecosystems: Above and below Ground. *Environ. Sci. Technol.* **2019**, *53*, 11496–11506. [CrossRef] [PubMed]
12. Yu, H.; Hou, J.; Dang, Q.; Cui, D.; Xi, B.; Tan, W. Decrease in Bioavailability of Soil Heavy Metals Caused by the Presence of Microplastics Varies across Aggregate Levels. *J. Hazard. Mater.* **2020**, *395*, 122690. [CrossRef] [PubMed]
13. Feng, X.; Wang, Q.; Sun, Y.; Zhang, S.; Wang, F. Microplastics Change Soil Properties, Heavy Metal Availability and Bacterial Community in a Pb-Zn-Contaminated Soil. *J. Hazard. Mater.* **2022**, *424*, 127364. [CrossRef] [PubMed]
14. Zhao, M.; Li, C.; Zhang, C.; Han, B.; Wang, X.; Zhang, J.; Wang, J.; Cao, B.; Zhao, Y.; Chen, Y.; et al. Typical Microplastics in Field and Facility Agriculture Dynamically Affect Available Cadmium in Different Soil Types through Physicochemical Dynamics of Carbon, Iron and Microbes. *J. Hazard. Mater.* **2022**, *440*, 129726. [CrossRef]
15. Shi, R.; Liu, W.; Lian, Y.; Wang, Q.; Zeb, A.; Tang, J. Phytotoxicity of Polystyrene, Polyethylene and Polypropylene Microplastics on Tomato (*Lycopersicon esculentum* L.). *J. Environ. Manag.* **2022**, *317*, 115441. [CrossRef]
16. Teng, L.; Zhu, Y.; Li, H.; Song, X.; Shi, L. The Phytotoxicity of Microplastics to the Photosynthetic Performance and Transcriptome Profiling of Nicotiana Tabacum Seedlings. *Ecotoxicol. Environ. Saf.* **2022**, *231*, 113155. [CrossRef]
17. Qi, Y.; Yang, X.; Pelaez, A.M.; Huerta Lwanga, E.; Beriot, N.; Gertsen, H.; Garbeva, P.; Geissen, V. Macro- and Micro- Plastics in Soil-Plant System: Effects of Plastic Mulch Film Residues on Wheat (*Triticum aestivum*) Growth. *Sci. Total Environ.* **2018**, *645*, 1048–1056. [CrossRef]
18. Machado, A.A.D.S.; Lau, C.W.; Kloas, W.; Bergmann, J.; Bachelier, J.B.; Faltin, E.; Becker, R.; Görlich, A.S.; Rillig, M.C. Microplastics Can Change Soil Properties and Affect Plant Performance. *Environ. Sci. Technol.* **2019**, *53*, 6044–6052. [CrossRef]
19. Gan, Q.; Cui, J.; Jin, B. Environmental Microplastics: Classification, Sources, Fates, and Effects on Plants. *Chemosphere* **2023**, *313*, 137559. [CrossRef]
20. He, D.; Guo, T.; Li, J.; Wang, F. Optimize Lettuce Washing Methods to Reduce the Risk of Microplastics Ingestion: The Evidence from Microplastics Residues on the Surface of Lettuce Leaves and in the Lettuce Washing Wastewater. *Sci. Total Environ.* **2023**, *868*, 161726. [CrossRef]
21. Huang, F.; Hu, J.; Chen, L.; Wang, Z.; Sun, S.; Zhang, W.; Jiang, H.; Luo, Y.; Wang, L.; Zeng, Y.; et al. Microplastics May Increase the Environmental Risks of Cd via Promoting Cd Uptake by Plants: A Meta-Analysis. *J. Hazard. Mater.* **2023**, *448*, 130887. [CrossRef] [PubMed]
22. Ding, L.; Huang, D.; Ouyang, Z.; Guo, X. The Effects of Microplastics on Soil Ecosystem: A Review. *Curr. Opin. Environ. Sci. Health* **2022**, *26*, 100344. [CrossRef]
23. Huang, C.; Ge, Y.; Yue, S.; Zhao, L.; Qiao, Y. Microplastics Aggravate the Joint Toxicity to Earthworm Eisenia Fetida with Cadmium by Altering Its Availability. *Sci. Total Environ.* **2021**, *753*, 142042. [CrossRef]
24. Wang, F.; Zhang, X.; Zhang, S.; Zhang, S.; Adams, C.A.; Sun, Y. Effects of Co-Contamination of Microplastics and Cd on Plant Growth and Cd Accumulation. *Toxics* **2020**, *8*, 36. [CrossRef]
25. Wen, X.; Yin, L.; Zhou, Z.; Kang, Z.; Sun, Q.; Zhang, Y.; Long, Y.; Nie, X.; Wu, Z.; Jiang, C. Microplastics Can Affect Soil Properties and Chemical Speciation of Metals in Yellow-Brown Soil. *Ecotoxicol. Environ. Saf.* **2022**, *243*, 113958. [CrossRef]
26. Hurley, R.R.; Nizzetto, L. Fate and Occurrence of Micro(Nano)Plastics in Soils: Knowledge Gaps and Possible Risks. *Curr. Opin. Environ. Sci. Health* **2018**, *1*, 6–11. [CrossRef]

27. Golia, E.E.; Angelaki, A.; Giannoulis, K.D.; Skoufogianni, E.; Bartzialis, D.; Cavalaris, C.; Vleioras, S. Evaluation of Soil Properties, Irrigation and Solid Waste Application Levels on Cu and Zn Uptake by Industrial Hemp. *Agron. Res.* **2021**, *19*, 92–99. [CrossRef]
28. Golia, E.E.; Chartodiplomenou, M.A.; Papadimou, S.G.; Kantzou, O.D.; Tsiropoulos, N.G. Influence of Soil Inorganic Amendments on Heavy Metal Accumulation by Leafy Vegetables. *Environ. Sci. Pollut. Res.* **2021**, *30*, 8617–8632. [CrossRef]
29. Golia, E.E.; Bethanis, J.; Ntinopoulos, N.; Kaffe, G.-G.; Komnou, A.A.; Vasilou, C. Investigating the Potential of Heavy Metal Accumulation from Hemp. The Use of Industrial Hemp (*Cannabis sativa* L.) for Phytoremediation of Heavily and Moderated Polluted Soils. *Sustain. Chem. Pharm.* **2023**, *31*, 100961. [CrossRef]
30. Dioses-Salinas, D.C.; Pizarro-Ortega, C.I.; De-la-Torre, G.E. A Methodological Approach of the Current Literature on Microplastic Contamination in Terrestrial Environments: Current Knowledge and Baseline Considerations. *Sci. Total Environ.* **2020**, *730*, 139164. [CrossRef]
31. Ren, X.; Tang, J.; Wang, L.; Liu, Q. Microplastics in Soil-Plant System: Effects of Nano/Microplastics on Plant Photosynthesis, Rhizosphere Microbes and Soil Properties in Soil with Different Residues. *Plant Soil.* **2021**, *462*, 561–576. [CrossRef]
32. Roy, T.; Dey, T.K.; Jamal, M. Microplastic/Nanoplastic Toxicity in Plants: An Imminent Concern. *Environ. Monit. Assess.* **2023**, *195*, 1–35. [CrossRef] [PubMed]
33. Zhou, W.; Wang, Q.; Wei, Z.; Jiang, J.; Deng, J. Effects of Microplastic Type on Growth and Physiology of Soil Crops: Implications for Farmland Yield and Food Quality. *Environ. Pollut.* **2023**, *326*, 121512. [CrossRef]
34. Golia, E.E.; Diakoloukas, V. Soil Parameters Affecting the Levels of Potentially Harmful Metals in Thessaly Area, Greece: A Robust Quadratic Regression Approach of Soil Pollution Prediction. *Environ. Sci. Pollut. Res.* **2022**, *29*, 29544–29561. [CrossRef] [PubMed]
35. Golia, E.E. The Impact of Heavy Metal Contamination on Soil Quality and Plant Nutrition. Sustainable Management of Moderate Contaminated Agricultural and Urban Soils, Using Low Cost Materials and Promoting Circular Economy. *Sustain. Chem. Pharm.* **2023**, *33*, 101046. [CrossRef]
36. Golia, E.E.; Papadimou, S.G.; Cavalaris, C.; Tsiropoulos, N.G. Level of Contamination Assessment of Potentially Toxic Elements in the Urban Soils of Volos City (Central Greece). *Sustainability* **2021**, *13*, 2029. [CrossRef]
37. Zhao, T.; Lozano, Y.M.; Rillig, M.C. Microplastics Increase Soil PH and Decrease Microbial Activities as a Function of Microplastic Shape, Polymer Type, and Exposure Time. *Front. Environ. Sci.* **2021**, *9*, 101–114. [CrossRef]
38. Degryse, F.; da Silva, R.C.; Baird, R.; Cakmak, I.; Yazici, M.A.; McLaughlin, M.J. Comparison and Modelling of Extraction Methods to Assess Agronomic Effectiveness of Fertilizer Zinc. *J. Plant Nutr. Soil Sci.* **2020**, *183*, 248–259. [CrossRef]
39. Alexakis, D.E.; Bathrellos, G.D.; Skilodimou, H.D.; Gamvroula, D.E. Land Suitability Mapping Using Geochemical and Spatial Analysis Methods. *Appl. Sci.* **2021**, *11*, 5404. [CrossRef]
40. Alexakis, D.E.; Bathrellos, G.D.; Skilodimou, H.D.; Gamvroula, D.E. Spatial Distribution and Evaluation of Arsenic and Zinc Content in the Soil of a Karst Landscape. *Sustainability* **2021**, *13*, 6976. [CrossRef]
41. Chen, L.; Han, L.; Feng, Y.; He, J.; Xing, B. Soil Structures and Immobilization of Typical Contaminants in Soils in Response to Diverse Microplastics. *J. Hazard. Mater.* **2022**, *438*, 129555. [CrossRef] [PubMed]
42. Zhao, M.; Liu, R.; Wang, X.; Zhang, J.; Wang, J.; Cao, B.; Zhao, Y.; Xu, L.; Chen, Y.; Zou, G. How Do Controlled-Release Fertilizer Coated Microplastics Dynamically Affect Cd Availability by Regulating Fe Species and DOC Content in Soil? *Sci. Total Environ.* **2022**, *850*, 157886. [CrossRef] [PubMed]
43. Rillig, M.C.; Lehmann, A.; de Souza Machado, A.A.; Yang, G. Microplastic Effects on Plants. *New Phytol.* **2019**, *223*, 1066–1070. [CrossRef]
44. Huang, D.; Wang, X.; Yin, L.; Chen, S.; Tao, J.; Zhou, W.; Chen, H.; Zhang, G.; Xiao, R. Research Progress of Microplastics in Soil-Plant System: Ecological Effects and Potential Risks. *Sci. Total Environ.* **2022**, *812*, 151487. [CrossRef]
45. Su, R.; Ou, Q.; Wang, H.; Dai, X.; Chen, Y.; Luo, Y.; Yao, H.; Ouyang, D.; Li, Z.; Wang, Z. Organic–Inorganic Composite Modifiers Enhance Restoration Potential of *Nerium oleander* L. to Lead–Zinc Tailing: Application of Phytoremediation. *Environ. Sci. Pollut. Res.* **2023**, *30*, 56569–56579. [CrossRef] [PubMed]
46. Esposito, M.P.; Domingos, M. Establishing the Redox Potential of *Tibouchina pulchra* (Cham.) Cogn., a Native Tree Species from the Atlantic Rain Forest, in the Vicinity of an Oil Refinery in SE Brazil. *Environ. Sci. Pollut. Res.* **2014**, *21*, 5484–5495. [CrossRef]
47. Xu, G.; Lin, X.; Yu, Y. Different Effects and Mechanisms of Polystyrene Micro- and Nano-Plastics on the Uptake of Heavy Metals (Cu, Zn, Pb and Cd) by Lettuce (*Lactuca sativa* L.). *Environ. Pollut.* **2023**, *316*, 120656. [CrossRef]
48. Pinto-Poblete, A.; Retamal-Salgado, J.; López, M.D.; Zapata, N.; Sierra-Almeida, A.; Schoebitz, M. Combined Effect of Microplastics and Cd Alters the Enzymatic Activity of Soil and the Productivity of Strawberry Plants. *Plants* **2022**, *11*, 536. [CrossRef]
49. Su, R.; Xie, T.; Yao, H.; Chen, Y.; Wang, H.; Dai, X.; Wang, Y.; Shi, L.; Luo, Y. Lead Responses and Tolerance Mechanisms of Koelreuteria Paniculata: A Newly Potential Plant for Sustainable Phytoremediation of Pb-Contaminated Soil. *Int. J. Environ. Res. Public Health* **2022**, *19*, 14968. [CrossRef]
50. Dou, C.-M.; Fu, X.-P.; Chen, X.-C.; Shi, J.-Y.; Chen, Y.-X. Accumulation and Detoxification of Manganese in Hyperaccumulator *Phytolacca americana*. *Plant. Biol.* **2009**, *11*, 664–670. [CrossRef]

51. Sun, H.; Lei, C.; Xu, J.; Li, R. Foliar Uptake and Leaf-to-Root Translocation of Nanoplastics with Different Coating Charge in Maize Plants. *J. Hazard. Mater.* **2021**, *416*, 125854. [CrossRef] [PubMed]
52. Hu, J.; He, D.; Zhang, X.; Li, X.; Chen, Y.; Wei, G.; Zhang, Y.; Ok, Y.S.; Luo, Y. National-Scale Distribution of Micro(Meso)Plastics in Farmland Soils across China: Implications for Environmental Impacts. *J. Hazard. Mater.* **2022**, *424*, 127283. [CrossRef] [PubMed]

soil systems

MDPI

Article

Tracking Soil Health Changes in a Management-Intensive Grazing Agroecosystem

Tad Trimarco [1,*], Joe E. Brummer [1], Cassidy Buchanan [1] and James A. Ippolito [1,2]

[1] Soil and Crop Sciences Department, Colorado State University, Fort Collins, CO 80523, USA; joe.brummer@colostate.edu (J.E.B.); cassidy.buchanan@colostate.edu (C.B.); ippolito.38@osu.edu (J.A.I.)
[2] School of Environment and Natural Resources, The Ohio State University, Columbus, OH 43210, USA
* Correspondence: trimarco@colostate.edu

Abstract: Management-intensive Grazing (MiG) has been proposed to sustainably intensify agroecosystems through careful management of livestock rotations on pastureland. However, there is little research on the soil health impacts of transitioning from irrigated cropland to irrigated MiG pasture with continuous livestock rotation. We analyzed ten soil health indicators using the Soil Management Assessment Framework (SMAF) to identify changes in nutrient status and soil physical, biological, and chemical health five to six years after converting irrigated cropland to irrigated pastureland under MiG. Significant improvements in biological soil health indicators and significant degradation in bulk density, a physical soil health indicator, were observed. Removal of tillage and increased organic matter inputs may have led to increases in β-glucosidase, microbial biomass carbon, and potentially mineralizable nitrogen, all of which are biological indicators of soil health. Conversely, trampling by grazing cattle has led to increased bulk density and, thus, a reduction in soil physical health. Nutrient status was relatively stable, with combined manure and fertilizer inputs leading to stabilized plant-available phosphorous (P) and increased potassium (K) soil concentrations. Although mixed effects on soil health were present, overall soil health did increase, and the MiG system appeared to have greater overall soil health as compared to results generated four to five years earlier. When utilizing MiG in irrigated pastures, balancing the deleterious effects of soil compaction with grazing needs to be considered to maintain long-term soil health.

Keywords: soil health; management-intensive grazing; irrigation; SMAF

Citation: Trimarco, T.; Brummer, J.E.; Buchanan, C.; Ippolito, J.A. Tracking Soil Health Changes in a Management-Intensive Grazing Agroecosystem. *Soil Syst.* **2023**, *7*, 94. https://doi.org/10.3390/soilsystems7040094

Academic Editor: Luis Eduardo Akiyoshi Sanches Suzuki

Received: 10 September 2023
Revised: 20 October 2023
Accepted: 21 October 2023
Published: 23 October 2023

1. Introduction

Sustainable intensification of agroecosystems has become an important topic for land managers, climate scientists, and numerous stakeholders looking to meet environmental goals while maintaining highly productive farms and ranches. One increasingly popular management strategy for meeting producer on-site goals is Management-intensive Grazing (MiG), which is a management scheme that may increase stocking rates, improve forage quality, and reduce negative environmental impacts from production [1,2].

Martz et al. [3] defined MiG as a "flexible approach to rotational grazing management whereby animal nutrient demand through the grazing season is balanced with forage supply and available forage is allocated based on animal requirements". This style of management may broadly apply to a variety of more specific practices, such as rotational stocking, strip grazing, creep grazing, limit grazing, and many other methods that require intense management techniques and frequent cattle movement to maintain high productivity in a pasture throughout a grazing season [4]. As management techniques improve and increased land efficiency of these intensive operations competes with fluctuating commodity prices in conventional cropland, producers may find benefit in transitioning from cropland to MiG systems, using existing irrigation infrastructure to support high forage productivity goals. However, there is little research on how this transition may impact soil health, a primary driver of long-term productivity and sustainability in MiG agroecosystems.

Research exists comparing annual cropland to perennial grassland, but much of this research is focused on conversion to Conservation Reserve Program land or compares long-term cropland sites to long-term grassland sites [5–10]. However, there is little research on converting to grassland and much less that examines conversion in systems that are characterized by intense management and frequent livestock rotation. Studies that examine not only the transition from annual cropland to perennial grassland but, more specifically, the transition to intensively grazed rotational systems may provide insight into the soil characteristics that promote healthy, productive agroecosystems that provide filtration of water, physical stability, and resistance to erosion, and adequate nutrient cycling to maintain soil fertility [11]. However, we may look to past research on perennial grasslands and pastures to guide our investigation of Management-intensive Grazing.

Within perennial pasture systems, soil biological activity associated with the healthy cycling of key plant nutrients to support sustainable crop production may be improved as compared to crop production systems [5,6]. Soil enzymatic activity has often been used as an indicator of soil health due to the high sensitivity of exoenzyme activity to management practice change and the important role that microbial enzymes play in nutrient cycling [11–14]. Moreover, the deep rooting systems and lack of intense tillage in perennial grasslands have been shown to increase soil organic carbon [15–18], an important source of energy for microbial communities. Microbial biomass carbon (MBC) has been shown to be greater in perennial grassland or pasture systems than in cropland systems [8,19], supporting the theory that these management systems may support larger and more active microbial communities. Grazing of varying intensities has also been shown to shift microbial community structure, activity, and diversity [20,21], indicating that the interactions of livestock, plants, and soil have a profound impact on soil microbial activity.

However, the potential soil health benefits of perennial grassland management are not limited to microbial activity. Manure inputs provide a way to supplement soil fertility and limit the need for inorganic fertilizers [22,23]. Decreased inorganic fertilizer use in perennial pasture systems may decrease fertilizer acidification and salinization of soils [8], both of which are threats to long-term sustainability. However, changes in soil pH and accumulation of salts in managed perennial grasslands tend to be relatively small and likely insignificant in terms of either directly or indirectly affecting plant growth and soil fertility.

Similarly, soil physical properties may be altered by the conversion from cropland to MiG systems. While perennial root systems may reduce bulk density over annual crop systems, the lack of regular tillage and the repeated hoof action of cattle and other livestock in MiG systems may increase bulk density (Bd) and disrupt soil aggregates in grazed soils [24], potentially limiting root growth, reducing available water content, and decreasing infiltration rate [25–27]. Thus, MiG agroecosystems need to be focused upon to further understand the tradeoffs between improvements versus degradation of soil biological, chemical, and physical properties, and, ultimately, the creation of resilient and sustainable agroecosystems.

The current study builds upon the work by Shawver et al. [1], which used SMAF (see Andrews et al. [11] for full framework details) to monitor soil health changes during the early transition from irrigated cropland to an irrigated MiG perennial pasture. While the Shawver et al. [1] study examined years 1 and 2 following the transition, we present the soil health measurements of years 5 and 6 to reflect upon the continued changes to soil health following cropland to MiG perennial pasture transition.

Based on current literature, we hypothesized that as this MiG pasture matures, the soil will experience (a) negative changes in physical soil health from continued trampling via hoof pressure, (b) positive changes in biological soil health from manure inputs with low soil disturbance, (c) no change in chemical soil health, as high $CaCO_3$ content present is likely to buffer pH change and EC is already low, with irrigation likely to push salts deeper into the soil profile, (d) a positive change in nutrient status from manure inputs of P and K, and an improvement in overall soil health from biological soil health and nutrient content improvements outweighing physical soil health impairments.

2. Materials and Methods

2.1. Site Description

This study was conducted under an 82 ha center pivot at the Colorado State University Agricultural Research, Development, and Education Center northeast of Fort Collins, CO USA (40°39′30.16″ N, 104°59′09.00″ W) at an altitude of 1554 m. Monthly mean temperatures tend to peak in July with an average high of 30 °C and reach a low in December with a minimum average temperature of −10 °C. Total average precipitation is approximately 340 mm, with much of this precipitation occurring between April and August [28]. The region is classified under the Köppen–Geiger classification system as a semi-arid, cold steppe [1,29]. Across this portion of Colorado, many producers graze cattle on unirrigated land, while cropland is dominated by irrigated commodity crops—namely corn and wheat. Maintaining the high productivity of intensively managed pasture consistent with this study requires additional irrigation inputs above typical climactic conditions. For example, the 2021 center pivot irrigation totals ranged from 460 to 590 mm, depending on in-field location and irrigation demands by the perennial grasses. Soil fertility requirements were supplemented by inorganic fertilizer inputs. For all years, fertilizer needs were determined by commercial soil testing and are as follows:

- 2017: No fertilizer added;
- 2018: ~14 kg N ha^{-1} and ~67 kg P ha^{-1} as monoammonium phosphate;
- 2019: no fertilizer added;
- 2020: no fertilizer added;
- 2021: ~90 kg N ha^{-1}, 22 kg P ha^{-1}, and 13 kg S ha^{-1} as a mix of monoammonium phosphate, urea, and ammonium sulfate;
- 2022: ~56 kg N ha^{-1}, 22 kg P ha^{-1}, and 13 kg S ha^{-1} as a mix of monoammonium phosphate, urea, and ammonium sulfate.

Before 2016, the 82 ha pivot field was managed as a sprinkler-irrigated, fully tilled crop rotation of grain corn, silage corn, dry beans, and alfalfa. Between the 2016 and 2017 growing seasons, the field was converted to a cool-season grass-forage mix of multiple bromes, clover species, and other common forages, such as alfalfa. The full details of the grass mix and livestock grazing pattern are available in Shawver et al. [1]. In the spring of 2017, the field was split into ~2.6 ha paddocks. Cattle were then introduced, and approximately 230 animal units consisting of cow-calf pairs, yearling heifers, and yearling steers were rotated through the paddocks depending on forage availability and dietary needs. Broadly, the system was comprised of intense grazing in small paddocks for 1–4 days to remove approximately 50% of forage biomass, with cattle subsequently being moved into adjacent paddocks delineated with mobile electric fences; GPS tools were used to precisely manage the entire area. Decisions to move cattle through paddocks were based on maintenance of stand health and to avoid over-grazing.

2.2. Soil Sampling and Processing

Soils were sampled in the same location as in the Shawver et al. [1] study, focusing sampling on the primary soil series in the pivot: Nunn clay loam (fine, montmorillonitic, mesic Aridic Argiustolls, 26 ha; [30]), Kim loam (fine-loamy, mixed, calcareous, mesic Ustic Torriorthents, 10 ha), and Garrett loam (fine-loamy, mixed, mesic Pachic Argiustolls, 7.5 ha; [30]). In order to maintain continuity between samples from 2017 to 2022, soil was not collected in the northwest quarter of the pivot due to poor initial forage establishment in the spring of 2017. Thus, while the pivot encompasses 82 ha, sampled soil only represented 62 ha. The sample locations were randomly chosen within the paddocks using ArcMap (Version 10.5.1, ArcMap GIS) in 2017, but the paddocks were chosen to represent both the primary soil textures in the field and the forage mixtures. For the current study, soil sampling occurred in late June of 2021 and late May of 2022.

Soil Syst. **2023**, *7*, 94

Soil samples were collected using a 2.5 cm diameter soil probe to a depth of 15 cm, with each core split into 0-to-5 and 5-to-15 cm depths. Approximately 25–30 cores were sampled within a 3 m radius centered around the GPS-located sampling point, composited per sample, and mixed in a plastic bucket before being transferred to a plastic bag, sealed, and placed in dry coolers. Within this sampling radius, an additional intact core at both depth increments was preserved in a metal can for gravimetric soil moisture and bulk density (Bd) determination.

Soils were returned to the laboratory the same day and stored at 4 °C until processing. Bulk density and moisture content were determined by immediately weighing moist cores stored in metal cans, drying at 105 °C for 24 h, followed by weighing. The composited soil samples were passed through an 8 mm sieve to remove rocks and large plant debris. Approximately 150 g of field-moist, 8 mm sieved soil was then stored at 4 °C prior to MBC analysis. An additional ~150 g of soil was passed through a 2 mm sieve and air-dried, while the remainder of the 8 mm sieved soil was also air-dried, both for further analysis. Both air-dried subsamples were returned to plastic bags and stored at room temperature.

2.3. Soil Health Analyses

The SMAF [11] is a Microsoft Excel-based tool used to score and provide relative interpretations of soil health measurements within the context of climactic conditions, cropping system, soil taxonomy, and texture. Selection of soil indicators may be based on an intended research goal, but are broadly split into four categories: soil physical indicators (Bd and water-stable aggregates (WSA)), soil biological indicators (soil organic carbon (SOC), MBC, potentially mineralizable nitrogen (PMN), and β-glucosidase activity (BG)), soil chemical indicators (pH and electrical conductivity (EC)), and soil nutritional indicators (plant-available P and K). Indicators are selected to represent key soil ecosystem services, agronomic needs, and sensitivity to changes in management [11]. The SMAF translates the raw measurements of these soil indicators into unitless scores from 0 to 1 (0 being "worst" and 1 being "best") based on algorithms accounting for soil texture, climate, and cropping system. The SMAF has been used previously to study cropland, pastures, cropland-to-pasture conversions, and various other management schemes [1,10,31,32].

2.3.1. Soil Physical Health Indicators

Soil moisture content and bulk density were determined using an intact soil core of known volume. The weight of the soil core was measured at field moisture and after 24 h at 105 °C until dried mass was consistent. Water-stable aggregates were determined using the method described in Kemper and Rosenau [33] using 100 g of 8 mm air-dried soil. The soil was placed on top of a stack of 23 cm diameter sieves (2.0, 1.0, 0.5, and 0.25 mm sized screens), which were attached to a Yoder sieving machine and submerged at 30 strokes per minute for 5 min. The soil remaining on all of the sieves was rinsed into an aluminum pan, and the water from the pan was evaporated until completely dry at 105 °C, at which point soil weight in the pan was determined.

2.3.2. Soil Biological Health Indicators

β-glucosidase activity was determined using the methodology published by Green et al. [34]. In triplicate, 1.0 g of air-dried 2 mm sieved soil was weighed into 50 mL Erlenmeyer flasks to create three sets of each sample. One set was treated as the control and contained an additional blank empty Erlenmeyer flask. The other two sets were treated as a sample set and a duplicate set. Following this, 4 mL of modified universal buffer (MUB) at pH 6.0 and 0.25 mL of toluene were added to all flasks, and 1 mL of 0.05 M ρ-nitrophenyl-β-glucopyranoside (PNG) was added to the sample flasks and duplicate flasks, but PNG is not yet added to the control flasks. All samples were swirled and incubated at 37 °C for 1 h, at which point 1 mL of 0.5 M $CaCl_2$ and 4 mL of 0.1 M TRIS (hydroxymethyl) aminomethane (THAM) buffer at pH 12 is added to all flasks and 1 mL of 0.05 M PNG is added to the control flasks. These soil suspensions were filtered through Whatman #2 filter paper, and the filtrate was diluted by adding 4 mL of 0.1 M THAM to 1 mL of sample. B-glucosidase activity was measured using a Genesys 10S UV-VIS spectrophotometer at 410 nm, using a standard curve of *p*-nitrophenol at 0, 10, 20, 30, 40, and 50 µg L^{-1} in 1 mL of 0.5 M CaCl$_2$ and 4 mL of 0.1 M THAM. Microbial biomass carbon was determined using the chloroform fumigation/non-fumigation method [35], which estimates MBC by measuring dissolved C analyzed on a TIC/TOC analyzer (Shimadzu TOC-L; Shimadzu Scientific Instruments, Inc., Kyoto, Japan), subtracting dissolved C extracted from unfumigated samples from dissolved C extracted from fumigated samples and using a ratio of chloroform-labile C to microbial biomass C (0.45).

Soil organic C was determined as the difference between total C and inorganic C. Total C was measured using a dry combustion VELP Dumas Elemental Analyzer (VELP Scientifica, Usmate Velate, Italy; [36]), while inorganic C was determined using the pressure transducer method [37].

Potentially mineralizable N was determined by subtracting NO$_3$-N and NH$_4$-N concentrations from non-incubated soils from those same soils that were allowed to incubate aerobically for 28 days [38]. Approximately 30 g of air-dried, 2 mm sieved soil was weighed into a 50 mL beaker, and the beaker was tapped gently to bring the soil to approximately 1.0 g cm^{-3} bulk density. The soil was then brought to 60% water-filled pore space using deionized water, and the flask was placed in a Mason jar with ~1 cm of water in the bottom of the jar to maintain soil moisture. This Mason jar was sealed and placed in a cool, dark cabinet for 28 days; every 7 days, the jars were opened briefly to allow air exchange. After 28 days, a 10 g subsample of the soil was removed and placed into a 125 mL plastic bottle. Concurrently, a 10 g sample of air-dried 2 mm sieved soil that was not incubated was weighed into a 125 mL plastic bottle to serve as the control. Both controls and samples were shaken for 30 min with 50 mL of 2M KCl and filtered through Whatman #1 filter paper. Following this, the filtrate was analyzed for NO$_3$-N and NH$_4$-N. NO$_3$-N was determined by a combination of 15 µL sample filtrate with 250 µL of Vanadium (III) Chloride reagent and 35 µL 2M KCl to force a Griess reaction and measure NO$_2$-N concentration colorimetrically on a Genesys 10S UV-VIS spectrophotometer at 540 nm. Similarly, NH$_4$-N was determined by combining 15 µL of sample with 25 µL citrate reagent, 50 µL salicylate-nitroprusside reagent, 25 µL hypochlorite reagent, and 160 µL 2M KCl and determining concentration colorimetrically on a Genesys 10S UV-VIS spectrophotometer at 610 nm. Both NO$_3$-N and NH$_4$-N concentrations were calculated by use of a standard curve containing 0, 0.1, 0.5, 1, 2, 5, 10, 20, and 40 mg L^{-1} of the respective analyte.

2.3.3. Soil Chemical Health Indicators

Soil pH and EC were both determined using a 1:1 soil solution (20 g air-dried 2 mm sieved soil:20 mL DI) ratio [39,40]. Soil-water slurries were shaken on low for 2 h in a 50 mL centrifuge tube. pH was read directly with a pH electrode, and EC was determined by centrifuging the samples and measuring the liquid phase in a conductivity meter.

2.3.4. Soil Nutrient Content

P and K concentrations were determined by the Olsen extraction method due to the high pH observed in all samples [41]. Briefly, 2 g of air-dried 2 mm sieved soil was shaken on low with 40 mL of 0.5 M sodium bicarbonate and filtered through the Whatman #2 filter paper. These filtrates were covered in parafilm and left out overnight to allow for release of CO$_2$ gas. The filtered solution was diluted at a 10:1 ratio in DI water and analyzed for P and K in a high throughput inductively coupled plasma-optical emission spectrophotometer (ICP-OES).

2.4. Statistical Analysis

Each composite soil sample was considered an individual replicate for statistical analysis. Analysis of variance (ANOVA) was performed using a linear fixed effects model with year and depth as interacting predictor variables. If the interaction term was significant at the $\alpha \leq 0.05$ level, a pairwise comparison of means was performed for the significance of depth in each year and of year at each depth. This perspective was used to acknowledge and account for the potential impact of manure deposition and trampling action occurring primarily in shallow soil depths. Analysis was performed using R (Version 4.2.2) in RStudio (Build 446) using the stats package (Version 4.2.2), emmeans package (Version 1.7.4-1), and ARTool package (Version 0.11.1). Raw measurements of soil health indicators and soil indicator scores were used as outcome variables in two separate tests to determine the effect of MiG transition on project years 5 and 6 within both depths. Significance was evaluated at $\alpha \leq 0.05$. If data was not normally distributed or otherwise violated model assumptions for ANOVA analysis, the outcome variable was log-transformed, and assumptions were rechecked. If model assumptions still failed to be met, data was examined using the Aligned-Rank Transformation in ANOVA test for non-parametric distributions [42]. The Aligned-Rank Transformation in ANOVA test was frequently required when analyzing index scores, as the curve-fitting algorithm often resulted in scores concentrated at one far end of the spectrum (e.g., for EC, where the distribution of actual measurements was normal, but the unitless soil health scores are curved to represent risk of salinity and a wide range of low electrical conductivity soils may all have a SMAF soil indicator score of 1.0).

Tables of measured values and SMAF scores are presented as the untransformed mean and standard error to clarify and contextualize measurements, but formal statistical analysis was performed on transformed models, where appropriate. The transformation performed is provided in each table below, alongside the results of the model analysis.

Due to data accessibility constraints, statistical analysis was only performed on the 2021 and 2022 sampling years. However, these results are semi-quantitatively compared to the same analyses performed on soils in 2017 and 2018 [1] to draw general conclusions on trends in soil health indices where statistical inferences are not possible. These soils are comparable, using the same methodologies and sampling locations, but this paper attempts to examine the longer-term state of soil health in the MiG system.

3. Results and Discussion

3.1. Soil Physical Indicators

All mean soil physical indicator characteristics are presented in Table 1. Bulk density increased from 2021 to 2022, particularly in the 5-to-15 cm depth, perhaps as a function of hoof action from grazing cattle [24,43]. The findings were similar to those observed by Shawver et al. [1] on this site. A meta-analysis of 64 studies by Byrnes et al. [24] found that grazing activities significantly increased Bd, yet rotational grazing had a smaller impact than continuous grazing. Byrnes et al. [24] also noted that rotationally grazed systems generally had lower Bd values than continuously grazed ecosystems, with increasing grazing frequency and intensity correlated to increased Bd. While the current field study is generally managed to minimize grazing activity on recently irrigated paddocks, trampling may still increase Bd, limiting root penetration and reducing available water content, particularly in heavy clay soils [25–27]. The average clay content of soil samples from the 5 to 15 cm depth was approximately 34%, and the soils were broadly classified as clays and clay loams, likely increasing the deleterious effect of compaction when this soil is wet and exceeds its plasticity index. Consequently, this increase in Bd at depth resulted in a significant change in the bulk density indicator score (Table 2). Compared to samples taken in 2017 and 2018 (Table A1), the Bd at all depths in 2021 and 2022 appears to have increased slightly over the initial study years, suggesting that prolonged grazing activity has continued to compact the soil over time after transitioning. Furthermore, plant root growth might become impeded at Bd values greater than 1.7 g cm^{-3} [44], as observed in the 5 to 15 cm depth in 2022. Bulk density should continue to be monitored with depth in the future.

The quantity of WSA did not change over year or depth in 2021–2022 (Table 1), yet aggregate stability appeared to be greater than in 2017–2018 (Table A1). Continuous grazing has been shown to reduce soil aggregate stability [25], in large part through the destruction of soil structure, particularly in wet, heavy (i.e., clayey) soils [27]. Results suggest that MiG does not act to the extent that continuous grazing does on WSA. Also playing a role in this system is tillage, or lack thereof. Reduced tillage has been shown to improve shallow soil aggregate stability [45,46], indicating that the opposing effects of lack of tillage and increased grazing action via MiG may result in a somewhat small (albeit positive) change in soil aggregate stability during the study period. An identical response in WSA was observed by Keshavarz et al. [32] in a furrow irrigated continuous corn grazed agroecosystem study under no-till as compared to conventional tillage.

Bulk density and water-stable aggregates are likely to be impacted by additional irrigation inputs in MiG ecosystems that would not be received in traditional, rain-fed pastures. Warren et al. [27] conducted a study of bulk density and soil aggregate stability under varying trampling rates and moisture regimes in silty clay soil, noting that aggregate stability was generally poorer in moist soils, particularly at higher stocking rates. Conversely, bulk density was less sensitive to change as a function of stocking rate in moist soils, perhaps due to the incompressibility of water helping to maintain soil structure during trampling [27]. However, the USDA-NRCS recommends avoiding heavy trampling and field operations while the soil is wet, citing concerns about compaction [26]. Consequently, MiG systems incorporating irrigation require additional planning and oversight to ensure that excessive trampling does not occur during wet conditions to avoid adverse soil physical damage. Thus, in MiG systems such as this, having a contingency plan for relocating grazing animals when fields are extremely wet is suggested, such as a sacrificial area within or near the field [47].

The soil physical health index (Table 3) is an average of the individual scores for Bd and WSA (from Table 2). The combined effect of increasing Bd and somewhat constant WSA resulted in a significant decrease in the soil physical health index from 2021 to 2022. The soil physical health index scores, with depth, in 2021 and 2022 are comparable to scores from 2017 and 2018 (Table A1), suggesting that overall changes in soil physical health have not been altered over the past four to five years. These findings do not support our hypothesis that negative changes in physical soil health would occur from continued trampling via hoof pressure.

Table 1. Soil indicator means (± standard error) in 2021 and 2022 with ANOVA results.

Soil Indicator	2021	2022	2021	2022	ANOVA	ANOVA	ANOVA	Transformation
	(0–5 cm Depth)		(5–15 cm Depth)		(Year)	(Depth)	(Year × Depth)	
Physical								
Bd (g cm^{-3})	1.42 ± 0.04	1.56 ± 0.04	1.41 ± 0.05	1.70 ± 0.03	**			
WSA (%)	59.2 ± 4.5	64.3 ± 3.3	61.3 ± 3.3	63.5 ± 3.4				
Biological								
BG (mg pnp kg^{-1} soil hr^{-1})	490 ± 34	839 ± 34	229 ± 21	300 ± 19	** at 0–5 cm	** for both years	**	
MBC (mg g^{-1})	219 ± 15	438 ± 16	137 ± 13	203 ± 10	** at both depths	** for both years	**	
SOC (%)	2.10 ± 0.10	2.38 ± 0.17	2.06 ± 0.09	1.86 ± 0.12		*		
PMN (mg kg^{-1})	15.5 ± 2.8	50.7 ± 4.2	16.1 ± 1.0	42.2 ± 3.4	**			
Chemical								
pH, 1:1	7.87 ± 0.02	8.07 ± 0.04	7.99 ± 0.03	8.05 ± 0.03	** at 0–5 cm		*	
EC, 1:1 (dS m^{-1})	1.27 ± 0.17	1.04 ± 0.16	1.59 ± 0.23	2.19 ± 0.22		** for both years	**	Aligned-Rank
Nutrient								
P (mg kg^{-1})	29.7 ± 3.8	27.0 ± 2.9	14.8 ± 2.0	9.9 ± 1.2		**		Logarithmic
K (mg kg^{-1})	487 ± 50	523 ± 63	281 ± 29	410 ± 59		**		

Significance is denoted with * if significant at 0.05 probability level and ** if significant at the 0.01 probability level. If the interaction term was significant, the result of pairwise comparisons is shown in the ANOVA (Year) and ANOVA (Depth) columns, comparing within a single depth or year, respectively. Transformation performed on outcome variables to fit model assumptions is noted but means and standard error are not transformed. Blank cells indicate a lack of significance or a lack of transformation, respectively. Bd = bulk density; WSA = water stable aggregates; BG = beta-glucosidase activity; MBC = microbial biomass carbon; SOC = soil organic carbon; PMN = potentially mineralizable nitrogen; EC = electrical conductivity; P = plant-available phosphorus; K = plant-available potassium.

Table 2. Mean soil indicator scores ($\pm$ standard error) in 2021 and 2022 with ANOVA results.

Soil Indicator	2021	2022	2021	2022	ANOVA	ANOVA	ANOVA	Transformation
	(0–5 cm Depth)		(5–15 cm Depth)		(Year)	(Depth)	(Year × Depth)	
Physical								
Bd	0.47 ± 0.06	0.31 ± 0.02	0.47 ± 0.07	0.24 ± 0.01	**			Aligned-Rank
WSA	0.94 ± 0.03	0.99 ± 0.01	0.96 ± 0.02	0.98 ± 0.02				Aligned-Rank
Biological								
BG	0.80 ± 0.06	0.98 ± 0.01	0.34 ± 0.08	0.48 ± 0.08	*	**		
MBC	0.38 ± 0.07	0.85 ± 0.03	0.18 ± 0.04	0.30 ± 0.05	**	**		Logarithmic
SOC	0.43 ± 0.06	0.53 ± 0.07	0.42 ± 0.06	0.37 ± 0.07				Logarithmic
PMN	0.69 ± 0.10	1.00 ± 0.00	0.89 ± 0.05	1.00 ± 0.00	** at 5–15 cm	* for 2022	*	Aligned-Rank
Chemical								
pH	0.03 ± 0.00	0.01 ± 0.00	0.01 ± 0.00	0.01 ± 0.00	** at 0–5 cm	** for 2021	*	Logarithmic
EC	0.86 ± 0.07	0.93 ± 0.06	0.74 ± 0.09	0.53 ± 0.08		** for both years	**	Aligned-Rank
Nutrient								
P	1.00 ± 0.00	1.00 ± 0.00	0.94 ± 0.03	0.81 ± 0.06	[a]	** for both years	**	Aligned-Rank
K	1.00 ± 0.00	0.99 ± 0.01	0.98 ± 0.01	0.95 ± 0.04				Aligned-Rank

Significance is denoted with * if significant at 0.05 probability level and ** if significant at the 0.01 probability level. If the interaction term was significant, the result of pairwise comparisons is shown in the ANOVA (Year) and ANOVA (Depth) columns, comparing within a single depth or year, respectively. Transformation performed on outcome variables to fit model assumptions is noted but means and standard error are not transformed. Blank cells indicate a lack of significance or a lack of transformation, respectively. Bd = bulk density; WSA = water stable aggregates; BG = beta-glucosidase activity; MBC = microbial biomass carbon; SOC = soil organic carbon; PMN = potentially mineralizable nitrogen; EC = electrical conductivity; P = plant-available phosphorus; K = plant-available potassium. [a] The ANOVA test for P SMAF scores at 0–5 cm was significant with respect to year. However, this is due to exact ties in every single data point, where the Aligned-Rank Transformation is seriously limited [42], and statistical analysis is not warranted. Almost every soil sample in the 0–5 depth scored 1.00 for both years.

Table 3. Mean soil health index scores ($\pm$ standard error) in 2021 and 2022 with ANOVA results.

Soil Indicator	2021	2022	2021	2022	ANOVA	ANOVA	ANOVA	Transformation
	(0–5 cm Depth)		(5–15 cm Depth)		(Year)	(Depth)	(Year × Depth)	
Physical	0.70 ± 0.04	0.65 ± 0.01	0.71 ± 0.04	0.61 ± 0.01	*			Logarithmic
Biological	0.58 ± 0.05	0.84 ± 0.02	0.46 ± 0.04	0.54 ± 0.04	** at both depths	** for both years	**	Aligned-Rank
Chemical	0.45 ± 0.03	0.47 ± 0.03	0.38 ± 0.04	0.27 ± 0.04		** for both years	**	Aligned-Rank
Nutrient	1.00 ± 0.00	1.00 ± 0.00	0.96 ± 0.02	0.88 ± 0.05	[a]	** for both years	*	Aligned-Rank
Overall	0.66 ± 0.02	0.76 ± 0.01	0.59 ± 0.03	0.57 ± 0.02	** at 0–5 cm	** for both years	**	Aligned-Rank

Significance is denoted with * if significant at 0.05 probability level and ** if significant at the 0.01 probability level. If the interaction term was significant, the result of pairwise comparisons is shown in the ANOVA (Year) and ANOVA (Depth) columns, comparing within a single depth or year, respectively. Transformation performed on outcome variables to fit model assumptions is noted but means and standard error are not transformed. Blank cells indicate a lack of significance or a lack of transformation, respectively. [a] The ANOVA test for P SMAF scores at 0–5 cm was significant with respect to year. However, this is due to exact ties in over 28 of the 30 data points, a condition where the Aligned-Rank Transformation is seriously limited [42], and statistical analysis is not warranted.

3.2. Soil Biological Indicators

Biological indicators of soil health have been purported to be particularly sensitive to field management and play an important role in soil biogeochemical cycling and other ecosystem services [11,14]. All mean soil biological indicator characteristics are presented in Table 1. B-glucosidase is an enzyme important for cellulose biodegradation, with BG activity often used as an indicator of general microbiome capacity for organic matter assimilation [11,13]. Significant BG activity differences existed between years, depths, and for the year by depth interaction. Generally, it appeared that 2022 had significantly greater BG activity than 2021 in the top 5 cm of soil (Table 1). This dynamic may be due to organic matter and manure inputs in the top portion of soil, which are thought to increase enzyme activity, though the impact of manure on exoenzyme activity has shown mixed results [12,48,49]. Notably, increased soil moisture has been shown to increase enzyme activity [50,51], but the average soil moisture at the time of sampling fell from 19.5% in 2021 to 14.0% in 2022, indicating that other factors were likely playing a role in the increase in BG in 2022 as compared to 2021. Temperature has been positively correlated to enzyme activity [51,52]. However, soils were sampled in late June and May of 2021 and 2022, respectively, and June was warmer than May (average maximum soil temperatures of 27.8 °C compared to 18.6 °C at a depth of 5 cm, respectively; Colorado State University—CoAgMet Station ftc03—CSU-ARDEC; available at: https://coagmet.colostate.edu/ (accessed on 1 July 2023); [53]). Thus, soil temperature certainly does not support the increased BG activity between years. It is possible that the continuation of current management practices was likely the driver of changes in BG activity.

The change in BG activity resulted in an increase in the BG indicator score (Table 2). The BG indicator score increased from 2021 and 2022 and was greater in the 0 to 5 versus 5 to 15 cm depth. The 2021–2022 change in the BG indicator score is noteworthy, but the more impactful story may be that the indicator score has drastically increased from that of the 2017–2018 study (Table A1). This clearly shows that enzyme activity continues to increase in this ecosystem and may have not yet reached a steady state. It is important to note that increases in BG activity are suggestive of both overall biological change [6] and potential increases in SOC accumulation [54].

Continuous increases in BG activity should lead to increases in MBC. Indeed, MBC was greater in 2022 than in 2021 and greater in the 0 to 5 cm versus the 5 to 15 cm depth (Table 1), and the MBC indicator scores responded identically (Table 2). Microbial biomass carbon is typically used as a measure of total microbial population size in soils, with healthier soils typically having larger microbial communities [11]. Both BG and MBC are thought to increase as a function of carbon inputs and manure inputs, particularly in pastures [5,8,48]. This may explain why both of these biological indicators were significantly greater in the 0 to 5 as compared to the 5 to 15 cm depth, as both manure inputs and plant biomass inputs are primarily deposited on the soil surface or in the shallow subsurface, with no tillage or significant soil mixing to move them deeper into the profile. The continued increase in MBC contributes to the evidence that the transition to MiG systems provides significant biological benefits to soils.

Supporting the contention that BG activity may eventually lead to increases in SOC [54], in conjunction with the MBC findings above, SOC has significantly increased in the 0 to 5 as compared to the 5 to 15 cm depth (Table 1). Furthermore, at both depths, SOC was 40 to 82% greater in 2021–2022 than in 2017–2018 (Table A1), indicating that continuous manure inputs under MiG have greatly increased SOC, a dynamic that has been shown by others [16,17,24,55]. The SOC indicator scores showed no significant differences (Table 2), which was similar to those found by Shawver et al. [1] (Table A1). Regardless, the SOC indicator scores in the current study are ~2 times those found in 2017–2018 (Table A1). This also suggests that this ecosystem is improving in terms of SOC accumulation. Future research should continue to monitor potential improvements in SOC content.

Potentially mineralizable nitrogen is a measure of the portion of nitrogen in soil organic matter that is susceptible to be mineralized to plant-available forms, supplementing fertilizer N requirements [56]. For this reason, PMN has been used as an indicator of soil health [11,57]. Potentially mineralizable nitrogen increased from 2021 to 2022 (Table 1), and the PMN indicator score followed an identical response (Table 2). As compared to 2017 and 2018 (Table A1), the 2021 PMN appeared to increase slightly, while the 2022 PMN increased drastically. These changes were likely due to additional factors other than the slow progressive transition to manure-fed systems. Mahal et al. [56] assessed the potential of conservation agriculture to increase PMN in soils, finding that manure-fed systems had higher PMN levels than fields with inorganic N inputs, and no-till systems had greater PMN than conventionally tilled systems. Manure is often rich in organic N that is not yet plant-available but is steadily made available by microbial activity over multiple years. Furthermore, intense tillage is likely to decrease measured PMN, often by increasing the organic matter degradation rate [58,59]. The combined effect of these changes to the no-till, manure-fed MiG system was likely responsible for the increase in PMN over the past several years.

The biological soil health index (Table 3) is an average of the BG, MBC, PMN, and SOC scores from Table 2. All individual indicator scores, except for SOC, improved from 2021 to 2022, resulting in a significant increase in the biological soil health index in 2022 as compared to 2021. Shawver et al. [1] found a similar increase between years, albeit lower than the biological soil health scores found in the current study. Furthermore, BG and MBC had greater indicator scores in the 0 to 5 versus the 5 to 15 cm depth, leading to a greater biological soil health index in the 0 to 5 as compared to the 5 to 15 cm depth. No differences between soil depth were observed in 2017 and 2018 [1]. This supports our hypothesis that biological soil health should continue to improve as manure inputs and minimal soil disturbance support large and active microbial communities.

3.3. Soil Chemical Indicators

All mean soil chemical indicator characteristics (i.e., pH and EC) are presented in Table 1. Soil pH is considered a master variable for biochemical reactions, nutrient availability and toxicity, and other important soil functions [11]. Although significant differences existed with respect to soil pH (Table 1), pH differed only by 0.1 to 0.2 pH units and thus may be inconsequential with respect to altering soil biogeochemical reactions. Given the relatively high $CaCO_3$ and clay content of these soils (~8 and 34%, respectively), there is a large buffering capacity to resist change in soil pH [1,60]. Shawver et al. [1] found similar soil pH values as in the current study (Table A1). Continuous grazing and perennial grasslands have been shown to have mixed effects on pH over time [61–63], likely as a function of initial pH and the effect of changing soil inputs. Moreover, these calcareous soils are typical of the region, and producers are well-acquainted with high-pH-tolerant forage varieties. The relatively high pH of these soils resulted in a relatively low pH indicator score (Table 2), yet this is more telling of the lack of SMAF algorithm curves to fit high-pH-tolerant forage varieties than actual soil health degradation. Further SMAF algorithm development for high pH soils, such as in this study, is warranted.

While high pH may not be a concern for most producers, EC is a significant risk in this region of the western US, particularly as high irrigation requirements may result in steady salinization of agricultural land [64]. Electrical conductivity differed across depth in the current study (Table 1), and subsequently, a depth effect was observed in the EC indicator score (Table 2). Perennial grassland has been shown to produce increased salinity, perhaps by increasing surface evapotranspiration and reducing salt leaching further into the soil profile [65]; manure additions have also been shown to increase EC due to manure-borne salts [22]. However, compared to EC measurements in 2017 and 2018 (Table A1), salinity has generally decreased across all depths, perhaps due to irrigation inputs that leached fertilizer and manure-borne salts deeper into the soil profile. Year-to-year measurement of EC may vary as a function of spatially heterogeneous manure inputs, but the general trend appears to be that the conversion from row crops with inorganic fertilizer inputs to this perennial, animal-based pasture system has decreased EC. It is important to note that all

EC measurements were well below those values that would be of concern for crop growth (~4.0 dS m^{-1}; [66]).

There was a significant increase in the chemical soil health index score as a function of depth and, to a lesser extent, year, driven primarily by the reduction in EC in the soil surface (Table 3). The soil chemical health index scores in the current study tend to represent an improvement over soil chemical health index scores determined in 2017 and 2018 (Table A1). This indicates a positive effect of the decrease in EC over time. In contrast to our hypothesis of no change in chemical soil health, this indicates that long-term decreases in EC may contribute to improving soil health in perennial pasture systems.

3.4. Soil Nutrient Indicators

All mean soil nutrient indicator characteristics (i.e., extractable P and K) are presented in Table 1. Nutrient indicators of soil health are key elements to understanding the capacity of soils to sustain highly productive forage biomass. Neither P nor K showed significant changes in plant availability between 2021 and 2022, but both showed significantly increased concentrations in the top 5 cm of soil compared to the 5 to 15 cm depth. The difference in concentration of both nutrients across depths may be indicative of the significant impact that manure inputs on the soil surface have on soil fertility, as manure and urine are known to be rich in both P and K and likely remain near the soil surface [22,23,67], though top-dressing of phosphate fertilizers likely also played a role. Phosphorus remaining near the soil surface led to greater plant-available P concentrations in the 0 to 5 versus 5 to 15 cm depth, leading to a significant increase in the P index score in the 0 to 5 cm as compared to the 5 to 15 cm depth (Table 2). Pairwise comparisons of years within the top 5 cm for the P index score did produce a significant difference between 2021 and 2022, but this was omitted from Table 2 for brevity's sake. Every soil sample in the top 5 cm of soil scored 1.00 for P index, and the detected difference was due to statistical limitations of assigning ranks under the Aligned-Rank Transformation procedure to a dataset that is entirely ties [42]. The amount of plant-available K in both depths was adequate for plant growth, resulting in no difference in the K indicator score (Table 2). In both years, 22 kg ha^{-1} of P was applied, while no K was applied, as K is generally plentiful in Colorado soils and thus is rarely a concern for plant growth [23]. Chemical P inputs from monoammonium phosphate supplement manure inputs of P to a degree that deficiency is not a concern, again indicated by high SMAF scores.

It is worth noting that the 0 to 5 cm soil P concentrations have increased four- to five-fold as compared to the 2017–2018 P concentrations (Table A1). Due to both manure inputs and the use of phosphorous fertilizer, it is difficult to elucidate if increasing P concentrations are due to fertilizer or manure inputs. However, as the same quantity of fertilizer P was applied in both 2021 and 2022 (22 kg ha^{-1}), and P concentrations decreased slightly, manure inputs alone may not be enough to support highly productive MiG systems, and producers should continue to regularly test soils to determine nutrient needs. Given the near 1% slope of the field, there is not a large concern over P mobilization risk to nearby waters, though mobilization risk was not directly measured in this study and was inferred from the SMAF curve algorithms that punish extremely high P concentrations on sloped fields (i.e., >70 mg kg^{-1}; [11]). While P runoff is not explicitly a concern in this field, the rapid increase in P concentration is a reminder that producers transitioning to systems with large manure inputs should manage herd movements thoughtfully to avoid localized deposition of nutrient-rich manure near receiving water bodies and to carefully balance the combination of fertilizer P and manure P to minimize mobilization risk [23,68].

The effect of decreasing P concentrations with depth led to a decrease in the nutrient content index score from 2021 to 2022 in the top 5 cm but an observable increase since 2017–2018 (Table A1); a similar finding was observed by Shawver et al. [1]. Shawver et al. [1] hypothesized that future monitoring would show that the nutrient status would increase over several years given further manure inputs; the current study proves that hypothesis as correct. However, due to multiple sources of P in the system, it is difficult to identify the source of nutrient status changes. Regardless, the decrease in nutrient content does not support our hypothesis of increasing nutrient content from manure inputs from 2021 to 2022.

3.5. Overall Soil Health

The SMAF averages together the above 10 indicator scores to produce an overall soil health index (Table 3). While biological scores tended to improve from 2021 to 2022 (Table 2) of the nonbiological indicators, only two—pH and Bd—changed from 2021 to 2022, and the change in Bd could be considered deleterious to soil health.

Meanwhile, though sampling depth had a significant impact on just four soil indicators (BG, MBC, EC, and plant-available P; Table 2), these differences led to a significant improvement in overall soil health in the 0-to-5 as compared to the 5-to-15 cm depth (Table 3). These differences were likely driven by manure inputs on the soil surface. A similar finding was noted by Shawver et al. [1] (Table A1). Every biological indicator, except for SOC, showed significant improvement from 2021 to 2022, particularly in the 0 to 5 cm depth. When comparing the current findings to those from 2017 and 2018, in addition to positive changes in biological characteristics mentioned above, SOC appears to have increased from ~1% to ~2% (40 to 82% greater), a change that was predicted via increasing BG activity and MBC by Shawver et al. [1]. Furthermore, compared to 2017 and 2018, the overall soil health scores seemed to increase, driven largely by biological soil health and nutrient status improvements. This supports our hypothesis that overall soil health would improve under MiG systems, as biological soil health improvements outweigh physical soil health degradation.

It is important to note that not all soil health changes are positive as a function of conversion from conventional agricultural practices to MiG. Bulk density increased significantly from 2021 to 2022, and Bd values were greater than in the initial 2017–2018 study. Trampling action, paired with removal of tillage from the management system, seemed to have had an expected negative effect on Bd. The management of MiG systems needs to consider removing animals from wet soils to lessen the effects of hoof pressure on soil bulk density. Opposite, the water-stable aggregates percentage appeared to increase from 2017/2018 to 2021/2022, indicating that perhaps the positive impact of the transition to perennial no-till pastureland under MiG may balance the negative effect of trampling in terms of physical soil health. One of the more interesting findings may be that changes to soil health and nutrient status in these systems are relatively quick, and significant changes occurred in just a few years. This finding provides additional insight into past research comparing intense cropping systems and managed grasslands decades into an established management scheme, highlighting the need for future work to study transitioning landscapes. Building upon the preliminary work by Shawver et al. [1], these overall results provide additional evidence that irrigated, perennial pasture MiG systems have the capacity to significantly improve soil health following decades of successive cropping, providing promising insight for future environmental sustainability efforts in livestock agriculture.

Author Contributions: T.T.: Conceptualization; Data curation; Investigation; Formal analysis; Software; Writing—original draft; Visualization. J.A.I.: Conceptualization; Methodology; Validation; Investigation; Resources; Writing—review and editing; Supervision; Project administration; Funding acquisition. C.B.: Validation; Writing—review and editing; Data curation; Investigation. J.E.B.: Conceptualization; Validation; Resources; Writing—Review and editing; Project administration; Funding acquisition. All authors have read and agreed to the published version of the manuscript.

Funding: This research received no external funding.

Institutional Review Board Statement: Not applicable.

Informed Consent Statement: Not applicable.

Data Availability Statement: Data available upon request.

Conflicts of Interest: The authors declare no conflict of interest.

Appendix A

Table A1. Raw soil indicator measurements, soil indicator scores, and soil health index scores from 2017 and 2018. Table adapted from Shawver et al. [1] (see publication for statistical analysis of these results).

Soil Health Indicators	2017 (0–5 cm Depth)	2018 (0–5 cm Depth)	2017 (5–15 cm Depth)	2018 (5–15 cm Depth)
Physical				
Bd ($g\,cm^{-3}$)	1.15 ± 0.05	1.52 ± 0.05	1.29 ± 0.04	1.59 ± 0.04
WSA (%)	40.1 ± 3.9	44.3 ± 6.1	54.3 ± 3.8	55.6 ± 5.6
Biological				
BG ($mg\,pnp\,kg^{-1}\,soil\,hr^{-1}$)	65.3 ± 2.8	84.9 ± 4.2	66.9 ± 3.3	70.2 ± 5.3
MBC ($mg\,g^{-1}$)	122 ± 6	355 ± 25	136 ± 9	271 ± 22
SOC (%)	1.24 ± 0.06	1.31 ± 0.08	1.21 ± 0.09	1.33 ± 0.06
PMN ($mg\,kg^{-1}$)	11.8 ± 1.9	17.3 ± 1.0	11.2 ± 1.7	14.9 ± 1.0
Chemical				
pH, 1:1	8.00 ± 0.02	8.17 ± 0.03	7.90 ± 0.02	8.05 ± 0.02
EC, 1:1 ($dS\,m^{-1}$)	1.96 ± 0.25	1.12 ± 0.25	2.94 ± 0.20	2.52 ± 0.23
Nutrient				
P ($mg\,kg^{-1}$)	11.8 ± 1.1	6.9 ± 0.9	8.0 ± 1.0	5.7 ± 0.9
K ($mg\,kg^{-1}$)	175 ± 9	351 ± 25	172 ± 13	186 ± 21
Soil Health Indicator Scores	**2017 (0–5 cm depth)**	**2018 (0–5 cm depth)**	**2017 (5–15 cm depth)**	**2018 (5–15 cm depth)**
Physical				
Bd	0.81 ± 0.06	0.37 ± 0.04	0.61 ± 0.07	0.31 ± 0.05
WSA	0.77 ± 0.05	0.76 ± 0.08	0.93 ± 0.02	0.88 ± 0.06
Biological				
BG	0.06 ± 0.01	0.07 ± 0.01	0.06 ± 0.01	0.06 ± 0.01
MBC	0.17 ± 0.05	0.67 ± 0.06	0.17 ± 0.03	0.49 ± 0.08
SOC	0.19 ± 0.04	0.20 ± 0.05	0.20 ± 0.06	0.21 ± 0.04
PMN	0.64 ± 0.10	0.97 ± 0.01	0.56 ± 0.09	0.90 ± 0.04
Chemical				
pH	0.01 ± 0.00	0.00 ± 0.00	0.02 ± 0.00	0.01 ± 0.00
EC	0.62 ± 0.09	0.87 ± 0.09	0.25 ± 0.06	0.43 ± 0.07
Nutrient				
P	0.92 ± 0.03	0.62 ± 0.07	0.71 ± 0.08	0.47 ± 0.08
K	0.92 ± 0.03	0.99 ± 0.00	0.93 ± 0.02	0.91 ± 0.04
Soil Health Indices	**2017 (0–5 cm depth)**	**2018 (0–5 cm depth)**	**2017 (5–15 cm depth)**	**2018 (5–15 cm depth)**
Physical	0.79 ± 0.04	0.56 ± 0.05	0.77 ± 0.04	0.59 ± 0.04
Biological	0.26 ± 0.04	0.48 ± 0.03	0.25 ± 0.03	0.42 ± 0.03
Chemical	0.32 ± 0.04	0.44 ± 0.04	0.14 ± 0.03	0.22 ± 0.04
Nutrient	0.94 ± 0.02	0.81 ± 0.03	0.82 ± 0.04	0.69 ± 0.05
Overall	0.51 ± 0.03	0.55 ± 0.02	0.45 ± 0.02	0.47 ± 0.02

All values presented as mean ± standard error. Bd = bulk density; WSA = water stable aggregates; BG = beta-glucosidase activity; MBC = microbial biomass carbon; SOC = soil organic carbon; PMN = potentially mineralizable nitrogen; EC = electrical conductivity; P = plant-available phosphorus; K = plant-available potassium.

References

1. Shawver, C.J.; Ippolito, J.A.; Brummer, J.E.; Ahola, J.K.; Rhoades, R.D. Soil health changes following transition from an annual cropping to perennial management-intensive grazing agroecosystem. *Agrosystems Geosci. Environ.* **2021**, *4*, e20181. [CrossRef]
2. Wang, T.; Jin, H.; Kreuter, U.; Teague, R. Expanding grass-based agriculture on marginal land in the U.S. Great Plains: The role of management intensive grazing. *Land Use Policy* **2021**, *104*, 105155. [CrossRef]
3. Martz, F.A.; Gerrish, J.; Belyea, R.; Tate, V. Nutrient Content, Dry Matter Yield, and Species Composition of Cool-Season Pasture with Management-Intensive Grazing. *J. Dairy Sci.* **1999**, *82*, 1538–1544. [CrossRef] [PubMed]
4. Hancock, D.; Andrae, J. What Is Management-Intensive Grazing (MiG) and What Can It Do for My Farm? 2009. Available online: http://www.caes.uga.edu/commodities/fieldcrops/forages/questions/023FAQ-grazmethods.pdf (accessed on 5 June 2023).

5. Acosta-Martínez, V.; Acosta-Mercado, D.; Sotomayor-Ramírez, D.; Cruz-Rodríguez, L. Microbial communities and enzymatic activities under different management in semiarid soils. *Appl. Soil Ecol.* **2008**, *38*, 249–260. [CrossRef]
6. Bandick, A.K.; Dick, R.P. Field management effects on soil enzyme activities. *Soil Biol. Biochem.* **1999**, *31*, 1471–1479. [CrossRef]
7. Huang, X.; Skidmore, E.L.; Tibke, G.L. Soil quality of two Kansas soils as influenced by the Conservation Reserve Program. *J. Soil Water Conserv.* **2002**, *57*, 344–350.
8. Moore, J.M.; Klose, S.; Tabatabai, M.A. Soil microbial biomass carbon and nitrogen as affected by cropping systems. *Biol. Fertil. Soils* **2000**, *31*, 200–210. [CrossRef]
9. Staben, M.L.; Bezdicek, D.F.; Fauci, M.F.; Smith, J.L. Assessment of Soil Quality in Conservation Reserve Program and Wheat-Fallow Soils. *Soil Sci. Soc. Am. J.* **1997**, *61*, 124–130. [CrossRef]
10. Veum, K.S.; Kremer, R.J.; Sudduth, K.A.; Kitchen, N.R.; Lerch, R.N.; Baffaut, C.; Stott, D.E.; Karlen, D.L.; Sadler, E.J. Conservation effects on soil quality indicators in the Missouri Salt River Basin. *J. Soil Water Conserv.* **2015**, *70*, 232–246. [CrossRef]
11. Andrews, S.S.; Karlen, D.L.; Cambardella, C.A. The soil management assessment framework: A quantitative soil quality evaluation method. *Soil Sci. Soc. Am. J.* **2004**, *68*, 1945–1962. [CrossRef]
12. Akhtar, K.; Wang, W.; Ren, G.; Khan, A.; Feng, Y.; Yang, G. Changes in soil enzymes, soil properties, and maize crop productivity under wheat straw mulching in Guanzhong, China. *Soil Tillage Res.* **2018**, *182*, 94–102. [CrossRef]
13. Chae, Y.; Cui, R.; Woong Kim, S.; An, G.; Jeong, S.W.; An, Y.J. Exoenzyme activity in contaminated soils before and after soil washing: ß-glucosidase activity as a biological indicator of soil health. *Ecotoxicol. Environ. Saf.* **2017**, *135*, 368–374. [CrossRef] [PubMed]
14. Dick, R.P. Soil Enzyme Activities as Indicators of Soil Quality. In *Defining Soil Quality for a Sustainable Environment*; Doran, J.W., Coleman, D.C., Bezdicek, D.F., Stewart, B.A., Eds.; Soil Science Society of America: Madison, WI, USA, 1994. [CrossRef]
15. Beniston, J.W.; DuPont, S.T.; Glover, J.D.; Lal, R.; Dungait, J.A.J. Soil organic carbon dynamics 75 years after land-use change in perennial grassland and annual wheat agricultural systems. *Biogeochemistry* **2014**, *120*, 37–49. [CrossRef]
16. Ledo, A.; Smith, P.; Zerihun, A.; Whitaker, J.; Vicente-Vicente, J.L.; Qin, Z.; McNamara, N.P.; Zinn, Y.L.; Llorente, M.; Liebig, M.; et al. Changes in soil organic carbon under perennial crops. *Glob. Chang. Biol.* **2020**, *26*, 4158–4168. [CrossRef] [PubMed]
17. Luo, Z.; Wang, E.; Sun, O.J. Soil carbon change and its responses to agricultural practices in Australian agro-ecosystems: A review and synthesis. *Geoderma* **2010**, *155*, 211–223. [CrossRef]
18. O'Brien, S.L.; Jastrow, J.D. Physical and chemical protection in hierarchical soil aggregates regulates soil carbon and nitrogen recovery in restored perennial grasslands. *Soil Biol. Biochem.* **2013**, *61*, 1–13. [CrossRef]
19. Zhang, C.; Liu, G.; Xue, S.; Song, Z. Rhizosphere soil microbial activity under different vegetation types on the Loess Plateau, China. *Geoderma* **2011**, *161*, 115–125. [CrossRef]
20. Cao, J.; Wang, H.; Holden, N.M.; Adamowski, J.F.; Biswas, A.; Zhang, X.; Feng, Q. Soil properties and microbiome of annual and perennial cultivated grasslands on the Qinghai–Tibetan Plateau. *Land Degrad. Dev.* **2021**, *32*, 5306–5321. [CrossRef]
21. Wang, Z.; Jiang, S.; Struik, P.C.; Wang, H.; Jin, K.; Wu, R.; Na, R.; Mu, H.; Ta, N. Plant and soil responses to grazing intensity drive changes in the soil microbiome in a desert steppe. *Plant Soil* **2022**, *491*, 219–237. [CrossRef]
22. Aarons, S.R.; O'Connor, C.R.; Hosseini, H.M.; Gourley, C.J.P. Dung pads increase pasture production, soil nutrients and microbial biomass carbon in grazed dairy systems. *Nutr. Cycl. Agroecosyst.* **2009**, *84*, 81–92. [CrossRef]
23. Whiting, D.; Card, A.; Wilson, C. *Understanding Fertilizers*; CMC Garden: Fort Collins, CO, USA, 2022; pp. 1–7.
24. Byrnes, R.C.; Eastburn, D.J.; Tate, K.W.; Roche, L.M. A Global Meta-Analysis of Grazing Impacts on Soil Health Indicators. *J. Environ. Qual.* **2018**, *47*, 758–765. [CrossRef] [PubMed]
25. Shah, A.N.; Tanveer, M.; Shahzad, B.; Yang, G.; Fahad, S.; Ali, S.; Bukhari, M.A.; Tung, S.A.; Hafeez, A.; Souliyanonh, B. Soil compaction effects on soil health and crop productivity: An overview. *Environ. Sci. Pollut. Res.* **2017**, *24*, 10056–10067. [CrossRef] [PubMed]
26. USDA NRCS. Soil Health-Guides for Educators: Soil Bulk Density/Moisture/Aeration. Soil Quality Kit-Guides for Educators, May, 1–11. 2019. Available online: https://www.nrcs.usda.gov/wps/portal/nrcs/detailfull/soils/health/assessment/?cid=nrcs142p2_053870 (accessed on 15 June 2023).
27. Warren, S.D.; Nevill, M.B.; Blackburn, W.H.; Garza, N.E. Soil Response to Trampling Under Intensive Rotation Grazing. *Soil Sci. Soc. Am. J.* **1986**, *50*, 1336–1341. [CrossRef]
28. Colorado Climate Center-Station Normals for Nunn, CO. Colorado Climate Center-Station Normals. 2023. Available online: https://climate.colostate.edu/station_normal.html?USW00094074 (accessed on 1 July 2023).
29. Kottek, M.; Grieser, J.; Beck, C.; Rudolf, B.; Rubel, F. World map of the Köppen-Geiger climate classification updated. *Meteorol. Z.* **2006**, *15*, 259–263. [CrossRef]
30. Soil Survey Staff, Natural Resources Conservation Service, United States Department of Agriculture. Web Soil Survey. Available online: https://websoilsurvey.nrcs.usda.gov/app/ (accessed on 12 July 2022).
31. da Luz, F.B.; da Silva, V.R.; Kochem Mallmann, F.J.; Bonini Pires, C.A.; Debiasi, H.; Franchini, J.C.; Cherubin, M.R. Monitoring soil quality changes in diversified agricultural cropping systems by the Soil Management Assessment Framework (SMAF) in southern Brazil. *Agric. Ecosyst. Environ.* **2019**, *281*, 100–110. [CrossRef]
32. Keshavarz, R.; Banet, T.; Li, L.; Ippolito, J.A. Furrow-irrigated corn residue management and tillage strategies for improved soil health. *Soil Tillage Res.* **2022**, *216*, 105238.

33. Kemper, W.D.; Rosenau, R.C. Aggregate stability and size distribution. In *Methods of Soil Analysis, Part 1—Physical and Mineralogical Methods Agronomy Monograph 9*, 2nd ed.; Klute, A., Ed.; Soil Science Society of America: Madison, WI, USA, 1986; pp. 425–442.

34. Green, V.; Stott, D.; Cruz, J.; Curi, N. Tillage impacts on soil biological activity and aggregation in a Brazilian Cerrado Oxisol. *Soil Tillage Res.* **2007**, *92*, 114–121. [CrossRef]

35. Hobbie, S.E. Chloroform Fumigation Direct Extraction (CFDE) Protocol for Microbial Biomass Carbon and Nitrogen. 1998. Available online: https://web.stanford.edu/group/Vitousek/chlorofume.html (accessed on 6 July 2023).

36. Nelson, D.W.; Sommers, L.E. Total C, organic C, and organic matter. In *Methods of Soil Analysis, Part 3—Chemical Methods*; Sparks, D.L., Ed.; Soil Science Society of America: Madison, WI, USA, 1996; pp. 975–977.

37. Sherrod, L.A.; Dunn, G.; Peterson, G.; Kolberg, R. Inorganic C analysis by modified pressure-calcimeter method. *Soil Sci. Soc. Am. J.* **2002**, *66*, 299–305.

38. Curtin, D.; McCallum, F.M. Biological and chemical assays to estimate nitrogen supplying power of soils with contrasting management histories. *Aust. J. Soil Res.* **2004**, *42*, 737–746. [CrossRef]

39. Thomas, G.W. Soil pH and soil acidity. In *Methods of Soil Analysis, Part 3—Chemical Methods*; Sparks, D.L., Ed.; Soil Science Society of America: Madison, WI, USA, 1996; pp. 475–490.

40. Rhoades, J.D. Electrical conductivity and total dissolved solids. In *Methods of Soil Analysis, Part 3—Chemical Methods*; Sparks, D.L., Ed.; Soil Science Society of America: Madison, WI, USA, 1996; pp. 417–435.

41. Olsen, S.; Cole, C.; Watanabe, F.; Dean, L. *Estimation of Available Phosphorus in Soils by Extraction with Sodium Bicarbonate*; United States Department of Agriculture: Washington, DC, USA, 1954; p. 19.

42. Wobbrock, J.O.; Findlater, L.; Gergle, D.; Higgins, J.J. The Aligned Rank Transform for nonparametric factorial analyses using only ANOVA procedures. In Proceedings of the Conference on Human Factors in Computing Systems–Proceedings, Vancouver, BC, Canada, 7–12 May 2011; pp. 143–146. [CrossRef]

43. Stavi, I.; Ungar, E.D.; Lavee, H.; Sarah, P. Grazing-induced spatial variability of soil bulk density and content of moisture, organic carbon and calcium carbonate in a semi-arid rangeland. *Catena* **2008**, *75*, 288–296. [CrossRef]

44. Bruand, A.; Gilkes, J.R. Subsoil bulk density and organic carbon stock in relation to land use for a Western Australian Sodosol. *Aust. J. Soil Res.* **2002**, *40*, 431–459. [CrossRef]

45. Devine, S.; Markewitz, D.; Hendrix, P.; Coleman, D. Soil aggregates and associated organic matter under conventional tillage, no-tillage, and forest succession after three decades. *PLoS ONE* **2014**, *9*, e84988. [CrossRef]

46. Zheng, H.; Liu, W.; Zheng, J.; Luo, Y.; Li, R.; Wang, H.; Qi, H. Effect of long-term tillage on soil aggregates and aggregate-associated carbon in black soil of northeast China. *PLoS ONE* **2018**, *13*, e0199523. [CrossRef]

47. Shawver, C.; Brummer, J.; Ippolito, J.; Ahola, J.; Rhoades, R. *Management-Intensive Grazing and Soil Health, Fact Sheet 0.570*; Colorado State University Extension: Fort Collins, CO, USA, 2020.

48. Elzobair, K.A.; Stromberger, M.E.; Ippolito, J.A.; Lentz, R.D. Contrasting effects of biochar versus manure on soil microbial communities and enzyme activities in an Aridisol. *Chemosphere* **2016**, *142*, 145–152. [CrossRef]

49. Zhang, Y.L.; Sun, C.X.; Chen, Z.H.; Zhang, G.N.; Chen, L.J.; Wu, Z.J. Stoichiometric analyses of soil nutrients and enzymes in a Cambisol soil treated with inorganic fertilizers or manures for 26 years. *Geoderma* **2019**, *353*, 382–390. [CrossRef]

50. Gómez, E.J.; Delgado, J.A.; González, J.M. Persistence of microbial extracellular enzymes in soils under different temperatures and water availabilities. *Ecol. Evol.* **2020**, *10*, 10167–10176. [CrossRef]

51. Steinweg, J.M.; Dukes, J.S.; Wallenstein, M.D. Modeling the effects of temperature and moisture on soil enzyme activity: Linking laboratory assays to continuous field data. *Soil Biol. Biochem.* **2012**, *55*, 85–92. [CrossRef]

52. Brzostek, E.R.; Finzi, A.C. Seasonal variation in the temperature sensitivity of proteolytic enzyme activity in temperate forest soils. *J. Geophys. Res. Biogeosciences* **2012**, *117*. [CrossRef]

53. Colorado State University. CoAgMet Homepage. CoAgMET. Available online: https://coagmet.colostate.edu/ (accessed on 1 July 2023).

54. Paudel, B.R.; Udawatta, R.P.; Anderson, S.H. Agroforestry and grass buffer effects on soil health parameters for grazed pasture and row-crop systems. *Appl. Soil Ecol.* **2011**, *48*, 125–132. [CrossRef]

55. Zhang, T.; Zhang, Y.; Xu, M.; Zhu, J.; Wimberly, M.C.; Yu, G.; Niu, S.; Xi, Y.; Zhang, X.; Wang, J. Light-intensity grazing improves alpine meadow productivity and adaption to climate change on the Tibetan Plateau. *Sci. Rep.* **2015**, *5*, 15949. [CrossRef]

56. Mahal, N.K.; Castellano, M.J.; Miguez, F.E. Conservation Agriculture Practices Increase Potentially Mineralizable Nitrogen: A Meta-Analysis. *Soil Sci. Soc. Am. J.* **2018**, *82*, 1270–1278. [CrossRef]

57. Moebius-Clune, B.N.; Moebius-Clune, D.J.; Gugino, B.K.; Idowu, O.J.; Schindelbeck, R.R.; Ristow, A.J.; van Es, H.M.; Thies, J.E.; Shayler, H.A.; McBride, M.B.; et al. *Comprehensive Assessment of Soil Health—The Cornell Framework Manual, Edition 3.1*; Cornell University: Geneva, Switzerland; Ithaca, NY, USA, 2016.

58. Martin-Lammerding, D.; Tenorio, J.L.; Albarran, M.; Zambrana, E.; Walter, I. Influence of tillage practices on soil biologically active organic matter content over a growing season under semiarid Mediterranean climate. *Span. J. Agric. Res.* **2013**, *11*, 232–243. [CrossRef]

59. Needelman, B.A.; Wander, M.M.; Bollero, G.A.; Boast, C.W.; Sims, G.K.; Bullock, D.G. Interaction of Tillage and Soil Texture Biologically Active Soil Organic Matter in Illinois. *Soil Sci. Soc. Am. J.* **1999**, *63*, 1326–1334. [CrossRef]

60. Wei, H.; Yang, J.; Liu, Z.; Zhang, J. Data Integration Analysis Indicates That Soil Texture and pH Greatly Influence the Acid Buffering Capacity of Global Surface Soils. *Sustainability* **2022**, *14*, 3017. [CrossRef]

61. Abdalla, M.; Hastings, A.; Chadwick, D.R.; Jones, D.L.; Evans, C.D.; Jones, M.B.; Rees, R.M.; Smith, P. Critical review of the impacts of grazing intensity on soil organic carbon storage and other soil quality indicators in extensively managed grasslands. *Agric. Ecosyst. Environ.* **2018**, *253*, 62–81. [CrossRef]
62. Dear, B.S.; Virgona, J.M.; Sandral, G.A.; Swan, A.D.; Morris, S. Changes in soil mineral nitrogen, nitrogen leached, and surface pH under annual and perennial pasture species. *Crop Pasture Sci.* **2009**, *60*, 975–986. [CrossRef]
63. Hao, Y.; He, Z. Effects of grazing patterns on grassland biomass and soil environments in China: A meta-analysis. *PLoS ONE* **2019**, *14*, e0215223. [CrossRef]
64. Bauder, T.A.; Waskorn, R.M.; Sutherland, P.L.; Davis, J.G. *Irrigation Water Quality Criteria*; Fact Sheet No. 0.506 Crop Series; Colorado State University Extension: Fort Collins, CO, USA, 2011; pp. 10–13.
65. Bremer, E.; Pauly, D.; McKenzie, R.H.; Ellert, B.H.; Janzen, H.H. Twenty-four years of contrasting cropping systems on a brown chernozem in Southern Alberta: Crop yields, soil carbon, and subsoil salinity. *Can. J. Soil Sci.* **2023**, *103*, 134–142. [CrossRef]
66. Waskom, R.M.; Bauder, T.; Davis, J.G.; Andales, A.A. *Diagnosing Saline and Sodic Soil Problems*; Fact Sheet No. 0.521 Crop Series; Colorado State University Extension: Fort Collins, CO, USA, 2012; pp. 1–2.
67. Early MS, B.; Cameron, K.C.; Fraser, P.M. The fate of potassium, calcium, and magnesium in simulated urine patches on irrigated dairy pasture soil. *N. Z. J. Agric. Res.* **1998**, *41*, 117–124. [CrossRef]
68. Sherman, J.F.; Young, E.O.; Coblentz, W.K.; Cavadini, J. Runoff water quality after low-disturbance manure application in an alfalfa–grass hay crop forage system. *J. Environ. Qual.* **2020**, *49*, 663–674. [CrossRef]

 soil systems

Article

Water Erosion Processes on the Geotouristic Trails of Serra da Bocaina National Park Coast, Rio de Janeiro State, Brazil

Guilherme Marques de Lima [1,*]**, Antonio Jose Teixeira Guerra** [1]**, Luana de Almeida Rangel** [1]**, Colin A. Booth** [2] **and Michael Augustine Fullen** [3]

1   Department of Geography, Federal University of Rio de Janeiro, Rio de Janeiro 21910-240, Brazil; antoniotguerra@gmail.com (A.J.T.G.); luarangel24@gmail.com (L.d.A.R.)
2   School of Engineering, University of the West of England, Bristol BS161QY, UK; colin.booth@uwe.ac.uk
3   Faculty of Science and Engineering, University of Wolverhampton, Wolverhampton WV11LY, UK; m.fullen@wlv.ac.uk
*   Correspondence: guilhermem.lima@ufrj.br

Abstract: Conservation units are strategic territories that have a high demand for public use, as they protect attractions of great scenic beauty, geodiversity sites, and numerous leisure areas. However, when carried out in an intensive and disorderly manner, tourist activity in these areas tends to catalyze environmental degradation, triggering, for example, water erosion processes caused by intensive soil trampling on the trails. In this sense, the aim of this study was to determine the soil's physicochemical characteristics, and to spatiotemporally monitor the microtopography of those areas degraded by erosion along two trails on Serra da Bocaina National Park coast of the Paraty Municipality. The findings verified that intensive trampling, the values of some soil physicochemical characteristics, and the specific meteorological conditions of the coastal region of this protected area were factors that contributed significantly to the evolution of erosion features monitored on these trails. Finally, strategies for appropriate management and recovery actions for these degraded areas are proposed in order to not only stop the erosive processes and re-establish the local ecosystem balance, but also avoid accidents involving the numerous tourists who visit the coastal region.

Keywords: environmental degradation; soil erosion; trail; protected areas; conservation units

Citation: Lima, G.M.d.; Guerra, A.J.T.; Rangel, L.d.A.; Booth, C.A.; Fullen, M.A. Water Erosion Processes on the Geotouristic Trails of Serra da Bocaina National Park Coast, Rio de Janeiro State, Brazil. *Soil Syst.* **2024**, *8*, 24. https://doi.org/10.3390/soilsystems8010024

Academic Editor: Luis Eduardo Akiyoshi Sanches Suzuki

Received: 16 November 2023
Revised: 31 January 2024
Accepted: 1 February 2024
Published: 17 February 2024

1. Introduction

Water erosive processes cause soil degradation through the breakdown, transport, and deposition of particles by the action of water from raindrops and runoff, which corresponds with ~72% of the land affected by soil erosion in the world [1]. Its occurrence is intensifying due to climate change and human actions; therefore, it is one of the main environmental and socioeconomic challenges that society faces today. When induced by anthropic actions, these processes, in addition to impacting biogeochemical cycles, promote biodiversity loss, declines in agricultural productivity, carbon storage, increases in hunger, food insecurity, poverty, and social inequality [2–6].

Soil trampling on trails for public use, for example, is a human activity that causes soil degradation and occurrence of these processes, especially when these actions are carried out in an intensive and disorderly way. These activities corroborate with the alteration of the characteristic physicochemical and biological changes in the soil, which, in turn, modify the hydraulic and edaphic dynamics, since, in addition to reducing organic matter levels, porosity values, and water infiltration rates, they cause an increase in bulk density, runoff volume, and soil loss due to erosion [7–10].

There are many networks of local, regional, national, and international trails maintained and built to offer leisure and recreation opportunities, especially in protected areas, such as conservation units (UCs). These territories have experienced considerable growth in the number of visitors, as their trails are spaces that allow access to areas of leisure and scenic beauty. Consequently, these places have become highly susceptible to soil degradation through water erosive processes [11–15].

Fonseca Filho [16], for example, monitored soil compaction and erosion on trails in the UCs of Minas Gerais State and identified intensive trampling as one of the causes of trail degradation. Rangel et al. [17] also demonstrated the influence of human trampling on soil erosion in two UCs in southeastern Brazil. Figueiredo and Martins [18] discussed the mechanisms that lead to trail degradation in protected areas, pointing out that a lack of planning, management, and maintenance is associated with the occurrence of erosion processes in these areas. Lima and Guerra [10] studied the Morro Dois Irmãos Trail, which is part of a protected area in Rio de Janeiro Municipality, and identified several erosion features along its path, pointing to the physical–chemical characteristics of the soil together with the intensive trampling by people as the main causes of the development of rills. Lima et al. [9] also identified the evolution of erosion features on trails on the Serra da Bocaina National Park (PNSB) coast, indicating intense and disorderly visitation as one of the factors that corroborates the dynamics of this behavior.

Monitoring and analyzing the microtopography and soil physical–chemical characteristics in time and space on trails with water erosion processes, especially those located inside UCs, are extremely useful for the management and planning of these protected areas, because surveys can not only support decision making by their managers in actions that aim to identify and minimize these processes, but also help in optimizing management practices in degraded areas, in addition to guiding the re-establishment of their ecosystem balance through erosion processes in these territories [14,19].

In this context, this study aims to conduct a spatial survey of the soil's physical and chemical characteristics on trails on the coast of the Serra da Bocaina National Park (PNSB) in Paraty Municipality of Brazil, in addition to evaluating the space–time evolution of water erosion processes through monitoring associations between soil microtopography and rainfall data. This protected area, located in one of the largest stretches of continuous and conserved forests in the Atlantic Forest, has numerous geotouristic attractions in its territory, which, in turn, makes it a scene of high demand for tourists, with an intense flow of people on its trails during the summer and holidays (vacations). Its trails and beaches receive an average of around eight thousand people per day [20], making the high demand for visits an agent that triggers the environmental degradation of the trails [9,17,20–22].

2. Materials and Methods

2.1. Study Area

The PNSB, covering ~106,000 ha, is a UC that was created in 1972 in São Paulo State (Areias, Cunha, São José do Barreiro and Ubatuba Municipalities) and Rio de Janeiro State (Paraty and Angra dos Reis Municipalities). It has the objectives of preserving and conserving forest formations and associated refuges; ensuring the maintenance of natural landscapes; protecting biodiversity and water resources; and providing opportunities for scientific research, leisure, and recreation [23,24]. This UC is one of the largest protected areas of the Atlantic Forest, the largest continuous extension, and one of the most significant Brazilian remnants of this biome, with several geotouristic attractions (Figure 1).

Figure 1. Geotouristic attractions on PNSB coast. Middle Beach (**left**), Caixa D'Aço Beach (**center**), and Middle Beach Tombolo (**right**). Photos: Lima.

On the PNSB coast, the trails which we will focus on in this study are Waterfall of Stone that Swallows (WSS) (red) and the Caixa D'Aço Natural Pool (CDN) (orange) (Figure 2).

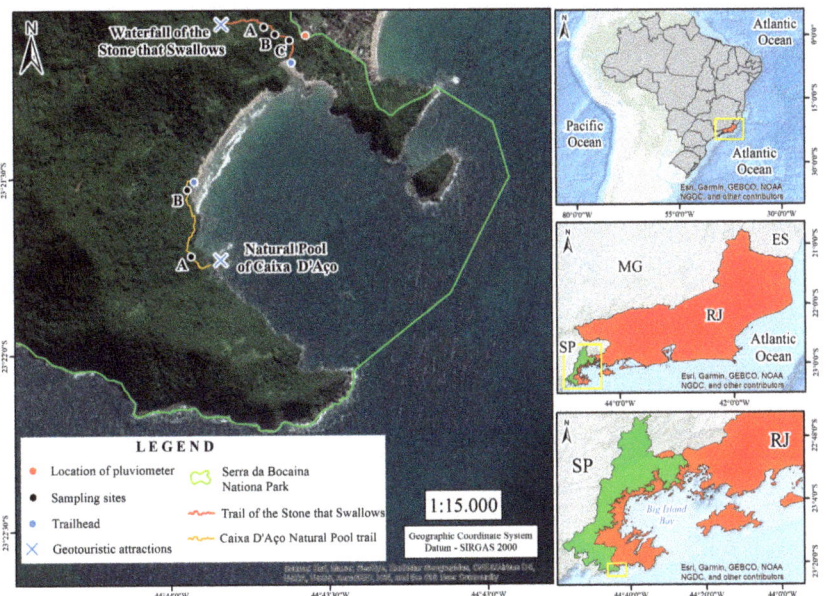

Figure 2. Location map of the study area on the Serra da Bocaina National Park (PNSB) coast. Author: Lima.

These trails are located in the relief of the mountains and hills, under soils and surface residuals, with rocky outcrops of granite Paraty-Mirim and colluvial–alluvial sediments. In general, they have a base saturation of less than 50%, which, in turn, gives them a dystrophic character, in addition to high acidity and high aluminum content. Also, according to mapping carried out at a scale of 1:250,000 [25], the soils that predominate the trails are dystrophic Haplic Cambisols. These classes, as well as Litholic Neosols, also occur in the mountainous region of the PNSB, on the escarpments and on the coastal strip between Serra do Mar escarpment and the plains. Oxisols, on the other hand, occur at the top and on smoother slopes [9,23,26].

The Swallows trail has a length of ~600 m, with an average slope of approximately 12.4% and an altitude range between 0 and 46 m (Figure 3), It gives access to the WSS (Figure 4). The CDN Trail is also ~600 m long, with an average slope of around 14.3% and an altitude range of 0 to 63 m (Figure 3), and it provides access to the CDN (Figure 4).

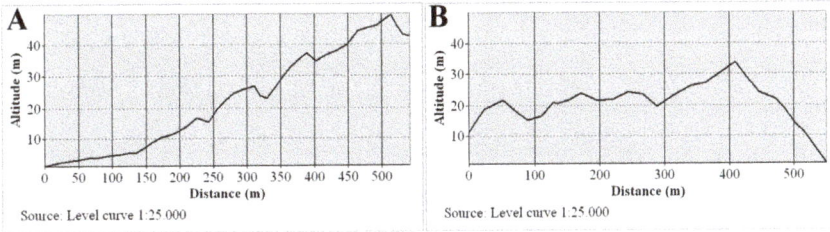

Figure 3. Topographic profiles of the trail to the Waterfall of Stone that Swallows (**A**) and the Caixa D'Aço Natural Pool Trail (**B**) on the coast of the PNSB. Author: Lima.

Figure 4. Geotouristic attractions on the trails studied on the PNSB coast: (**left**) Waterfall of Stone that Swallows and (**right**) Caixa D'Aço Natural Pool. Photos: Lima.

Geologically, the rocks are from the Rio Negro Magmatic Arch (orthogneisses) and the Parati-Mirim Granite Suite (granites), as well as later fluvio-marine sediments (Cenozoic). Regarding climate, this is influenced by the orographic effect of the Serra do Mar, as the compartmentalization of the relief and the altimetric unevenness of the mountain range. The mountain range extends from sea level to altitudes exceeding 2000 m and acts as a barrier to the air masses responsible for the behavior of meteorological phenomena. The escarpments of Serra do Mar and its mountainous plateau make passage difficult for frontal systems (cold fronts) coming from the South Atlantic/Antarctica, and, as such, are responsible for regional rainfall, generating temporal space–time discontinuities in these phenomena [24,27]. Therefore, the Serra do Mar slopes facing the ocean (south) generate a strong seasonality in the precipitation on PNSB coast, making the frontal systems (cold fronts) and the rains frequent in the summer, while winter precipitations are minimal [23,24,27].

2.2. Methodology

The determination of the soil physicochemical properties (e.g., granulometry, porosity, bulk density, pH, and organic matter content) were chosen because they are related to erodibility and soil use and management, given that their values are influenced and/or modified by intensive trampling of the ground, mainly by the flow of tourists where the surveyed trails are located [9,10,17,22].

Soil samples were collected at depths between 0 and 20 cm at strategic points along the trails: in places on the beds of the trails where there are erosive features, that is, areas that suffer trampling, and those areas immediately adjacent (edge) to the trails, where there is no passage of visitors (Figures 5 and 6). Therefore, it was possible to infer the impact of trampling by people and to spatially compare soil quality.

Undisturbed samples were collected to determine bulk density and porosity, and disturbed samples to determine pH, granulometry (texture), and organic matter content. Concerning the depth of these collections, these are justified because they are the main factors that suffer from the impact of trampling during public use activities.

All soil physical–chemical parameters were assessed at the Maria Regina Mousinho de Meis Geomorphology Laboratory of the Geography Department of the Federal University of Rio de Janeiro (UFRJ). Our techniques were in accordance with the methods of Teixeira et al. [28], that is, the texture was assessed by total dispersion of individual soil particles (pipette method); the organic matter content using the indirect method (oxidation); the bulk density by collecting samples in a cylinder of known volume (100 cm^3); the mineral density via volumetric flask and ethyl alcohol; the porosity according to its relationship with density values; and the pH using a digital meter.

Figure 5. Soil collection sites (**A–C**) on the trail to the Waterfall of Stone that Swallows on the PNSB coast. The red arrow shows the direction of main flow during periods of intense rainfall. Photos: Lima.

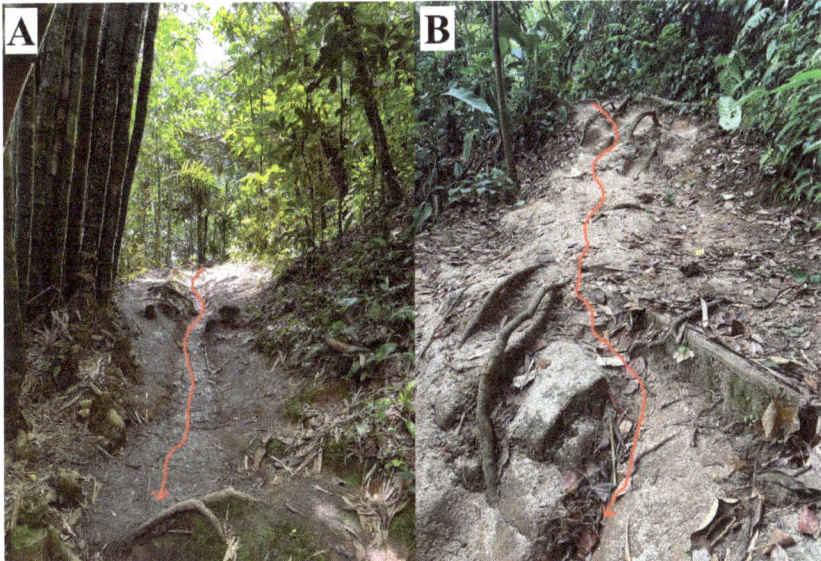

Figure 6. Soil collection sites (**A,B**) on Caixa D'Aço Natural Pool Trail on the PNSB coast. The red arrow shows the direction of main flow during periods of intense rainfall. Photos: Lima.

Determination of the texture (granulometry) took place via crushing, sieving, and mixing the soil samples with sodium hydroxide and distilled water [28], while the classification was based on the textural triangle of the United States Department of Agriculture (USDA) [29]. Wet organic matter was obtained by mixing soil samples with potassium dichromate, silver sulfate, orthophosphoric acid, diphenylamine, and ferrous ammonium sulfate [28]. Bulk density was obtained by collecting samples in a cylinder with a known volume (100 cm^3), while particle density was obtained by mixing samples with ethyl alcohol [28]. Finally, the porosity was determined by the ratio between the particle density (g cm^{-3}) and bulk density (g cm^{-3}), while the pH was obtained by dissolving the soil in distilled water and its respective reading after equipment calibration [28].

Each physical–chemical parameter was assessed in triplicate. At each site studied on the trails, three collection repetitions were performed, and the presented results refer to the averages of these triplicates. In addition to these averages, the values of the standard deviation and the coefficient of variation were also calculated in order to obtain the statistical variations using Microsoft Excel software.

The soil microtopography was determined through the installation of erosion bridges (PDE) with stakes from one edge to the other in the cross sections of sites and on the beds of the trails that presented water erosion processes, following the method developed by Shakesby [30] and adopted by Ferreira [31] in studies on water erosion in soils and subsequently applied to trails [9,22,32]. Thus, the evolution of erosion features was monitored between August 2021 and March 2022 and between August 2022 and March 2023. Therefore, we were able to obtain significant and contrasting variations in the temporal data on rainfall in the region. It is important to note that the monitored sites were the same sites from which the soil samples were collected to obtain the physical–chemical characteristics (Figures 5 and 6).

The model developed by Shakesby [30] and adapted for trails by Silva and Botelho [32] was used, whereby 50 cm wooden stakes/rods (useful for leveling), 2 m slats (erosion bridge), and a 1 m iron rod (measuring rod) were used with a measuring ruler offering 100 holes (analysis points) at 2 cm intervals. Two stakes were installed at the edges of the cross section (sites) so that they could be leveled between the edges of the trails and the values could be measured (Figure 7).

Figure 7. Example of the PDE leveling on sites (**A**,**B**) the trail to the Waterfall of Stone that Swallows on the PNSB coast. Photos: Lima.

From this monitoring program, graphs were created using Excel software to demonstrate the space–time evolution of erosion features during the four monitoring periods. We aimed to identify sites of removal and accumulation of sediments and organic matter and places where there was intense trampling and removal of organic and inorganic particles, in addition to estimating areas of lost soil along the beds of the trails. After preparing and analyzing these graphs, it was possible to estimate the area of the monitored cross-sections and quantify the evolution of the lost soil areas at the monitoring sites.

Rainfall data were obtained using a manual rain gauge installed at a campsite near the PNSB coast (latitude 23°21′5.28″ S and longitude 44°43′34.47″ W) at a height of ~1.5 m on a wooden platform, away from obstacles, so that it was possible to store and record daily rainfall throughout the monitoring period. Finally, it should be noted that the records of these data correspond to the dates of the beginning and end of the monitoring program of the five erosion bridges (August 2021, March 2022, August 2022 and March 2023), and the readings were recorded daily at 11:00 am.

3. Results

3.1. Rainfall Data

Daily rainfall data allowed total monthly volumes to be calculated (Figure 8). Moreover, rainfall occurred throughout less than half of the total monitoring period (219 days), with daily values higher than the total average, which was 11 mm, being recorded on more than half of these days (Figure 8). It was also possible to identify that this precipitation was concentrated in December (2021), January (2022), April (2022), November (2022), and January (2023), which in turn were the months with the highest averages of rainfall (21 mm, 19 mm, 21 mm, 17 mm, and 17.1 mm, respectively) (Figure 8). In addition, the highest monthly total volumes were also recorded in these months, that is, 647.5 mm in December 2021, 593 mm in January 2022, 630 mm in April 2022, 510 mm in November 2022, and 530 mm in January 2023 (Figure 8).

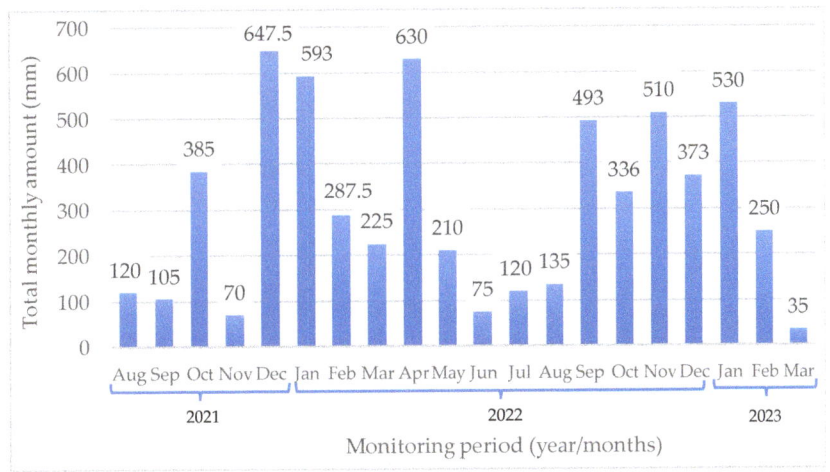

Figure 8. Total monthly values of precipitation on PNSB coast during the four monitoring periods.

When considering the intervals of days between each monitoring period, it was also possible to identify that, from the first monitoring (2021) to the second monitoring period (2022/1), the daily rainfall average was 11 mm, and these precipitations were concentrated within just 85 days. From the second monitoring period (2022/1) to the third (2022/2), despite the daily average rainfall being lower than the total average (11 mm), that is, equal to 8.3 mm, rainfall was concentrated within only 33 days. From the third monitoring period (2022/2) to the fourth (2023), the daily average rainfall was greater than 13 mm, that is, greater than the total average, with rainfall concentrated within 95 days (Figure 8). In this way, in addition to the rainfall concentration, in specific months, we noted that, between the intervals of the monitoring periods, with the exception of the second to the third, the daily average rainfall was equal to or greater than the total average precipitation during the entire monitoring period. This indicates an ideal scenario for the evolution of erosion features due to splash erosion and the increase in runoff amount on the beds of the trails.

3.2. Spatial Variation of Soil Physical–Chemical Characteristics

The survey of the spatial variation in the physical–chemical soil characteristics in the degraded sites due to water erosion processes along the WSS Trail and the CDN Trail on the PNSB coast allowed us to identify disparities between the values of the attributes that were obtained, as well as the relationship between these parameters and the evolution of erosion features. In general, porosity and bulk density values, as well as granulometry, pH, and organic matter content, showed significant relationships with the presence or absence of rills along the two trails.

Analyzing specifically the data from the WSS Trail (Table 1), it was possible to identify that the values of total porosity (%) and organic matter (%) were, without exception, lower on the bed of the trail compared to those obtained on the edges of the trail, given that the average values of these attributes were, respectively, 39% and 0.6% on the bed and 59% and 2.4% on the edges (Table 1). On the other hand, the values of bulk density (g cm^{-3}) and pH were, also without exception, higher on the bed of the trail and lower at the edges, so the average values of these attributes were, respectively, 1.1 g cm^{-3} and 5.3 on the edges and 1.6 g cm^{-3} and 5.5 on the beds of areas degraded by erosion processes along the trail (Table 1).

Table 1. Values of the physical–chemical characteristics of soils obtained from the Waterfall of Stone that Swallows Trail on the coast of the PNSB.

Trail to the Waterfall of Stone That Swallows	Position on the Trail	Pore Arrangement		Granulometry (%)					Chemical Analysis	
		Total Porosity (%)	Bulk Density (g cm^{-3})	Coarse Sand	Fine Sand	Silt	Clay	Textural Classification	pH	Organic Matter (%)
Point A	Bed	39	1.6	52	8	16	24	Sandy clay loam	4.8	0.6
	Edge	51	1.2	60	8	19	14	Sandy loam	4.5	1.2
Point B	Bed	34	1.7	49	8	24	19	Sandy loam	5.7	0.4
	Edge	64	1.0	52	6	17	25	Sandy clay loam	5.5	3.6
Point C	Bed	44	1.6	41	13	31	15	Sandy loam	5.9	0.9
	Edge	62	1.1	42	3	23	32	Sandy clay loam	5.8	2.5
Mean	Bed	39	1.6	47	10	24	19	-	5.5	0.6
	Edge	59	1.1	51	6	19	24	-	5.3	2.4
Standard deviation	Bed	4.7	0.1	5.7	2.9	7.4	4.6	-	0.6	0.3
	Edge	7.0	0.1	8.9	2.5	2.9	9.5	-	0.7	1.2

Regarding the values of soil granulometry, it was possible to identify the domain of the sand fractions, especially fine sand and silt, on the bed of the trail, since the average values were 57, 10, and 24%, respectively. Those on the edges of 57, 6, and 19%, respectively (Table 1). The only exceptions occurred at site A, where the fine sand content was the same in both the soil sample from the edge and the soil from the trail bed (8%), as well as in the silt content, which was lower in the trail bed in comparison with that obtained at the edge (16 and 19%, respectively) (Table 1). It was also possible to identify that the clay contents in the border soil samples were higher than those obtained on the beds, with average values of 24 and 19%, respectively (Table 1). The only exception was at site A, where the clay content was higher in the trail bed samples compared to those obtained at the edges (24 and 14%, respectively) (Table 1). This granulometric composition, rich in sand, indicates a textural classification of soils that vary between sandy loam and sandy clay loam (Table 1).

Regarding the data obtained on the CDN Trail (Table 2) regarding the soil's physical and chemical characteristics, it was possible to identify that the values of total porosity (%) and organic matter (%) were lower on the trail bed compared to those that were obtained at the edges, as was the case on the WSS Trail. This is because on the bed, the average values of these attributes were 33% and 1.0%, respectively, and on the edges they were 59% and 1.4%, respectively (Table 2). The values of bulk density (g cm^{-3}) and pH were higher on the trail bed and lower at the edges, so the average values of these attributes, respectively, were 1.7 g cm^{-3} and 5.6 on the beds of the trails and 1.1 g cm^{-3} and 5.4 in the soil along the edges of the trails (Table 2).

Table 2. Physical–chemical parameters of the soils obtained from the Caixa D'Aço Natural Pool Trail on the coast of the PNSB.

Caixa D'Aço Natural Pool Trail	Position on the Trail	Pore Arrangement		Granulometry (%)					Chemical Analysis	
		Total Porosity (%)	Bulk Density (g cm^{-3})	Coarse Sand	Fine Sand	Silt	Clay	Textural Classification	pH	Organic Matter (%)
Point A	Bed	33	1.8	41	14	16	29	Sandy clay loam	5.6	1.2
	Edge	56	1.1	30	9	26	36	Clay loam	5.4	1.5
Point B	Bed	36	1.7	10	51	28	11	Sandy loam	5.5	0.8
	Edge	62	1.1	14	39	17	31	Sandy clay loam	5.4	1.4
Mean	Bed	35	1.7	25	33	22	20	-	5.6	1.0
	Edge	59	1.1	22	24	21	33	-	5.4	1.4
Standard deviation	Bed	2	0.1	22.0	26.2	8.3	12.8	-	0.1	0.2
	Edge	4	0.0	11.6	21.2	6.4	3.4	-	0.0	0.1

Regarding the granulometric analysis of the soils, it was noted that, on the bed of the trail, there was a predominance of fractions of silt and sand, especially fine sand. The average values of these fractions were 22%, 58%, and 33%, respectively (Table 2), whereas, at the edges, the values of these granulometric fractions were 21%, 46%, and 24%, respectively (Table 2). It is still possible to observe that clay contents were, without exception, higher on the border than on the bed of the trail; the average values of these attributes were 33% and 20%, respectively (Table 2). These compositions of the granulometric fractions of the soils in this trail indicate the predominance of sand fractions; thus, with the exception of the soil on the edge of site A, the other soils were classified between sandy loam and sandy clay (Table 2).

3.3. Spatiotemporal Dynamics of Soil Microtopography

Soil microtopography monitoring of degraded sites with erosive features along the WSS Trail and the CDN Trail (Figures 9–13) indicated a space–time evolution of erosion processes, as well as deposition of organic and inorganic materials on the beds of these trails. This may be associated with intensive trampling and soil compaction, removal and deposition of vegetation cover and soil organic matter, and the rainfall indices recorded in the region. The evolution of the cross-sectional area at the monitored site was evident when observing that the soil surface from the last monitoring (black line) was in a lower position than the soil surface during all other periods (orange, blue, and yellow lines) at all monitored sites (Figures 9–13).

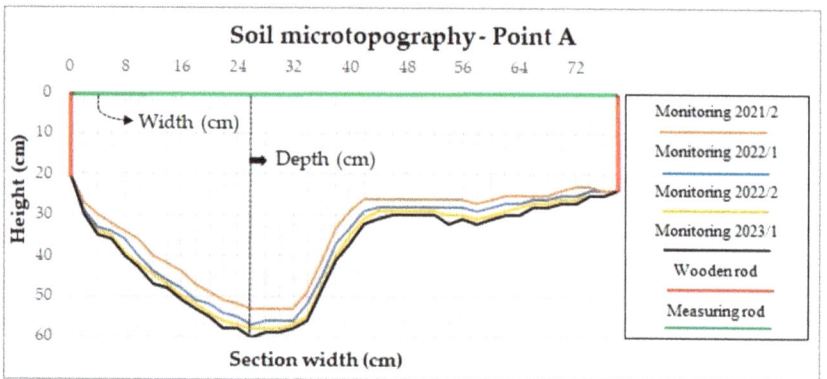

Figure 9. Soil microtopography at site A of the Waterfall of Stone that Swallows Trail on the PNSB coast.

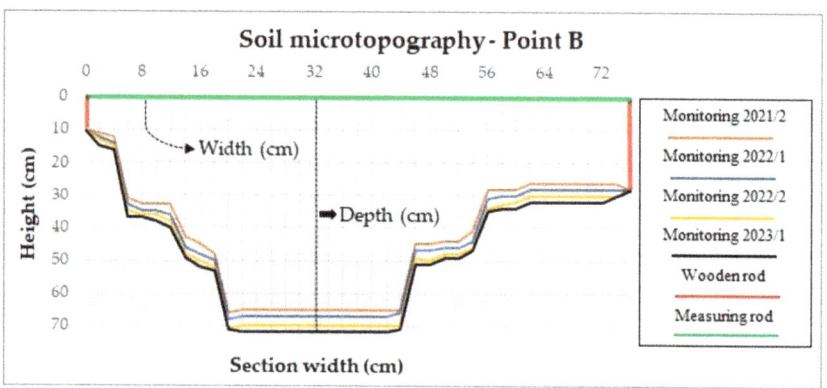

Figure 10. Soil microtopography at site B of the Waterfall of Stone that Swallows Trail, on the PNSB coast.

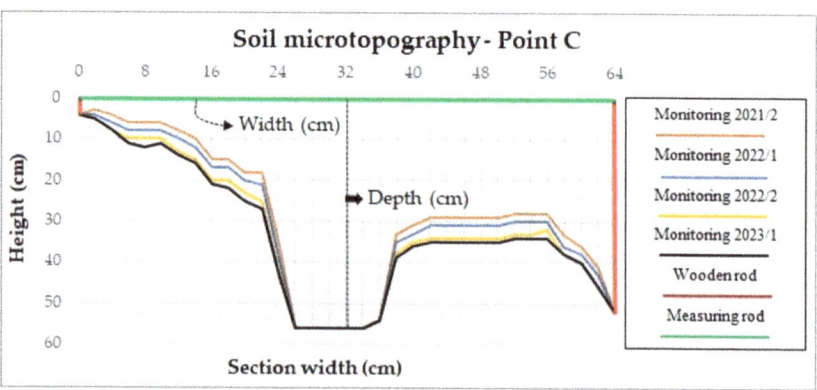

Figure 11. Soil microtopography at site C of the Waterfall of Stone that Swallows Trail on the PNSB coast.

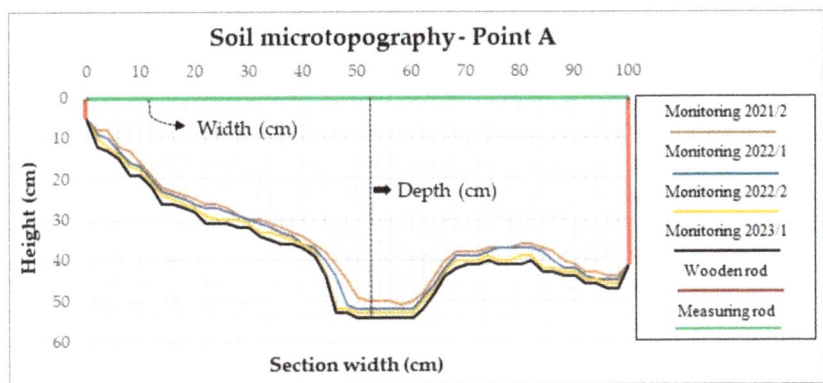

Figure 12. Soil microtopography at point A of the Caixa D'Aço Natural Pool Trail on the coast of the PNSB.

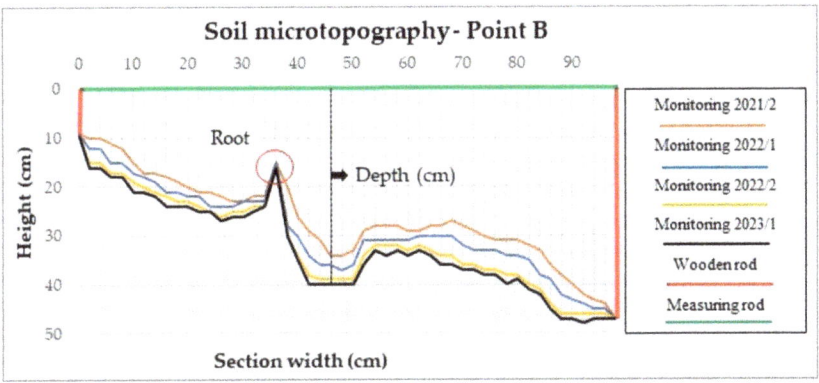

Figure 13. Soil microtopography at site B of the Caixa D'Aço Natural Pool Trail on the PNSB coast.

Figure 9 shows that, during the first monitoring period of this site (2021), for example, the maximum depth of this rill was approximately 25 cm, while in the last monitoring period (2023), the depth increased by ~8 cm. Furthermore, it should be noted that the width of this rill also showed space–time evolution, since the value in the last monitoring period was ~4 cm greater than that in the first period (Figure 9). This space–time evolution corroborates the increase in the total cross-sectional area between the monitoring periods, since this area was 0.265 m² in the first monitoring period (2021), 0.284 m² in the second period (2022/1), 0.295 m² in the third period (2022/2), and 0.304 m² in the last period (2023). Therefore, this indicates an increase of approximately 0.039 m² in the cross-sectional area from the beginning of monitoring.

Figure 10, which shows the soil microtopography at site B of the WSS Trail, also indicates a space–time evolution of the monitoring period erosion feature. In the central portion of this rill, for example, it was possible to identify that in the first monitoring period (2021), the maximum soil depth surface was ~65 cm, while in the last monitoring period (2023), this depth was greater than 70 cm (Figure 10). This space–time evolution was reinforced by the increase in the total cross-sectional area between the monitoring periods, since the total area was 0.327 m² in the first monitoring period (2021), 0.336 m² in the second period (2022/1), 0.258 m² in the third period (2022/2), and 0.372 m² in the last one (2023), i.e., an increase of about 0.044 m² in the total cross-sectional area from the beginning of the monitoring period (Figure 10).

Figure 11, which shows the microtopography of the soil at site C of the WSS Trail, also shows a space–time evolution of the monitored erosion feature, especially in the lateral portions of the cross section (Figure 11). The exception was found in the central portion of this erosive feature. This occurred because the soil loss was severe at this site, and, in turn, reached the rocky layer underlying the soil layer, as can be seen in Figure 7B. Although the central portion of this rill did not indicate a space–time evolution at this point due to the rock outcrop, which is more resistant to the erosion process, it was possible to identify lateral soil loss in the other portions of this cross section. We considered the total cross-sectional area at this site, which was 0.177 m^2 in the first monitoring period, 0.192 m^2 in the second, 0.205 m^2 in the third, and 0.210 m^2 in the last, that is, an increase of 0.032 m^2 from the beginning of the monitoring period (Figure 11).

Figure 12 indicates the data associated with the monitoring of soil microtopography at site A of the CDN Trail. It is possible to identify that, in the central portion of this cross section, that is, inside the erosion feature, the maximum depth during the first monitoring period (2021) was approximately 14 cm. In the last period, this depth was approximately 4 cm greater in relation to the first. This space–time evolution is corroborated by data on the total area of the cross-section, which was 0.336 m^2 in the first monitoring period, 0.355 m^2 in the second, 0.37 m^2 in the third, and 0.381 m^2 in the fourth. This indicates an increase of approximately 0.045 m^2 in the total area from the beginning of the monitoring period.

Figure 13, which presents the soil microtopography data from site B of the CDN Trail, also shows a space–time evolution of the monitored erosion feature that is similar to the behavior of the other monitoring sites,. At this site, it was possible to identify a significant space–time evolution of soil loss by exposing the roots, the greater distance between the lines that indicate the soil surface, or the increase in the total value of the cross-sectional area (Figure 13). In the first monitoring period, for example, the total cross-sectional area was approximately 0.259 m^2, while in the second monitoring period, it was 0.284 m^2. In the third monitoring period, the total value of this area was 0.307 m^2, and in the last monitored period, 0.318 m^2. This indicates an increase of about 0.059 m^2 from the beginning of the first monitoring period.

4. Discussion

When jointly considering the obtained data, it was possible to identify that these factors are related to each other, since certain physical-chemical parameters, for example, when subjected to certain dynamics of use and precipitation, tend to condition the evolution of erosive features such as those that were evidenced during the soil microtopography monitoring period (Figures 9–13) [9,14,17,18].

Knowing that water erosion is a process of mobilization, transport, and deposition of soil sediment through the action of rainwater, the concentration of rainfall in the PNSB region, especially in specific periods, tends to intensify aggregate breakdown and the detachment of soil particles from the bed of the trails, corroborating the evolution of the erosive features that were identified (Figures 9–13). In this sense, the impact of the kinetic energy of raindrops when they reach the soil surface causes its sealing, compaction, and detachment through splash erosion. Consequently, this material is lost and transported through runoff, therefore leading to the space–time evolution of the erosion features monitored. This culminates in an increase in the area of the soil surface and its cross-sections, as observed during the monitoring periods [9,14,17,18].

It is worth mentioning that Rangel and Guerra [22], as well as Lima et al. [9], have already identified that the evolution of cross-sectional areas on trails during monitoring periods of soil microtopography can also be associated with the presence or removal of organic material/litter by surface runoff. Therefore, this evolution is not strictly associated with the loss of mineral material from the soil. This behavior reinforces a possible influence of physical–chemical characteristics on this evolution, given that the organic matter content, for example, can accentuate or delay the evolution of the erosion processes.

Considering that the beds of the trails are the places with the most trampling and least vegetation cover compared to their edges, the values of bulk density, total porosity, organic matter, and pH, for example, tended to be different from those of soil samples that were collected on the edges, as evidenced in the data in Tables 1 and 2. In the soils on the trail beds, the minimal vegetation cover and the intensive trampling of the soil tend to contribute to the decrease in porosity values and organic matter content, in addition to favoring the increase in bulk density values and soil loss. This explains not only the evolution of the monitored erosion features (Figures 9–13), but also the values of porosity, organic matter, and bulk density (Tables 1 and 2). At the trail edges, the presence of vegetation cover and low trampling tend to reduce bulk density and affect pH values, while also directly affecting total porosity values and organic matter content (Tables 1 and 2).

The lack of vegetation cover on the trail beds also contributes to the dynamics of this behavior, in addition to exposing the soil to the impact of the kinetic energy of raindrops (splash), which in turn contributes to its compaction. Its respective loss also influences the levels of organic matter [8,33–36].

As the lack of soil trampling on the edges favors a greater development of vegetation compared to the beds of the trails, the mechanical action of roots during the growth and retraction processes increases the amount of empty space inside the soil that can be filled by air and water, that is, the value of total soil porosity [35–37]. In addition, because it is composed predominantly of plant and animal residues, organic matter provides nutrients for the endopedonic fauna, which, together with the agents secreted by the roots, favor an increase in microbial activity and total soil porosity [33,36,38].

The soil's characteristic chemical conditions influence the formation of aggregates, retention, and water infiltration. They also increase aggregate stability, the number of pores, and biological activity, in addition to reducing the density [8,14,38–41]. By acting as a cementing agent that unites soil particles, organic matter forms organic substances that, when released and made available to the soil, bind and agglutinate the particles, increasing aggregate stability and reducing their erodibility [39,42]. Therefore, the ability of organic matter to affect the structural integrity of the soil, form aggregates, and stimulate the development of its biota favors aeration and increased permeability, which in turn increases water infiltration, reduces runoff, and increases soil resistance to erosion. Its absence, on the other hand, tends to trigger water erosion processes, remove soil material, and both cause and reduce nutrient retention [8,41,43,44].

In addition to vegetation cover influencing the disposition of organic matter, it also influences the total soil porosity, which in turn controls percolation, infiltration of water inside the soil, the penetration of air, and the movement of roots. The modifications to the distribution, volumes, sizes, and shapes of these pores through use and soil cover, for example, affect permeability, aeration, hydraulic conductivity, and, consequently, the dynamics of water erosion processes [8,44–46].

Considering that porosity values are inversely proportional to their density, and that intensive trampling by people contributes to reducing soil porosity, this behavior implies that the soils on the bed of the trail are more susceptible to erosion compared to the soils on the edge, either due to low porosity values, low levels of organic matter, or high bulk density values. This justifies not only the values found in Tables 1 and 2, but also the evolution of soil microtopography (Figures 9–13). On the other hand, the edge soils, which are not trampled, have low erodibility due to the maximum values of porosity and organic matter and the low values of bulk density, thus favoring the dynamics of the values shown in Tables 1 and 2 [8,45].

Erodibility tends to increase with high rainfall in the PNSB region, which, in turn, tends to be concentrated in specific periods, especially during the beginning of summer and the holiday (vacation) period (December and January) (Figure 13). In addition to the absence of vegetation covering the low levels of organic matter in the trails' soils and, in turn, increasing the instability of its aggregates, it facilitates the rupture of its particles due to the impact of raindrops, thus favoring the formation of crusts in the topsoil layer, reducing water infiltration, and increasing the soil loss via runoff. Therefore, when concentrated in specific periods, the rain tends to intensify the compaction of soils due to trampling by people on the beds of the trails, favoring the formation of preferential paths of water that culminate in the development of erosion features. This influences not only the values of some physical–chemical parameters (Tables 1 and 2), but also the loss of organic and mineral particles, as evidenced in the monitoring period of soil microtopography with the evolution of the cross-sectional area (Figures 9–13).

The pH values were all acidic (pH < 7), with the highest values found at the beds of the trails and the lowest associated with the trail edges (Tables 1 and 2). The fact that the edges have greater vegetation cover and, consequently, higher levels of organic matter may indicate the excretion of certain acidic substances, either by the action of the roots or by the decomposition of vegetable and animal matter. Soils obtained on the beds of the trails, on the contrary, lack vegetation cover and materials for decomposition (Tables 1 and 2). The more acidic soils at the trail edges tend to present colloidal complexes deficient in chemical elements that confer greater stability, that is, they lead to reductions in erodibility and increase the shear resistance of soil particles, since some of these elements result not only in greater biological activity, but also in stability of the aggregates [47–49].

Soils on the edges, despite having lower pH values, are less susceptible to erosion processes because, in addition to not suffering from trampling or the direct impact of raindrops, these soils are more porous, less dense, less compacted, and have higher levels of organic matter, which, by producing and releasing humus into the soil, favor their aggregate stability. In contrast, the soils on the trails, despite having less acidic pH values than the soils on the edges, in addition to being subjected to intensive public use and the impact of raindrops, have higher bulk density values, low porosity, and low levels of organic matter. This justifies our findings (Tables 1 and 2), and the evolution of erosion was evidenced by monitoring the soil microtopography (Figures 9–13).

The granulometry and textural classification of the soils indicate the predominance of clayey, medium-texture soil (clay-sandy loam), in which clay and sand fractions predominate in the granulometric composition of the soils. The predominance of coarse sand fractions and clay fractions causes the soils to have low erodibility, considering that the coarse sand fractions, due to their diameters, weights, and fast decanting speeds, make their removal and transport difficult due to the action of water. The clay fractions, in turn, have high aggregation capacity, the colloids and specific surfaces with cohesive strength, and a significant presence of fractions of silt and fine sand, especially in soils on the beds of trails. These favor the formation and loss of soil through erosion, since these fractions are easily removed by low cohesion and insufficient weight in the face of detachment and transport caused by the action of water [50–53].

Further, these soils have granulometric compositions of solid particles with minimal cohesion, so these characteristics together can be fundamental conditioning factors for soil loss, which became evident due to the evolution of the area of the cross sections recorded during the monitoring of the soil microtopography (Figures 9–13). On the other hand, the soils on the edges had the highest values of porosity and organic matter, the presence of vegetation cover, low values of bulk density and pH, a lack of trampling, and greater presence of granulometric fractions of clay. These factors increase the cohesion of soil particles, hinder soil loss, and elucidate the data (Tables 1 and 2) [8,10,14,17,20].

In this sense, the joint action of soil physical and chemical characteristics, mainly the values of bulk density, total porosity, granulometry, and organic matter content associated with rainfall indices in the coastal region of the PNSB, as well as the intense public use of the trails of this protected area, were assessed. These conditions culminate in the development and evolution of erosion features along the trails, as evidenced in cross sections of the soil microtopography (Figures 9–13). This is because public use tends to influence some of these physical–chemical characteristics, while the dynamics of regional rainfall together with other intrinsic characteristics of the soil (granulometry) favor the evolution of erosion features.

5. Conclusions

The obtained data indicate the intensive trampling of the soil resulting from public use is associated with the intrinsic characteristics of the location, such as the granulometric fractions of high erodibility and concentrated rains, especially in periods that coincide with greater local visitation. These conditions are leading to the environmental degradation of the trails of this conservation unit, as evidenced by the evolution of the erosion features in the cross-sectional areas monitored during the survey.

Monitoring the soil microtopography has demonstrated a loss of solid organic and inorganic particles which culminates in a degradation process, given that in all the cross sections monitored, there were increases in the areas of erosion features from the beginning of the monitoring. Such behavior, therefore, evidences the surface runoff process, which has a high capacity to remove solid materials. In addition, the reduction in soil protection against the impact of raindrops is enhanced by the trampling of the ground resulting from the intense flow of people along the trails.

The physical–chemical characteristics of the soil also showed an association with the degradation process, given that compaction, evidenced by the values of bulk density, especially on the bed of the trails, contributes to the reduction in water infiltration and to the increase in surface runoff. This, in turn, culminates in soil loss along the trails. Organic matter content also contributes to the dynamics of this process; all results were higher at the edges of the trails compared to those obtained in the respective beds, and these contents affect soil erodibility.

Intrinsic pedological and climatic factors, such as soils with coarse texture and intense rainfall in concentrated periods, also contribute to the evolution of the monitored erosion processes. The presence of sand and silt fractions and low levels of clay in the soils from the bed of the trails, for example, contribute to soil loss, especially when accompanied by compacted soils and by concentrated rainfall in specific periods.

This collective evidence shows the need for management and recovery actions for degraded areas along these trails through strategic management actions and improvement of their physical–chemical attributes, whether through the incorporation of organic matter or the construction of steps.

Finally, it is concluded that the obtained data can be fundamental tools for identifying environmental weaknesses or potential, as well as supporting actions aimed at planning and managing trails for public use in conservation units, especially when these protected areas are in high demand for public use and have intrinsic environmental characteristics that favor the triggering of environmental degradation processes.

Author Contributions: Data curation, G.M.d.L.; formal analysis, G.M.d.L., A.J.T.G., L.d.A.R., C.A.B. and M.A.F.; funding acquisition, A.J.T.G.; investigation, G.M.d.L., A.J.T.G., L.d.A.R., C.A.B. and M.A.F.; methodology, G.M.d.L.; project administration, A.J.T.G.; supervision, A.J.T.G., L.d.A.R., C.A.B. and M.A.F.; validation, G.M.d.L., A.J.T.G., L.d.A.R., C.A.B. and M.A.F.; visualization, G.M.d.L., A.J.T.G., L.d.A.R., C.A.B. and M.A.F.; writing of the original draft, G.M.d.L., A.J.T.G., L.d.A.R., C.A.B. and M.A.F.; writing—review and editing, G.M.d.L., A.J.T.G., L.d.A.R., C.A.B. and M.A.F. All authors have read and agreed to the published version of the manuscript.

Funding: This research was funded by FAPERJ (Fundação Carlos Chagas Filho de Amparo à Pesquisa do Estado do Rio de Janeiro), grant number E-26/201.369/2023.

Institutional Review Board Statement: Not applicable.

Informed Consent Statement: Not applicable.

Data Availability Statement: Data are contained within the article.

Acknowledgments: The authors of this research would like to thank the Post-Graduate Program in Geography (UFRJ), CAPES, and FAPERJ for the financial support needed to carry out the field work, and the Chico Mendes Institute of Biodiversity (ICMBio) for the necessary authorization to carry out the research.

Conflicts of Interest: The authors declare no conflicts of interest. The funders had no role in the design of the study; in the collection, analyses, or interpretation of data; in the writing of the manuscript; or in the decision to publish the results.

References

1. Centeri, C. Soil Water Erosion. *Water* **2022**, *14*, 447–501. [CrossRef]
2. Boardman, J.; Poesen, J.; Evans, M. Slopes: Soil erosion. In *The History of the Study of Landforms or the Development of Geomorphology: Volume 5: Geomorphology in the Second Half of the Twentieth Century*; Geological Society of London: London, UK, 2022. [CrossRef]
3. Ferreira, C.S.S.; Seifollahi-Aghmiuni, S.; Destouni, G.; Ghajarnia, N.; Kalantari, Z. Soil degradation in the European Mediterranean region: Processes, status and consequences. *Sci. Total Environ.* **2022**, *805*, 150106–150123. [CrossRef]
4. Golubović, T.D. Environmental Consequences of Soil Erosion. In *Advances in Environmental Engineering and Green Technologies*; Milutinović, S., Živković, S., Eds.; IGI Global: Hershey, PA, USA, 2022; pp. 112–131.
5. Yin, C.; Zhao, W.; Pereira, P. Soil conservation service underpins sustainable development goals. *Glob. Ecol. Conserv.* **2022**, *33*, 1–8. [CrossRef]
6. Guerra, A.J.T.; Bezerra, J.F.R.; Jorge, M.C.O. Recuperação de voçorocas e de áreas degradadas, no Brasil e no mundo—*Estudo* de caso da voçoroca do Sacavém—São Luís—MA. *Rev. Bras. Geomorfol.* **2023**, *24*, 1–20. [CrossRef]
7. Dragovich, D.; Bajpai, S. Managing Tourism and Environment—Trail Erosion, Thresholds of Potential Concern and Limits of Acceptable Change. *Sustainability* **2022**, *14*, 4291–4307. [CrossRef]
8. Pereira, L.S.; Rodrigues, A.M.; Jorge, M.C.O.; Guerra, A.J.T.; Booth, C.A.; Fullen, M.A. Detrimental effects of tourist trails on soil system dynamics in Ubatuba Municipality, São Paulo State, Brazil. *Catena* **2022**, *216*, 106431. [CrossRef]
9. Lima, G.M.; Rangel, L.A.; Guerra, A.J.T. Monitoramento da microtopografia do solo em trilhas de uso público no litoral do Parque Nacional da Serra da Bocaina. *Rev. Bras. Geomorfol.* **2023**, *24*, 1–17. [CrossRef]
10. Lima, G.M.; Guerra, A.J.T. Áreas degradadas por processos erosivos hídricos na Trilha do Morro Dois Irmãos, no município do Rio de Janeiro (RJ). *Rev. Ciência Geográfica* **2023**, *27*, 376–395. [CrossRef]
11. Costa, N.M.C.; Oliveira, F.L. Trilhas: "caminhos" para o geoturismo, a geodiversidade e a geoconservação. In *Geoturismo, Geodiveridade e Geoconservação: Abordagens Geográficas e Geológicas*; Guerra, A.J.T., Jorge, M.C.O., Eds.; Oficina de Textos: São Paulo, Brazil, 2018; pp. 201–223.
12. Bhammar, H.; Li, W.; Molina, C.M.M.; Hickey, V.; Pendry, J.; Narain, U. Framework for Sustainable Recovery of Tourism in Protected Areas. *Sustainability* **2021**, *13*, 2798–2808. [CrossRef]
13. Zhang, X.; Zhong, L.; Yu, H. Sustainability assessment of tourism in protected areas: A relational perspective. *Glob. Ecol. Conserv.* **2022**, *35*, 1–14. [CrossRef]
14. Marion, J.L. Trail sustainability: A state-of-knowledge review of trail impacts, influential factors, sustainability ratings, and planning and management guidance. *J. Environ. Manag.* **2023**, *340*, 117868. [CrossRef] [PubMed]
15. Spernbauer, B.S.; Monz, C.; D'antonio, A.; Smith, J.W. Factors influencing informal trail conditions: Implications for management and research in urban-proximate parks and protected areas. *Landsc. Urban Plan.* **2023**, *231*, 104661. [CrossRef]
16. Fonseca Filho, R.E.; Varajão, A.F.D.C.; Castro, P.T.A. Compactação e erosão de trilhas geoturísticas de parques do Quadrilátero Ferrífero e Serra do Espinhaço meridional. *Rev. Bras. Geomorfol.* **2019**, *20*, 825–839. [CrossRef]
17. Rangel, L.A.; Jorge, M.C.; Guerra, A.J.T.; Fullen, M.A. Soil Erosion and Land Degradation on Trail Systems in Mountainous Areas: Two Case Studies from South-East Brazil. *Soil Syst.* **2019**, *3*, 56–70. [CrossRef]
18. Figueiredo, M.A.; Martins, J.V.A. Erosão em trilhas e sua relação com o turismo em áreas protegidas: Uma breve discussão. In *Turismo em Áreas Protegidas*; Sutil, T., Ladwig, N.I., Silva, J.G.S., Eds.; UNESC: Criciúma, Brazil, 2021; pp. 173–195.
19. Wolf, I.D.; Croft, D.B.; Green, R.J. Nature Conservation and Nature-Based Tourism: A paradox? *Environments* **2019**, *6*, 104–126. [CrossRef]
20. ICMBIO. Instituto Chico Mendes de Conservação da Biodiversidade. ICMBio Realiza Operação de Ordenamento da Visitação nas Praias do Meio e Caixa D'aço em Trindade no Feriado de Carnaval de 2022. Available online: https://www.icmbio.gov.br/parnaserradabocaina/destaques/190-icmbio-realiza-operacao-de-ordenamento-da-visitacao-nas-praias-do-meio-e-caixa-d-aco-em-trindade-no-feriado-de-carnaval-de-2022.html (accessed on 20 July 2023).
21. ICMBIO. Instituto Chico Mendes de Conservação da Biodiversidade. Parque Nacional da Serra da Bocaina Promoveu Ação de Ordenamento Turístico na Trindade. Available online: http://www.icmbio.gov.br/parnaserradabocaina/destaques/152-parque-nacional-da-serrada-bocaina-promoveu-acao-de-ordenamento-turistico-na-trindade.html (accessed on 20 July 2023).
22. Rangel, L.A.; Guerra, A.J.T. Microtopografia e compactação do solo em trilhas geoturísticas no litoral do Parque Nacional da Serra da Bocaina—Estado do Rio de Janeiro. *Rev. Bras. Geomorfol.* **2018**, *19*, 391–405. [CrossRef]
23. MMA (Ministério do Meio Ambiente). *Plano de Manejo do Parque Nacional da Serra da Bocaina*; Instituto Brasileiro de Meio Ambiente; Ministério do Meio Ambiente: Brasília, Brazil, 2002. Available online: https://www.icmbio.gov.br/parnaserradabocaina/extras/62-plano-de-manejo-e-monitorias.html (accessed on 19 July 2023).
24. Leuzinger, M.D.; Santana, P.C.; Souza, L.R. *Parques Nacionais do Brasil: Pesquisa e Preservação*; CEUB: Brasília, Brazil, 2020; pp. 1–748.

25. Carvalho Filho, A.; Lumbreras, J.F.; Wittern, K.P.; Lemos, A.L.; Santos, R.D.; Calderano Filho, B.; Oliveira, R.P.; Aglio, M.L.D.; Souza, J.S.; Chaffin, C.E.; et al. *Mapa de Reconhecimento de Baixa Intensidade dos solos do Estado do Rio de Janeiro*; Escala 1:250.000; Embrapa Solos: Rio de Janeiro, Brazil, 2003.
26. Guerra, A.J.T.; Jorge, M.C.O.; Fullen, M.A.; Bezerra, J.F.R. The geomorphology of Angra dos Reis and Paraty municipalities, Southern Rio de Janeiro State. *Rev. Geonorte* **2013**, *9*, 1–21.
27. Kamino, L.H.Y.; Rezende, E.A.; Santos, L.J.C.; Felippe, M.F.; Assis, W.L. Atlantic Tropical Brazil. In *The Physical Geography of Brazil: Environment, Vegetation and Landscape*; Salgado, A.A.R., Santos, L.J.C., Paisani, J.C., Eds.; Springer: New York, NY, USA, 2019; pp. 41–74.
28. Teixeira, P.C.; Donagemma, G.K.; Fontana, A.; Teixeira, W.G. *Manual de Métodos de Análise de Solos*; Embrapa Solos: Rio de Janeiro, Brazil, 2017; pp. 1–574.
29. USDA. United States Department of Agriculture. Soil Texture Calculator. Available online: https://www.nrcs.usda.gov/resources/education-and-teaching-materials/soil-texture-calculator (accessed on 20 July 2023).
30. Shakesby, R.A. The soil erosion bridge: A device for micro-profiling soil surfaces. *Earth Surf. Process. Landf.* **1993**, *18*, 823–827. [CrossRef]
31. Ferreira, C.G. Erosão hídrica em solos florestais. Estudo em povoamentos de Pinu spinaster e Eucalyptus globulus em Macieira—Alcôba. *Rev. Fac. Let. Geogr.* **1996/7**, *12/13*, 145–244.
32. Silva, A.O.; Botelho, R.G.M. Diagnóstico das condições ambientes e de uso público na trilha do Peito do Pombo por meio do Protocolo de Avaliação Rápida (Sana—Macaé—RJ). *Rev. Iberoam. Tur.* **2021**, *11*, 177–195. [CrossRef]
33. Igwe, P.U.; Ezeukwu, J.C.; Edoka, N.E.; Ejie, O.C.; Ifi, G.I. A Review of Vegetation Cover as a Natural Factor to Soil Erosion. *Int. J. Rural Dev. Environ. Health Res.* **2017**, *1*, 21–28. [CrossRef]
34. Xia, L.; Song, X.; Fu, N.; Cui, S.; Li, L.; Li, H.; Li, Y. Effects of forest litter cover on hydrological response of hillslopes in the Loess Plateau of China. *Catena* **2019**, *181*, 104076. [CrossRef]
35. Prescott, C.E.; Vesterdal, L. Decomposition and transformations along the continuum from litter to soil organic matter in forest soils. *For. Ecol. Manag.* **2021**, *498*, 119522. [CrossRef]
36. Sayer, E.J.; Rodtassana, C.; Sheldrake, M.; Bréchet, L.M.; Ashford, O.S.; Lopez-Sangil, L.; Kerdraon-Byrne, D.; Castro, B.; Turner, B.L.; Wright, S.J. Revisiting nutrient cycling by litterfall—Insights from 15 years of litter manipulation in old-growth lowland tropical forest. In *Advances in Ecological Research*; Holzer, J.M., Baird, J., Hickey, G.M., Eds.; Elsevier: Amsterdam, The Netherlands, 2022; Volume 62, pp. 173–223.
37. Giweta, M. Role of litter production and its decomposition, and factors affecting the processes in a tropical forest ecosystem: A review. *J. Ecol. Environ.* **2020**, *44*, 11. [CrossRef]
38. Gmach, M.R.; Cherubin, R.; Kaiser, K.; Cerri, C.E.P. Processes that influence dissolved organic matter in the soil: A review. *Sci. Agric.* **2020**, *77*, e20180164. [CrossRef]
39. Zhang, X.; Li, Z.; Nie, X.; Huang, M.; Wang, D.; Xiao, H.; Liu, C.; Peng, H.; Jiang, J.; Zeng, G. The role of dissolved organic matter in soil organic carbon stability under water erosion. *Ecol. Indic.* **2019**, *102*, 724–733. [CrossRef]
40. Wiesmeier, M.; Urbanski, L.; Hobley, E.; Lang, B.; von Lützow, M.; Marin-Spiotta, E.; van Wesemael, B.; Rabot, E.; Ließ, M.; Garcia-Franco, N.; et al. Soil organic carbon storage as a key function of soils—A review of drivers and indicators at various scales. *Geoderma* **2019**, *333*, 149–162. [CrossRef]
41. Cotrufo, M.F.; Lavallee, J.M. Soil organic matter formation, persistence, and functioning: A synthesis of current understanding to inform its conservation and regeneration. *Adv. Agron.* **2022**, *172*, 1–66. [CrossRef]
42. FAO. *Global Status of Black Soils*; Food and Agriculture Organization of the United Nations: Rome, Italy, 2022; pp. 1–176. [CrossRef]
43. Fernández-Raga, M.; Palencia, C.; Keesstra, S.; Jordán, A.; Fraile, R.; Angulo-Martínez, M.; Cerdà, A. Splash erosion: A review with unanswered questions. *Earth-Sci. Rev.* **2017**, *171*, 463–477. [CrossRef]
44. Yadav, G.K.; Dadhich, S.K.; Bhateshwar, M.C. *Recent Innovative Approaches in Agricultural Science*; Bhumi Publishing: Maharashtra, India, 2022; pp. 1–218.
45. D'acqui, L.P.; Certini, G.; Cambi, M.; Marchi, E. Machinery's impact on forest soil porosity. *J. Terramechanics* **2020**, *91*, 65–71. [CrossRef]
46. Meadema, F.; Marion, J.L.; Arredondo, J.; Wimpey, J. The influence of layout on Appalachian Trail soil loss, widening, and muddiness: Implications for sustainable trail design and management. *J. Environ. Manag.* **2020**, *257*, 109986. [CrossRef] [PubMed]
47. Totsche, K.U.; Amelung, W.; Gerzabek, M.H.; Guggenberger, G.; Klumpp, E.; Knief, C.; Lehndorff, E.; Mikutta, R.; Peth, S.; Prechtel, A.; et al. Microaggregates in soils. *J. Plant Nutr. Soil Sci.* **2017**, *181*, 104–136. [CrossRef]
48. Matsumoto, S.; Ogata, S.; Shimada, H.; Sasaoka, T.; Hamanaka, A.; Kusuma, G.J. Effects of pH-induced changes in soil physical characteristics on the development of soil water erosion. *Geosciences* **2018**, *8*, 134. [CrossRef]
49. Schlatter, D.C.; Kahl, K.; Carlson, B.; Huggins, D.R.; Paulitz, T. Soil acidification modifies soil depth-microbiome relationships in a no-till wheat cropping system. *Soil Biol. Biochem.* **2020**, *149*, 107939. [CrossRef]
50. Ker, J.C.; Curi, N.; Schaefer, C.E.G.R.; Vidal-Torrado, P. *Pedologia: Fundamentos*; Sociedade Brasileira de Ciência do Solo: Viçosa, Brazil, 2015; pp. 1–343.
51. Nguyen, V.B.; Nguyen, Q.B.; Zhang, Y.W.; Lim, C.Y.H.; Khoo, B.C. Effect of particle size on erosion characteristics. *Wear* **2016**, *348–349*, 126–137. [CrossRef]

52. Guerra, A.J.T.; Fullen, A.; Jorge, M.C.O.; Bezerra, J.F.R.; Shokr, M.S. Slope processes, mass movements and soil erosion: A review. *Pedosphere* **2017**, *27*, 27–41. [CrossRef]

53. Jorge, M.C.O. *Solos: Conhecendo sua História*; Oficina de Textos: São Paulo, Brazil, 2021; pp. 1–62.

MDPI AG
Grosspeteranlage 5
4052 Basel
Switzerland
Tel.: +41 61 683 77 34
www.mdpi.com

Soil Systems Editorial Office
E-mail: soilsystems@mdpi.com
www.mdpi.com/journal/soilsystems

www.ingramcontent.com/pod-product-compliance
Lightning Source LLC
LaVergne TN
LVHW070200100526
838202LV00015B/1972